Books are to be returned on or before
the last date below

# ONE BEST WAY?

# ONE BEST WAY?

*Trajectories and Industrial Models
of the World's Automobile Producers*

*Edited by*
MICHEL FREYSSENET
ANDREW MAIR
KOICHI SHIMIZU
GIUSEPPE VOLPATO

from the
International GERPISA Programme
Emergence of New Industrial Models

OXFORD UNIVERSITY PRESS
1998

Oxford University Press, Great Clarendon Street, Oxford OX2 6DP
Oxford New York
Athens Auckland Bangkok Bogota Buenos Aires Calcutta
Cape Town Chennai Dar es Salaam Delhi Florence Hong Kong Istanbul
Karachi Kuala Lumpur Madrid Melbourne Mexico City Mumbai
Nairobi Paris São Paulo Singapore Taipei Tokyo Toronto Warsaw
and associated companies in Berlin Ibadan

Oxford is a registered trade mark of Oxford University Press

Published in the United States
by Oxford University Press Inc., New York

© M. Freyssenet, A. Mair, K. Shimizu, and G. Volpato 1998

The moral rights of the author have been asserted

First published 1998

All rights reserved. No part of this publication may be reproduced,
stored in a retrieval system, or transmitted, in any form or by any means,
without the prior permission in writing of Oxford University Press.
Within the UK, exceptions are allowed in respect of any fair dealing for the
purpose of research or private study, or criticism or review, as permitted
under the Copyright, Designs and Patents Act 1988, or in the case of
reprographic reproduction in accordance with the terms of the licences
issued by the Copyright Licensing Agency. Enquiries concerning
reproduction outside these terms and in other countries should be
sent to the Rights Department, Oxford University Press,
at the address above

This book is sold subject to the condition that it shall not, by way
of trade or otherwise, be lent, re-sold, hired out or otherwise circulated
without the publisher's prior consent in any form of binding or cover
other than that in which it is published and without a similar condition
including this condition being imposed on the subsequent purchaser

British Library Cataloguing in Publication Data
Data available

Library of Congress Cataloging in Publication Data
One best way? : trajectories and industrial models of the world's
automobile producers / edited by Michel Freyssenet . . . [et al.].
p. cm.
Includes bibliographical references (p. ).
1. Automobiles—Design and construction. 2. Industrial
engineering. 3. Technology transfer. I. Freyssenet, Michel.
TL278.054 1998
629.2′31—dc21              98-22476

ISBN 0-19-829089-6

1 3 5 7 9 10 8 6 4 2

Typeset in Hong Kong
by Graphicraft Limited
Printed in Great Britain
on acid-free paper by
Bookcraft (Bath) Ltd,
Midsomer Norton, Somerset

# FOREWORD

GERPISA (Groupe d'Etude et de Recherche Permanent sur l'Industrie et les Salariés de l'Automobile: Permanent Group for the Study of the Automobile Industry and its Employees) was formed in 1981 as a multidisciplinary group of researchers from economics, history, management, and sociology with interests in the automobile industry. The network was initially directed by Michel Freyssenet and Patrick Fridenson at the Ecole des Hautes Etudes en Sciences Sociales (School for Advanced Social Science Studies) in Paris.

In the early 1990s, discussions within the Group came to focus on debates about the emergence of 'new industrial models'. The influential book, *The Machine that Changed the World*, by Womack, Jones, and Roos (directors of the International Motor Vehicle Programme (IMVP) at the Massachusetts Institute of Technology) argued that a new industrial model had been born in Japan. This model, which the authors called 'lean production', was said to be universally superior and transferable to other countries. It was set to become the industrial model for the twenty-first century, just as 'mass production' had been for the twentieth century.

The leaders of GERPISA therefore decided to invite colleagues from many different countries to participate in an international programme, The Emergence of New Industrial Models, which would seek a response to questions about the nature and dynamics of industrial models. GERPISA did not seek to imitate the centrally directed and costly research structures of the IMVP. Instead it launched an open network of international cooperative research, based upon common interests and the free exchange of ideas, focused on debates and discussions at a series of international meetings and seminars.

Almost two hundred researchers from twenty countries participated in the programme, to varying extents. Four working groups were formed: trajectories of automobile producers, transplantation and hybridization of industrial models, variety and flexibility of production, and teamwork and employment relations. The most active researchers in these groups became their moderators and the editors of the resulting books. Along with a representative of the French automobile producers, Jean-Claude Monnet, they formed the programme's steering group. The scientific directors of the programme, Robert Boyer and Michel Freyssenet, were responsible for the overall coordination and leadership of the scientific debate. An annual meeting was held to present the results, to clarify theory and methodology, and to discuss successive theoretical analyses. The University of Evry and the Ministry for National Education and Research provided GERPISA with support for a secretary and two research staff, making it possible to establish a secretariat. These two institutions, along with the European Union (DGXII, Human Capital and Mobility Programme), the Committee of French Automobile Producers, PSA, Renault, and France Télécom provided the financial and material support necessary to organize the meetings and international colloquia, and coordinate the research.

At the end of the programme the participants reached the shared conclusion that both theoretically and in practice there have been, there remain today, and there will probably be tomorrow, several successful industrial models. The reasoning behind this conclusion is presented and discussed in the four collective books produced by the four working groups, which represent different elements of the integrated project.

Two of the books are being published together by Oxford University Press. *One Best Way? Trajectories and Industrial Models of the World's Automobile Producers*, edited by Michel Freyssenet, Andrew Mair, Koichi Shimizu, and Giuseppe Volpato, analyses the trajectories of fifteen automobile producers since the 1960s and reveals the variety of solutions adopted and possible conditions of success. *Between Imitation and Innovation: The Transfer and Hybridization of Productive Models in the International Automobile Industry*, edited by Robert Boyer, Elsie Charron, Ulrich Jürgens, and Steven Tolliday, analyses more than a dozen subsidiaries of Japanese and American and European manufacturers in nearly ten different countries. It reveals the particular historical conditions in which transplantation succeeds and shows in particular how the creation of a subsidiary most often gives rise to hybridization and sometimes to a new industrial model as a result of the constraints and opportunities offered by the host region.

Two further books are planned. *Coping with Variety: Product Variety and Production Organization in the World Automobile Industry*, edited by Yannick Lung, Jean-Jacques Chanaron, Takahiro Fujimoto, and Daniel Raff, defines, dates, and compares the evolution of product variety and analyses how firms have adopted different ways of effectively managing this variety in production, design, and relations with suppliers and distributors. *Teamwork in the Automobile Industry: Radical Change or Passing Fashion*?, edited by Jean-Pierre Durand, Juan José Castillo, and Paul Stewart (Basingstoke: Macmillan, 1998), reveals the wide diversity of practices, objectives, and outcomes hidden by the general adoption of teamwork, through case studies of twenty-five automobile factories.

Each book has its own particular focus, but all explain the plurality of industrial models. The thesis of convergence towards a single model is based on the idea that success comes from combining the methods which appear to give the best results, assuming that the environment is largely common to all firms. But reality suggests otherwise. Successful techniques are so only under certain economic and social conditions. Although growing liberalization of international trade and economic deregulation in many countries may have led to a convergence in competitive conditions, other factors are creating fresh sources of differentiation in both demand and cost structures. Indeed industrial models emerge from these partly unintended processes which result in coherence between strategies, organizational forms and practices, and the fit between these and the economic and social environment. It is the process of achieving internal coherence and external fit which makes companies successful, because it enables them to reduce the uncertainties in work and markets in the time and place in which they operate. Yet their very success often modifies the environment and the conditions which made their models viable. At this point, a new era commences in which firms must seek out new forms of coherence. Manufacturers in a common environment cannot simply copy the most successful company, since by definition the latter has a competitive advantage in having developed the appropriate strategy and model earlier than its rivals. On the contrary, the other producers must try to find a strategy which permits them to compete effectively yet avoid direct confrontation. Accordingly, not only is there no global model, but there is also no national industrial model which firms are obliged to adopt. There are, instead, a limited number of possible models in a given historical period.

While the members of GERPISA, and especially the members of its steering committee, reached agreement on these conclusions, they did not reach agreement on precisely how to characterize the various models. Time and resources did not permit full development of the debates, and, therefore, the contributors to the four books have adopted their own characterizations of industrial models, leaving the door open to further theoretical work in this area. The scientific directors of the programme, Robert Boyer and Michel Freyssenet, have, nevertheless, developed an analysis of the industrial models which have been used in the twentieth century automobile industry, based on the research undertaken during the programme and their own work, in an effort to create an appropriate theory. This will appear in a further book, *The World that Changed the Machine*, which follows the origin, development, diffusion, and crisis of industrial models and offers an explanation based on the evolution of markets and work in different contexts.

The results that GERPISA presents in these books are the fruit of the patient and coordinated work of their fifty or so authors, as well as of the various contributions made by all the participants in the programme. Other material published by the programme includes over a hundred issues of the GERPISA newsletter, *La Lettre du GERPISA*; the publication of articles in the *Actes du GERPISA*; and the work-in-progress papers presented to the annual meetings. Many people have contributed to this large-scale international cooperative project and we thank them all. We would also like to thank the representatives of the French Automobile Manufacturers, particularly Annie Beretti of PSA, Frédéric Decoster and Jean-Claude Monnet of Renault and Beatrice de Castelnau and Christian Mory of the Committee of French Automobile Producers, who were not content merely to follow the GERPISA programme on behalf of their companies or institution, but actively participated in the network's theoretical development. The programme would never have been successfully completed without the liaison, documentation, and editorial activities and the organization of annual meetings undertaken by the members of the GERPISA secretariat, Carole Assellaou, Kémal Bécirspahic, and Nicolas Hatzfeld, under the direction of Michel Freyssenet, the interpretation of Jacqueline Colombat's team, the translations of

Sybil Hyacinth Mair and Jennifer Merchant, and the organization of working group meetings at Lower Slaughter by Andrew Mair, Bordeaux by Yannick Lung, Venice by Giuseppe Volpato, Berlin by Ulrich Jürgens, Lyon by Jean-Jacques Chanaron, Paris by Robert Boyer and Elsie Charron, and Madrid by Juan José Castillo. The publication of these books bears witness to their contributions.

Further information on continuing GERPISA activities can be obtained from:

GERPISA International Network
Université d'Evry-Val d'Essonne
4 Boulevard François Mitterand
91025 Evry Cedex
France

*Tel*: 33 (1) 69 47 70 23
*Fax*: 33 (1) 69 47 70 07
*Email*: contact@gerpisa.univ-evry.fr
*Webpage*: http//www.gerpisa.univ-evry.fr

The steering committee of the GERPISA programme, The Emergence of New Industrial Models:

Robert Boyer (CEPREMAP, CNRS-EHESS, Paris), Juan José Castillo (Complutense University, Madrid), Jean-Jacques Chanaron (CNRS, Lyon), Elsie Charron (CNRS, Paris), Jean-Pierre Durand (University of Evry), Michel Freyssenet (CNRS, Paris), Patrick Fridenson (EHESS, Paris), Takahiro Fujimoto (University of Tokyo), Ulrich Jürgens (WZB, Berlin), Yannick Lung (University of Bordeaux IV), Andrew Mair (Birkbeck College, University of London), Jean-Claude Monnet (Research Department, Renault), Daniel Raff (University of Pennsylvania), Koichi Shimizu (University of Okayama), Paul Stewart (University of Wales, Cardiff), Steven Tolliday (University of Leeds), Giuseppe Volpato (Ca'Foscari University, Venice).

Participants in the GERPISA programme, The Emergence of New Industrial Models:

*Argentina*

Martha ROLDÁN (Flacso University, Buenos Aires), Miguel ZANABRIA (Ministry of Industry, Buenos Aires).

*Australia*

Greg BAMBER, (Griffith University, Brisbane), Russel LANSBURY (University of Sydney).

*Belgium*

Michel ALBERTIJN (Tempera, Antwerp), Leen BAISIER (Flemish Foundation of Technology Assessment, Brussels), Rik HUYS (University of Leuven), Geert VAN HOOTEGEM (University of Leuven), Johan VANBUYLEN (European Centre for Work and Society, Brussels), André VANDORPE (Flemish Foundation of Technology Assessment, Brussels).

*Brazil*

Ricardo ALVES de CARVALHO (Minas Gerais Federal University, Belo Horizonte), Nadya Araujo CASTRO (São Lazaro Federal University), Jussara CRUZ de BRITO (Cesteh, Rio), Edna CASTRO (Federal University, Belem), Afonso FLEURY (University of São Paulo), Roberto MARX (University of São Paulo), Mario Sergio SALERNO (University of São Paulo), Rosa Maria SALES de MELO SOARES (IPEA, Brasilia), Mauro ZILBOVICIUS (University of São Paulo).

*Canada*

Daniel DRACHE (York University).

*Colombia.*

Elba CÁNFORA de ZALAMEA (National University, Bogotá), Anita WEISS de BELALCÁZAR (National University, Bogotá).

*France*

Délila ALLAM (University of Paris I), Annie AMAR (IREP-D, Grenoble), Michel ARIBART (Ministère de l'Industrie, Paris), Etienne de BANVILLE (CNRS, Saint-Etienne), François BEAUJEU (University of Paris IX), Kémal BÉCIRSPAHIC, known as Bécir, (University of Evry), Marie-Claude BÉLIS-BERGOUIGNAN (University of Bordeaux IV), Muriel BELLIVIER (University of Marne-la-Vallée), Annie BERETTI (PSA Peugeot-Citroën), Géraldine de BONNAFOS (France Télécom, Paris), Gérard BORDENAVE (University of Bordeaux IV), Robert BOYER (CEPREMAP, Paris), Christophe CARRINCAZEAUX (University of Bordeaux IV), Béatrice de CASTELNAU (CCFA, Comité de Constructeurs Français d'Automobile, Paris), Sylvie CÉLÉRIER (University of Evry), Jean-Jacques CHANARON (CNRS, Lyon), Elsie CHARRON (CNRS, Paris), Bertrand CIAVALDINI (PSA Peugeot-Citroën), Yves COHEN (CRH, Centre de Recherches Historiques, Paris), Guy CORNETTE (University of Evry), Emmanuel COUVREUR (Renault), Isabel DA COSTA (CEE, Centre d'Etudes de l'Emploi, Paris), Frédéric DECOSTER (Renault), Gabriel DUPUY (CNRS, Paris), Jean-Pierre DURAND (University of Evry), Joyce DURAND-SEBAG (University of Evry), Béatrice FAGUET-PICQ (INTEC, Evry), Simone FEITLER (Renault), Michel FREYSSENET (CNRS, Paris), Patrick FRIDENSON (EHESS, Paris), João FURTADO (University of Paris XIII), Christophe GALLET (University of Lyon II), Annie GARANTO (University of Paris X), Gilles GAREL (CRG, Centre de Recherche en Gestion, Paris), Patrick GIANFALDONI (University of Provence), Armelle GORGEU (CEE, Paris), Nathalie GREENAN (INSEE, Paris), Françoise GUELLE (IAO-MRASH Lyon), Dominique GUELLEC (OCDE), Cândido GUERRA FERREIRA (University of Paris XIII), Christian GUIBERT (France Télécom, Paris), Armand HATCHUEL (Ecole des Mines, Paris), Nicolas HATZFELD (University of Evry), Helena Sumiko HIRATA (CNRS, Paris), Jean-Paul HUBERT (University of Paris XIII), Marie-Noëlle HUME (University of Evry), Hee-Young HWANG (University of Paris X), Didier IDJADI (University of Paris XIII), Bruno JETIN (University of Rouen), Bernard JULLIEN (University of Bordeaux IV), Alex KESSELER (CRG, Paris), Gerson KOCH (University of Lille), Alain KOPFF (Paris), Daniel LABBÉ (Renault), Anne LABIT (University of Rouen), Lydie LAIGLE (LATTS, Paris), Pascal LARBAOUI (University of Paris XIII), Marc LAUTIER (Paris), Jean-Bernard LAYAN (University of Bordeaux IV), Nathalie LAZARIC (University of Compiègne), Yveline LECLER (IAO-MRASH, Lyon), Danièle LINHART (CNRS, Paris), Jean-Louis LOUBET (University of Evry), Yannick LUNG (University of Bordeaux IV), Jean-Loup MADRE (INRETS, Paris), Olivier MARÉCHAU (Rectorat de Créteil), Claire MARTIN (Renault), René MATHIEU (CEE, Paris), Alain MICHEL (EHESS), Christophe MIDLER (CRG, Paris), Jean-Claude MONNET (Renault), Christian MORY (CCFA, Paris), Aimée MOUTET (University of Paris X), Jean-Philippe NEUVILLE (CSO Paris), Jean-Pierre ORFEUIL (INRETS, Paris), Alfredo PENA-VEGA (Paris), Monique PEYRIÈRE (University of Evry), Jean-Marc POINTET (University of Paris XIII), Emmanuel QUENSON (GIP Mutations Industrielles, Paris), Jean-Philippe RENNARD (Paris), Luiz ROTHIER BAUTZER (University of Paris IX), Patrick ROZENBLATT (University of Paris X), Frédérique SACHWALD (IFRI, Paris), Laurence SAGLIETTO (University of Nice), Jean SAUVY (Paris), Benoît SCHLUMBERGER (University of Paris IX), Klas SODERQUIST (ESC, Grenoble), Jean-Claude THENARD (GIP Mutations Industrielles, Paris), Benoît WEIL (Ecole des Mines, Paris).

## Germany

Peter AUER (WZB, Berlin), Bob HANCKÉ, (WZB, Berlin), Jörg HOFMANN (IG Metall), Peter JANSEN (WZB, Berlin), Ulrich JÜRGENS (WZB, Berlin), Martin KUHLMANN (University of Göttingen), Steffen LEHNDORFF (Institut Arbeit und Technik, Gelsenkirchen), Roland SPRINGER (Daimler-Benz), Frank WEHRMANN (Volkswagen).

## Italy

Giovanni BALCET (University of Turin), Arnaldo CAMUFFO (Ca'Foscari University, Venice), Aldo ENRIETTI (University of Turin), Massimo FOLLIS (University of Turin), Stefano MICELLI (Ca'Foscari University, Venice), Giuseppe VOLPATO (Ca'Foscari University, Venice).

## Japan

Tetsuo ABO (University of Tokyo), Hisao ARAI (University of Shiga), Takahiro FUJIMOTO (University of Tokyo), Masanori HANADA (Kumamoto Gakuen University), Masayoshi IKEDA (Chuo University, Tokyo), Yasuo INOUE (Nagoya University), Osamu KOYAMA (Sapporo University), Kazuhiro MISHINA (JAIST Institute, Tokyo), Yoichiro NAKAGAWA (Chuo University, Tokyo), Hikari NOHARA (Hiroshima University), Masami NOMURA (Tohoku University, Sendai), Ichiro SAGA, (Kumamoto Gakuen University), Shoichiro SEI (Kanto-Gakuen University, Yokohama), Koichi SHIMIZU (Okayama University), Koichi SHIMOKAWA (Hosei University, Tokyo).

## Korea

Myeong-Kee CHUNG (Han Nam University, Taejon), Hyun-Joong JUN (Seoul University).

## Mexico

Jorge CARRILLO (Frontera Norte College, Tijuana), Patricia GARCÍA-GUTIERREZ (Autonomous University of Mexico), Sergio Fernando HERRERA-LIMA (Autonomous University of Mexico), Yolanda MONTIEL (CIESAS, Mexico).

## Netherlands

Ben DANKBAAR (Catholic University of Nijmegen), Frank DEN HOND (Free University, Amsterdam), Winfried RUIGROK (Erasmus University, Rotterdam), Rob VAN TULDER (Erasmus University, Rotterdam).

## Portugal

Paulo ALVES (University of Lisbon), Antonio BRANDÃO-MONIZ (Uni-Nova University, Lisbon), Ilona KOVÁCS (University of Lisbon), Maria Leonor PIRES (University of Lisbon), Marinùs PIRES de LIMA (University of Lisbon), Pedro PIRES de LIMA (University of Lisbon), Mario VALE (University of Lisbon).

## Spain

Ricardo ALÁEZ (University of País Vasco, Bilbao), Javier BILBAO (University of País Vasco, Bilbao), Vicente CAMINO (University of País Vasco, Bilbao), Juan José CASTILLO (Complutense University, Madrid), Juan Carlos LONGÁS (Public University of Navarra), Javier MÉNDEZ (Complutense University, Madrid), Manuel RAPUN (Public University of Navarra).

## Sweden

Christian BERGGREN (University of Linköping, Stockholm), Per Olav BERGSTRÖM (Metallförbundet, Stockholm), Anders BOGLIND (Volvo Car Corporation), Göran BRULIN (Institutet för Arbetslivsforskning, Stockholm), Kajsa ELLEGÅRD (Göteborg University), Tomas ENGSTRÖM (Chalmers University of Technology, Göteborg), Henrik GLIMSTEDT (Göteborg University), Nils KINCH (Uppsala University), Lars MEDBO (Chalmers University of Technology, Göteborg), Tommy NILSSON (The Swedish Institute for Working Life Research, Stockholm), Lennart NILSSON (Göteborg University), Åke SANDBERG (The Swedish Institute for Working Life Research, Stockholm).

## Switzerland

Ronny BIANCHI (Bellinzona).

## Turkey

Lale DURUIZ (University of Marmara), Nurhan YENTÜRK (Istanbul Technical University).

## United Kingdom

Philip GARRAHAN (University of Northumbria), John HUMPHREY (University of Sussex, Brighton), Arnoud LAGENDIJK (University of Newcastle), Andrew MAIR (University of London), Mari SAKO (London School of Economics), Elizabeth Bortolaia SILVA (University of Leeds), Paul STEWART (University of Wales, Cardiff), Joseph TIDD (Imperial College, London), Steven TOLLIDAY (University of Leeds).

## USA

Paul ADLER (University of Southern California), Steve BABSON (Labor Studies, Detroit), Bruce BELZOWSKI (University of Michigan), Richard FLORIDA (Harvard University), Michael FLYNN (University of Michigan), Susan HELPER (Case Western Reserve University, Cleveland), Harry KATZ (Cornell University), Ruth MILKMAN (University of California Los Angeles), Frits PIL (Pittsburgh University), Daniel RAFF (Pennsylvania University), Saul RUBINSTEIN (New Brunswick University), Harley SHAIKEN (University of California).

# CONTENTS

| | |
|---|---|
| List of Contributors | xv |
| List of Figures | xvi |
| List of Tables | xvii |
| List of Statistical Appendices | xviii |

Introduction     1
*Michel Freyssenet*

1. Intersecting Trajectories and Model Changes     8
   *Michel Freyssenet*

2. Models, Trajectories, and the Evolution of Production Systems: Lessons from the American Automobile Industry in the Years between the Wars     49
   *Daniel M. G. Raff*

## Part I    Only One Model in Japan?     61

3. A New Toyotaism?     63
   *Koichi Shimizu*

4. Nissan: Restructuring to Regain Competitiveness     91
   *Masanori Hanada*

5. The Globalization of Honda's Product-Led Flexible Mass Production System     110
   *Andrew Mair*

6. The Unique Trajectory of Mitsubishi Motors     139
   *Koichi Shimizu and Koichi Shimokawa*

7. Hyundai Tries Two Industrial Models to Penetrate Global Markets     154
   *Myeong-Kee Chung*

## Part II    Three Distinct Trajectories at North America's Big Three     177

8. The General Motors Trajectory: Strategic Shift or Tactical Drift?     179
   *Michael S. Flynn*

9. Globalization at the Heart of Organizational Change: Crisis and Recovery at the Ford Motor Company     211
   *Gérard Bordenave*

10. Reinventing Chrysler 242
    Bruce M. Belzowski

**Part III  Europe's Dilemma: Transfer, Adapt, or Innovate?** 271

11. The Development of Volkswagen's Industrial Model, 1967–1995 273
    Ulrich Jürgens

12. Making Manufacturing Lean in the Italian Automobile Industry: The Trajectory of Fiat 311
    Arnaldo Camuffo and Giuseppe Volpato

13. Peugeot Meets Ford, Sloan, and Toyota 338
    Jean-Louis Loubet

14. Renault: From Diversified Mass Production to Innovative Flexible Production 365
    Michel Freyssenet

15. From British Leyland Motor Corporation to Rover Group: The Search for a Viable British Model 395
    Andrew Mair

16. A Second Comeback or a Final Farewell? The Volvo Trajectory, 1973–1994 418
    Christian Berggren

17. Lada: Viability of Fordism? 440
    Jean-Jacques Chanaron

18. Conclusion: The Choices to be made in the Coming Decade 452
    Michel Freyssenet, Andrew Mair, Koichi Shimizu, and Giuseppe Volpato

Index 463

# LIST OF CONTRIBUTORS

BRUCE BELZOWSKI is Senior Research Associate in Sociology at the University of Michigan, Ann Arbor, USA.

CHRISTIAN BERGGREN is Professor of Sociology at the University of Linköping, Sweden.

GERARD BORDENAVE is Lecturer in Economics at the University of Bordeaux IV, France.

ARNALDO CAMUFFO is Professor of Business Economics at Ca'Foscari University, Venice, Italy.

JEAN-JACQUES CHANARON is Research Director in Economics at the CNRS, Lyon, France.

MYEONG-KEE CHUNG is Professor of Economics at Han Nam University, Taejon, Korea.

MICHAEL FLYNN is Associate Director in Sociology at the University of Michigan, Ann Arbor, USA.

MICHEL FREYSSENET is Research Director in Sociology at CNRS-Paris, and Co-Director of GERPISA and First International GERPISA Programme, University of Evry, Paris, France.

MASANORI HANADA is Professor of Economics at Kumamoto Gakuen University, Japan.

ULRICH JÜRGENS is Senior Researcher in Sociology at Wissenschaftszentrum, Berlin, Germany.

JEAN-LOUIS LOUBET is Professor of History at the University of Evry, Paris, France.

ANDREW MAIR is Lecturer in Management at Birkbeck College, University of London, United Kingdom.

DANIEL RAFF is Professor of History at Wharton School, University of Pennsylvania, USA.

KOICHI SHIMIZU is Professor of Economics at Okayama University, Japan.

KOICHI SHIMOKAWA is Professor of Business History at Hosei University, Tokyo, Japan.

GIUSEPPE VOLPATO is Professor of Business Economics at Ca'Foscari University, Venice, Italy.

# LIST OF FIGURES

| | | |
|---|---|---|
| 3.1. | The hierarchy of employee ranks at Toyota | 66 |
| 3.2. | Toyota market share, 1950–1991 | 72 |
| 3.3. | Factors in the variation of Toyota per-vehicle production cost | 73 |
| 3.4. | Profits as a proportion of turnover at Toyota | 74 |
| 4.1. | Evolution of rate of Nissan operating profit | 100 |
| 6.1. | Mitsubishi Motors Corporation: market share | 142 |
| 6.2. | Profits as a proportion of turnover at Mitsubishi Motors Corporation | 143 |
| 6.3. | Mitsubishi Motors Corporation: organizational hierarchy | 147 |
| 8.1. | Selected North American light vehicle market shares, 1967–1994 | 188 |
| 8.2. | Light trucks as a proportion of North American total sales, 1967–1994 | 194 |
| 9.1. | Vehicle factory sales and turnover of Ford | 212 |
| 9.2. | Turnover and costs per Ford vehicle | 227 |
| 10.1. | Chrysler US market share: cars/trucks | 257 |
| 10.2. | US vehicle import sales | 259 |
| 13.1. | Peugeot-Citroën product range, 1982–1996 | 355 |
| 14.1. | Total production of Renault passenger cars by model | 368 |
| 14.2. | Total production at Renault by category of vehicle: passenger cars, light commercial vehicles, trucks and buses | 370 |
| 14.3. | Renault's worldwide production of passenger cars and light commercial vehicles by place of production and sale | 372 |
| 14.4. | Renault employees by category | 375 |
| 17.1. | USSR/CIS and Avtovaz car production, 1970–1994 | 442 |
| 17.2. | Avtovaz profitability in 1994 | 443 |
| 17.3. | Shareholding of Avtovaz in 1994 | 447 |
| 17.4. | Avtovaz trajectory | 448 |

# LIST OF TABLES

| | | |
|---|---|---:|
| 7.1. | Hyundai product variety | 160 |
| 7.2. | Evolution of Korean producers' market share and registration | 161 |
| 7.3. | Export of Hyundai by region | 161 |
| 7.4. | Annual percentage change in labour productivity, unit labour cost, and unit labour output of Hyundai | 164 |
| 7.5. | Structure of Hyundai workforce 1988–1994 (per cent) | 166 |
| 7.6. | The wage structure of Hyundai in 1994 | 169 |
| 9.1. | Comparison of average annual growth rates in selected indicators for 1968–1979 and 1982–1994 at Ford | 218 |
| 11.1. | The development of VW's model range 1962–1972: VW and Audi in Germany and 'overseas' | 279 |
| 11.2. | The development of VW's model range 1977–1992: VW and Audi in Germany and 'overseas' | 286 |
| 11.3. | Car manufacturing divisions in the VW group, 1994 | 301 |
| 12.1. | Fiat Auto vertical integration | 323 |
| 12.2. | Fiat Auto components design | 324 |
| 16.1. | The wages of automobile workers in relation to the manufacturing average: a comparison of hourly wages in selected countries, 1970 and 1980 | 425 |
| 17.1. | Labour productivity at Avtovaz in 1994 | 443 |
| 17.2. | Strategic versus operational variables | 444 |
| 17.3. | Characteristics of a revived Fordist model | 450 |

# LIST OF STATISTICAL APPENDICES

Statistical Appendix 3: Toyota     88–9
Statistical Appendix 4: Nissan     108
Statistical Appendix 5: Honda     136–7
Statistical Appendix 6: Mitsubishi     151–2
Statistical Appendix 7: Hyundai     174
Statistical Appendix 8: General Motors     208–9
Statistical Appendix 9: Ford     239–240
Statistical Appendix 10: Chrysler     268
Statistical Appendix 11: Volkswagen     308
Statistical Appendix 12: Fiat     336
Statistical Appendix 13: Peugeot–Citroën     363
Statistical Appendix 14: Renault     391–2
Statistical Appendix 15: Rover     416
Statistical Appendix 16: Volvo     438

# Introduction

MICHEL FREYSSENET

The 'received wisdom' about the history of the automobile industry is common to those who study it and those who work in it. Between 1920 and 1950, following a primarily 'craft'-based phase, both American and European automobile companies are said to have adopted the mass production system, characterized by a search for economies of scale, a sequential and segmented organization of design and production, and the utilization of an unskilled workforce. Because of this system's inherent rigidity, it is said to have been thrown into crisis by the transition to a diversified and unstable replacement market increasingly open to international competition. Conversely, Japanese companies are deemed to have learned early how to rapidly adapt design and production to market changes and make profits in any conditions, thanks to a trained workforce who participate in cost reduction and quality improvement. Accordingly, the Japanese are purported to have invented an industrial model that can succeed in the market conditions which all automobile companies now share. Systematized under the name 'lean production' by the authors of *The Machine that Changed the World*, (Womack et al. 1990), this system—said to have proved its universality with the success of Japanese transplants in the USA and United Kingdom—is now supposedly obligatory for all American and European companies if they wish to survive into the twenty-first century. They are all said to be trying to adopt it, as revealed by the borrowing of 'Japanese methods': notwithstanding delays, resistance, and mistakes, the production systems of these companies will therefore converge towards 'lean production', just as they converged on mass production during the first half of the twentieth century.

Historical simplifications are useful and legitimate when they reveal the underlying essentials. They are dangerous in theory as in practice when they support widespread prejudices. The claim summarized above is problematic precisely because of the presuppositions upon which it is based, the conceptual ambiguities it supports, the facts it overlooks, the observational and analytical methods it encourages, and lastly the errors it can lead to on the part of company managers and union leaders alike.

In the first place, it presupposes that the market has evolved, and has, in the main, evolved along similar lines throughout the world: that is to say, from an elite market into a replacement market by way of a mass market. It overlooks the fact that the volume and structure of the automobile market depends upon the forms of mobility characteristic of each country, and the way national income is redistributed among social groups, which in turn is influenced by incomes policies and by the economic significance of the automobile industry in the countries in which it is located. Lastly, it implies that there is only a single viable solution when conditions are similar.

The three supposedly successive production systems, craft, mass production, and lean production, are actually the result of historic amalgams and conceptual ambiguities. The so-called craft system includes craft-based companies, but also genuine industrial companies, even if they continue to produce vehicles at fixed work stations. These companies did not disappear from the USA in the interwar years because of a lack of competitiveness, but due to a lack of liquidity, as revealed in Chapter 2 by Dan Raff. 'Mass production', otherwise known as the 'Taylor–Ford' system, encompasses different models, even if they do share certain principles in common. The Fordist model that can be seen today is different from Henry Ford's initial conception, and should not be confused either historically or conceptually with the Taylorist model or with the Sloanist model, for instance. The Fordist model has not been as universal as is often made out. Not only has it failed in countries where the conditions it required did not exist, but neither has it been automatically adopted in countries, like Japan, where mass production was certainly possible (Fujimoto and Tidd 1994, Tolliday 1998). The so-called lean production model is a combination of characteristics borrowed from various Japanese companies (Mair 1994). Finally, Toyota has been questioning the core components of its system at the very moment when it was being projected as the paradigm for the twenty-first century, as Koichi Shimizu reveals in Chapter 3.

The belief in a 'one best way' for each major economic era leads the researcher, and even the practitioner, to interpret and analyse companies in terms of how faithful they are to the model considered to be the best, and to interpret their performance, whether superior, equal, or inferior to what it is supposed to be, as a function of secondary, temporary, or contextual factors which have intervened to modify the presumed cause and effect relationship. This approach frequently obscures the essentials, that is, the existence of different profit strategies which require appropriate and coherent means if they are to be implemented effectively.

Lastly, several company leaders, having internalized the reassuring image of the history of the automobile industry presented above, and newly converted to the excellence of 'lean production', forget to properly verify whether their own market, work organization, and institutional contexts correspond to the conditions in which the apparently superior model is in fact valid and viable. It is scarcely surprising that they then experience failures, which are too quickly and conveniently explained away in terms of incomprehension of the system, errors of implementation, and resistance to change on the part of certain groups of employees.

For these reasons, the authors of the chapters in this book, together with the entire group of researchers who have participated in GERPISA's international programme The Emergence of New Industrial Models, have accepted a requirement to start from empirical research, in this case focused on the trajectories of the automobile producers from the 1970s through to 1995, prior to drawing conclusions about convergence or plurality of models. Two pitfalls had to be avoided: fastidious description, which too often leads to the conclusion that every case is unique, and superficial comparison which is content to classify companies according to perceived similarities and differences. In contrast, the analytical framework adopted here is based on studying the changes that companies have made to the particular profit strategies and the socio-production systems that characterized them in the 1960s, in the context of the shared or unique problems that they have encountered since the 1970s, and which help interpret what their evolution actually means. The authors of this book have then examined the attempts of the companies to make their changes coherent and, where this has been achieved, to characterize the models that have thus been constituted.

The identification of the problems encountered has focused on problems likely to affect the validity of the existing model: slow changes, new constraints, internal contradictions, severe crises, and so on. When the model adopted by a company required continuous market growth owing to regular increases in the consumer purchasing power and acceptance by the majority of workers of unskilled work, research was oriented towards internal or external economic and social processes likely to change these conditions. When a model includes a profit strategy based on selling innovative, quality, and high price products to a limited clientele able to pay the asking price, then it was necessary to investigate the longevity of this clientele and the company's capacity to respond to its expectations. If a model only functions properly when workers participate thoroughly in the realization of company objectives, it was necessary to look for the factors that have removed the preconditions for such involvement.

The empirical research on the trajectories was supported by an effort at conceptual clarification and methodological guidance undertaken by the programme's founders and other network members (Boyer and Freyssenet 1995). Hence industrial models were defined as the periodic processes of making the internal organizational and social practices and systems of companies coherent in order to implement chosen profit strategies, taking into account market and work transformations and the compromises that govern them. In practice, it is never certain that products offered by a given company will find sufficient buyers at the asking price. Neither is it guaranteed that the company's employees will be sufficiently involved in their work to realize output at the desired levels of quantity, quality, costs, and lead times. At the national or indeed international level, the uncertainties surrounding the market and work are to a certain degree foreseeable owing to various practices, agreements, laws, and social and economic policies, which represent so many compromises between economic, social, and political forces. The viability of a model therefore depends first on its compatibility with these compromises and secondly on the coherence between its production organization and its employment relationships.[1]

The procedure that has been adopted here has permitted the authors of this book to follow very similar analyses of automobile producer trajectories and present them in very similar ways. The socio-production systems of producers are first described prior to the appearance of significant problems which required them to be modified, that is, the problems which appeared during the late 1960s or early 1970s, depending on the case. Next, the company's history is periodized as a function of the solutions that were tried, along with their foreseen and unforeseen consequences. Then, the new socio-production systems that emerged from the changes undertaken are described, and their level of internal coherence is evaluated to find out whether a new model was created. To conclude, the likely challenges facing the company after 1995 are presented.

The trajectories of fifteen companies have been studied. These encompass the major foreign subsidiaries, such as Opel, Vauxhall, Saab, for General Motors (GM) and Ford of Europe for Ford, as well as the brands absorbed or taken over since the 1970s, such as Seat and Škoda for Volkswagen, Lancia and Alfa-Romeo for Fiat, Citroën and Chrysler-Europe for PSA, Mazda for Ford, and so on. Not all the independent companies that deserved to be studied in the book could be included, notably Mercedes and BMW, owing to an absence of academic experts. The companies have been grouped by the three regional poles of the global automobile industry, East Asia, North America, and Europe, given the relative similarity of markets and work contexts facing companies within each pole. This permits the differences of trajectory and models in relatively similar contexts to emerge clearly. However, this principle is not adhered to in two cases: Hyundai, which is grouped with the Japanese companies, and Lada, where production conditions at the time of the Soviet Union were quite unrelated to those in Western Europe. Other parts of the world are not entirely ignored. The subsidiaries of the major producers present there are mentioned within their trajectories, although they are not analysed directly. Another book resulting from the GERPISA programme studies them from the perspective of the transfer and hybridization of industrial models (Boyer *et al.* 1998), of the teamworking (Durand *et al.*, 1998), and of product policy and organization (Lung *et al.*).

Although to begin with the authors of this book had divergent views, as did many participants in the GERPISA programme (as can be read in the Foreword), by the end of the process they agreed that convergence towards a single successful industrial model is no more likely today than in the past.

The trajectories of the Japanese companies clearly reveal the diversity of their strategies and their methods. Koichi Shimizu underlines the significance of the hidden face of Toyota's production system, namely, its wage system, and the depth of the transformation the company has had to undertake in its production organization and employment relationships following the crisis of work it experienced at the end of the 1980s. The differences with Nissan, Honda, and Mitsubishi become even clearer. Masanori Hanada reveals Nissan's loss of competitiveness, already apparent in the domestic market, and now emerging internationally. The elimination of the dual (management–union) management system which characterized Nissan for many years has not enabled the company to return to an adequate level of profitability. Andrew Mair recalls the originality of Honda's strategy, which succeeded in creating a place for the company in the national and international market by providing high quality products that were innovative both technically and commercially and adapted to different markets, thanks to an internal system that valorized creative and expert individualism, a flexible mass production system, and a precocious and active strategy of international investment. He none the less questions the longevity of aspects of this model related to technical prowess. Mitsubishi, here analysed by Koichi Shimizu and Koichi Shimokawa, has ended up pursuing a similar strategy, after numerous setbacks, although it has been prevented from achieving Honda's success owing to high interest repayments on loans incurred to finance significant investments, particularly in automation.

The trajectory of the Korean producer, Hyundai, presented by Myeong-Kee Chung, shares few points of reference with the Japanese producers. Basing itself on the mass production of a few models sold for export at low prices, thanks to low wages, very long annual working hours, and favourable exchange rates, the company experienced repeated failure. The long social and political crisis at the end of the 1980s, the significant increase in wages, changes in economic policy, and the recognition of unions have radically changed the context. Hyundai is now seeking to establish a system that combines flexible automation with employee participation, despite industrial relations that remain mired in conflict.

The American producers, who were hard hit by competition from their Japanese counterparts in the 1970s and the 1980s, have recovered financially. While they have all cut costs drastically, particularly by closing factories, by refocusing on the automobile sector, and by obtaining lower prices from their suppliers, they have certainly not all opted for the same path to regain long-term profitability. As Michael Flynn shows, General Motors still has important decisions ahead in terms of volume, variety, flexibility, diversification, and form of internationalization, and it remains impossible to say to what extent these will be influenced by the experiences of its joint venture with Toyota (NUMMI), its subsidiary Saturn (designed together with the United Auto Workers (UAW) union), or even its European subsidiary, Opel. Ford, on the other hand, has clearly decided to globalize its product range and its organization, thus linking up again with the initial ambitions of Henry Ford, as Gérard Bordenave explains. The so-called 'Japanese methods' that the company borrowed were reinterpreted within the context of an organization which remained strongly hierarchical, with employment relationships still based on external mobility. As for Chrysler, this company has reinvented itself, as Bruce Belzowski shows, after various ill-starred attempts to resemble the other two American producers. The company relaunched itself with a strategy of commercial innovation and a low break-even point, which enabled it to achieve high levels of profitability with far lower volumes than its direct competitors.

European companies have had contrasting trajectories, in part at least owing to their division into generalists and specialists. Among the generalists, Volkswagen experienced two crises: the first occurred during its transition away from the Fordist model, of which it was the greatest exponent in Europe, to the Sloanist model during the early 1970s; the second came during its international expansion and its organization into five brands in the early 1990s. These crises were both overcome, as Ulrich Jürgens emphasizes, in a unique way thanks to a management structure which closely links the IG Metall union and the government of Lower Saxony with the major strategic choices and the management of the company. Arnaldo Camuffo and Giuseppe Volpato explain how Fiat has forged its own concept of organization. The 'Integrated Factory' aims to make each individual's practices transparent, inspired, they argue, by Toyotism in terms of the objective, yet different in terms of the means, which in the case of Toyota have been far more gradual. PSA (Peugeot Société Anonyme) was able, after a crisis which almost proved fatal, to integrate two other producers it had purchased, Citroën and Chrysler Europe, into one coherent entity. As Jean-Louis Loubet shows, Peugeot succeeded by returning to its long-term strategy of rigorous financial and management planning and by developing a systematic policy of using platforms for more than one model. Renault, which at the start of the 1980s believed itself capable of increasing its capacity, expanding its range, automating its manufacturing, investing in new countries, all the while lowering its break-even point, eventually owed its recovery to a complete change in strategy, as I described. After having drastically reduced its costs and changing its employment policies, the company focused on quality and innovation with the aim of positioning itself at the top end of each market segment and creating new market niches, in the hope that these would become significant market segments. Rover pursued a similar strategy, by taking advantage of its know-how in terms of four-wheel drive vehicles and urban cars, particularly in foreign markets. Andrew Mair retraces the chaotic trajectory which eventually propelled Rover in this direction, and contemplates whether the company will continue along this path following its purchase by BMW.

In contrast to the USA and Japan, several specialist producers were able to survive in Europe and remain independent and strong. Among them, Volvo, analysed in this book by Christian Berggren, symbolized the European attempt to 'humanize work' and to make the production process flexible during the 1970s. The closure of the emblematic factories of Kalmar and Uddevalla might lead one to believe that this path was now closed off. Yet the reopening of Uddevalla, and Toyota's borrowing from the Kalmar system for its new factories, warns us against drawing definitive conclusions too quickly.

The final producer studied here is Avtovaz, the company which produces Lada vehicles. At first glance Avtovaz appears to be the perfect incarnation of the Fordist model in terms of its product strategy, vertical integration, and organizational form. In practice, however, it never achieved this during the existence of the Soviet Union, because its employment relationships were incompatible. Jean-Jacques Chanaron explains why the appropriate conditions for Fordism might now be emerging for Avtovaz.

The authors of this book have had neither the time nor the financial means to arrive at shared definitions of the new models that they have discovered in the trajectories they have studied. Hence each author has been free to characterize in his own way the new model of the company he has studied. None the less, in the chapter that follows this introduction, I endeavour to present a coherent overview of the trajectories of companies in tabular form on the basis of the trajectories that have been analysed in the book and of the interpretations that Robert Boyer and I have drawn from the history of industrial models in the automobile sector since its inception (Boyer and Freyssenet forthcoming). In the concluding chapter, the editors of this book have attempted to highlight the challenges that confront automobile producers at the close of the twentieth century and the tensions that their industrial models will face. Just as history has not ended, neither has the evolution of industrial models. No one industrial model can guarantee companies the capability to withstand environmental changes and internal tensions under any circumstances.

## NOTES

Translated by Sybil Hyacinth Mair.
1. See Chapter 1, note 2.

## BIBLIOGRAPHY

Boyer, R., Charron, E., Jürgens, U., and Tolliday, S. (eds.), *Between Imitation and Innovation: The Transfer and Hybridization of Productive Models in the International Automobile Industry* (Oxford, 1998).
—— and Freyssenet, M., 'The Emergence of New Industrial Models. Hypotheses and Analytical Procedure', *Actes du GERPISA*, 15 (1995).
—— —— *The World that Changed the Machine* (forthcoming).
Durand, J. P., Castillo, J.-J., and Stewart, P. (eds.), *Teamwork in the Automobile Industry. Radical Change or Passing Fashion?* (Basingstoke, 1998).
Fujimoto, T., and Tidd, J., 'The UK & Japanese Automobile Industries: Adoption and Adaptation of Fordism', in *Actes du GERPISA*, 11 (1994).
Lung, Y., Chanaron, J.-J., Fujimoto, T., and Raff, D. (eds.), *Coping With Variety. Product Variety and Productive Organization in the World Automobile Industry* (forthcoming).
Mair, A., *Honda's Global Local Corporation* (Basingstoke, 1994).
Tolliday, S., 'The Diffusion and Transformation of Fordism: Britain and Japan Compared', in Boyer *et al.*, *Between Imitation and Innovation* (Oxford, 1998).
Womack, J. P., Jones, D. T., and Roos, D., *The Machine that Changed the World* (New York, 1990).

# 1
# Intersecting Trajectories and Model Changes
## MICHEL FREYSSENET

Has there been one, or have there been several successful industrial models in the period since the 1974 oil crisis, and, if so, which? The aim of this chapter is to identify and characterize the successful model or models and to investigate the conditions under which it or they are viable and function well. It is clear from the trajectories of the automobile manufacturers examined in this book that they have certainly pursued different profit strategies; some have emphasized volume, some product diversity, others quality, and so forth. Analysis of these strategies reveals that they in fact represent combinations of the five fundamental sources of company profits: scale effects, scope effects, quality, product innovation, and the reduction of costs at constant volume. Each strategy combines these sources of profit in precise proportions to make them structurally compatible. Accordingly, it appears to be impossible for a firm to simultaneously attain, and maintain over the long term, maximum volume, maximum diversity, perfect quality, permanent innovation, and the reduction of costs at constant volume. This finding stands in stark contrast to the assertions made in *The Machine that Changed the World* (Womack *et al*. 1990), with its presentation of 'lean production' as the optimal industrial model; one which, moreover, also apparently guarantees the fulfilment of employees, the satisfaction of consumers, and economic development worldwide!

Empirical analysis of the world's automobile producers reveals that five profit strategies were pursued during the period between 1974 and 1990–2. We have labelled these strategies 'volume and diversity', 'continuous reduction of costs at constant volume', 'innovation and flexibility', 'quality and specialization', and 'volume'. For these strategies to be profitable they had to fit their external environments and be implemented by industrial models that were consistent with them. Of the manufacturers studied in this book, only three entirely met these twin criteria during the period in question: Volkswagen, Toyota, and Honda. All three also maintained break-even points below their added value. This is an important indicator, since it not only reveals the profitability of automotive sector activities (by excluding financial activities, which distort the real situation in one direction or another), but also show the behaviour of profit margins when demand falls. Volkswagen, Toyota, and Honda adopted strategies respectively based on 'volume and diversity', 'continuous reduction of costs at constant volume', and 'innovation and flexibility', and each was based in a country where the growth and redistribution of national income was already based on competitiveness in export markets. Volkswagen was able to adapt the Sloanist model to the new international context, while Toyota and Honda created their own models. Volvo comes into the picture alongside the other three companies over a shorter period, from 1981 to 1990, when it pursued the 'quality and specialization' strategy by exploring a path which opened up onto a new potential industrial model, 'reflexive production' (Boyer and Freyssenet forthcoming).

In practice, the viability of each profit strategy presupposes a particular type of demand for automobiles and labour market conditions which cannot be assured under every mode of national economic growth. Hence, for as long as they continued in the countries which adopted them, post-war investment-oriented modes of growth did not allow the creation of the mass automobile markets which are indispensable to the 'volume and diversity' strategy. This was the case in Japan until the early 1960s. Modes of national growth can in fact be identified in terms of the source, criteria, and forms of national income redistribution. Immediately preceding the 1974 crisis there were three principal modes in the countries which had built their own automobile industries, if one excludes the Soviet Union which had its own particular economic system. The first, the 'autocentred' mode of economic growth, consisted of redistributing productivity increases through the purchasing power of a moderately unequal distribution of wages. This was the case in the USA, France, and Italy. The second, the 'export' mode, made wage growth a function of competitiveness in export markets. Competitiveness was gained either by specialization in products not (or rarely) manufactured in the other countries, as was the case for the German Federal Republic and Sweden, or by lower prices for more widely shared products, as was the case for Japan. The third mode of growth, observed in Great Britain, the most open country of all, was characterized by an absence of links between productivity, external competitiveness, and the setting of wage levels. In the early 1950s, the old system of free wage negotiations at the local level among the numerous trades unions and employers was resumed, recreating marked disparities in wage levels among the different categories of employees, among companies, and between regions, as well as wide variations from one period to another. Consequently, mass consumption, which requires regular, general, and moderate growth in wage levels could not properly take root and create conditions favourable to mass production (Fujimoto and Tidd 1994, Tolliday 1998). Each mode of national growth is therefore characterized by particular competitive inter-firm relations (practices and rules) and labour relations (a phrase used in this text in a broad sense, including especially the mode of determination and distribution of wages and national insurance benefits, work rights, general working practices, industrial relations, and so on). These relations are notably made by compromises (formal and in practice) between companies and between companies and employees at national level.[1]

The three firms which were the most successful between 1974 and 1992 were each based in countries which were already 'exporters' prior to 1974. However, not all the manufacturers in these countries were as successful. Their socio-productive configuration still had to be consistent with the profit strategy they pursued. To be profitable, a strategy must not only be relevant in the country where it has been adopted, but it must also be implemented with adequate and mutually compatible resources. It is hard to imagine, for example, being able to implement a strategy of 'continuous reduction of costs at constant volume' in a replacement vehicle market without methods to oblige employees to accept its consequences for them. The means to implement a profit strategy in a coherent way may vary according to the period and context, thus creating a number of distinct industrial models. The means concern four essential components of an industrial model: the form of power (types of capital and management), product policy (favoured market segments, types and designs of cars), production organization (design, purchasing, production, distribution), and employment relationships (phrase used in this text to indicate recruitment, work, working time, wage, promotion, and social benefits systems at company level). A 'company compromise' between management and employees is necessary to make coherent the production organization and employment relationships.[2] An industrial model can be thrown into crisis when the conditions which had previously made it viable and workable no longer exist. A period of great uncertainty then arises, in which companies must find a new form of coherence by betting on future forms of economic and political regulation and on the profit strategies that will be appropriate to these forms. Hence a battle breaks out as firms attempt to impose the environment which best suits the industrial model that each company is attempting to make work.

There have been four phases during the process of changing industrial models which the automobile industry has been going through since the mid-1960s. The first lasted until 1973. The model which served as a point of reference for most automobile companies at this time was the Sloanist model, embodied by General Motors (GM). This company had implemented a profit strategy which prioritized volume and diversity, within the context of the American mode of national economic growth, where a significant share of productivity increases was redistributed as purchasing power for employees. However, in the USA this mode of growth reached limitations due to its own success. In other countries, not all the conditions to make the profit strategy viable were met by the companies which tried to adopt it, notably as a result of the crisis of work which emerged at the end of the 1960s. In parallel, certain car manufacturers were choosing other profit strategies, notably the one giving priority to 'the continuous reduction of costs at constant volume' as well as the one based on 'innovation and flexibility'. They created innovative industrial models compatible with domestic modes of economic growth; these were the Toyota model and the Honda model, the principal features of which were already developed by the late 1960s.

The second period extended from 1974 to 1983 or 1985 (depending on the country). Floating exchange rates and oil crises brought global economic growth to a halt and led to confrontation between modes of national growth and industrial models. Confrontation in turn created crises for the labour compromises of countries where growth had been based on the redistribution to employee purchasing power of internal productivity gains: principally affected were the USA, France, and Italy. This accentuated wage disparities in Great Britain and dashed the hopes of some developing countries which were trying to acquire their own automobile industries. All the car manufacturers in the USA, France, and Italy experienced financial crisis. Conversely, companies in countries where the growth of purchasing power had long been linked to competitiveness in export markets, such as the German Federal Republic, Japan, and Sweden, were able to benefit. The Sloanist model was successfully transplanted to Volkswagen. The Toyota and Honda models, blithely subsumed under the category 'Japanese' in the USA and Europe, emerged into the open and revealed their effectiveness. Other Japanese firms, however, which had not successfully created a coherent industrial model, experienced difficulties. A fourth potential industrial model began to emerge at Volvo in Sweden. This developed into the 'reflexive production' model over the course of the following period, although it was never to characterize the entire firm.

This third period, lasting from 1983–5 to 1990–2, was marked by the recovery of the American economy, the oil 'counter-crisis' and the 'speculative bubble'. The explosive growth of automobile markets during this period enabled American, French, and Italian firms which were previously in crisis to profit as much as possible from the lower break-even points they had been forced to adopt. They then redefined their profit strategies and endeavoured to make coherent all the changes they had made. Far from a process of convergence of industrial models, there was a diversification of profit strategies and production configurations, although no coherent new industrial models were created. Conversely, the context of the period pushed the Sloan, Toyota, Honda, and 'reflexive production' models to their limits in the countries where they had been able to develop. Thus just as they were reaching their apogee, towards the end of this period the firms which embodied these models—Volkswagen, Toyota, Honda, and Volvo—experienced their own crises: crises of profitability for Volkswagen and Volvo, of work for Toyota, of innovative capacity for Honda.

The fourth period began with a dip in the automobile market in 1991–3 and transformations in patterns of demand provoked by the bursting of the 'financial bubble' and restrictive budget policies introduced in some countries. The period was marked by renewed confrontation between firms and modes of growth. But conditions had changed since 1974–82. The mode of national economic growth which involved redistributing national income as a function of internal productivity growth had been abandoned by all. Gaps in competitiveness between firms had been reduced. Previously successful models had been transformed. New countries had emerged. Global economic and political space was being reordered. Automobile firms now had to make their wagers on the trends they believed would prevail in terms of globalization or regionalization; they had to confirm their existing profit strategy or select a new one; and they had to adapt or create an industrial model capable of implementing it.

## 1.1. CONVERGENCE TOWARDS THE SLOANIST MODEL IS THWARTED, 1965-73

By the mid-1960s, Ford and Chrysler in the USA, Fiat, Renault, Peugeot, and Volkswagen in Europe, and Nissan in Japan had all selected the profit strategy of volume and diversity, and they attempted to mimic the industrial model which had successfully implemented this strategy in the automobile sector, the Sloanist model, named after the president of General Motors, Alfred Sloan. These companies were doing this at the very moment when the Sloanist model was reaching its limits in the USA. None of them had succeeded in becoming Sloanist by the eve of the first oil crisis, owing to their own previous trajectories and their particular economic and political environments. At the time, their attempts to adopt this model gave an impression of strong convergence, and it led to a belief that firms which did not do the same were doomed to fail. Yet two firms which followed different paths, Toyota and Honda, were in the process of creating new industrial models which were to prove very successful indeed.

### 1.1.1. The Sloanist Model and the Conditions that Make it Viable

The Sloanist model resolves the contradiction, once considered to be insurmountable, between economies of scale and product diversity. GM designed different car models on the basis of shared platforms. It was therefore able to offer models for each level of income, and versions of each model to meet different customer requirements. Moreover, these were fitted with extra equipment which enabled the company to demand slightly higher prices than competitors. The conditions which made the Sloanist model viable were based on a 'labour compromise' involving the redistribution of national productivity increases through the purchasing power of a moderately unequal distribution of wages. This compromise was created by GM following a 113-day strike in 1946. The unions renounced all demands related to the organization of work and technological change, which had been frequent before the war, and forbore to strike during the period covered by the contract signed with the company. In exchange they received regular reviews of wages against the cost of living, programmed increases in employee purchasing power, and improved social benefits. This compromise was then made national through various rules, conventions, and laws. Accordingly, the companies no longer sought to compete against one another through their wage levels. Furthermore, the compromise increased the incomes of a growing proportion of the population, thus creating the conditions for a diversified mass consumption and a relatively predictable level of demand. Hence the Sloanist model required companies to be attentive to changes in the demand for automobiles through marketing research, and it required recognition of unions in order to put employment relations on a contractual footing. Product variety and frequent replacement of models required significantly larger design departments, a relative polyvalency of machinery, assembly lines, and workforce, and a flexible integration of manufacturing within industrial groups which included components-making subsidiaries. In fact, the commonization of platforms, the definition of coherent and complementary product ranges, and control of the value chain across all the Group's companies and suppliers required not just coordination between divisions and subsidiaries but a true centralization of strategic choices, product and purchasing policies, together with their rigorous implementation across the companies which formed the Group.

If the Sloan model was to endure over time, it had to resolve two basic difficulties born of its own success: how to continue to increase volumes when the market had reached replacement level, and how to evolve the terms of the company compromise when the workforce became better educated and more demanding. The first difficulty could be overcome in a number of ways: by further increasing the ratio of models to platforms, by taking market share from competitors, by acquiring competitors and commonizing the platforms of the models retained, or by finding new markets in countries where economic growth was sufficiently well organized and income shared to give rise to regular growth in the demand for automobiles. The second difficulty required the content of work to be developed and career prospects to be created.

*1.1.2. The Problems of the Sloanist Model in the USA*

GM, Ford, and Chrysler ran into precisely this first problem between 1965 and 1973. The American automobile market became a replacement market at an average level of 12–14 million new vehicle registrations (cars and light trucks) per year. But at the same time the automobile parc continued to grow owing to used car sales. Air pollution, noise, accidents, and urban congestion all led the authorities to mandate increasingly strict safety and pollution norms. Without waiting for these to take effect, a proportion of the clientele was purchasing imported smaller cars which were less polluting and less expensive. Imports reached 15.2 per cent in 1973. With volume growth having stopped and new expenditure on research required, there were fewer resources to redistribute as salary increases and less new hiring, reducing promotion opportunities for existing employees. Working conditions became more difficult, and lay behind waves of 'wildcat' strikes which were neither authorized nor supported by the union (Bardou *et al.*, 1982).

We have seen above that an initial solution to the emergence of a replacement market was to increase platform commonization. The three American producers did just this, but not sufficiently to increase average volumes. GM was at its apogee, embodying the Sloanist model in its entirety. It controlled nearly half the domestic market. Its domestic production (USA and Canada) of passenger cars surpassed 5 million units in 1965. GM had long mastered product diversity, both in terms of design and from the manufacturing perspective. Moreover, its product diversity continued to increase, with the number of models rising from twenty-three to thirty-one even as the rate of commonization (the number of models per platform) rose from 2 to 2.4. Despite this, the average volume per platform remained approximately 350,000, because sales were stagnating (Bélis-Bergouignan and Lung 1994, Jetin 1998). Although it remained at a high average level of 7 per cent during the 1965–73 period (de Mautort 1980), the company's after-tax profit rate (as a proportion of turnover) began to decline. Ford and Chrysler, on the other hand, were still trying to make the transition to Sloanism when the domestic market reached its replacement level. Ford had not yet made a clean break with its original Fordist model. While it had long ceased to offer a single product (as during the 1920s), it had not commonized the platforms of the six models it offered at the beginning of the 1960s. Ford's product range progressively increased to twelve models, but the rate of commonization only rose to 1.5. Ford's profitability declined further than GM's, with average profits of 4 per cent. Its market share was eroded from 26.2 to 24.5 per cent and its domestic passenger car production stabilized at 2.7 million during this period. Chrysler had succeeded during the inter-war years in finding a niche alongside Ford and General Motors by designing vehicles which were innovative both technically and in terms of style, meeting new expectations and the needs of particular clienteles. Accordingly, Chrysler offered a relatively high number of models (ten in 1960). It then abandoned this innovation strategy in an attempt to become a generalist producer like GM and Ford. Engaged in a programme to expand its capacities, to vertically integrate, and to establish operations in Europe and Japan largely financed by debt, Chrysler was hit hard when the domestic market stopped growing. Heavily indebted, it had to lay off 12,000 employees in 1970 and call a halt to the construction of new factories. With domestic production of about 1.6 million passenger cars and light trucks, Chrysler controlled 15 per cent of the US market.

Other potential methods of increasing volumes were in fact difficult for the American manufacturers to implement. Antitrust laws prohibited the takeover of, or merger with, competitors. The remaining small producers had marginal outputs. Exporting products to rapidly expanding European markets and an exploding Japanese market was neither possible nor desired. American cars were not adapted to driving conditions in these countries. Moreover, GM, Ford, and Chrysler had subsidiaries in Europe which designed and produced models that met local requirements. At the same time, they sought to take control of Japanese firms, an operation made possible by Japan's membership of the OECD and GATT from the early 1960s. The signature of an 'Autopact' between the USA and Canada in 1965, to create a free trade zone for automobiles between the two countries, had only limited impact given the restricted size of the Canadian market. The Big Three could still try to win back domestic market share captured by foreign brands, notably in the small car segment. However, they believed that the production of this type of vehicle under American conditions would be barely profitable, and that small car sales resulted only from a passing infatuation. They preferred to import compact cars from their European subsidiaries as a means to limit the penetration of foreign brands. The American balance of trade in automobiles entered into deficit for the first time. Having failed to increase volumes by one or other of these methods, it was still possible for the Big Three, following the logic of Sloanism, to increase productivity by increasing automation. But they preferred to create new production facilities in the south and south-east regions of the USA, to follow geographical shifts in demand for automobiles, and to profit from cheaper, non-unionized labour (see Flynn, Chapter 8).

### 1.1.3. Apparent Convergence towards the Sloanist Model in Europe Belies a Diversity of European Company Trajectories

The situation in Europe was entirely different from that in the USA. Automobile markets were growing rapidly and becoming more diverse at the same time. A unified market was created encompassing Belgium, France, Italy, Luxembourg, the Netherlands, and the German Federal Republic, all signatories of the 1957 Treaty of Rome which foresaw the elimination of customs duties in ten years. By 1971 these countries had been joined by Denmark, Great Britain, and Ireland. The market for passenger cars and commercial vehicles in Western Europe (seventeen countries) grew continuously, from 7 million in 1965 to 10.56 million in 1973. Economic growth was so rapid that labour shortages began to appear, particularly in the automobile industry, which had become less attractive than other sectors owing to the nature of work. A crisis of work therefore broke out, manifested in problems of recruitment, high turnover, increased absenteeism, and strikes, all with degrees of intensity that varied by country. While in the USA this crisis was caused by a slowing of growth and productivity, in Europe it was ignited by their acceleration. The 'labour compromises' underlying growth in these countries were questioned by employees, who rejected the types and organizational forms of work that were imposed upon them in exchange for increases in purchasing power. European automobile producers therefore had to choose strategies to cope with both diversifying demand and the crisis of work. Fiat, Renault, Volkswagen, and Peugeot attempted to adopt Sloanism but without success. Fiat and Renault were thwarted by the crisis of work, Volkswagen encountered difficulties in trying to move from the Fordist towards the Sloanist model, while Peugeot could not find a partner. The two other European producers studied in this book, British Leyland Motor Corporation (BLMC) and Volvo, sought other paths.

Fiat and Renault increased their product ranges from nine to fourteen and four to eight models respectively, while maintaining a commonization rate of 2. The average volume per platform reached 195,000 for Fiat and 275,000 for Renault. Fiat broadened its range in 1968 and 1969 by taking over three Italian producers: Autobianchi, Lancia, and Ferrari, thus becoming a multi-brand generalist and specialist producer. Renault achieved the same through internal growth. Moreover, both companies adopted strategies of exportation and active foreign investment in order to increase their volumes. Global production of cars and light commercial vehicles reached about 1.6 million for Fiat and 1.4 million for Renault.

Fiat encouraged large numbers of workers to migrate from southern Italy to its factories concentrated in the Turin region. However, poor housing and transport conditions, overtime work, and lack of social dialogue lay at the root of 'l'autumno caldo' (the hot autumn) of 1969, which was the starting-point for a complete and long-lasting reversal of power relations in favour of the unions. Renault brought in immigrant workers, who, far from releasing labour market pressures, added new dimensions to social conflicts. Besides steady increases in wages, reductions in working time, and the institutionalization of the union presence in the workplace, both companies had to accept restrictions on polyvalency: at Fiat, by having to negotiate any changes to work posts and by limiting the reduction of cycle times; Renault by abandoning the wage system based on classification of the work post in favour of a salary set according to employee category.

In 1968 Volkswagen had decided to offer a range of products. Up to this point, the company had embodied the Fordist model: a single model (the cumulative production of which reached 15 million units in 1972), production that was vertically integrated (more than 50 per cent) and mainly concentrated on the Wolfsburg site, a long and continuous moving assembly line, short cycle times, and finally a wage compromise which guaranteed social peace and high wages. This compromise was buttressed by employee representation on the governing board of the company, by a single and powerful union, and by political consensus on the social market economy (see Jürgens, Chapter 11). Volkswagen experienced major difficulties in adopting the Sloanist model, demonstrating once again that, contrary to appearances, it is not a natural extension of the Fordist model. Having been restricted to improving a single product, Volkswagen's engineering department had prepared no projects and did not possess the competencies required to design other models. Accordingly, Volkswagen attempted to acquire these by purchasing the foundering firm NSU in 1969. Hesitation and conflict delayed the launch of new models, to the point of placing the firm at risk and leading the Deutsche Bank to demand a reduction of the workforce, which took effect in 1973, before the oil crisis (Streeck 1992). Worldwide production reached 2.33 million vehicles, 71.2 per cent of which were exported or produced abroad.

Peugeot's trajectory was different again. After World War II the company had followed a strategy of 'continuously reducing costs at constant volume', as Toyota was doing at the same time. More concerned with margins than with volume, Peugeot avoided taking risks, whether in terms of capacity, innovation, export markets, or foreign investments. Geographically concentrated and strongly rooted in a predominantly rural region, the company remained conservative financially, so as to create adequate profits, finance its own investments and remain independent, and also to guarantee employment to its workforce, from whom loyalty was expected in return (see Loubet, Chapter 13). Yet in contrast to Toyota, Peugeot decided to change its direction in the mid-1960s. The company adopted a strategy based on 'volume and diversity'. After several attempts to ally itself with other car makers, Peugeot expanded its product range on its own, reaching five models by 1973 (but with a low commonization rate of 1.3). Peugeot also experienced conflicts, precursors to the 'May 1968' events in France, which disrupted production though at lower cost to employees. The company immediately denounced the wage agreements drawn up nationally with the unions, refusing to establish contractual relations with them.

In Great Britain, the creation of BLMC in 1968 resulted from the gradual merger since the early 1950s of two generalist brands, Austin and Morris, four specialist brands, Jaguar, MG, Triumph, and Rover, and one commercial and industrial vehicles brand, none of which was either Fordist or even Taylorist. They produced different models for limited and quite distinct clienteles, which corresponded to distinctions in income and social status in Great Britain. Work was organized by work groups, the representatives of which were members of particular trades unions, who negotiated with factory managers the piece-rate for the parts to be produced, and to a certain extent guaranteed that production would be undertaken (see Mair, Chapter 15). The pace of work was controlled neither by organization nor by a technical system. It was negotiated through the setting of piecework rates. Reaching a plateau at 0.8 million vehicles, and under increasing competitive pressures, at the time of its formation BLMC decided to continue expanding its product range but to change its system of work. It hoped to increase the productivity of its 195,000 employees, without reducing their number, by replacing the piece-rate by timing the tasks to be carried out during an eight-hour working day; in short to move towards the Taylorist system. Not only did the new car models fail, but the foremen, upon whom depended the success of the new wage and work system, were unable to impose it on the shop stewards.

Volvo, the sole specialist producer analysed in this book, had a system quite similar to that of BLMC in the 1950s. Work groups of about fifteen members paid per piece assembled small runs of numerous car models on short lines. In accordance with the policy of the social-democratic government to seek growth and redistribution of national income through competitiveness in export markets on the basis of specialized products, Volvo ceased attempting to satisfy the whole breadth of the domestic market and decided to produce no more than two models in larger volumes, destined for an international clientele of affluent families with three children living in suburban or rural areas who wished to own a robust, durable, and safe vehicle (Ellegård 1995). To this end, Volvo built two new factories designed along 'American lines': with a continuous assembly line, time and motion studies, and short cycle times. Introduced during a period of labour shortages, these methods were immediately contested in practice, with particularly high rates of turnover and absenteeism, despite the introduction of an immigrant workforce. The crisis was sufficiently serious and the political and social context sufficiently favourable for Volvo to decide in the early 1970s to build two new factories where the assembly line would be replaced by automated guided vehicles (AGVs) on which operators could build subassemblies over long cycle times (see Berggren, Chapter 16). These two factories began operating in the year of the first oil crisis.

## 1.1.4. The Creation of Toyotaism and Hondaism in Japan

Japan experienced a mode of national economic growth very different from that in post-war Europe. This mode favoured investment over consumption until the early 1960s. Accordingly, in 1960 new passenger car registrations were still only 140,000, with commercial vehicle registrations twice as great. The overall market of 400,000 vehicles was shared among no less than nine producers, which were protected by customs duties of 35 per cent. Japan's membership of OECD and GATT in 1962 and then the shift of economic policy to the benefit of consumption shook up the automobile market and industry. In anticipation of growth in imports, the government asked certain sectors, the automobile industry in particular, to prepare to export and therefore to be competitive in overseas markets so that it could contribute to the balance of trade when the time came.

Domestic demand and exports both grew strongly. The market for passenger cars reached 2.93 million new registrations in 1973 and that for light commercial vehicles 1.98 million. Exports rose from several thousand in the early 1960s to 2.07 million in 1973. As for vehicle production (passenger cars and commercial vehicles combined), this leaped to 7.08 million vehicles in 1973. Over a very short period of time, both the market and production system became mass and diversified, starkly accentuating urban congestion, pollution, and road safety problems. The Japanese government introduced regulations that constrained both consumers and producers. The labour market soon became tight, forcing the car makers to call upon recruits who were increasingly well educated.

Contrary to the idea that later prevailed in Europe and the USA and which is still widely believed, the various Japanese producers neither pursued the same profit strategy, nor developed the same industrial model. Toyota chose the strategy giving priority to the continuous reduction of costs at constant volume, and invented an original industrial model to achieve this. Nissan opted for the strategy of volume and diversity, without, however, succeeding at becoming Sloanist. Honda, initially a maker of motorbikes, succeeded in becoming an automobile producer after all others, in the mid-1960s, by implementing a strategy of innovation and flexibility and by constructing an industrial model enabling it to avoid or limit the risks peculiar to that strategy. Mitsubishi, committed to the same path as Honda, was not successful.

Toyota's output grew between 1960 and 1973 from 42,000 to 1.63 million cars and from 112,000 light commercial vehicles to 580,000. The number of passenger car models offered rose from two to nine, at a moderate commonization rate of 1.6. However, Toyota's distinguishing mark was its break-even point, which remained remarkably low despite a marked increase of investments and wages. The Toyotaist model was the result of a long process which was only completed in the 1970s. Although its founder held up Fordism as the ideal, by the 1950s Toyota had to find the resources to be profitable in a limited, diversified and highly competitive automobile market, while at the same time guaranteeing employment, as it was committed to do following the social conflicts that had been provoked by redundancies in 1950. The solution to these apparently contradictory demands was to continuously reduce costs at constant volume and require the participation of employees in this process. This was the inspiration for the organizational and wages innovations of Taiichi Ohno: just-in-time to reveal and eliminate the problems which generate waste, a system to make wages and promotion dependent upon the reduction of standard times by the relevant work group, and a working day which can be extended by overtime if so required (see Shimizu, Chapter 3). The joint declaration issued by the management and the union in 1962 made the company compromise official in that it exchanged participation in improved performance for a guarantee of employment and career advancement. Toyota might have abandoned its profit strategy, which was highly demanding for employees, suppliers, and managers, with the development of a mass market. Economies of scale, scope effects, and improved quality might well have replaced the model. But Toyota preferred to benefit from these other profit sources as the icing on its cake; it only relied upon volume to the extent that the market and its self-financing policy permitted, and it diversified its range and improved the quality of its models to the extent that this was useful commercially. At the same time, the company did not seek to innovate in terms of products, owing to the risks this might run. Instead it preferred to copy or buy patents, once an innovation undertaken by another company had been validated by the market.

Like all other models, the Toyota model had its own conditions of viability, and so was not viable in all circumstances and in all places as is sometimes claimed. Employee participation can be threatened when the labour market is tight and young people turn to less demanding companies and sectors. The participation of suppliers becomes less sure when they have opportunities to diversify their client base and become more independent. The other manufacturers and countries involved may erect barriers in order to contain the firm or firms which have been made expansionist by the very adoption of the model. Finally, the firm may be destabilized by a radical innovation in product design which it has been unable to purchase because it has been monopolized by the inventor/manufacturer or which the firm is unable to copy owing to a lack of the requisite competencies. While Toyota has yet to encounter the last situation, it certainly had to deal with the first three situations during the 1990s, as we shall see below.

In 1973, Nissan produced 1.42 million passenger cars and 0.56 million commercial vehicles. It had diversified its range, notably by acquiring the manufacturer Prince in 1966. On the eve of the first oil crisis, Nissan offered ten models, each with its own platform, for an average volume per platform of 142,000 units, compared to 300,000 for Toyota the same year. The company had oriented itself towards export markets at an early stage in order to attain higher volumes than were possible in the domestic market. It had applied American management methods, notably in terms of quality, in order to become competitive on the international market. Productivity increases were sought through technical advance. The wage system was based on the post occupied and age, only weakly influenced by the factory output coefficient and evaluation by the hierarchical superior. In the mid-1960s, Nissan had been particularly affected by labour shortages owing to its location in the Tokyo region. It had been necessary to recruit workers who were leaving high school after twelve years' schooling and accordingly to pay higher wages and grant a five-day week from 1971. To counteract absenteeism and the reduced interest of employees in improving quality, 'quality circles' had been launched in 1966 with union support. But Nissan had not succeeded in establishing a stable compromise. Like Toyota, it had experienced long and difficult social conflicts in the early 1950s, but these had ended differently. The company's survival had been the result of the creation of a second union which had been able to supplant the initial revolutionary union. On the basis of its role in resolving the crisis, this cooperative union had acquired a powerful position in issues ranging from promotion, changes in working hours, length of working hours, working conditions, to the overall strategy of the company. A kind of dual management system had been established, to the detriment of the unity and coherence of command (see Hanada, Chapter 4).

Honda was unusual in having already created an industrial model by the time it entered the automobile industry. Twelve years after it was founded in 1948, Honda had become the world's largest motorcycle manufacturer, on the basis of a strategy which focused on product innovation and production flexibility and on the mass production of products which had in effect opened new market segments. The firm's success owed much to the mechanical and commercial imagination of Soichiro Honda himself. His associate, Takeo Fujisawa, who was in charge of the organization and its finances, had been concerned from the start to find the resources needed to overcome the difficulties inherent in this profit strategy. Besides loss of capacity to innovate over the long term, the risks of the strategy included of course the inevitable failures, the over- or under-estimation of demand and the refusal of investors in banks to finance projects. Industrial models which are to be consistent with part of a strategy of 'flexibility and innovation' must therefore give the firm the resources to counter these risks or reduce their impact.

As a means to encourage creativity and get flexibility accepted, Honda and Fujisawa developed a company compromise which was not dependent upon group spirit and loyalty as at Toyota but rather on the recognition and gratification of individual talents as well as good work and employment conditions (see Mair, Chapter 5). Inventiveness and expertise were first valued by a promotion path and wage scale, named the expert system, which ran in parallel to the traditional lines and scales. Although first conceived in the 1950s, this system could not be established until 1967, due to union opposition. Research activity was separated from product development so that it would not depend closely upon the requirements of the design. Each engineer had the freedom to submit his projects to an evaluation committee, and to receive a budget and form his own team if one of them was accepted. Salesmen were integrated into the process of innovation, by being given the task, in addition to selling, of discovering client expectations. In order to retain its independence, Honda did not attempt to join a *keiretsu* with linked banks and industrial firms, nor to maintain close ties with the political world as Nissan had done. Neither did it form an 'association' of suppliers, to which it would have obligations. The company used Toyota's and Nissan's suppliers, and was able to profit from their experience and prices. Above all, Honda was careful to be self-financing and not become financially dependent upon the banks.

By 1967 Honda had become a proper car manufacturer. It opted for an innovative automobile niche and exportation in order to create a place for itself among Japanese producers. It marketed a mini front-wheel-drive car with a small but powerful air-cooled engine. The model's commercial success in Japan propelled Honda into third place behind Toyota and Nissan, with an annual production of 277,000 passenger cars in 1970. However, the second model launched was a failure: too expensive and too polluting. The engineers at Honda then refined a water-cooled engine, the CVCC, which reduced pollution at source (at the point of combustion) and consumed less petrol. It was the first to respect the norms of the American Clean Air Act, two years before they came into force, and equipped the Civic, a small and somewhat sporty model, which was spacious, reliable, and was sold in five versions: two, three, four and five doors and estate. This was precisely the model to have at the time of the first oil crisis, notably in the American market.

Like Honda, Mitsubishi Motors Company (MMC) pursued an innovation-based strategy, but unlike Honda it was not a success story. As a manufacturer of trucks, buses, and jeeps, Mitsubishi Heavy Industry attempted to become a producer of passenger cars in the early 1960s (see Shimizu and Shimokawa, Chapter 6). The company carved out a small niche for itself by launching a minicar, but the failure of its two other models, despite their advanced technology, drove the company to link up with Chrysler (which took 15 per cent of its capital) in 1970 after having rejected a government proposal to merge with Isuzu. This agreement was an important opportunity for Mitsubishi, since it was planned to use the two sales networks jointly, with Mitsubishi designing and producing a compact car for Chrysler. Although Mitsubishi's market share in Japan dipped from 6.0 to 3.7 per cent between 1970 and 1973, its exports enabled it to double its output. Mitsubishi counted on technology to increase productivity, equipping its factories with the most modern machinery (without recourse to *kaizen* and *kanban*), as well as relying on a skilled workforce. From 1970 the company recruited workers who had finished their secondary education and trained them for a year before assigning them to a work post. The wage and promotion systems were based on assessments of the individual's competence and on results. The adaptation of volume to demand was achieved, as at Toyota, by a system of two shifts per day, but with the possibility of overtime limited to an hour and a half and annual working hours restricted to 1,960 hours in 1973. Mitsubishi had formed an association of suppliers, which were certified for quality, in 1968.

## 1.1.5. Attempts to Transplant Fordism to Countries Outside the Three Global Poles

During the 1950s most countries elsewhere in the world had adopted policies of import-substituting industrialization. This policy required foreign manufacturers to invest in the country and to buy locally an increasing proportion of components if they wished to sell their cars there. Countries which adopted this policy hoped that it would permit technology transfer and the development of the local network of suppliers that would be vital to the establishment of an autonomous national industry. Car makers which accepted this policy, mostly European companies, hoped that these countries' domestic markets would grow over the medium term. In the short term, they could at little cost transplant out-of-date equipment, prolong the life of products that had already been renewed in their home markets, and lengthen production runs for the components they were permitted to import. In practice, then, they set up production lines to assemble one or two old models. The policy was successful in terms of the numbers of car makers setting up factories, but soon reached its limitations. While demand increased substantially in Brazil, for instance (0.57 million in 1973), markets remained narrow and production costs high. National income distributions remained very unequal and for some countries national income varied greatly according to variations in raw material prices. Accordingly, mass demand could not develop. Moreover, the balance of trade in automobiles remained in deficit, owing to importation of machinery and components and the payment of royalties (Maxcy 1981). With few exceptions, the car makers were unable to develop Fordist production. With markets protected, they nevertheless made substantial profits. This explains why, by the start of the 1970s, several countries, notably Argentina, Brazil, and Mexico, required companies to compensate for their imports of machinery and parts through exportation, giving fiscal incentives to encourage this.

From the mid-1960s, the Eastern European countries had already adopted a compensation policy in the form of their contracts with Western European producers. Desiring to accelerate the production of consumer goods, the Soviet Union, Poland, and Romania, in particular, purchased licences to make models which had reached the ends of their lives from Fiat, Renault, and Citroën, as well as the tooling and equipment necessary to produce them. To compensate, these companies had to buy components and assembled cars. The largest such operation was undertaken by Fiat in the Soviet Union, with the construction of the Avtovaz factories at Togliattigrad and the sale of a licence to manufacture a family car. This totally integrated production complex, with a capacity to produce 700,000 vehicles per year, was opened in 1970. Given significant household savings, there was mass demand in the domestic market (see Chanaron, Chapter 17). All the ingredients for Fordist production seemed to be present.

South Korea wished to develop a national automobile industry. From 1962, importation was forbidden and only companies with majority Korean ownership were authorized to manufacture passenger cars. Ten years later, total annual output had only reached 12,400 vehicles. Since 1968, Hyundai had been assembling a few thousand units of first one, and then two of Ford-Europe's cars. Although the cars were assembled on an assembly line, there was no real cycle time (see Chung, Chapter 7).

On the eve of the first oil crisis, a convergence of the majority of automobile producers towards the Sloanist model had not in fact taken place. Moreover, GM, which had embodied this model up to this point, had been unable to adapt it to a context of slow growth and a national market which had reached the replacement level. This was not to imply that the difficulties being experienced could not be overcome in the context of existing national modes of growth. The US companies believed that the penetration of the domestic market by foreign firms was basically a result of the undervaluation of their currencies.

## 1.2. THREE SUCCESSFUL INDUSTRIAL MODELS EMERGE FROM THE CONFRONTATION BETWEEN FIRMS AND NATIONAL MODES OF ECONOMIC GROWTH: 1974–1983/5

The new monetary policies of the USA finished, through a chain of unanticipated events, by breaking the rhythm of global economic growth and provoking a confrontation between modes of national growth and industrial models, via the medium of the high levels of competition in which firms were forced to engage. The firms which succeeded during this period were all based in countries where the mode of growth was already rooted in competitiveness in export markets: Toyota and Honda in Japan, Volkswagen in the German Federal Republic, and Volvo (only during the 1980s). Yet to be based in these countries was not a sufficient condition for success, as revealed by the difficulties experienced by Toyo Kogyo (the future Mazda), Nissan, Mitsubishi, Isuzu and Suzuki in Japan, and Saab in Sweden. Conversely, and without exception, firms in the countries where growth had been based on the autonomous mode, and which were or sought to become Sloanist, experienced a financial crisis. Each went on to reduce its break-even point drastically and then completely change its employment relationships. Simultaneously, the home countries of these firms changed their modes of national economic growth. Lastly, the countries which had adopted the 'import substitution' gave this policy up in favour of, first, a policy to compensate for imports with exports, and then, insertion into the international division of labour. Only South Korea continued to try to create an autonomous automobile industry.

### 1.2.1. Global Growth Falters

The USA attributed the overvaluation of the dollar, which gave imported products a price advantage, to fixed exchange rates and to the inflation raging in a number of countries. The trade deficit, in which the automobile sector was a significant factor, was one reason for abandoning the gold standard in 1971 and freeing exchange rates in 1973. These measures effectively reduced the imports from the generalist European producers, but not those of the Japanese and specialist European producers. Above all, they reduced the incomes of the countries which exported raw materials. The oil producing countries reacted by quadrupling the price per barrel of oil in constant dollars during a period of political crisis in the Middle East.

The first oil crisis broke the rhythm of economic growth. In 1974 automobile markets were in decline everywhere (–23.3 per cent in the USA, –22.0 per cent in Japan, –14.0 per cent in Europe). Markets recovered their 1973 levels by 1976 in Europe, but not until 1978 in the USA and 1979 in Japan. Measures to save energy, speed limits, and new norms for fuel consumption exerted their influence on demand. Demand shifted towards small and medium-sized cars, which were less expensive and more economical in terms of fuel consumption. However, the first oil crisis also dramatically increased total payments for oil, and therefore inflation. All countries seriously endeavoured to reduce their energy dependence, but they were all constrained to increase their exports in order to pay for more costly imports and therefore to find abroad the sales which their domestic markets no longer offered. Those which were already oriented towards export markets had to increase their export volumes, notably those deficient in raw materials, such as Japan. Others were suddenly confronted with questions about their international competitiveness and their ability to resist the penetration of foreign firms into their domestic markets. Both European and Japanese producers were confronted with the problem of remaining profitable in diversified replacement markets, like American producers had been ten years earlier, and now had to face stiff international competition. Italy had applied an import quota on Japanese cars, set at a few thousand units per year, during the 1960s. In 1975 Great Britain negotiated a restriction of Japanese market share to 11 per cent of the domestic market. In 1977, France unilaterally set a level of 3 per cent. These barriers did not hinder Japanese competition in third markets, however, or competition within the European Community between European firms.

The unanticipated consequences of the depreciation of the dollar clearly had an impact contrary to what had been hoped for on the US balance of trade; not only had the price of raw materials increased, but so too had the volume of imported manufactured goods. It also contributed to the growth of inflation, which became the fundamental problem in the eyes of the American monetary authorities. In order to rein in inflation, in 1978 the Federal Reserve Board began to increase interest rates. The resulting appreciation of the dollar brought about a second oil shock and crisis for many companies. The price of oil doubled in constant dollars between 1979 and 1981. The rise in petrol prices due to dollar appreciation was amplified by the building up of precautionary stocks and by the revolution in Iran. Contrary to expectations, global inflation was reignited. Recovery of the automobile market, which had seemed well in train, was once more thwarted. Demand did not regain its 1979 levels until 1984 in the USA, 1985 in Europe, and 1987 in Latin America. While demand had recovered in Japan by 1982, it then stagnated until 1987.

In the USA, the Reagan administration at first adopted a more relaxed attitude, persuaded that businesses would only restructure themselves when threatened by crisis and that their employees would only accept revisions to working rules under threat of unemployment. The share of the American market controlled by Japanese producers rose from 6.2 per cent in 1973 to 19.4 per cent in 1982. However, in 1981, the American automobile producers and the UAW obtained a concession from the Reagan administration, that a 'self-limitation' agreement restricting exports to the USA to 1.76 million vehicles per year should be negotiated with Japanese producers. The European Community, fearing that Japanese producers would redirect their exports to Europe, then negotiated a restriction to 9 per cent of the Community market. Japanese producers were now obliged to invest in the USA and Europe if they wished to increase their sales there.

*1.2.2. The Successful Firms from the 'Exporting' Countries and the Others*

Companies in countries where the mode of national economic growth and income redistribution was based on competitiveness in export markets were better prepared to deal with and profit from an accelerated interpenetration of economies. Their company compromises could be adapted without major political and social conflicts. The unions accepted wage moderation in exchange for job preservation. Of the firms studied here, Toyota, Honda, Volkswagen, and Volvo (to a lesser degree) were the firms which did not experience a financial crisis and which managed to maintain a break-even point permanently below added value (between 70 and 20 per cent). This low break-even point allowed them to remain profitable throughout, with the exception of Volkswagen in 1982–3 (owing to losses at its electronics subsidiary and Latin American subsidiaries).

One reason for the success of these companies was that their strategies were consistent with the mode of growth in their country of origin. Following the first oil crisis, Toyota's strategy to 'reduce costs at constant volume' was the perfect strategy to reinforce the Japanese mode of growth based on price-based competitiveness in export markets. In accordance with this strategy, the firm progressively broadened its model range (from 9 to 19 between 1973 and 1985), the commonization of its platforms (from 1.5 to 1.9), and its average volume per platform, at the same time as improving the quality of its cars. In 1985 Toyota exported or manufactured abroad 57.5 per cent of its global production. Conservative, but also conscious of the conditions necessary for its model to function, it was the last of the three big Japanese producers to invest in the USA. Honda also had a strategy consistent with the Japanese mode of growth, although for other reasons. As seen above, Honda had chosen to respond to new types of demand and to export its products as the means to stake its place as an automobile manufacturer. Its export success was due in part, of course, to competitive prices and to the quality of its cars, but more so to its innovative car model, the Civic, of which 3 million units had been produced by 1982. Honda was able to expand its range to six models without commonizing their platforms, to give each its own clear personality. By 1985 it exported or produced abroad 63.4 per cent of its worldwide production. Before being obliged to invest in the USA to increase its sales, Honda had already decided to do so. The company made its first cars in the USA in 1982. For its part, Volkswagen's strategy remained consistent with the German mode of growth through specialist exports, making the transition from a volume-based strategy to a strategy of 'volume and diversity'. Two parallel ranges were launched, one Volkswagen and one Audi, each covering the four principal market segments. By increasing the commonization rate to 2, Volkswagen was able to increase its average volume to 300,000 per platform. By 1985 it had produced 2.4 million vehicles worldwide, 77.3 per cent of which were exported from the German Federal Republic or produced abroad. Like Sweden in general, Volvo found it more difficult to maintain its growth path as a specialist exporter in the wake of the first oil crisis. Volvo initially tried to move downmarket, following demand, with its 1974 purchase of the crisis-ridden Dutch producer Daf which had two models in the small to medium segment. In doing this Volvo not only faced the difficulties related to all growth abroad, but also lost its position as a specialist, exposing itself directly to competition from the generalists. The results were disappointing, and Volvo refocused on its particular strengths: upmarket family cars which were strong and safe. With the successful launch of the 700 series, and a 26 per cent devaluation of the Swedish krona in 1982 at a period when the dollar was at its highest level, Volvo's worldwide production of passenger cars reached 400,000 in 1985, with 85 per cent of cars exported or produced abroad.

A second reason for the success of these four firms was the consistency between their production organization, their employment relations, and their profit strategy. For these companies, the new phase was an opportunity to reinforce these linkages. Toyota deepened its system by diffusing continuous cost reduction to design, administration, and product distribution, also requiring first-tier suppliers to adopt the Toyota production system. For its part, Honda continued to seek ways to maintain and indeed increase its capacity to innovate and to make its production apparatus more flexible. In particular, Honda decided always to seek two solutions to the same problem, so as not to become prisoner to a premature decision. Conversion of its production lines to other models was made easier, and the principle of producing homogeneous lots of thirty to sixty vehicles was retained. Unlike Toyota and Honda, Volkswagen and Volvo still had to bring their production organization and employment relations into line with their new profit strategies. Volkswagen had to introduce polyvalency and to develop an appropriate social compromise in order to make its 'volume and diversity' strategy viable in a market where growth was weak. The restructuring plan, with its elimination of 43,000 jobs, which the Deutsche Bank had imposed prior to the first oil crisis because of the company's delay in making the transition from a single model to a full range, permitted both goals to be attained. Management and union agreed to postpone wage increases in order to preserve employment. Overtime was preferred to employment expansion. Prior to potential recourse to redundancies, all efforts were to be made to utilize internal mobility, early retirement, and, from 1983, reductions in working time. Acceptance of internal mobility led the wage and classification system to be modified, which increased the polyvalency of employees. In 1974, Volvo opened its Kalmar factory, which was to become the symbol of the 'socio-technical' path. Along with similarly inspired reorganizations of the older factories, this factory was to permit Volvo to offer high quality products at acceptable costs in a replacement market and in an environment of labour shortage. The goal of reducing absenteeism rates was achieved, but that of reducing turnover was not (Auer and Riegler 1990). Above all, the announced revolution in the content of work went unnoticed by Kalmar's workers. The principle of the production line remained intact in practice, with centralized and programmed control of AGVs, even if the system was applied in a less constraining way. The strong growth of Swedish exports, and those of Volvo in particular after 1982, led to a renewed shortage of labour even as elsewhere in Europe unemployment was rising rapidly. Volvo was once again obliged to improve the attractiveness of work in its factories, and in 1985 decided to 'go further' than Kalmar in the new assembly factory that was to be built at Uddevalla.

Other firms in the countries where economic growth and income redistribution were based on export competitiveness were not as successful as the four firms discussed above. Their production organization and/or their employment relations were not consistent with their profit strategy. Toyo Kogyo (the future Mazda), which had followed an innovation strategy, had become overly reliant on the rotary engine. The rise in the price of oil militated strongly against this type of engine, which used too much fuel, but the company was not flexible enough either in design or in production to reorganize itself easily (Shimokawa 1994). It had to rely on support from Ford in 1974 and again in 1979, selling Ford 25 per cent of its capital. Mitsubishi continued to innovate, notably with the 1982 launch of an off-road recreational vehicle which was to prove very successful. However, its growth was financed by borrowing. Heavy debt repayments reduced its profitability and led to losses when some of its models fared badly, especially in the domestic market. Nissan's profit rate plunged in 1983 and became negative in 1986 despite the fact that 62.3 per cent of its vehicles were exported or made abroad. While Nissan was competitive overseas, it was less strong in the domestic market, losing market share continuously from 1974 (from 33 per cent to 28 per cent in 1985). The product range was broadened from 10 to 18 models, but the platforms were not commonized (a rate of 1.3). Accordingly, the average volume per platform remained very low, at about 130,000. Nissan tried to regain lost market share with a programme to increase productivity through automation and employee participation, and by introducing the *kanban* method. The failure of this programme led to tension between management and union, each blaming the other. The conflict emerged into the open when the union opposed a plan to build a factory in the United Kingdom. The factory which opened in the USA in 1983 had indeed resulted in a reduction of exports and above all a fall in employment in Japan (from 59,600 in 1983 to 51,200 in 1987). The dual management system was in open crisis.

### 1.2.3. The English Case

All the producers located in Great Britain, whatever their form of production organization, ran into difficulties. The national mode of fixing wages, based on local negotiations involving separate employee categories as a function of particular power relations without national level regulation, was unable to ensure the long-term competitiveness of firms, particularly in the case of the British Leyland Motor Corporation.

BLMC, barely profitable during the growth years, had to be nationalized after the first oil crisis to avoid disappearing altogether. It then lurched from crisis to crisis. Neither the considerable number of models offered, nor automation, nor the progressive reduction of the workforce sufficed to re-establish competitiveness. The reappearance of financial losses and the arrival of a Conservative government under the leadership of Margaret Thatcher, who was prepared to break the power of the unions, encouraged management to require an employee vote on its plan to close a number of factories, to reduce the workforce by half, and to prohibit shop stewards and trade unions from involvement in the organization of work. The restoration of managerial authority and the partial replacement of the product range with support from Honda did not, however, prevent a new crisis. BLMC was taken over by British Aerospace and a new strategy adopted in 1986. The break-even point was reduced to 400,000 vehicles, the workforce reduced to 40,000, the production of all automobile components with the exception of engines, gearboxes, and body parts abandoned. Bus and truck activities were sold to Daf. For the first time in many years, BLMC, which had now become Rover, had defined a coherent product policy: a single range of passenger cars, restricted to four models, positioned at the top of each market segment, coupled with the transformation of Land Rover models into recreational vehicles. Following a partial swap of ownership with Honda UK, Rover entrusted Honda with the design of new models.

### 1.2.4. Firms in Countries with the Autocentred National Economic Growth Model All Experience Crisis

Firms in countries which had adopted an autocentred mode of growth based on the redistribution of productivity increases into purchasing power, like the USA, France, and Italy, all experienced crisis at some point between 1974 and 1986: Chrysler, Citroën, Ford, Fiat, PSA, Renault, and GM. All lost market share. Chrysler passed below the 10 per cent threshold in 1978. Ford's sales in the USA had dropped by half between 1979 and 1982. Its market share shrank to 17 per cent, a loss of 7 points in three years. Fiat lost 13 points in its domestic market and 5.4 in the European Community market between 1973 and 1979. PSA's losses between 1979 and 1982 were of the same order. Renault initially benefited from the misfortunes suffered by its European competitors; it even briefly became the largest European producer, although it still lost money on every vehicle sold. The collapse was even more severe when it came, in 1984. GM too at first profited from the disintegration of Chrysler and Ford, prior to losing market share in its turn (from 43.6 per cent to 35.0 per cent between 1982 and 1986). The breakeven point of all these firms exceeded their added value over periods lasting several years. Some were only able to make profits, and only fended off crisis, with the aid of income from their foreign subsidiaries or their financial activities. The American firms diversified their product ranges (including importing small vehicles manufactured by Japanese firms in which they held a share of the capital: Isuzu and Suzuki for GM, Mazda for Ford, Mitsubishi for Chrysler), while substantially increasing the commonization of their platforms (from 2.4 to 3.0 for GM, from 1.5 to 2.7 for Ford, from 1.3 to 2.6 for Chrysler between 1973 and 1984). Average volumes nevertheless declined, or stagnated in the case of GM, owing to falling sales, and they were unable to avoid financial crisis. The French and Italian firms also diversified their product ranges, but without significantly commonizing their platforms. Average volume per platform declined at Fiat and PSA, while at Renault it remained stable between 1973 and 1984.

The fundamental cause of these crises was the contradiction between the mode of redistribution of national income as a function of productivity increases and the new necessity for firms to be competitive internationally. The previous labour compromises were now under attack by employers, but were defended in terms of social progress by employees, even though they had been denounced by employees and supported by employers during the previous phase. These conflicts were not simply short-term conflicts of interest. In practice, global confrontation was not unavoidable, and there were in fact attempts to avoid it. The first attempt to do so involved trying to relaunch growth on the basis of the sudden enrichment of countries endowed with raw materials and energy resources. Some automobile producers hoped that these markets would take up the slack in demand in the industrialized countries. Fiat invested in an assembly factory in Brazil in 1976. Renault and PSA signed (in 1975 and 1977, respectively) agreements with Iran to each produce 100,000 vehicles per year. However, market growth in these countries was a short-term phenomenon. The rise in interest rates added greatly to their debt repayments. The second attempt to avoid global confrontation was made by member countries of the European Community, which tried to accelerate its development by coordinating economic policies, by reducing variations in exchange rates, and by harmonizing regulations. However, swings in the political control of European countries made it more difficult for them to overcome divergences in their interests and structural differences. The third attempt was made by the Reagan administration. Greater military spending was to provide American companies with the means to acquire advanced technologies for their sector. The final attempt to avoid confrontation was pursued by companies which transferred capital and labour towards new activities with strong growth potential outside the automobile sector. This strategy was attempted by the majority of firms in these countries, as well as by Volkswagen and Volvo, which diversified into sectors such as electronics, aeronautics, leisure activities, pharmaceutical products, or even food and oil production. With few exceptions, this diversification did not deliver the anticipated results. It even reduced the availability of the investment funds needed to replace car models as fast as some of the car producers desired and to raise the quality of their products to the levels now required by the market.

The second cause of continuing problems was failure to adapt supply to demand. Demand had become redirected towards, on the one hand, compact vehicles which were less expensive and more economical in terms of petrol consumption, and on the other hand, similarly priced vehicles that were more reliable and of higher quality. The inability to adapt products particularly affected the American producers, which were essentially supplying only large cars. The new trends in demand, reinforced by new norms governing safety, environmental impact, and fuel economy enacted by the public authorities, obliged the companies to change their automobile designs. Not only had they to design new engines, but they also had to reduce the weight and size of their vehicles, to improve their driving characteristics, and, in order to do these things, to change to front-wheel drive technologies. The French and Italian generalist firms did not have to revise the fundamental design of their cars, but they did indeed have to improve their quality. They had also only just launched new market models which often then failed as demand for large cars declined. Citroën, already quite indebted, was unable to support the poor sales of a new luxury model. Its principal shareholder, Michelin, was obliged to sell it to Peugeot in 1974.

Finally, the programmable automation and external growth which were supposed to improve the performance of the firms that followed these paths frequently, on the contrary, contributed to their deterioration. Programmable automation ought to have increased productivity while simultaneously improving quality, making production equipment more flexible, and eliminating arduous unskilled work. The significant investments made did not deliver the anticipated results, in part because they were too ambitious, and in part because of the economic and social assumptions implicit in the particular forms of automation adopted (Freyssenet 1997). Acquisition of a competitor or merger with it might be the way of finding new economies of scale: assuming, of course, a rapid commonization of platforms across brands. A number of firms which started off down this path did not succeed, given that it was a long, difficult, and costly operation. The average volume per platform at PSA fell to 73,000 in 1981, following the acquisition of Citroën in 1974 and then Chrysler Europe in 1978. Ford's purchase of part of the capital of Toyo Kogyo in 1974 and 1979 did not permit Ford to modify the Japanese manufacturer's product range. In 1982 Renault had to take over the American Motors Company (AMC), with which it had formed an alliance in 1979. As a result, Renault became even more indebted just as interest rates began to rise fast.

None of the companies in countries which had previously followed the auto-centred mode of national economic growth, then, was able to avoid a financial crisis. All of them then embarked on a drastic lowering of their break-even points through the immediate reduction of their workforce, the closure of factories, the externalization of some of their manufacturing, and the restriction of investment to the replacement of existing models. They reduced their debt by selling shares, notably in sectors outside the automobile industry and abroad. They demanded price reductions from their suppliers and upfront payments from their distributors. Chrysler was the first to be affected, reducing its workforce by 49,000 employees between 1973 and 1975. It sold its European subsidiaries (Simca, Barreiros, and Rootes) to Peugeot in 1978. Even this was not enough, and Chrysler was obliged to drastically lower its break-even point once more from 2.4 million vehicles to 1.1 million, reduce wage levels, close twenty-one factories, and eliminate 75,000 jobs between 1978 and 1982. By 1979 Ford was also in crisis. It closed nine factories, including three assembly plants, and reduced employment by 25 per cent over four years. Fiat reduced its break-even point to 1.2 million cars by laying off (with priority in rehiring) 20,500 employees. It withdrew from countries where it had not been successful: the USA, Argentina, and Spain. PSA eliminated 98,000 jobs worldwide between 1979 and 1986, closing factories in the United Kingdom, Spain, and elsewhere. Renault reduced its workforce in France by 26,000 between 1984 and 1988. It sold off numerous subsidiaries, notably AMC to Chrysler. Lastly, GM, having underestimated Ford and Chrysler's capacity to recover and the early success of the Japanese transplants, in 1986 announced the closure of eleven factories and the reduction of its workforce by 25 per cent over five years.

For these rescue plans to be implemented, they had to be accepted by the employees and financed by the state. The agreement signed between Chrysler and the UAW marked a turning-point in industrial relations in the USA. For the first time, in 1979, the union made important concessions to ensure the firm's survival, notably by accepting wage reductions. At Ford the UAW went even further. It became deeply involved in the implementation of teamworking and in the improvement of productivity and quality. It accepted wage reductions in exchange for fewer redundancies than had been desired by management. GM and the UAW directly collaborated in experiments with new forms of industrial relations and new ways to manage the company. The president of GM and the president of the union personally committed themselves to the creation of Saturn, a subsidiary producing a model of the same name, which was co-managed at every level from the boardroom to the shopfloor (Pil and Rubinstein 1998). In France and Italy, on the other hand, the drastic reduction in the break-even point was attained after direct confrontation with the unions led to their defeat. In every case state aid was called upon: loans to permit the purchase of a failing company (Citroën by Peugeot), finance for product renewal (Chrysler's K-Car front-wheel-drive platform), contributions to debt reduction (Renault), or funding to take care of redundant employees or early retirees.

### 1.2.5. Problems Implementing Fordism in Other Countries

Increasing raw materials prices brought about economic growth in some of the producing countries, fuelled by the loans which international banks were quick to make. This economic boom did not last long. Interest rate rises greatly increased indebtedness. The Brazilian market, for instance, the largest among these countries, which had doubled in size during the 1970s, stagnated during the 1980s at around 0.6 million passenger cars per year, shared among four producers: Volkswagen, Fiat, Ford, and GM. Vehicle exports, which had been demanded by the host countries first to compensate for automobile sector imports and then to help the balance of trade after the oil price rises, could only be sent to other developing countries, given the characteristics of the models being produced.

After the failure of the first attempt to create an automobile industry in Korea, in 1974 the government announced a new plan to raise production to 500,000 in 1980 and 1,000,000 in 1985 by exporting models which were basically Korean. In 1982, however, production remained at a level of 94,500 units (after peaking at 113,600 in 1979). The lack of development of a domestic market prevented the main automobile producers, Hyundai and Daewoo, from obtaining the economies of scale that were indispensable to competitiveness on the international market (see Chung, Chapter 7). In 1979 there was a price gap of $1,700 between the Hyundai Pony and the Toyota Corolla (Roos and Altshuler 1984). In practice, domestic demand was restrained by the heavy taxes on vehicles and petrol, as well as by low and relatively egalitarian incomes.

As for Eastern Europe and in particular the Soviet Union, Fordist automobile production proved impossible in the absence of the necessary labour relations. While Avtovaz attained very high volumes per model with a technical organization system based on that of Fiat at Mirafiori, productivity remained very low. The fact that the factories were obliged to give jobs to the working age population, the lack of control over investments and employee wages and benefits, and the irregularity of components deliveries owing to the great difficulties of coordinating the overall production system made it impossible to exert the discipline over the workforce that the moving assembly line could not impose by itself.

## 1.3. THE RECOVERY OF AMERICAN AND EUROPEAN FIRMS: THE SLOAN, TOYOTA, AND HONDA MODELS PUSHED TO THEIR LIMITS OF VIABILITY, 1984–1992

The oil 'counter-crisis' and the 'speculative bubble' relaunched and fed demand in the automobile markets, especially in Europe and Japan. These exceptional circumstances helped to rescue the firms which had previously been in difficulty. Several of them changed their profit strategy. They all modified their production organization and employment relationships. As during the 1960s, there appeared to be convergence towards the same model: towards what the authors of *The Machine that Changed the World* were to call 'lean production' (Womack *et al.* 1990). However, to reduce the break-even point did not automatically entail a long-term strategy to continuously reduce costs at constant volumes. The implementation of this or that technique, whether just-in-time or teamwork, did not necessarily signify the adoption of the Toyota model, in the absence of the particular employment relationships that were required to make the techniques effective. If proof was required, it came with the rapid rise of break-even points again as soon as demand began to fall off. As for the firms which embodied a model, they appeared to reach their apogee. Yet their successes pushed them to their limits: limits which were social for Toyota, economic for Volkswagen and Volvo, and related to innovation capacity for Honda.

*1.3.1. The Decline of the Dollar, the Oil 'Counter-Shock' and the 'Speculative Bubble' Change the International Environment and the Global Automobile Market*

The Reagan economic recovery, which was supposed to take place through investment, paradoxically became a recovery through consumption. Tax reductions were supposed to encourage households to save and give companies the financial resources to invest. Households consumed the resources instead. The American automobile market began to recover in 1984. The USA reduced interest rates and permitted its currency to depreciate. Between 1985 and 1994, the dollar declined from 251 to 103 yen. The dollar's decline, the opening up of new oil reserves, energy savings, and the collapse of OPEC led to a gradual reduction of oil prices in real terms (declining 2.9 times in constant dollars between 1982 and 1987). This reduction had the effect of an oil 'counter-shock'. In combination with the de-indexation of wages against prices and the reduction of average labour costs in a number of countries, it led to a significant fall in inflation. In order to boost growth further, governments wanted to relaunch investment so that companies could achieve their plans to restructure and create new jobs. However, insufficient savings in the USA and Europe led companies to compete for the floating capital created by the trade surpluses of countries such as Japan, West Germany, and some of the oil producing countries. The liberalization of capital flows and credit facilities favoured speculation, giving rise to what became known as the 'financial bubble'.

This new context had three implications for the global automobile industry: strong recovery of automobile markets in the industrialized countries, reduced exports of Japanese cars in favour of production abroad, and a change in the types of vehicle demanded. In the USA, purchases of new cars and light trucks rose from 10.53 million in 1982 to 15.74 million in 1986 and maintained this level until 1988. In Japan they climbed from 5.87 million in 1987 to 7.78 million in 1990. In Western Europe (seventeen countries) purchases of new cars and light commercial vehicles (under 5 tonnes) rose from 11.72 million in 1985 to 15.03 in 1990, maintaining this level until 1992 under the influence of the impact of German reunification. Alone among the industrializing countries, South Korea also experienced a boom in domestic demand, which tripled between 1987 and 1993 (from 0.4 to 1.44 million). Protectionist measures against the Japanese firms led these to invest in the USA and Europe, either by creating their own subsidiaries (Nissan) or by allying themselves with local producers (Mazda and Mitsubishi) or by doing both (Toyota and Honda). The American producers and the UAW hoped that when the Japanese transplants were subjected to the same environment they would lose their competitive advantage. They also believed that their customers would return to large cars as the cost of oil declined and the population aged. None of this transpired; the market share of the Japanese producers continued to rise and the balance of trade remained strongly in deficit as a result of increased imports of components. Less attractive in terms of price, Japanese cars were still attractive in terms of quality and reliability. Not only did demand in the three global poles rise rapidly, but it shifted towards higher quality, up-market cars and towards new types of vehicle: the minivan and recreational vehicle. These vehicles, together with a proportion of pick-ups, replaced the family car for consumers with an active leisure life and with multiple transport needs. The phenomenon started and was particularly significant in the USA. Minivans and 'compact utilities', which only accounted for 2.1 per cent of 'utility vehicles' or 56,000 units in 1982, accounted for 38.6 per cent or 1.68 million ten years later.

### 1.3.2. The Recovery of the Manufacturers Previously Experiencing Difficulties, with the Exception of GM and Nissan, and their Apparent Convergence towards 'Lean Production'

After having drastically reduced their break-even points, the manufacturers which had previously been experiencing difficulties became even more profitable as automobile demand rose substantially. On the other hand, GM and Nissan, which were the last to implement recovery measures, were not able to profit as much from this period, and suffered most from the next downturn. All these companies tried to change their production organization and to build the foundations for employment relations. The American and European companies had sent numerous missions to observe the Japanese producers since the early 1980s to discover the basis for their success. Some of them created joint subsidiaries or concluded agreements not only to produce models which were missing in their product ranges but also to study 'Japanese methods'. Some of these they borrowed: *kanban*, teamwork, quality control by workers themselves, just-in-time, preventive maintenance, selection, partnership, and the hierarchical tiering of suppliers. Yet empirical research reveals that underneath all the declarations of adoption of Japanese methods, the real practices and actual content of the methods adopted have varied widely (Lung et al., forthcoming, Durand et al., 1998). The underlying reason for this variation was that the profit strategies being followed, far from converging, were becoming more diverse.

Indeed five different profit strategies can be identified between 1983–5 and 1990–2. GM, Nissan, and Fiat continued to follow their 'volume and diversity' strategy. Ford moved away from this by attempting to implement a volume-based strategy at the global level. Chrysler and Mitsubishi returned to the 'innovation and flexibility' strategy. BLMC and Renault tried to position themselves at the top of each market segment with quality-oriented strategies. PSA was the only firm to adopt a strategy of 'continuous cost reduction at constant volumes', although this was temporary and was not pursued in the same way as Toyota. Measures taken by GM to reduce capacity and employment even while demand was strong were met by incomprehension on the part of employees. Moreover, confused and poorly coordinated implementation reduced their credibility in the eyes of employees and suppliers alike (see Flynn, Chapter 8). The number of models at GM remained stable at 39–40 between 1984 and 1988 before dropping suddenly to 33 in 1990. But platforms were not commonized even as demand declined. The average volume per platform therefore halved between 1983 and 1990 (from 410,000 to 200,000). GM's break-even point remained close to its value added and rose above it in 1990. Nissan's managers, having won a test of strength with the union, not only opened the factory planned for the United Kingdom in 1986 but also took control over personnel management, notably issues of promotion. Each employee's tasks were strictly redefined. The classification system was changed to permit greater mobility. The evaluation by hierarchical superior in the determination of the individual salary was enhanced. In the context of the speculative bubble, production rose again in Japan, likewise employment (to 55,000 in 1989). Yet market share continued to fall, passing under the 20 per cent threshold. Employment relations were not the only area of difficulty. Nissan continued to extend its product range (from eighteen to twenty-four models between 1986 and 1989). Average volume per platform remained low, at 130,000 units. The quality of vehicles was below average. Fiat had perhaps the best results. It increased the number of its models from nine to fourteen following its mergers in 1987 with Alfa Romeo, the number two Italian producer by volume (200,000 vehicles per year) and in 1990 with Innocenti. It immediately commonized platforms, attaining a rate of two, with an average volume of 280,000 in 1989, the highest level reached. Fiat actively pursued an automation strategy. It reduced absenteeism markedly, strikes practically disappeared, and the company obtained the required flexibility from a favourable power relationship. However, the cost and problems of automation, the disappointing results of some models, and inadequate product quality drove Fiat's break-even point up towards its value added (see Camuffo and Volpato, Chapter 12).

In 1983 Ford became profitable once more, and continued to improve its results until 1988. It increased purchases from abroad and imposed a quality certification on its suppliers. Ford began systematically comparing its performance with that of its competitors and attempting to surpass these. Nevertheless, the most important factor was a return to a volume-based strategy. In contrast to the firms discussed above, Ford reduced the number of its models from seventeen to thirteen between 1983 and 1987. Its average volume per platform exceeded that of GM. Taking up the idea of a global car once more after having previously failed to realize it with the Escort, Ford announced its intention to make each of its three global poles responsible for a segment of the global market: small cars for Mazda, the middle of the range for Ford of Europe, large cars and light trucks for the US operations. The first manifestation of this global volume strategy was the Mondeo, launched in 1993. Ford simultaneously proceeded to share suppliers across its global poles and signed cooperative accords with Volkswagen in Brazil and Europe. This volume-based strategy was implemented by following typically Fordist procedures. The successful experience with the Taurus served as a matrix for reorganization. The simultaneous and interactive design of product and process was codified in detail, including the time allotted, and then generalized throughout the company in the early 1990s. Transversal groups were charged with controlling and reducing costs in both design and production (see Bordenave, Chapter 9).

Chrysler and Mitsubishi rescued themselves by returning to their original product innovation strategy. Chrysler innovated by creating the minivan segment in 1984, followed by the 'compact sports utility' segment thanks to AMC's Jeep, after Renault sold its American subsidiary to Chrysler. It obtained greater flexibility in the utilization of its workforce, thanks to a new agreement signed with the UAW in 1985 which accepted a teamwork form of organization with a reduction in the number of employee categories. Chrysler repurchased its own shares, in order to become more independent financially. However, yet again, prosperity having returned, the company launched into an expensive and disorderly expansion and diversification of operations. With the downturn in the USA after 1989, the company began to lose money again. Once again without the resources to design new platforms, Chrysler was again forced into debt, selling its aeronautical and electronics subsidiaries and its own shares for less than it had bought them (see Belzowski, Chapter 10). Car and light truck production collapsed from 1.72 million in 1988 to 0.96 million in 1992. Mitsubishi gradually improved its profitability thanks to the increasing success of its light trucks, in particular its sports-utility vehicles launched in 1982, then by giving its passenger cars their own 'personalities' to distinguish them from their competitors, and lastly by relaunching the minicar segment. Simultaneously, Mitsubishi greatly increased the flexibility of its three assembly plants, in particular by organizing the systematic rotation of tasks between production operators every two to four hours.

Renault and Rover sought to profit through improved quality (not being able to do so through greater volumes) and by trying to control costs. Renault deliberately reduced its rate of platform commonization from 2.3 to 1.2 in 1992, designing each model as a coherent but separate whole. It deliberately reoriented its sales towards the Northern European countries, which were used to paying and were better able to pay for quality, and it formed an alliance, and then drew up merger plans, with Volvo, with the aim of linking the two complementary product ranges. Cost reductions were initially approached by strictly controlling the size of the workforce: leavers were not replaced, and temporary workers were hired when demand required. Renault attempted to introduce a new employment relationship. But the participation of employees in the economic recovery of the firm was obtained more through fear of unemployment than through the proposed compromise. As soon as they believed that Renault had recovered, the workers went on strike in 1991 to demand their share of the progress that had been made, in the form of higher wages and an end to job losses. Rover also attempted to position itself at the top of each market segment. The success of its three off-road vehicles permitted Rover to be profitable for four years in succession for the first time in many years between 1987 and 1990. The strategy that Rover and Renault had adopted of matching better quality with higher prices was threatened by the market downturn of the early 1990s and the displacement of demand towards less expensive models. Rover's management and unions agreed to a company compromise, largely inspired by Honda, which broke with traditional British practices. Local negotiations and negotiations by category of employee were replaced by centralized negotiation. The multiple categorical divisions were eliminated and there were now only three unions. All employees were henceforth governed by the same rules, which facilitated internal mobility. Previously excluded from discussion, the organization of work was again introduced as a matter of negotiation with employees, although only at the level of the factories and individual work areas.

PSA had been the only firm to follow a strategy of 'continuous cost reduction at constant volume' during the period when demand was rising. Now in a healthy financial situation, it had continued to make savings everywhere, not just in terms of the workforce. It was thus able to achieve a break-even point lower than that of any other European producer and close to that of the Japanese. It had reorganized its components subsidiaries into the ECIA group, thus preserving both its competencies and control over its value chain. Having abandoned the Talbot brand (the former Simca-Chrysler) which had not been successful, PSA reduced its model range from eighteen to eleven in 1990 and decided to create two parallel model ranges, Peugeot and Citroën, thus offering two models sharing the same platform in each of the four principal market segments. It attained a rate of two models per platform in 1992 with an average volume of 300,000. The PSA trajectory temporarily intersected with that of Toyota. Yet its employment relations were quite different, and PSA underwent a long strike in 1989 at its Sochaux factory.

### 1.3.3. *The Firms Embodying the Three Successful Models Reach their Apogee but Simultaneously Reach their Limits*

Volkswagen, Toyota, and Honda reached their historic peak in terms of profits, production, exports, production abroad, and employment. Volkswagen methodically pursued its Sloanist strategy. It retained its commonization rate of 2.0, which permitted it to reach an average volume per platform of 350,000 with the boom in demand, the highest level for any producer during the 1985–9 period. Volkswagen had to close its American factory, but it had taken control of SEAT in 1986 and Škoda in 1991. Its worldwide production reached 3.5 million in 1992, with 1.93 million in Germany where there were 164,000 employees, thus reproducing the peak outputs of the Fordist period. Toyota's worldwide production peaked in 1991 with 4.8 million vehicles, 3.24 million of which were exported or manufactured abroad. Its rate of commonization remained 2.0, and its average volume 270,000. The same year Toyota employed 72,900 in Japan. After having used its joint subsidiary with GM, NUMMI, to test the possibilities of adapting its production system to the USA, in 1988 Toyota opened its own North American factories, one in Kentucky, the other in Ontario in Canada. Production abroad, which was 136,000 in 1985, reached 677,000 in 1990 and 1.05 million in 1994. In 1992, Honda's worldwide production had more than doubled since 1979, reaching 1.96 million vehicles, which made Honda the world's eighth largest producer, ahead of Chrysler and Renault. Half of its sales came from North America, 35 per cent from Japan, 9 per cent from Europe, and 6 per cent from elsewhere in the world. Production abroad accounted for 39 per cent of output. The workforce in Japan grew to 31,000. Honda's investments in North America had been successful. They were founded on an egalitarian and democratic approach, hiring a workforce composed primarily of young employees of German stock who lived in rural parts of the state of Ohio, rather than basing itself on employee individualism as in Japan or on agreement with the UAW as at Toyota at NUMMI. In Great Britain Honda had focused on removing obvious symbols of hierarchy.

Yet these remarkable achievements occluded a rise in the break-even point at each of the three firms and a reduction in their profitability. Indeed they had been pushed to their limits in certain respects. Volkswagen had seen its costs rise in terms of purchases as well as production costs and debt repayments. Toyota had been destabilized by its recruitment problems in Japan during the period when demand was rising rapidly. During a period of labour shortage, young people refused jobs which they judged too hard. Toyota's employees and those of some of its suppliers had refused to undertake further overtime work, and supervisors could no longer deal with the multiple problems caused by turnover among young employees and the inexperience of temporary workers. Instead of attaining the target 50 per cent domestic market share, Toyota fell back to 43 per cent. Toyota's management, in concert with the union, began to question the three basic pillars of the system: the wage and promotion system, the system of unplanned overtime hours, and the assembly lines without intermediate stocks (see Shimizu, Chapter

3). They had to significantly reduce the part played by the coefficient of production efficiency of each work group in the determination of wages, create two day shifts without the possibility of overtime work after the eight statutory hours, reduce annual working hours, and divide assembly lines into segments separated by buffer stocks. Honda did not experience the same crisis of work as Toyota. Better working conditions, the lowest annual working hours among the automobile producers, planned overtime hours, mobility based primarily on the personal competency of the employee, all helped Honda avoid the same problems of recruitment. Honda's difficulties appeared where they were least expected: its capacity to understand the new requirements of its customers. Honda believed that the 'speculative bubble' would create an enduring shift in demand towards sporty and luxury cars in the image of increasingly wealthy young professionals. The product range grew from six models in 1985 to seventeen in 1992, reducing the average volume per platform to 107,000, compared with 240,000 in 1985 and 350,000 in 1979. From being a manufacturer of innovative models, Honda had evolved towards the production of niche models. As long as the bubble continued, this strategy appeared justified. Focused since the origin of the company on sporty vehicles and on the improvement of their engines, Honda's engineers and managers chose not to believe in the emergence of a new market segment oriented towards recreational vehicles. This historical error reveals that the system adopted by Honda to retain an innovative capacity which was commercially sensitive was not adequate to overcome the gradual specialization of the designers in a single type of vehicle.

### 1.3.4. The Emergence of a Possible 'Reflexive Production' Model at Volvo

As we have seen, Volvo had to contend with a new crisis of work as a result of its expanding output and a shortage of labour in Sweden. It responded in two ways. On the one hand, production capacity was expanded abroad where there was no labour shortage. The proportion of cars manufactured abroad rose from 46 per cent in 1979 to 58 per cent in 1989. On the other hand, in Sweden Volvo decided to go further than the system adopted at Kalmar (see Berggren, Chapter 16). One of the most significant results was the new factory at Uddevalla. While Kalmar retained the principle of the assembly line and the cycle time, Uddevalla introduced the assembly of whole vehicles at fixed stations by two-to-four workers, with times allocated equivalent to those on the assembly line, on the basis of small subassemblies and parts delivered on AGVs and organized on trays according to their position and function in the finished vehicle. The memorization of operations with no logical links between them which was required on the assembly line and was the source of numerous errors, not to mention lack of interest in work, was replaced by understanding based on the very logic of the product's construction and on the normal cognitive capacity of people (Ellegård et al. 1991). Experience showed that it was possible to attain times below those possible on the assembly line given the absence of the wasted time which the latter generated structurally: the impossibility of filling up the cycle time, which became a greater problem to the extent that products became more varied, the need to reorganize tasks with each change in output and in demand, the stoppage of the whole line if there was an incident or breakdown at any point along it, the need to keep replacement workers to fill in for absences. Uddevalla also revealed that 'reflexive production' permitted production to be adjusted to demand by simply withdrawing or reutilizing the assembly stations and permitted variety to be further increased without disrupting production. Lastly, this mode of assembly gave the production workers a vision of the whole and an overall control over the quality of the vehicle which was not permitted by the assembly line. On the contrary, reflexive production, at least at the stage of development it had reached, did not in itself lead to reductions in the time allocated and to participation in the improvement of performance. It needed to be linked to an employment relationship which would encourage these activities. To become a true industrial model, the Uddevalla system still had to develop a company compromise coherent with its production organization (Freyssenet, 1998).

*1.3.5. Fordism succeeds in Korea*

The Korean automobile industry took off in 1985. In 1992 South Korea overtook Italy and Great Britain and its output reached a spectacular 1.81 million passenger cars in 1994, 35.9 per cent of which were exported. Indeed exports had pulled production along until 1988, a year when exports had accounted for 64.7 per cent of output. The collapse of Korean exports in 1989, owing to problems of quality, was counter-balanced by rapid growth in the domestic market. Consuming 1.14 million passenger cars in 1994, the domestic market had probably reached the replacement phase. This explained a new export offensive on the part of the Korean producers, notably in Europe after their withdrawal from North America. A new plan was launched to increase production capacity to 4 million vehicles per year by the year 2001 and make South Korea the fourth biggest vehicle-producing country.

On the eve of the bursting of the speculative bubble, the firms which had previously been experiencing difficulties, with the exception of GM and Nissan, had greatly reduced their indebtedness, selected clear and different profit strategies, and modified their production organization. On the other hand, none of them had been able to establish a new and stable company compromise and therefore to constitute and embody a new industrial model. As for the three successful industrial models of the years 1974–92, they had reached their limits in the firms which had embodied them. Far from converging on a single industrial model, anticipated to correspond to the market conditions of the twenty-first century, the automobile manufacturers had therefore followed diverging trajectories, which the recomposition of global economic and political space under way during the 1990s would only accentuate.

## 1.4. THE SEARCH FOR STRATEGIES AND INDUSTRIAL MODELS APPROPRIATE TO A CONTEXT OF NEWLY EMERGING ECONOMIC SPACES AFTER 1993

The bursting of the 'financial bubble' led to renewed confrontation between firms and national economies. However, the terms of confrontation were no longer those of the years between 1974 and 1985. Competitive gaps between firms had been reduced. Modes of growth in the USA, Great Britain, France, Italy, and Sweden had also changed, whi le those in Japan, Germany, and Korea had been destabilized. With income distribution having become more unequal than previously, the structure of automobile demand in the industrialized countries had changed both quantitatively and qualitatively. With the purchasing power of employees having stagnated or declined depending upon the country, and with gaps in competitiveness having been reduced, the producers had to reduce their prices. Some of them became involved in a price war which greatly reduced their profit margins. They all sought new sources of economies of scale by considerably increasing the commonization of their platforms and seeking to invest in so-called emerging countries.

Simultaneously, some countries had changed their economic system and other markets had emerged. The communist regimes of Eastern Europe had imploded. In the following period of disorganization, in the absence of the institutions necessary to make a market economy viable, automobile production in these countries collapsed (from 3.56 million in 1988 to 1.66 million in 1994). Several producers were purchased by Western European car makers. China and to a lesser extent Vietnam attempted to manage a transition to a market economy, and began to grow rapidly. The countries of South-East Asia and Latin America had taken off economically, with strong demand for automobiles. The passing of the baton of growth from the saturated automobile markets of the industrialized countries to the developing countries which the producers had expected to occur during the second half of the 1970s seemed to be taking place twenty years later. Yet it remained unclear what was the most appropriate model range to offer in these countries, in the sense that the distribution of income remained very unequal and varied greatly. Political and economic stability was far from certain. All the producers were becoming involved, creating overcapacity and hypercompetition. Was it appropriate to sell models designed for industrialized countries, or to adapt them, or to strip them down in order to make them accessible in price terms, or to design a specific model on the basis of a global platform, or even a number of models with their own platforms?

This new international context provoked a clash between a number of tendencies in the recomposition of global and political space. A first tendency was the 1993 liberalization of global trade under the umbrella of the World Trade Organization, strongly supported by the USA. A second tendency was the constitution of regional spaces which favoured internal trade, relatively autocentred growth, and in some cases the creation or defence of their own social model; this was the case notably of the European Union, NAFTA, Mercosur, ASEAN. In the first two groups the emerging countries were treated as satellites, while the two others were groups of emerging countries which were attempting to create autonomous poles despite the centripetal forces exerted upon them by the USA and Japan. The third tendency was that of the countries which believed they could create their own industry, such as Russia, China, or India, or which hoped to retain autonomy like South Korea.

The automobile producers now had to place their bets on the type of spatial recomposition that would carry the day in order to select the most appropriate profit strategies, modes of internationalization, product ranges, production organization, and employment relations: in short, their industrial models. The choices they made, strongly influenced by their past trajectories, would in turn have an impact on the recomposition of the global economy itself.

The question of which industrial model to select is now being posed in an unanticipated manner. How is it possible to choose a profit strategy and an industrial model while the environment is being restructured and its future form unknown? How is it possible to perhaps integrate different models in the same globalized or regionalized firm? Up until now, international firms have had autonomous subsidiaries at the regional, and even national, levels, each with its own product range and own industrial model, yet all pursuing the same profit strategy. Current internationalization tendencies are modifying this spatial form and are posing the question of the articulation of industrial models within a single firm.

Ford appears to believe that trade will be increasingly open over the long term, and that automobile demand will become homogeneous. Ford is pursuing the objective of establishing a single range composed of world cars for each of the principal market segments. It has reiterated its intention that its three geographic poles will each specialize in one of the three segments. In contrast, the company seems less entrepreneurial when it comes to investing in emerging markets. How viable is this approach? The homogenization of global demand is highly unlikely, even if the World Trade Organization is able to harmonize national regulations (Freyssenet and Lung 1996). A world car which is a compromise between the expressed desires of different markets has every chance of satisfying none. It is perhaps possible to define world products situated in the middle of each market segment, sold on the basis of price and reliability. Is this the path Ford is following? While the homogenization of demand presupposes a homogenization of the labour relationship across the world, this pattern of global firm may, during a phase of transition, find itself with different forms of production organization and employment relations in the three world poles, and therefore taking into account regional specificities, in so far as each of the poles is responsible for design as well as production and distribution. This form of internationalization, which may perhaps integrate the different industrial models dominant in the three world poles, requires effective free trade across the world as one of its preconditions. An alternative way to produce world cars is the one which Hyundai and Daewoo appear to be following, and which Avtovaz may adopt (see Chanaron, Chapter 17): produce a small range of standard models in countries where the cost of labour is low.

GM, Fiat, Volkswagen, and Nissan have oriented themselves towards a regional diversification of products combined with a global commonization of platforms. In the Sloanist tradition, built on realism and pragmatism, they are betting on an enduring differentiation of demand related to income differentiation and historical and cultural differences between regions and countries. They are also betting on the possibility of globalizing mechanical and other parts not visible to the driver, the manufacture of which can be shared across countries. This form of internationalization presupposes low customs duties on components and relatively stable exchange rates, but it can easily accommodate variations in regulations and demands for local production of some vehicle components. It is compatible with the relative liberalization of world trade and the development of regional spaces. This system offers the possibility of optimizing the division of responsibilities between various global sites as a function of differences in the evolution of demand and costs of production. Yet it is already provoking moves by unions to coordinate their activities in order to impose shared rules.

GM and Nissan are active in each of the three world poles, as well as in several emerging countries. In the USA, GM has reduced its model range, has obtained the flexibility that was previously lacking, and has reduced its costs to become competitive. It may become Sloanist at the global scale, rejecting the alternative industrial models sketched out by its two experiments, NUMMI and Saturn. Volkswagen has renewed its strategy of 'volume and diversity', especially at the European level, and has made each of its four brands responsible for emerging markets. Active in China and Latin America, it is possible that it will eventually develop unique models there. Fiat is doing the same. PSA, which ought logically to follow the same approach, probably does not have the resources to match such ambitions unless it links up with another producer.

Honda, Chrysler, Mitsubishi, and Renault, in line with their profit strategies, their previous trajectories, and what is actually possible for them, appear to believe that new market segments will appear more or less regularly. For this strategy to succeed, the regionalization and heterogenization of demand would have to prevail over globalization and homogenization. This hypothesis may suit the producers which remain concentrated on a single region, taking advantage of their detailed knowledge of their markets just as producers which internationalize ensure that they have the means to understand consumer expectations in the regions in which they invest. In these cases, regional design offices do more than simply restyle base models: they modify them in response to local desires, and they may even design specific models. Regional subsidiaries have a broad autonomy because they have to detect emerging local requirements. To do so, they rely heavily on local managers, designers, engineers, and distributors. The function of the company at the global level is to take responsibility for financial control, distribution of investments, particularly to new regions, and ensuring that knowledge drawn from experience is circulated. Chrysler seems in a good position to follow this path and to eventually create an industrial model consistent with its 'innovation and flexibility' strategy. The flexibility it previously lacked in the USA is provided by the deregulation of the labour market and new flexibility in job rules. It is maintaining its innovative capacities by regularly bringing in imaginative outside designers. It has created the resources to become highly reactive by developing a technocentre which permits designers, engineers, and suppliers to work in parallel. Honda, which is active in each of the three world poles, has to make a choice. Either it becomes a specialist producer of upper range and niche vehicles at the global level, or it returns to its original approach of detecting new consumer expectations. Its links with Rover offered it two choices, both of which were also open to the English firm: either a complete range of models situated at the top of each market segment, or a focus on Rover's successful recreational vehicles. However, the surprise sale of Rover by British Aerospace to BMW in 1994 closed off this possibility. Honda has since entered the recreational vehicle market in Japan with great success, but has only caught up with a trend which it ought to have anticipated. Mitsubishi's success has been due to its off-road vehicles. It has bridgeheads in the USA and Europe. But it appears to have retained its ambition to offer a broad range of products, and it remains heavily indebted. As for Renault, it does not seem to have made a clear choice as yet. Renault hopes to follow a strategy involving a traditional range plus a range of innovative vehicles, with the first protecting it against failures in the second while benefiting from the image of the latter. But its difficulties in responding to demand for its successful mid-range minivan, the Scenic, which are giving its competitors the opportunity to copy it quickly, reveal that it has neither the resources nor the necessary production flexibility. It possesses the design system required to create its future models rapidly, with a technocentre similar to that of Chrysler, but is prevented from pursuing a strategy of 'innovation and flexibility' by the absence of a process to develop internally, or find outside, innovative personnel, and the absence of a sales network capable of analysing the needs of its actual and potential clients. Renault is also hampered by the absence of utilization of expertise at all levels, an appropriate company compromise, and financial independence.

Toyota is continuing to reduce costs at constant volumes, but is now following other paths to meet this goal. It has retained the same strategy, but in Japan it has probably already changed its industrial model. Toyota no longer relies upon its production operators to reduce production times, but on special teams which seek wastage and opportunities for savings. It is converging around the system it has implemented in the USA and Great Britain, where it did not transfer its wages, hours, and promotions systems and where it introduced segmented lines which leave teams with autonomy (Mishina 1998). Yet Toyota still has to find a new way to involve its employees in the elimination of wastage that is only indirectly related to time and work and in the improvement of quality. Active in all three geographical poles, Toyota can continue to be the redoubtable competitor it has been in the past, destabilizing the other industrial models. But this will be more difficult to achieve than before.

Volvo remains a specialist producer, with new questions to consider for its future. Its European competitors, BMW and Mercedes, have decided to shift towards smaller cars and to invest in the new market segments, minivans and recreational vehicles: BMW by purchasing Rover, and Mercedes by itself. Rover's purchase by BMW is perhaps a measure of the former's success in finding a stable and viable profit strategy and creating a coherent industrial model. Mercedes is innovating by itself, both in terms of products and in terms of processes. In association with the watchmaker Swatch, Mercedes is launching a two-seat urban car, designed as an assemblage of a dozen modules which are produced by suppliers located on the same site as the final assembly factory. Only time will tell whether this linking of a 'quality and specialization' strategy with an 'innovation and flexibility' strategy is viable, and whether the modular design and production of cars prefigures a new era for the automobile industry.

## CONCLUSION

There is no 'one best way'. There was none in the past, and there is none today. In all likelihood, there will be none in the future. 'Lean production', an amalgam of profit strategies and industrial models which are different and incompatible, cannot be the industrial model of the twenty-first century. History moves fast. The Toyota system, having reached its apogee, was forced to change significantly, just at the moment when other firms were trying to draw their inspiration from it. A similar phenomenon appears to have occurred in the 1960s. In the context of autonomous national growth models, the Sloanist model appeared to be the only path to success. Many experts claimed that all producers ought to adopt it, on pain of disappearing. Indeed several automobile makers tried to do just this, without succeeding, while others invented new models which were to prove effective following the first oil crisis, in a context where autonomous and export-oriented national modes of growth clashed. Three industrial models, which put into practice three distinct profit strategies ('volume and diversity', 'continuous reduction of costs at constant volumes', and 'innovation and flexibility') coexisted and succeeded between 1974 and 1990–2: the Sloanist model adapted to an export-oriented national mode of growth, the Toyota model, and the Honda model. They were embodied in three firms, Volkswagen, Toyota, and Honda, which shared only the fact that they followed their own profit strategy, which fitted their own socio-economic context, and created a production organization and employment relations coherent with this profit strategy. Yet at the end of the 1980s these three producers encountered the social, organizational, or economic limits of the models which they embodied. They too experienced crisis. Far from a convergence among automobile firm trajectories, what can be observed today is their divergence, an outcome of the different profit strategies they selected and followed during the 1980s, and the uncertainties over the restructuring of global economic and political space which have characterized the 1990s.

## NOTES

Translated by Sybil Hyacinth Mair.

The analysis in this chapter owes much to the research undertaken by the authors involved in the book. Since all the information concerning the various automobile producers is drawn from their chapters, they have not been cited on each occasion such information has been used. On the other hand, their analyses have been cited when these have been utilized. Any complementary information gleaned from other authors has also been cited. As will be evident to the reader, the data on volumes per platform, product variety, and break-even points have played a significant role in characterizing the firms. These data have been drawn from research undertaken by Marie-Claude Bélis Bergouignan, Bruno Jetin, Yannick Lung, Robert Boyer, and myself, as indicated in the reference list. This chapter could not have been written without the theoretical and methodological analyses that I have been undertaking with Robert Boyer on industrial models in the automobile industry, the results of which will be published in a forthcoming book. These analyses have been stimulated by the discussions to which all members of GERPISA have contributed to a greater or lesser degree, as part of the programme 'The Emergence of New Industrial Models'. Patrick Fridenson gave me useful comments. My thanks go to all those mentioned.

1. The concept of 'rapport salarial' is translated in English by the phrase 'labour relations'. Moreover, it includes not only the modalities of relations between employers and employees at national level, but also the conditions and the results of these relations: that is the social composition of employers and employees, types of association of each of them, labour market, dominant forms of work organization, work practices and rights, modalities and rules of wage growth, distribution of national insurance benefits, the macroeconomic link between productivity and consumption. By extension, the concept of 'compromis salarial' is translated by 'labour compromise', in other words, compromise (formal and in practice) between employers and employees at national level over work practices, rules and rights, the link between macroeconomic or company performances and salaries and social benefits (Boyer and Saillard 1995).

2. The concept of 'relation salariale', translated by 'employment relationships', includes also the conditions and the results of these relations, but at company level. More precisely, it includes the type of owners, management, workforce and unions, the local labour market, the systems of recruitment, job classification, promotion, training and career development, working time, work organization, modes of participation, employment adjustment, wage and social benefits. We have used the phrase 'company compromise' to indicate the compromise (formal and in practice) between management, unions, and employees of the company, concerning these work and employment conditions (Boyer and Freyssenet 1995).

## BIBLIOGRAPHY

Auer, P., and Riegler, C. H., *Le post-taylorisme. L'entreprise comme lieu d'apprentissage du changement organisationnel* (Paris, 1990).

de Banville, E., and Chanaron, J.-J., *Vers un système automobile européen* (Paris, 1991).
Bardou, J. P., Chanaron, J.-J., Fridenson, P., and Laux, J. M., *The Automobile Revolution* (Chapel Hill, 1982).
Bélis-Bergouignan, M.-C., and Lung, Y., 'Le mythe de la variété originelle. L'internationalisation dans la trajectoire du modèle productif japonais', *Annales*, Histoire, Sciences Sociales (HSS), 3 (1994).
Bhaskar, K., *The Future of the World Motor Industry* (New York, 1980).
Boyer, R., Charron, E., Jürgens, U., and Tolliday, S. (eds.), *Between Imitation and Innovation: The Transfer and Hybridization of Productive Models in the International Automobile Industry* (Oxford, 1998).
—— and Freyssenet, M., '*The Emergence of New Industrial Models. Hypotheses and Analytical Procedure*', *Actes du GERPISA*, 15 (1995).
—— —— *The World that Changed the Machine* (forthcoming).
—— —— and Jetin, B., 'Les Stratégies de profit des firmes automobiles' (forthcoming).
—— and Saillard, Y., *La Théorie de la régulation. Un état des savoirs* (Paris, 1995).
Chanaron, J.-J., and Lung, Y., *Economie de l'automobile* (Paris, 1995).
Durand, J. P., Castillo, J. J., and Stewart, P. (eds.), *Teamwork in the Automotive Industry. Radical Change or Passing Fashion?* (London, 1998).
Ellegård, K., 'Automobile Industry In Sweden', *Proceedings of Fourth International Colloquium of GERPISA* (Paris, 1995).
—— Engström, T., and Nilsson, L., *Reforming Industrial Work. Principles and Realities* (Stockholm, 1991).
Freyssenet, M., 'The Current Social Form of Automation and a Conceivable Alternative', in Shimokawa *et al.*, *Transforming Automobile Assembly* (Berlin, 1997).
—— 'Reflective Production: An Alternative to Mass Production and Lean Production?' in *Economic and Industrial Democracy*, 19 (1998), 91–117.
—— and Lung, Y., 'Between Globalization and Regionalization: What Future for the Automobile Industry?' *Actes du GERPISA*, 18 (1996).
Frybourg, M., and Prud'homme, R., *L'Avenir d'une centenaire: l'automobile* (Lyon, 1984).
Fujimoto, T., and Tidd, J., 'UK & Japanese Automobile Industries: Adoption and Adaptation of Fordism', in *Actes du GERPISA*, 11 (1994).
Jetin, B., 'The Historical Evolution of Supply Variety: An International Comparative Study', in Lung *et al.*, *Coping with Variety* (forthcoming).
Keiser, B., and Kenigswald, L., *La Triade économique et financière* (Paris, 1996).
Kochan, T. A., Katz, H. C., and McKersie, R. B., *The Transformation of American Industrial Relations* (New York, 1994).
Laux, J. M., *The European Automobile Industry* (New York, 1992).
Lung, Y., Chanaron, J.-J., Fujimoto, T., and Raff, D. (eds.), *Coping With Variety. Product Variety and Productive Organisation in the World Automobile Industry* (forthcoming).
de Mautort, L., 'Concurrence internationale et norme de production dans l'industrie automobile', *CEPII* (Paris, 1980).
Maxcy, G., *The Multinational Motor Industry* (London, 1981).
Mishina, K., 'Revealing the Essence of Toyota's Manufacturing Capability: The Kentucky Transplant, 1986–1994', in Boyer *et al.*, *Between Imitation and Innovation* (Oxford, 1998).
Pil, F. K., and Rubinstein, S., 'Saturn: A Different Kind of Company', in Boyer *et al.*, *Between Imitation and Innovation* (Oxford, 1998).
Roos, D., and Altshuler, A. (eds.), *The Future of the Automobile* (Boston, 1984).

Sandberg, Å. (ed.), *Enriching Production* (Aldershot, 1995).
Shimokawa, K., *The Japanese Automobile Industry* (London, 1994).
—— Jürgens, U., and Fujimoto, T. (eds.), *Transforming Automobile Assembly* (Berlin, 1997).
Shiomi, H., and Wada, K. (eds.), *Fordism Transformed* (Oxford, 1995).
Streeck, W., *Social Institutions and Economic Performance* (London, 1992).
Tolliday, S., 'The Diffusion and Transformation of Fordism: Britain and Japan Compared', in Boyer *et al.*, *Between Imitation and Innovation* (1998).
Volpato G., *L'industria automobilistica internazionale* (Padua, 1983).
Whisler, T. R., *At the End of the Road* (Greenwich, 1995).
White, L. J., *The Automobile Industry since 1945* (Cambridge, Mass., 1971).
Williams, K., Haslam, C., Johal S., and Williams, J., *Cars: Analysis, History, Cases* (Providence, 1994).
Womack, J. P., Jones, D. T., and Roos, D., *The Machine that Changed the World* (New York, 1990).

# 2

## *Models, Trajectories, and the Evolution of Production Systems: Lessons from the American Automobile Industry in the Years between the Wars*

DANIEL M. G. RAFF

First came the tinkerers, lonely visionaries, and then the artisans. Then came Ford. Some years later, in a far away and economically very differently situated country, the Toyoda firm asked Mr Ohno to consider how the Americans manufactured cars (Cusumano 1985). The fruits of this became the Toyota production system. This, the MIT report tells us, is the be all and end all of manufacturing systems and should be implemented forthwith everywhere (Womack *et al.* 1990). Is there nothing but managerial myopia or sloth behind the slow and uncertain process of its diffusion? What of the experiments in neo-craft production, for example in the Volvo works at Uddevalla? What of the varied and incomplete adaptions in North America, in Europe, and in transplant operations even farther afield? What of Toyota's own recent systematic retreat in its newer plants in Japan?

The GERPISA Programme on the Emergence of New Industrial Models was organized to make sense of this rich but disordered experience. Early on in our discussions, two concepts—those of industrial models and of firm trajectories—themselves emerged as crucial in any systematic analysis. The empirical questions they suggest seem in themselves straightforward. Are there really alternative industrial models for automobile manufacturing? Are there really different trajectories firms might follow? Indeed, are these just the same question seen from the different perspectives of the external analyst and the internal decision-maker?

The first two questions are closely and, in fact, intimately, related to one another. But they are not the same question. Becoming clear about how they are related is an important first step in understanding the broad sweep of the industry's recent history and, from there, in recognizing the most useful questions to ask of the experiences of individual firms.

To begin at the beginning, the only interesting version of the question about alternative models at a strictly formal level is whether there are distinct alternative models—that is, alternative means of organizing work, logistics, and physical production and sales—that are actually viable in some relevant competitive environment. One might want to assess viability in the context of a war of all against all or in some more limited but still realistic context. Product markets may be segmented, after all. Shareholders may not demand the absolutely maximal rate of return on invested capital. The point remains that our subject-matter here is business, not some sort of physics of the factory, and therefore that no system that fails to turn a reasonable profit will persist. If a system loses too much money—more precisely, too much money relative to the costs of scrapping it and starting over in some other way—then it represents a tax and transfer system rather than a viable investment vehicle and, sooner or later, the owners will simply insist upon change of one sort or another.

It is worth being explicit that the criterion identified in the preceding sentence is not the naive economic one that all systems produce identical rates of return in equilibrium. It is not even the somewhat more sophisticated simple idea that the rates of return should be equal, adjusting for risk appropriately defined. It is that incremental change seem profitable over some specific horizon given the operating circumstances (the states of the world) likely to arise in that interval.

All this being so, the question about models is a question about steady states (or at least about something like the ergodic states of a stochastic system). It is a question about where, in a parametrized space of possible systems, actual firms might settle. It is a question, posed in a somewhat abstract way, about what the stable alternatives are. Let us suppose that a careful examination of this question reveals that there really are distinct possible alternatives. Firms might in principle attain different destinations. The question about trajectories is then a question about how firms get to wherever they might be going. Given the opportunities and costs as these will be perceived by firm decision-makers, is it true that all viable end-points are attainable from all actual starting-points?

The main thesis of this paper is that there may be a logic to the economics of trajectories, or at least to the interaction between operating circumstances and that logic, that constrains which potentially viable end-states are in fact possible. Given the motives of firm decision-making, it may well be true of some pairs of starting-points and potential end-points that, as the old joke has it, 'You can't get there from here'. In order to say something about practical future possibilities, one must understand this logic and what it might rule out.

I propose to make this point not as a piece of the high theory of dynamic systems but rather in terms of the history of the American automobile industry during the years between the two world wars. That history is less transparent than it is generally taken to be. The usual analysis of the explosion of output in American motor vehicle manufacturing between, say, the manufacturing censuses of 1909 and 1929 refers to the appearance of mass production techniques at Ford and says little more. The usual story of the Depression years that followed, one of abrupt and stark contraction followed by a slow return to something like pre-Crash conditions, is told in representative firm language, as if the industry was one big factory adjusting at the margin. But in 1929, industry was not Ford writ only slightly larger. As I will discuss below, production techniques were quite heterogeneous in 1929. And there was a significant amount of entry into the population of establishments even over the next six years. The patterns of exit were quite complex. There was a lot of activity going on behind the aggregates. Furthermore, manufacturers were not stuck with techniques in the same way that a farmer is stuck, at least inframarginally, with the quality of his land. The story of the diffusion of mass production methods turns out to be a much more complicated and thought-provoking one. I will argue that the story of what is to come next seems likely to involve the same issues and principles.[1]

2.1. THE MASS PRODUCTION OF AUTOMOBILES

In exposing this hidden history, it is helpful to begin with the product. Automobiles are at once familiar and strange to us. Few among the many who use cars on a daily basis have thought much about how the car parts fit together, never mind how they are made or about the relationship between the two. We might divide the car's components into three broad groups. The car is a vehicle that moves forward by virtue of various parts moving round and round. The power for this comes from an engine in which linear motion is converted into rotary motion. The drive train transmits this rotary motion to the wheels. Thus there is an entire category of parts that rotate. These parts are entirely metal. They are both dense and heavy. The second broad category is coach and interior work. This is the part of the car visible to the customers. The materials are fabric, leather, wood, light metal, and paint and varnish. The residual category we might call the frame. It comprises all the parts that hold the components and assemblies of the first two together. Its elements are rigid but invisible to the casual consumer.

A reasonable caricature of the old production process is as follows. The firms that assembled the cars did not make the parts. The parts may or may not have been made with American System methods; but in general the tolerances were not very tight. Parts were delivered to the assembly firm in relatively large quantities. The parts sat until they were needed. Sub and final assembly took place in poorly lit general purpose rooms. There were typically workbenches around the perimeter, on which could be found vices, mallets, files, and other instruments useful in making recalcitrant pieces of metal fit together. There were also general purpose machine tools around the perimeter, used for the same purpose. (Fancier equipment on site appears to have belonged to the workers rather than the firm.)

The assemblers of metal parts were highly skilled machinists. They needed to be. While there was a single basic design to the product they were turning out, heterogeneity in the actual shapes of the parts rendered their work highly unroutinized in its detail. Their jobs required a great deal of judgement and discretion. Those employees who worked in leather or wood faced a similar idiosyncrasy in material shapes and pliability. Their jobs also were judgement- and discretion-intensive. In this, the three groups of components were all being treated in the same way. The process was one of fitting.

Monitoring this process was quite difficult. Light was not always good. Caches of raw materials and work in process obscured sight lines. The evolution of these caches was so irregular that there was no indirect monitoring through them. More fundamentally, control was difficult because the primary object was not pace *per se*. It was extraordinarily difficult to judge fit from a distance. In a broad sense, judging fit was a lot of what these workers were being paid for. Given this, mechanisms for controlling the process were inevitably also very crude. Almost all establishments tried to hire skilled artisans who took pride in their work and then simply paid piece-rates.

The development of mass production methods had three salient elements. The first and most difficult involved procuring interchangeable parts. It is well known that ideas and methods that were in principle practical for manufacturing rifles with interchangeable parts date from the early nineteenth century. But the development of methods and organizational infrastructure for actually doing it, and doing it on a large scale, is really a phenomenon of the twentieth century. The first car company to do this on any scale left no records about its parts sourcing and in any case did not do it for very long.[2] The idea lay dormant for another decade until bursting into flower at Ford in the years 1912–14.

The second key element was progressive assembly. Instead of leaving the men to go to the work—at the sawhorses at the centre of the room when the parts fit easily, off at the perimeter fitting when they did not, hunting down some part or subassembly hoard when more were required, and occasionally far from the sawhorses sharpening tools and the like—the men were fixed in position and the work came to them. The cycle time of individual jobs dropped radically. The man who worked the piece was not the man who kept the input stocks, nor was he the man who kept in good working order any machines or tools that might be required. Work tasks themselves were routinized and rendered more easily monitored. Being more easily monitored, they were more easily controlled.

If the pace of work tasks could be controlled centrally, through job design and standard setting, so too could the pace of part requirements. Enhanced logistical control was the third element of the innovation. This was not carried anywhere near so far as Ohno and his associates at Toyota did in these early years; but control certainly was improved and tasks certainly were consciously more tightly coupled. This was in fact what made progressive assembly an easy vehicle for monitoring.

Designing and controlling the parts production process so that the parts came out truly interchangeable was difficult. And the word 'truly' here can be overemphasized. The parts could be far from identical and still not require fitting: this was simply a matter of design tolerance. Anyone who has ridden in, listened to, or even watched a Model T under its own power will know that the ride was not smooth. And the standard of coachwork and interiors was intentionally not fine. None the less, the significance of assemblers having adequately interchangeable parts was simple. Assembly ceased being skilled work. *A fortiori*, it ceased being judgement- and discretion-intensive. Relatively unskilled employees could do it. Managers knew what they could expect each worker to do and how long it ought to take. Given that degree of knowledge of the basic production process tasks, implementing progressive assembly seems, at least in the few documented cases, to have been a relatively minor task. Logistical control does not appear, at least in the surviving sources, to have in itself been a major trial. The real practical difficulties on the manufacturing end seem to have lain in conceiving the complementarities between these activities and in implementing them more or less all at once.

The output consequences at Ford were certainly dramatic. Production was soon an order of magnitude larger than that of his closest competitors and two orders larger than the industry norm. The operational consequences of the new system were both more profound and more complex than the shopfloor ones identified above. Those costs, after all, were almost entirely fixed. There were sunk costs too. Machines and tools needed to be designed. Ford made large-scale investments in machine and tool-building expertise and facilities as well as in machines and tools to get the parts it needed. The machines and tools it developed were sufficiently revolutionary that they were the subject of extensive description and photography in the leading metalworking trade journals as well as in those of the automobile industry itself. To the extent that the company was not to be backward integrated, relationships with vendors needed to be set up. A new human resource management system needed to be designed and implemented. Perhaps most important, a marketing niche needed to be created. Demand for a branded product is not discovered but created. The public had to be educated as to what a Ford car was and convinced that that particular bundle of attributes and price was a desirable one. A network of dealers needed to be set up and an infrastructure to support them designed and implemented. In the event, these last two tasks were, respectively, easy and not so difficult. But no one in the organization seems to have thought them trivial matters.

2.2. DIFFUSION HISTORY

Ford himself became, as Marshall would surely want to note, publicly extremely wealthy. The techniques were widely thought to have made the money. It is natural to ask whether there was swift entry into the use of mass production methods. The famous outward signs of the new practices, at least in the public's eye, were conveyor belts. It is indisputable that these became common in American automobile plants within a year or two of the publicity showered on Ford methods in connection with the announcement of the Five Dollar Day compensation scheme in January 1914. On the surface, then, the diffusion of mass production methods in automobiles has a very different pattern from the slow and halting nineteenth-century history chronicled by Hounshell (Hounshell 1984). It is, however, extremely unclear from this publicity—even down to articles on the subject in trade journals—whether the factories in question had adopted the underlying Ford production practices identified above. Such sources and the small number of intra-company sources offering glimpses into the operating and investment decisions suggest, rather, that the conveyors were simply a substitute for unskilled labour.[3]

The traditional literature on the history of the American automobile industry leaps from these traces and the output data recorded in the Manufacturing Censuses conducted in 1909, 1914, and 1919 to the conclusion that Ford methods took the industry by storm. The operational end of Ford methods certainly was better known by 1919, at least in all shops that spent the war years manufacturing standard issue products in large volumes. But the fact that the strikes that so threatened the industry in 1919 appear to have been craft strikes did not register.[4] Statistics about the industry's growing output never separate out Ford, whose giant plant at the River Rouge, in its time the largest industrial establishment on the continent, was constructed during this period. There are no estimates of the growth of per-firm output exclusive of Ford, nor are data collected on capital stock during 1919 used to estimate productivity growth exclusive of Ford in order to compare this with the Ford performance. The capital stock data are probably not to be taken too literally. But taking them as given and making the most straightforward intermediate assumptions about the timing of investment, the rest of the industry looks nothing like Ford.

The overt history of diffusion is sketched in Hounshell's book and is given more detail in an article of mine (Raff 1991). The first large-scale implementation was, in relative isolation, in the Chevrolet Division of General Motors from about 1924.[5] The executive behind it had been hired away from Ford for the purpose. The methods began their diffusion through the product lines of the Corporation starting in 1926 with the design and production of the Pontiac. Sloan's challenge to the Pontiac designers was to create a new car, to as large an extent as was feasible, out of Chevrolet and Oldsmobile parts with already existing dies and tooling. Internal GM documents reveal his exuberant belief that they had succeeded handsomely. The common parts strategy diffused slowly through the Corporation's other lines up until 1929. The company's main operational response to Depression conditions was to increase the extent to which it pursued the strategy. The competitive value of this approach was so obvious to Walter Chrysler when organizing the company that now bears his name that he wrote it into the organizational diagram.[6] GM had a separate engineering staff for each of its five divisions. Sloan had to entreat them to share designs. Chrysler cut this Gordian knot by establishing a single engineering office for his entire multi-divisional corporation.

This is a story of slow but purposive diffusion. It took fifteen years; but on the other hand, it seems to be simply a matter of superior (actually, inferior) unit costs steadily winning out. Yet such an interpretation follows only from assuming that in 1929 Ford, GM, and Chrysler were the industry. The overt history is the only history there was. It is in fact true that the Big Three were responsible for about three-quarters of the 1929 output of the Census motor vehicle industry.

Yet it comes as something of a shock to most people to discover that the Big Three operated only one-quarter of the establishments in that year (Bresnahan and Raff 1991). The non-Big Three establishments appear to have been organized according to very different principles: the modal car was not produced in the modal type of factory. Most motor vehicle manufacturers, and indeed most car manufacturers, ran single establishment firms with relatively short production runs. On most of the measures derivable from the 1929 Census and contemporary Moody's data, their establishments look very different from those of the Big Three. A glance at a randomly selected issue of *Automotive Quarterly* will reveal that their output looked quite different also.

This would have been obvious enough to their customers. Details from the Census forms reveal that these firms sold principally to two markets. The first was for utilitarian vehicles specific to fairly idiosyncratic purposes. These were unlovely but distinctive. They were made substantially by hand because the market was too small to support the infrastructure of interchangeable parts. The second was for relatively luxurious vehicles. The purchasers of these paid for visible hand work in the bodies and interiors and for the consequences of hand fitting in the less visible parts. That work was patient and extremely careful balancing of all rotating parts and then dynamic balancing of components and subassemblies. While this work was not visible to the silk-sleeved customer, anyone could feel its consequences riding in the car. The following is from a trade journal, not advertising literature.

To attain the degree of refinement, length of life, freedom from noise, and absence of vibration required in a car selling in the highest price range, it is essential to work to unusually close limits, and to finish many parts all over that otherwise might be left rough on some surfaces. Grinding, polishing, and lapping operations represent a considerable portion of the total machining costs of the Cadillac engine; and only a few of those operations can be done [i.e. to the requisite standard] on fully or even semi automatic machines.[7]

The requirements of selling such output had imposed strong constraints on factor inputs and so on the structure of costs. Much of the equipment required tended to be general purpose rather than special purpose. This was in itself a major contrast to the Ford-style system. But the human capital required to make this sort of output was very different too. The workers needed very sophisticated skills. They rapidly developed highly firm-specific (indeed, design-specific) knowledge.

It is natural, comparing the production processes, to speculate about the shape of cost curves. This raises the question of which process had the greater fixed costs, which raises the subtler question of which fixed costs were sunk (that is, irrecoverable) and which were not. The Ford-style operations had large investments in plant and equipment. But much of that was dedicated to specific products and therefore sunk: if demand for Fords was slack, there were few other uses for this equipment. The equipment in the artisanal-type plants tended not to be sunk: there was, even in the Depression, a secondary market for general purpose machine tools. Those plants' labour forces, on the other hand, did represent sunk capital investment. There was no real secondary market in the early 1930s for knowledge about how swiftly to balance the rotating parts of Packard engines or gearboxes, and there were, in any case, no labour market institutions to enable Packard to sell the knowledge.[8]

Bresnahan's and my recent study of a panel of establishment data drawn from the Censuses conducted in 1929, 1931, 1933, and 1935 mobilizes these ideas to develop systematic clues to the covert history of how and why this population of heterogeneous techniques and firms evolved as it did (Bresnahan and Raff 1996). We do this by estimating a model of the plant shut-down decision in which we can distinguish the effects on those decisions of cash flow (and its drivers) from those of recoverable (i.e. unsunk) fixed costs. Since we work from an establishment-level database, we can also match the estimated values of unsunk fixed costs to particular establishments and correlate those values to estimates of, for example, the speed with which short-run labour demand responds to shocks.

Over the course of the first half of the 1930s, the industry did have a shake-out. The representative firm vision suggests a Marshallian (survivorship) explanation, with an adverse demand shift as the animating mechanism. The sector of this industry whose production process was relatively intensive in highly skilled workers was indeed essentially wiped out during the Depression. But it was not wiped out, as the traditional automobile history literature suggests, because its average cost curve was too high. High end plants had better, not worse, margins than the mass production facilities. (Presumably product differentiation protected them.) Shut-down was presumably driven by total revenues being insufficient to cover the recoverable fixed costs. On our calculations, those costs were in fact highest for the relatively artisanal plants. Our investigation concluded that the economic source of those costs was the treatment of blue collar labour as a quasi-fixed asset. The mass production plants behaved very differently, of course.

What actually happened then was this. The Depression struck. Most of the costs which the mass production firms treated as fixed were already sunk. Most of their labour costs, on the other hand, represented unskilled labour and were avoidable; and avoid them the mass production firms did. The firms contracted radically at the margin but did not close: they rode out the storm. The artisanal firms could not do this. Their labour costs were, roughly speaking, not avoidable: to have let their skilled workforce go would have been to dissipate many of the assets on the basis of which they competed. No one thought the Depression would last as long as it did. Optimistic, the artisanal firms hung on to their workforces as long as they could. But their cash flows were small and their retained earnings limited. The firms proved unable to hold on for long enough. Eventually, they ran out of money and went out of business. Other sunk investments of the Big Three had created serious barriers to entry. Once out, the artisanal firms were never able to return.[9]

Why did the non-Big Three producers not switch over when demand was booming? To ask the question so bluntly is to suggest they faced the same choices as the Big Three did. But they did not. They had not made the Big Three's sunk investments in manufacturing. They had made very different ones downstream, for brand names are also capital investments. Cadillac had invested so heavily in its brand identity and distribution infrastructure that it could not significantly change its manufacturing methods in the early 1930s without causing any radical revolt from customers or dealers.[10] Launching a new model or moving into a new market required new and substantial sunk expenditures. Knute Rockne's name and a well-produced vehicle were not sufficient to get consumers to take a mass-produced Studebaker seriously.[11] Packard also tried to launch a car for a more popular market in 1937. Their dealers did not know how to attract the new target buyers; and the old ones viewed the old lines with new suspicion. The two groups of producers thus faced fundamentally different choices.

## 2.3. IMPLICATIONS

The meaning of this story, it seems to me, is that the history of industrial organization in automobiles is not simply a history of the coming of machine production, played out in terms of fixed and average costs. The history is one of path dependency, which turns on the degree of sunkness—an attribute not often studied—of investments which themselves are not often focused upon by historians of the industry. This feature explains the absence of continuous adaption and change by incumbent firms.[12] This absence makes for differences over time and between early- and late-comers. It therefore makes for heterogeneity within the cohorts of firms active at any particular time. To a set of decentralized decision-makers in a world with sunk costs, there may well be more than One Best Way. This seems on the face of things as likely to be true for the lean production model as it was for Fordist mass production.[13]

The second main point seems to me to be that while there are important sources of this sunkness in the realm of operations in which students of the industry usually look in manufacturing, there may be equally important sources in the much less frequently surveyed domains of upstream supply and of distribution. The impulse to theorize explicitly, to measure, and to test could in principle add a great deal to the writing of the business and industrial organization history of the automobile. But to address the current ferment in thinking about the nature of firms—their competencies and their opportunities—and the evolution of markets and competition, researchers will have to take such sources seriously. This suggests that an illuminating history of industrial organization and of organizational innovation must cast its evidentiary net far more widely than we are used to doing. The usual statistical data is restricted to counts and values of inputs and outputs, and this does not even come close to measuring the scope of the sort of relationships we now recognize to be critical.

My third concluding point is methodological on a level intermediate between these two. If the first was that sunk costs can matter and the second that we are not yet surveying them—never mind measuring them—systematically, the third is that some sunk costs may matter more than others and we need to sort out which ones are which. To understand the evolution of the population of these production systems is to understand the incentives facing those who make decisions about change. Those decisions are made in the context of a set of expectations about future operating and competitive conditions. The decisions have whatever selection consequences they do in the context of the set of conditions that actually do come about. One important characteristic of operating systems is therefore their relative robustness to the range of these conditions that might come about. The sector of the American industry that largely died in the 1930s did so because its systems were not robust in the conditions it confronted. In doing company-level research and seeking to identify alternative possible models, researchers should contrast the sorts of shocks and market changes each finds advantageous or is at least relatively robust to. We need to understand the differential adaptive abilities and opportunities of each. This is the aspect of a technological trajectory that makes it an interesting way of studying the future.

# NOTES

Precursors of this Chapter were presented at several GERPISA seminars and meetings and at the 1993 Annual Meeting of the Economic History Association. Robert Boyer and Ross Thomson made unusually valuable comments on those occasions and later. Queries and suggestions from virtually the whole of the GERPISA Trajectories group, through most especially Michel Freyssenet and Patrick Fridenson, were also very helpful. For longstanding encouragement and useful advise, I thank Alfred Chandler and David Landes, Sidney Fine and David Hounshell, Kim Clark and Robert Hayes, and Richard Nelson. I also owe thanks to my collaborator Timothy Bresnahan and to Donald Critchlow, Thomas De Fazio, Takahiro Fujimoto, Bruce Greenwald, Mr Jozlin the Packard buff, Frederic Scherer, Steven Tolliday, and Daniel Whitney. The underlying research has been supported by the Harvard Business School, the Colombia Business School, and the National Science Foundation. The usual disclaimer applies.

1. The idea of extending the discussion that follows in order to comprehend the consequences of that lesser but still powerful shock, the downturn of 1920–1, is intriguing. Unfortunately, a parallel discussion is not possible; broadly based statistical data analogous to those that underlie the analysis of the 1930s Depression put forward below do not survive. Worse, the population of firms deploying the new technology at that time was extremely small, making generalization hazardous in any case. Parallels in the behaviour of (and outcomes for) specific firms are certainly striking, however.
2. This was the Olds Motor Company *circa* 1904. All the records and most of the capital stock disappeared in a freak fire. When the works went back into production, it made a very different sort of vehicle with very different methods.
3. For two examples, see (reading closely) 'Conveyor System Aids Big Production', *The Automobile*, 20 July 1916: 100–4, and (completely on the surface) Minutes of the

Studebaker Corporation Finance Committee for 9 February 1914, in the Studebaker National Museum, South Bend, Indiana.
4. See the articles on the strikes in *Automotive Industries* for 25 April, 1 May, and 8 May, 1918 (pp. 841, 972, and 1028, respectively).
5. The process is described in convincing detail in William S. Knudsen (1927).
6. Personal communication from Bruce Thomas, Chrysler Engineering, 1991.
7. Herbert Chase (1930).
8. And these production costs were not necessarily the artisanal firm's greatest sunk investments. See below.
9. An intriguing implication of this analysis is that the artisanal sector might have survived longer, and even survived the Depression, if its accumulated financial resources at the beginning had been larger or had the Depression itself been shorter.
10. Steve Babson of Wayne State University, Detroit, Michigan, USA, has interviewed a man who claims to have assembled Cadillac engines by hand in 1931. By the mid-1930s, Cadillac was involved in the common parts strategy described above.
11. Studebaker's bankruptcy was a strictly financial event, caused by a stubborn dividend policy and set off by the Bank Holiday. The receivership passed smoothly and swiftly precisely because there was plenty of demand for the cars.
12. My interpretation of the abstract concept of 'sunkness' has been entirely in terms of investments in productive capabilities and in marketing. But the concept is equally useful in analysing organizational politics.
13. It is natural to wonder how this analysis relates to the discussion of firm responses to 'environmental jolts' put forward by the organization theorist Alan Meyer (e.g. Alan D. Meyer 1982). The main thrust of his analysis, derived from a case study of actions by hospitals during an anaesthesiologists' strike, is that ideational ('ideological' and 'strategic') aspects of organizations can be more important predictors of their responses to an extreme change in operating environments than objective measures of their organizational structures and resources. That ideational influences can be important is certainly true. But the case Meyer studied was a very special one, and one should be cautious about drawing from it inferences about what must generally be true. The shock in question was not big (some of his hospitals survived it by ignoring it; there was no exit whatever). The industry is one in which competition between firms was quite limited (doctors typically have admitting privileges in relatively few local hospitals). And it is not at all clear that the hospitals were differently situated *ex ante*. The more general case is the one under consideration here, in which the shock is large enough to demand some response, in which there is real competition between firms, and in which heterogeneous commitments by firms to dedicated resources give the firms heterogeneous choices. Here, economic factors were of first-order importance.

# BIBLIOGRAPHY

Bresnahan, T. F., and Raff, D. M. G., 'Intra-Industry Heterogeneity and the Great Depression: The American Motor Vehicle Industry 1929–35', *Journal of Economic History*, 51 (1991), 317–31.

Bresnahan, T. F., and Raff, D. M. G., 'Technological Heterogeneity, Adjustment Costs, and the Dynamics of Plant Shut Down Behaviour: The American Motor Vehicle Industry in the Time of the Great Depression', *Wharton School Reginald Jones Centre for Management Policy, Strategy, and Organization Working Paper Series*, Working Paper 96–05 (Philadelphia, 1996).

Chase, Herbert H., 'Practice in Precision by Cadillac', *American Machinist*, 73 (1930), 359–61.

Cusumano, M., *The Japanese Automobile Industry: Technology and Management at Nissan and Toyota* (Cambridge, Mass., 1985).

Hounshell, D. A., *From the American System to Mass Production: The Development of Manufacturing Technology in the United States* (Baltimore, 1984).

Knudsen, W. S., 'For Economical Transportation: How the Chevrolet Motor Company Applies its Own Slogan to Production', *Industrial Management* 76 (1927), 65–8.

Meyer, A. D., 'Adapting to Environmental Jolts', *Quarterly*, 27 (1982), 515–37.

Raff, D. M. G., 'Making Cars and Making Money in the Inter War Automobile Industry: Economies of Scale, Economies of Scope, and the Manufacturing that Lay behind the Marketing', *Business History Review*, 65 (1991), 721–53.

Studebaker papers, Studebaker National Museum, South Bend, Indiana, USA.

Womack, J. P., Jones, D. T., and Roos, D., *The Machine that Changed the World* (New York, 1990).

# PART I
# ONLY ONE MODEL IN JAPAN?

# 3

# A New Toyotaism?

## KOICHI SHIMIZU

Toyota introduced innovations in industrial organization and labour force management throughout the 1950s and 1960s. By the early 1970s these innovations had evolved into the coherent and successful Toyota Production System (TPS). Prior to the first oil crisis, the system was mainly geared towards mass production. The company had to find creative solutions in order to diversify its product range and make its production process more flexible during the turbulent 1970s. Emerging strengthened from these trials, Toyota revealed its capacity to achieve even higher performance levels during the 1980s. The company's success led numerous managers, observers, and researchers to consider that its system had become the new best industrial model which all companies ought to adopt (Womack *et al.* 1990). Yet by the end of the 1980s, at the very point at which it was being promoted as the industrial model for the twenty-first century, the system was encountering difficulties at Toyota itself. Toyota was having to modify its production organization to make work more attractive and humane. Its share of the Japanese market was declining, and it was suffering the adverse consequences of trade conflicts between Japan and several other countries.

### 3.1. THE TOYOTA MODEL BEFORE THE FIRST OIL CRISIS

In 1973, Toyota had four assembly factories. Honsha had been constructed in 1938, Motomachi in 1959, Takaoka in 1966, and Tsutsumi in 1970. Toyota produced 1,631,000 private and 676,000 commercial vehicles, of which around 40 per cent were assembled by other automobile assembly companies belonging to the Toyota group (Toyota Auto Body, Kanto Auto Works, Hino, Daihatsu, Toyoda Automatic Loom, Araco and Central Jidosha). About 70 per cent of Toyota's components were delivered by suppliers according to just-in-time principles. In 1970, Toyota's four distribution networks, Toyota, Toyopet, Corolla, and Auto, which were managed by Toyota's sister company, Toyota Motor Sales (TMS), included 247 dealerships. The number of passenger car models rose from four to nine between 1965 and 1971, yet in general each model was assembled on a single assembly line with the aid of specialized production equipment. Toyota had still not arrived at the stage of diversified mass production.

### 3.1.1. The Toyota Production System

The two main organizational principles of the TPS were just-in-time production and autonomization (Ohno 1978). The principle of just-in-time, conceived by Kiichiro Toyoda, the founder of Toyota, arose from the idea that it was in a company's interest to buy the exact quantity of parts it required and to encourage suppliers to deliver these parts precisely when needed. Taiichi Ohno sought to apply this idea within the company by advocating just-in-time production, without intermediary stocks. He was first inspired, in 1948, by a supermarket system; downstream activities fetched parts and components from upstream. However, it was the use of tickets, or *kanbans*, to manage this system, from 1954, which made it efficient. *Kanbans* were used in all the factories by 1962, and were extended to suppliers starting in 1965. The system was accompanied by a smoothing of production to avoid disruptions or overproduction.

The idea of autonomization (*ji-do-ka*) was first conceived of by Sakichi Toyoda, father of Kiichiro; it consisted of equipping machines with an automatic system to stop them in case of problems. Autonomization also applied to operators, who were supposed to stop their production line if they ran into problems, so that product quality could be guaranteed and the problem solved on the spot. Thus autonomization was a means of managing anomalies (Monden 1991*a*). This type of management was carried out by 'eyes', because, if a problem arose, the machine or worker would emit an optical signal on an *andon* board. Given that the machines were autonomous, workers were able to supervise several machines at once as long as they were positioned in such a way that the operator, who would have to be polyvalent, could move easily from one machine to another. Hence production lines were built in a 'U' shape. The primary goal of autonomization was to increase productivity by reducing the workforce (Ohno 1978). To achieve this, the number of workers per line was diminished through the *kaizen* (continuous improvement) of production processes. This was the Toyota method of increasing productivity.

So just-in-time production was organized by the autonomization of production lines and by the circulation of *kanbans*, which synchronized lines producing components with assembly lines. However, to be successful, the system required operators and managers to become actively involved in production and in *kaizen* activities.

### 3.1.2. Methods of Involving Workers

Lifetime employment, seniority-based wages, and the company union are considered to form the triptych of 'Japanese-style industrial relations'. Yet this triptych does not help us understand the active involvement of Toyota workers, nor its results. Up until 1974 lifetime employment coexisted with the extensive use of temporary and seasonal workers (34 per cent of the workforce in 1960,

9.2 per cent in 1974, and under 5 per cent in 1975). Pay rises occurred from one year to the next, but they did not increase automatically as a function of seniority. Toyota's union was a company union, but different from its counterparts elsewhere.

The relationship between the union and management had been based on reciprocal trust, as embodied in the 1962 Management–Union joint declaration. This had been signed twelve years after a major labour conflict in 1950 which followed a financial crisis at the company. In fact, the declaration made official a cooperative relationship which had existed between the two parties since 1954. The declaration defined the company as a community composed of administrators and workers. Increased productivity was considered to be a precondition both for increasing company profits and for improving the living standards of workers. Management and union declared themselves ready to cooperate in building mutual trust in order to obtain these increases (Shimizu and Nomura 1993). There were no more strikes after 1954. Collective bargaining was replaced by a summit conference organized by company administrators and members of the union's executive committee. While no labour agreement was signed between 1951 and 1974, the commitment to mutual trust prevented management from taking decisions which might undermine this trust, such as redundancies.

The basic work unit in the factory was the group (*kumi*, comprised of about fifteen workers divided into three *han*). Group leaders established and organized the tasks of their group members. Standard tasks and standard times were introduced in 1953; the standard task indicated the procedure and the time allotted for performing operations, the order and timing of movements, and the cycle time within which the operations and movements had to be finished. To establish standard tasks, group leaders analysed and timed all operations undertaken by their group. They were advised by management to have their subordinates participate in the determination of standard tasks so that they would be able to perform their tasks with the feeling of having designed them. Toyota accorded a great deal of importance to the standard task. It allowed production operators to assure product quality and to work in safety; moreover, it became the focus of *kaizen* activities. Groups were expected to improve the way operations were distributed and carried out, with the aim of eliminating possible defects as well as reducing the number of individuals needed to carry out the tasks. The standard time served as the point of reference for manufacturing a product on a given line as well as for calculating the 'remuneration of production' (see below). It was considered to be the strategic parameter for managing production efficiency.

Until 1990, the monthly wage of employees classified below rank 2B (see Fig. 3.1), that is, below section leader (*kacho*) was principally composed of basic pay, remuneration for production, and overtime pay (Nomura 1993). The basic salary of an employee was the salary at which he was hired plus annual increases. There were three levels of salaries for newly hired workers, reflecting their level of education. The average increase in basic salary was negotiated between the union and management in March of each year, during the *shunto* negotiations.

| RANK | QUALIFICATION | AGE | | POST | |
|---|---|---|---|---|---|
| | | mininum | Average 1993 | Management | Professional post* |
| 1B | Vice director | | 56.0 | Ji-cho | |
| 2A | Superior section chief | | 53.0 | | |
| 2B | Section chief | | | Ka-cho | |
| 30 | Subsection chief | 41 | 47.3 | | |
| 40 | Superior instructor | | | Ko-cho | CX (chief expert) |
| 50 | Group leader | 35 | 43.5 | Kumi-cho | SX (superior expert) |
| 60 | Team leader (1st rank) | 33 | 40.4 | | |
| 7A | Team leader (2nd rank) | 29 | 36.8 | Han-cho | EX (expert) |
| 7B | Instructor | 27 | 33.0 | | |
| 80 | Instructor (2nd rank) | 24 | 28.2 | Staff/ engineer | |
| 9A | Superior rank operator | 21 | 23.4 | | |
| 9B | Middle rank operator | 19 | 19.5 | | |
| 9C | Inferior rank operator | 18 | 18.0 | Operator | |

FIG. 3.1. *The hierarchy of employee ranks at Toyota*
*Note*: The professional posts were created in 1991.
*Source*: Toyota.

This average increase was distributed according to the hierarchical rank, and as a function of the *satei* (skill and ability evaluation) score of each employee; the basic salary was therefore individualized. The remuneration of production was the product of the basic salary and a remuneration of production coefficient (RPC). This had been designed by Taiichi Ohno to give workers an incentive to increase production efficiency. The RPC was calculated for each shift unit (*hakari*). It was determined by the standard time, production volume, and paid working time (the product of hours worked and number of production workers). In principle, an increase in production volume and a reduction in paid working time for the same volume would increase the RPC, while a reduction in standard time would decrease it. Overtime pay was the product of the paid overtime coefficient (POC) and the sum of basic salary plus remuneration of production. The monthly salary (minor allowances excepted) was therefore calculated by the following formula:

monthly salary = basic salary $(1 + \text{RPC})(1 + \text{POC})$.

The management of production efficiency therefore focused on paid working time for a given volume, and on standard time. The production efficiency of a work unit was increased through the *kaizen* of its tasks and production processes, which was carried out with the encouragement of management. Its production efficiency coefficient (PEC) rose, and so did the remuneration of its members, since the PEC in part determined the RPC; productivity increases were therefore shared. However, when the PEC of a work unit surpassed the average PEC of the best performing units (30 per cent of all work units) its standard time was reduced in such a way that its PEC equalled this average. The reduction did not affect the remuneration of its members. Yet since other units were also pursuing their *kaizen* activities, its PEC was then classed below that of the best performing units. The work unit would have to initiate the *kaizen* processes once more. Toyota's wage system thus incorporated incentives for workers to engage in *kaizen*. Admittedly, this kind of *kaizen* initiative was started off by the supervisory staff, who alone were informed of the PEC classification. Consequently, wages varied from month to month owing to fluctuations in the RPC (as well as variations in overtime).

In addition to monthly wages, there were also bonuses paid twice yearly, and a retirement bonus. The average sum of annual bonuses was negotiated during the *shunto*. It was of the order of six months' standard wages, a proportion which remained stable from 1968, owing to Toyota's high level of profitability. The retirement bonus that employees received at the end of their career was the product of their final basic salary and a coefficient for years of service (the coefficient of an employee with thirty years of service was 90).

The individual wage rose as the employee rose in rank. Newly hired workers were classed hierarchically according to their level of education: individuals with secondary school diplomas were positioned at rank 9C; individuals with the equivalent of two years of university study were positioned at 9B, while those holding full university degrees (four years) were positioned at 9A. Salaries for ranks 9B and 9A were approximately the sum of the basic salary of a rank 9C worker plus annual increases corresponding to the additional number of years of education. Workers were then promoted up to rank 7B, from which point a selection process began. The company's personnel management department set the minimum age for supervisory staff and presented the factory with the number of promotions allotted per rank. In the factory, it was effectively the section leaders who had the right to nominate workers for group or team leader positions. Candidates, who therefore had a minimum level of seniority, were selected on the basis of their *satei* scores. Their promotion was decided following a training period. What counted most was the *satei* score.

The wage system, promotion, and the *satei* were the three institutional instruments designed to get workers to participate actively in their work and in *kaizen*. However, management did not consider them sufficient. Following the major labour conflict of 1950, Toyota had developed various activities aimed at turning the company into a coherent working collective. These activities comprised basically the suggestions system and quality circles (QC) on the one hand, and human relations activities on the other (Nomura 1993). Improvements made to production efficiency were essentially carried out by supervisory staff, engineers assigned to the factory, and the department of production technology. Those made by production workers through the suggestions system and QCs were of minor importance from an economic perspective. In reality, these activities fulfilled other functions in the eyes of management. First, they increased the ability of workers by encouraging them to propose solutions to certain problems. Secondly, they gave them a feeling of ownership over their workplace through participation in its improvement. Thirdly, they reinforced bonds among workers by encouraging them to discuss and reflect upon their work. Lastly, they permitted leaders to emerge and developed their ability to communicate. Human relations activities were also aimed at promoting the integration of the team ('personal touch', 'mentor' system) and a feeling of belonging to the company. Toyota's workers were organized into eight 'corps', as a function of their rank upon entering the company; a corps for holders of a Toyota school of technology diploma, a corps for holders of secondary school diplomas, a corps for university graduates, and so on. The corps were set up to reinforce human relations between their members and to reveal those who might be leaders. Employees over 30 years of age became leaders of their respective corps, and were made responsible for organizing activities (sports competitions, parties, excursions, and so on) in consultation with their superiors. Their leadership activities therefore went beyond the workplace, but were definitely part of Toyota's industrial relations system, in the sense that they encouraged employees to involve themselves in their work and *kaizen* activities and to develop a sense of community.

### 3.1.3. Supplier Relationships: Partnership and Competition

Suppliers delivered to Toyota according to just-in-time principles. They contributed to cutting production costs and increasing the quality of company products. Toyota's relationship with its suppliers was a long-term investment. In practice, Toyota treated them as if they were subsidiaries, sharing profits as well as risks with them. However, this did not mean that they were protected. Indeed Toyota exerted various forms of control over them, in terms of their production processes, work organization, production costs, and product quality. Furthermore, the company exerted pressure on its suppliers to reduce component prices, according to the following system (Asanuma 1984; Ueda 1989).

In general, Toyota chose suppliers for each new model, the supplier contract being concluded as the product was launched. Roughly speaking, the price of a component was determined by applying a profit rate ($r$) to the cost of unit production ($C$), according to the following equation: price = $C(1+r) + d$, $d$ being the depreciation per unit of the production tools needed. When sales of a model were substantially lower than forecast in the contract, some suppliers were unable to recuperate tooling costs. In this case, Toyota reimbursed them their sunk costs, by raising the price for parts supplied in future. Profit rates having been agreed, price was fixed during negotiations between Toyota and the supplier over unit production costs ($C$). If the supplier were able to reduce the cost of producing a part at the design stage, the difference between the previous cost and the new cost became an extra profit which Toyota accepted the supplier should make for six months or one year. Moreover, if in month t+1, the supplier had achieved $C_{t+1} = C_t - a$ by reducing costs, then this $a$ ($>0$) gave an extra profit, $a(1+r)$. However, since price negotiations were conducted every six months, Toyota would at that point require its suppliers to lower the price of the part by $a\,(1+r)$. According to Asanuma (1984), there were two categories of supplier: those capable of designing components (design-approved suppliers), and those without this capability, to which Toyota supplied the design (design-supplied suppliers). Generally speaking, suppliers in the first group possessed know-how which Toyota did not wholly possess. This gave them a hidden $a$. However, they could not impose any price, since Toyota ordered the same part from several suppliers and carried out cost studies. Conversely, the cost of parts from 'design-supplied' suppliers was well known to Toyota. It was difficult for these suppliers to conceal an $a$. They nevertheless participated in Toyota's research and development activities, either to develop new parts, or to improve the quality of parts.

Toyota informed its suppliers of its quarterly production plan so that they could prepare themselves. The monthly order volume was given to them before the end of the preceding month, although it might still be modified, within a margin of about 5 per cent of expected volume, as a result of daily orders made through *kanbans*. The cooperative relationship between Toyota and its suppliers was given a dynamic by competition among suppliers (parallel orders), and the price system incorporating incentives to *kaizen*. By creating competent suppliers, Toyota was able to improve its price/quality competitiveness. However, in the early 1970s suppliers had not yet begun to apply the TPS; they supplied Toyota just-in-time by holding stocks in their own factories.

### 3.1.4. Distribution Network and Production Planning

The final element of the TPS was the distribution network, the interface between Toyota and its customers. The network was organized by Shotaro Kamiya, president of TMS (Toyota Motor Sales, separated from Toyota in 1950). The priority was satisfaction for customers, followed by dealers; both came before Toyota. The relationship between TMS and its dealerships mirrored that between Toyota and its suppliers. With the knowledge TMS had of their accounts, it could both control and help them. Dealer contracts prohibited them from selling other marques. Toyota gave them financial support on condition that they remained loyal. Each dealer signed an annual contract indicating sales per model, including retention of a small permanent stock. Toyota believed that the stock compelled them to try to increase sales. Eighty per cent of sales were made by the 'visiting' salesmen of the dealerships, the remainder being sold at the sales points, of which a dealer had several. The role of the dealers was not limited to sales, for they also diligently noted the complaints, requirements, and desires of existing as well as potential customers, so that these factors could be taken into account in the design of new models.

Toyota established its production plan on the basis of the orders it received, as well as its own sales forecasts. In 1965, Toyota implemented a system whereby orders were sent in every ten days, which permitted orders to be delivered within sixteen to thirty days. In 1970, it adopted a daily ordering system for the Celica model, which allowed customers to choose between twenty-eight versions equipped with multiple options. The vehicles were produced on command and delivered between eight and eleven days afterwards. However, not only was this system not expanded, but it was abandoned following the first oil crisis. Contrary to what many in the West have believed, therefore, Toyota has never produced only on command, with the exception of the Celica model over a brief period. Ordering on a daily basis actually contradicted the principles of the TPS, which required detailed planning production (annual, quarterly, monthly, and every ten days) in order to smooth monthly and daily volumes and thereby avoid stocks.

### 3.1.5. Design and the Management of Production Costs

In the final instance, the viability of an industrial model depends on the competitiveness of its products. This in turn results on the one hand from the innovative design of the vehicle and its components, and on the other hand from quality and price. From the mid-1970s, each model of middle-range cars was completely replaced every four years. The management of production costs and *kaizen* activities started in the 1960s (Monden 1991b; Tanaka 1991).

When a complete model replacement or new model launch was decided upon, TMS proposed certain characteristics and a price to Toyota, which then selected a style and target production cost. A chief engineer, responsible for the model's design, would then organize a team of engineers from the various relevant design departments. Components were manufactured and assembled into prototypes three times, allowing the factory to ask designers for modifications to the product, and suppliers to establish and negotiate the terms of their contracts. Meanwhile, a production cost committee, composed of senior managers and the chief engineer, oversaw changes to the estimated cost of production, seeking to reduce any gap between it and the target cost. At the end of this process, the definitive plans were adopted. The next stage unfolded in the production department. After having reorganized the production lines, the department of production technology determined the reference production cost by production line. The basic unit in this calculation was the working group, each group therefore having its own reference production cost. Then, after having commenced mass production, the production department instituted the *kaizen* process in order to achieve and improve upon the reference production cost.

In practice, management imposed the objective on the production department of reducing production costs every six months in order to achieve targeted profits. If anticipated profits were lower than targeted profits, the gap had to be made up. This took place as follows: half had to be absorbed by increased sales (increased economies of scale), and the remainder by reduced variable costs (raw materials, energy, and workforce). Hence half the difference was the sum to be obtained through *kaizen* in the production department. The production department succeeded in obtaining lower costs for raw materials and energy by reducing their consumption and by making suggestions for the modification of parts, the utilization of cheaper parts, and so on. Part of the sum thus saved would be paid to the factories or to the workers who had made the suggestions. This form of *kaizen* was controlled by a production cost committee which convened once a month at all hierarchical levels, from management down to work groups. Yet the key element was the management of production efficiency through the wage system, as discussed above. This was the *kaizen* process to reduce the cost of the workforce, an activity scrupulously supervised by management. The TPS would not have been so successful without this *kaizen* process.

### 3.1.6. The Coherence and Dynamism of the Toyota Industrial Model

Beyond certain general socio-production principles, such as additive production on assembly lines, the components of the Toyota model were all innovative compared to those of the canonical Fordist model. It was the management of production costs that orchestrated them and provided the dynamic. This was the core of 'Toyotaism', which therefore had five basic characteristics. The first was mutual trust among Toyota, its employees, its suppliers, and its dealers. The second was competition among the members of each of these categories. The third was the research and development capacity of the Toyota group as a whole, including its subcontract or assembly companies (Toyota Auto Body and Kanto Auto Works, for instance) and its suppliers, which shared technical information with Toyota. The fourth was the continual effort to reduce production costs, to increase production efficiency, and improve quality. Lastly, improvements obtained through *kaizen* were shared, for a limited period, with those who had created them. The management of production costs and of production efficiency thus constituted both a means of controlling group members and an incentive to *kaizen*. This was the heart of Toyota management.

## 3.2. THE ERA OF TOYOTAISM: THE 1970S AND 1980S

Toyotaism continued into the late 1980s without significant modification. It was able to overcome the various difficulties the company faced, all of which arose externally, by further developing its principles. This was the case with the diversification of product variety and the flexibilization of production, two major challenges which followed the first oil crisis. Thus it was that the Toyota model appeared to be a post-Fordist model, and was named 'lean production'.

### 3.2.1. Long-Term Evolution, 1974–1994

The sale of Toyota vehicles in the Japanese market was flat during the period from 1973 to 1985, except for a drop following the first oil crisis when Toyota

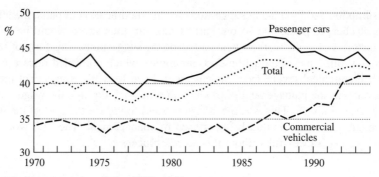

FIG. 3.2. *Toyota market share, 1950–1991*
*Sources*: Toyota (1987) and *Nikkan Jidosha Shinbun* [Automobile Industry Handbook] (1995).

lost market share (between 1975 and 1977). It then took seven years, until 1984, for the company to recover its 1975 market share (see Fig. 3.2). Registrations increased again with the advent of the bubble economy, before falling again after 1992 when the 'bubble' collapsed.

On the other hand, sales turnover continued to rise until 1993 (with the exception of 1987 when turnover was particularly affected by the appreciation of the yen). The rise in turnover in fact reflected not only the sale of vehicles, including exports, but also Toyota's other activities. Increased turnover was less a result of housing construction (Toyota entered the construction business in 1976 but this only accounted for 0.48 per cent of turnover in 1994) or the production of industrial vehicles (which began in 1983, and accounted for 1.38 per cent of turnover in 1994), than of increased sales of components, including those destined for overseas transplants (the contribution of these sales to turnover increased from 18.7 per cent in 1980 to 33.5 per cent in 1994).

Toyota's production also rose constantly between 1970 and 1991, with the exception of a small dip in 1974 and a flattening out between 1980 and 1983. The growth of output between the first oil crisis and 1985 was due to rising exports. In 1985, Toyota began production in the USA, which led to a structural decline in exports. This fall was more than compensated by a strong upturn of Toyota sales in the Japanese market until 1991. However, after 1992 exports and registrations in Japan declined together, causing domestic production to contract by 16.3 per cent between 1991 and 1994.

Toyota's workforce, including temporary workers, only increased by 12.1 per cent between 1970 and 1979, whereas total production (passenger cars and commercial vehicles) rose by 88.7 per cent. By contrast, the 1980s were marked by an inverse tendency: the workforce in Japan expanded by 66.4 per cent between 1979 and 1992, even though total production in Japan only rose by 40.9 per cent. The increased workforce can partially be explained by the merger of TMS and Toyota in 1982. Yet even if the whole rise in the workforce between 1982 and 1983 is imputed to this merger, the increase due to Toyota alone would still be

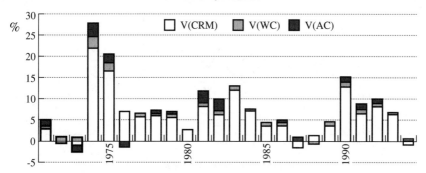

FIG. 3.3. *Factors in the variation of Toyota per-vehicle production cost*

*Notes*: V(WC) is the weighted effect of unit wage costs.
V(AC) is the weighted effect of unit administrative costs and depreciation.
V(CRM) is the weighted effect of unit raw material and purchased parts costs.

*Sources*: Toyota, financial reports.

51.3 per cent. Hence during this period, the workforce grew more rapidly than output. The workforce began to decline in 1993, a year after the fall in production. This was accomplished through the non-renewal of temporary worker contracts and the partial non-replacement of workers taking retirement. Toyota had thus avoided redundancies and could continue to take on employees, although in much smaller numbers. Indeed the company had not made anyone redundant since 1951, thanks to the 'mutual trust' expressed in the joint declaration of 1962, and to the fact that after 1950 Toyota experienced no major crises.

### 3.2.2. *How Toyotaism Responded to Obstacles: 1974 to 1980*

During the first half of the 1970s, the Japanese automobile industry had to deal both with issues of pollution and the first oil crisis. Rising air pollution led the Japanese government to adopt American standards of acceptable levels of toxicity for exhaust fumes, and to use tax exemptions to encourage research and development in this domain. From 1976, the engines of new Japanese car models had to comply with these standards. However, Toyota was slow to develop engines conforming with the regulations. Moreover, the first oil crisis increased the cost of energy used in manufacture (see Fig. 3.3). Toyota was forced to increase the price of its vehicles, and then new taxes were levied on automobiles from 1974. Sales fell off and operating profits declined markedly (see Fig. 3.4). At that point Toyota decided to go on the offensive with a sales strategy, known as 'T23', designed to sell 230,000 vehicles in the two months from June to July 1974. To achieve this Toyota modified its order system in March; it was at this point that the system of daily orders for the Celica model was eliminated. At the same time the company introduced into its ten-day-order system the opportunity for clients

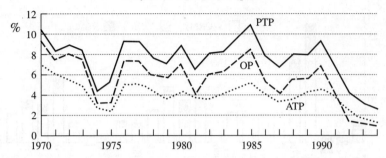

FIG. 3.4. *Profits as a proportion of turnover at Toyota*
*Notes*: PTP: pre-tax profit; OP: operating profit; ATP: after-tax profit. The difference between operating profit and pre-tax profit indicates profits from financial activities.
*Sources*: Toyota, financial reports.

to change certain technical specifications of a car ordered up to six days before production, and the company was committed to delivering the car within an average of under ten days.

However, the company's slowness to develop low pollution engines led to a fall in market share (see Fig. 3.2). While the company did not lose money in 1974, it did decide to modify its production system and its sales strategy.

Once the technical problems with low pollution engines were resolved, Toyota reinforced its range by replacing six models and launching four more models with the objective of regaining its position in the market (for a discussion of product diversification in detail, see Bélis-Bergouignan and Lung 1994). The concept of diversification was also modified. From 1972 to 1976, Toyota had increased the number of variants per model. After 1977, the company designed models which shared the same platform: hence the Mark II and the Chaser in 1977, the Tercel and the Corsa in 1978, the Corolla and the Sprinter in 1979, the Camry and the Vista in 1982 all shared platforms. This diversification of products, even if limited with platforms shared, required a flexibilization of the production system as well as efforts to contain production costs.

Three types of modification were made to the production system (Toyota 1978). Specialized equipment was replaced by equipment which could be rapidly adapted to a change of models or even be used to produce different models (mixed model production). This was concretized through quick tool and die change; so that stamping dies, for instance, could be replaced in under ten minutes. However, this flexibilization was not introduced into the engine factories or body shops at this stage. The second reorganization consisted of creating linked U-shaped production lines so that the number of operators could be adjusted according to fluctuations in production, and the number of operators be reduced in general (Monden 1991*a*). Thirdly, preventive maintenance was re-emphasized in order to increase the reliability of equipment.

Meanwhile, Toyota had to consolidate its management of production costs. In 1974, the company formed a team responsible for lowering the production costs of the Corolla model (Toyota's best-selling car). The team surpassed its target by 28 per cent. Their method differed from previous methods in that the team was comprised of members from all departments concerned: design, production technology, manufacturing, purchasing, general affairs, and accounting (Toyota 1978). It was at this stage that the management of production costs was systematized all the way from design through to production. Moreover, Toyota now required all its first-tier suppliers to utilize the TPS and produce without stocks.

Toyota was able to re-establish its competitiveness and profitability through these strategies of diversification and production cost reductions. The company was able to build up significant financial reserves to the point that borrowing was eliminated from its balance sheet in 1977, since when Toyota has become known for its debt-free management and internal reserves which have brought considerable financial benefits to the company.

On the strength of these excellent results, in 1978 Toyota launched a production plan to reach 3.5 million vehicles and set a target for domestic sales of 2 million vehicles (a 40 per cent share of the passenger car market). To this end, the company increased its production capacity by building a factory at Kinuura and a second factory at Shimoyama in 1978, along with three factories at Tahara between 1978 and 1981. Toyota now owned two transmissions factories (Tsutsumi and Kinuura), two engine factories (Kamigo and Shimoyama), and nine assembly factories (not including those of its subcontractor assemblers). From 1979 onwards, the company also undertook a programme of robotization.

### 3.2.3. The Zenith of Toyotaism in the 1980s

The growth programme was maintained despite a change in the economic environment. Like the other Japanese producers, Toyota had to confront the yen's appreciation in 1978, followed by the second oil crisis in 1979, and finally US pressure to voluntarily limit exports in 1981 (with a quota of 1.68 million Japanese cars per year, of which 516,659 were allocated to Toyota). The domestic market stopped growing, exports contracted in 1981–2, and profit rates dipped in 1981. Toyota continued to strengthen its management so that it would remain profitable even at 80 per cent capacity utilization (Toyota 1987).

The company began designing cars that used fewer raw materials and less energy (with engines designed to consume less fuel), incorporated lighter raw materials (aluminium, plastics, ceramics), and included electronic technologies and front-wheel drive. The components supply system had also been reorganized, with the production of certain parts outsourced. Suppliers and subcontractor assemblers were also involved in production management and now had to implement Toyota's methods of planning production costs.

A new distribution network was created in 1981 for the Vista model. The five Toyota networks each sold three passenger car models. Toyota merged with

TMS in 1982 in order to be able to react more rapidly in a context of fierce global competition. The dealers were obliged to undertake *kaizen* activities and start quality circles, and to open on Sundays. The ordering system was modified once more, so as to reduce delivery lead-times and to permit the technical specification of the vehicle ordered to be changed up until five days prior to its manufacture. To resolve quality problems raised by customers (scratches, traces of glue, loose screws, and so on), the company introduced tools which prevented certain assembly errors, and created inspection posts on the assembly lines.

Between 1980 and 1985 Toyota replaced ten models and launched seven others. Some models had their own platform: Starlet, Carina, Soarer, Supra, Corona, Crown, and Century. Others shared platforms: Corolla and Sprinter; Mark II, Chaser, and Cresta; Celica and Carina ED; Camry and Vista; Corsa, Tercel, and Corolla II. Toyota's strategy appeared to be to offer models with their own platform at the top and bottom of the range in order to make their production profitable, and to share platforms among models in the middle of the range, where competition was greatest. The renewal and broadening of the range was accompanied by an increase in the quality and specification of models. Despite the stability of wholesale prices (0.6 per cent average annual increase) between 1981 and 1984, production costs increased by an average of 10.3 per cent (see Fig. 3.3). Between 1982 and 1985 Toyota's profit rate was the highest in its history. The company reached a 44.9 per cent share of the domestic passenger car market (Figs. 3.2 and 3.4). 'Operation T50' was launched in 1986, with goal of obtaining a 50 per cent share of the passenger car market.

However, Toyota's plans were compromised by the rapid appreciation of the yen (over 70 per cent from 1985–7), which brought about two years of economic recession. Output stagnated in 1987 as exports declined despite increased sales on the domestic market, and operating profits fell from 8.3 per cent in 1985 to 4.1 per cent in 1987. As ever, Toyota focused on reducing production costs in order to recover its profit level. This was done by reducing the costs of raw materials, purchased components, and wages (Fig. 3.3). The company also increased the flexibility of its production system. In 1984, an assembly line was installed in one of the engine factories which was capable of mixed production but in which production costs were the same as for a specialized line (Toyota 1987). In 1985, the company installed a Flexible Body Line (FBL) in one of its body-welding factories, which enabled different bodies to be welded on the same line. To control the assembly process and robots in real time, a new Assembly Line Control (ALC) system was adopted in 1989. Moreover, in order to adjust the workforce to fluctuations in production with increased precision, the reorganization of U-shaped lines into linked U-shape lines was extended, the sending of workers from overstaffed areas to understaffed area was systematized, and the hiring of temporary workers was resumed. In 1986, the installation of a computerized network (Toyota Network System) embracing Toyota and its dealers enabled the company to control orders, production, and deliveries in real time. The TPS was being made more flexible, and being computerized from upstream to downstream.

## A New Toyotaism?

Having reached this level of development, it served as a model for the construction of Toyota transplants in North America and the United Kingdom (see below).

A second reason Toyota was able to regain its former level of profitability was the strong recovery of domestic and global markets under the impact of the economic bubble which began in Japan in 1987. That year, the company announced its 'Global 10' objective, signifying its intention to control 10 per cent of the global market. To this end, Toyota diversified its range even further, by launching luxury models (the Celsior in 1989). It was also during this period that Toyotaism appeared to be the world's most successful and robust industrial model. And yet certain statistics already suggested that the Toyota model was entering a difficult period. Despite the rapid growth of its sales, Toyota's share of the domestic passenger car market fell from 46.2 per cent in 1987 to 42.9 per cent in 1991, running counter to the objective of 50 per cent. Profitability did not grow as rapidly as during previous phases of market growth, and in fact fell in 1991 even as output continued to rise.

### 3.2.4. The Construction of Transplants and the Transferability of the Toyota Model

Trade conflict between the USA and Japan created a structural problem for the Japanese automobile industry. Toyota's exports to North America had increased rapidly between 1975 and 1986. With a desire to avoid conflict, in 1981 Toyota followed Honda and Nissan and decided to produce in the USA. Hesitating to manage a factory alone, and having failed to form a joint venture with Ford, in 1982 Toyota accepted an offer by General Motors (GM) to reopen one of the latter's closed factories, where each company could produce a model. GM's Fremont factory became the joint subsidiary New United Motor Manufacturing Inc. (NUMMI), the management of which was entirely entrusted to Toyota. When this experience showed Toyota that its system was transferable, the company decided to build its own factories in North America: Toyota Motor Manufacturing USA Inc. (TMM) in Kentucky, Toyota Motor Manufacturing Canada (TMC) in Ontario, both in 1988, and Toyota Motor Manufacturing United Kingdom (TMUK) in the United Kingdom in 1992. These subsidiaries were true 'transplants', since Toyota effectively transferred its industrial model, requiring only a few adaptations to accommodate local industrial relations.

Toyota's strategy for transfer can be summarized in the following statement: 'risk is truly great if both the workforce and the equipment are new' (TMM, cited by Suzuki 1991). For NUMMI, Toyota had to use some equipment from the old factory. However, a new stamping plant was built, and the body plant was replaced with the installation of two welding lines identical to a standard Toyota line. In the paint and assembly areas, Toyota divided the old lines (two kilometres long) into several sections, thereby creating intermediate stocks which would allow operators and supervisors the time to resolve problems without stopping the entire line (Shimada 1988). By contrast, when the TMM factory was constructed, Toyota installed equipment similar to that used in its Japanese factories, in particular the Tsutsumi factory (Mishina 1995). By and large, the TMM and TMUK factories resembled the Tsutsumi factory. Both engineers and supervisory staff were sent out from Japan to train maintenance workers for the machines they knew well.

The greatest difficulties encountered during the implementation of the TPS at the transplants lay in industrial relations and components supply. Toyota had to discover a way to forge relations based on mutual trust with local workers, whether unionized (NUMMI, TMUK) or not (TMM, TMC), and with suppliers. As far as industrial relations were concerned, NUMMI has been studied extensively (Parker and Slaughter 1988; Shimada 1988; Suzuki 1991). The Fremont factory was known for having one of the worst records for industrial relations at GM. Toyota was none the less able to create a model of cooperative industrial relations in North America, despite the fact that 85 per cent of the workforce had worked at the factory previously, and were therefore members of the United Automobile Workers (UAW) union. The work relationships Toyota wished to develop at its transplants were clearly expressed in NUMMI's collective agreement, which was accepted by the UAW and management in 1985. This American version of Japan's 'joint declaration' stipulated that the parties would make every effort to create the most innovative industrial relations in the USA, on the one hand to deliver to customers vehicles of the highest quality in the world at the lowest possible cost, and on the other hand to assure equitable wages to the employees. According to Suzuki (1991), the principal clauses of this agreement were: the classification of blue-collar workers into three categories, direct workers, indirect workers, and maintenance workers, in contrast to the 84 categories at GM; hourly wages and bonuses fixed according to scales negotiated by the UAW and GM; redundancies to be avoided at all cost (not one worker was made redundant despite a loss of $100m. in 1987–8); management to discuss with the union its intentions regarding production planning and changes in the distribution of the workforce as well as working hours; lastly, the union to accept the rules governing work under the TPS: teamwork, quality circles, operating according to standard tasks, and so on. This collective agreement contained the minimal conditions necessary for implementation of the TPS. In order to reconstruct, within a different social context, the sense of mutual trust which characterized industrial relations in the TPS, NUMMI's management guaranteed employment and planned to meet with union representatives to explain company strategy and discuss any problems that might arise.

Work organization at the transplants is characterized by teamwork, task rotation, *kaizen* activities conducted by quality circles, and flexible assignment of workers. The training of supervisors was particularly important, to enable them to fulfil their role of establishing standard tasks and encouraging *kaizen* activities. Future team leaders were sent to one of Toyota's Japanese factories, and even to TMM and TMC in the case of TMUK, for on-the-job training in TPS principles, quality circle activities, and teamwork. Japanese instructors were dispatched to help managerial staff at the transplants. However, the Toyota wage system was not transferred. In the North American transplants an hourly wage system was adopted, whereas at TMUK in the United Kingdom wages were calculated on an annual basis. In both cases Toyota opted to respect local practice.

As far as components supplies were concerned, NUMMI selected 104 American suppliers, nine of which were joint ventures with Japanese companies. In 1987, these suppliers permitted NUMMI to reach 60 per cent local content for the Nova and 50 per cent for the Corolla. In 1987 TMM had already selected 60 American suppliers, permitting it to attain 60 per cent local content. Suppliers belonging to the Toyota group also transplanted operations to North America (Denso, Toyota Gosei, Aisin, and members of the Kyoho Association). TMUK achieved 70 per cent local content, with parts and materials provided by 200, mainly European, suppliers. The principal criterion for choosing suppliers was their cooperative attitude and their acceptance of TPS rules: quality, respect for delivery lead-times, and use of *kanbans*. Once selected, suppliers were helped by the Toyota transplants to improve their technology and to improve the quality of their products. This helped establish mutual trust as the basis for long-term cooperation. Frequent deliveries in small batches were not necessarily required by the Toyota transplants; the company displayed a degree of flexibility in its thinking, and just-in-time production could assume different forms according to the industrial context. What remained important was that suppliers delivered components and materials in compliance with the required level of quality and lead-times previously established. The rest of the system was not going to be established immediately.

Above all, however, the success of the transplants depended upon the commercial success of the models produced. To begin with, NUMMI manufactured the Nova for GM and the Corolla FX16 for Toyota. But the Nova was a failure and had to be replaced by the Prism in 1988, while the four-door version of the Corolla replaced the three-door version (Suzuki 1991). These changes permitted adequate sales to be made. TMM quickly benefited from its 1990 adoption of the Camry, one of the best-selling models in the USA. However, the Carina E, produced by TMUK, did not succeed in the European market and so TMUK was also allocated production of the Corolla from 1998 in order to realize full capacity production (200,000 units per year).

In short, Toyota's transplants performed better than had been expected. Their output steadily replaced Toyota exports from Japan. The Toyota model had proved that it could be transferred outside Japan. However, this did not mean that all of the model's components should be transplanted to the host country. With their different social structures, the way the system was applied could vary across host countries, so long as the minimum necessary preconditions for the TPS to function were met.

### 3.2.5. *The Crisis of Work at the end of the 1980s*

A crisis of work emerged in Japan during the so-called bubble economy period between 1987 and 1991, paradoxically at the very moment Toyota seemed to have reached its apogee. The crisis was the result of structural change in the labour market and a growing rejection of assembly-line work by young employees. While the demand for automobiles grew rapidly, by the end of the 1980s companies found it difficult to recruit the necessary labour force. A declining birth rate had reduced the younger active working population. Young high school graduates wanted to avoid jobs characterized by the 3Ks (*kitanai*: dirty; *kitsui*: difficult; *kiken*: dangerous). Moreover, turnover of newly hired young employees in the automobile industry had increased, since employees were rejecting assembly-line work, which was monotonous, repetitive, and fast, in contradiction with the reputation of Japanese workers as polyvalent and motivated.

In Toyota's case, the way production efficiency was managed meant that the factories worked with the minimum number of operators necessary and a fast work pace. While this was viable during a phase of regular growth in demand, these working conditions were not consistent with rapid growth. The bubble economy pushed the TPS to the limit, revealing that it had become too lean. Toyota was incapable of responding to rising and diversifying demand. To fill the gaps in its labour force, from 1987 Toyota was hiring temporary employees on a large scale; by 1991, they constituted 10.38 per cent of the direct labour force. Far from resolving the nascent crisis of work, the recruitment of temporary workers precipitated it. The fact that temporary workers were less competent, combined with the growing complexity of tasks owing to the wider variety of components, disrupted production, requiring more overtime to produce the planned output. Annual working hours in the factories rose from 2,224 in 1987 to 2,315 in 1990. Fully one-quarter of new recruits left Toyota during their first year in 1990, unable to cope with the heavy workloads. Supervisory staff, the hard core of the TPS, who had to intervene directly to resolve all the problems, themselves became exhausted (Shimizu 1995). The shortage of labour had been transformed into a crisis for the whole working collective owing to the characteristics and logic of the TPS.

The management and the union at Toyota therefore began to question the Toyota way of managing work. Simultaneously, management concluded that it had pursued product variety too far, and decided to reduce the number of variants and reorganize design into four departments, each specializing in one domain: rear-wheel drive, front-wheel drive, commercial vehicles (including electric vehicles), and new components.

### 3.3. A NEW TOYOTAISM AND GROWING UNCERTAINTY

The reorganization of the Toyota model began by a questioning of work relations and by the publication of a new set of 'guiding principles at Toyota', of which there were seven: be a company of the world, serve the greater good of people everywhere by devoting careful attention to safety and to the environment, assert leadership in technology and customer satisfaction, become a contributing member of the community in every nation, foster a corporate culture that honours individuality while promoting teamwork, pursue continuing growth through efficient global management, and build lasting relationships with business partners around the world. The reorganization of the management of human resources and work appeared to be the most advanced and promising transformation. These changes could justifiably be labelled a 'new Toyotaism'. Yet, by the mid-1990s there were already uncertainties over its future, since Toyota's financial results did not appear to be recovering satisfactorily.

### 3.3.1. Rethinking the Toyota Approach to Work Relations

In 1990, a committee composed of union representatives and company management was formed in order to consider methods that might make factory work more attractive. The notion of 'humanizing work' appeared to this committee to be the sole means of resolving the crisis of recruiting and involving workers. Between 1990 and 1992 the group met to discuss a number of problems, issues related to the very foundation of the Toyota model: the management of production efficiency, the wage system, training, and assembly work (Shimizu 1995).

It was decided that the target for labour cost reduction, hence the number of operators needed, would be set according to the results obtained during the three months following the launch of a model, and not according to the best results obtained with the preceding model. The target for reducing the variable costs would be set according to objectives fixed by each individual factory, with top management restricting itself to ensuring the coherence of the various objectives proposed. Thus management withdrew from unilateral management of production costs, and above all production efficiency, in order to grant greater autonomy to the factories. Moreover, the company committed itself to think about reducing production costs as a whole, instead of always compelling the factories to increase their production efficiency. This effort focused more on the design stage, in which it was possible to make substantial savings in terms of raw materials and components.

The method used to evaluate production efficiency was also modified. The classification of the coefficient of production efficiency (PEC) was now to be carried out with reference to homogeneous groups, taking account of the specificities of work in different production processes: with a 'foundry, forging, stamping, and welding' group, a 'mechanical components' group, a 'bodywork, painting, and plastic moulding' group, and an 'assembly' group. Moreover, the determination of standard time was to take into consideration the time needed by older and female operators. A reduction in annual working hours was also planned, by 300 hours between 1991 and 1993. In practice, hours declined from 2,284 in 1989 to 1,915 in 1993. In short, the management of production efficiency became less constraining and more reasonable.

The wage system was revised simultaneously. In 1990 one form of change had already occurred. The introduction of two new criteria for remuneration, age (hence age-related pay: AP) and ability (hence ability-based pay: ABP), each accounting for 10 per cent of the average standard wage, had reduced the significance of the remuneration of production from 60 per cent to 40 per cent, with the remaining 40 per cent corresponding to the basic salary (BS). Making ability an autonomous criterion, whereas it was formerly integrated into the basic salary, meant there were now two distinct types of evaluation: a *satei* for competence which affected increases in the basic salary and promotion, and a *satei* for the results of worker activities upon which the non-cumulative increase in their ABP depended. However, reform did not extend to methods of calculating the remuneration of production.

The new system established in 1993 again reduced the significance of this criterion in determining the wage. It was simply eliminated for engineers and salaried staff (section S), who considered production efficiency irrelevant to them. Remuneration based on ability (ABP) now accounted for 40 per cent of the standard wage. As for other employees—those in production and engineering sections (sections P and E)—the proportion of the remuneration of production, renamed remuneration of productivity (RP) in the wage was cut from 40 per cent to 20 per cent, while the AP and ABP each rose from 10 to 20 per cent. The method of calculating the RP was also modified; the RP was the product of its coefficient (CRP) and a sum determined by the employee's rank within the hierarchy.

Hence for section P and E employees, the standard wage, excluding minor allowances and overtime pay, was calculated by the following equation:

Wage = BS (40%) + RP (20%) + ABP (20%) + AP (20%)

For section S employees, it was calculated as follows:

Salary = BS (40%) + ABP (40%) + AP (20%)

When in the 1950s Taiichi Ohno applied the RP to salaried staff and engineers too, it was because he wanted to mobilize them too in *kaizen* activities to increase production efficiency and to involve them in managing working time and workforce numbers (Toyota 1958). The new wage system implied that Ohnoism no longer applied to white-collar workers. Furthermore, workers of the same rank in a work section (*ka*) received the same RP sum, and this constituted a truly collective incentive to increase productivity. The ABP was determined by hierarchical rank and was increased according to the *satei* score. Because the increase was not cumulative, it was possible for employees with poorer evaluations to catch up by working harder. In addition, the new system aimed to modify the progression of wages during an employee's working life, since 30- to 40-year-old employees were relatively less well off than employees aged over 50, given the expenses the younger workers incurred: house purchase, school fees, and so on.

The break with the Ohnoist wage system was even clearer at Toyota Motor Kyushu (TMK). This subsidiary of Toyota, located on the southern Japanese island of Kyushu and founded in 1991, began production of the Mark II in 1992 with a production capacity of 160,000 vehicles per year. The monthly salary of all employees was comprised solely of the BS (60 per cent) and the ABP (40 per cent) (with the exception of minor allowances). As a consequence, the monthly salary remained the same all year, and was only revised each April, after negotiations between management and the union. As the wage did not include the RP, TMK's employees were not directly encouraged to increase production efficiency as they were at Toyota. Instead, TMK introduced PIT (production incentive of TMK), a second bonus paid every six months based on the results of *kaizen* activities evaluated by work section (*ka*).

To resolve the problem of high turnover among new production operators, in 1993 Toyota also modified its system of post-recruitment training. Training was now entirely devolved to the factories, and its duration was prolonged to nine weeks for workers assigned to stamping, welding, body, and assembly areas, and six weeks for others. The factories organized two weeks of general training, and then assigned workers to work areas for on-the-job training by rotating them among work posts where they did half an operator's task. Their first work post was decided after this training. Moreover, in 1991, a new form of professional training had been introduced to instil in all operators a more systematic know-how and a 'delight in producing'. Management instituted four certificates of professional competence based upon the results of tests following training. These certificates objectively stated the skills and know-how of the holder. The holder of the highest level certificate would possess the skills and know-how necessary to assemble a whole car. To help establish this training, task rotation was also systematized.

Reforms of production cost management, the wage system, and training were therefore pursued in an attempt to overcome the crisis of work and motivate workers more by granting them greater autonomy, reducing the pressure to reduce standard time, reducing the number of hours worked, and enriching training. Of even greater importance was the new conception of the assembly line and group work.

### 3.3.2. New Assembly Lines and the 'Humanization of Work'

The first experiment was undertaken at the Tsutsumi factory in 1990. One of the two assembly lines was divided into four segments, between which were placed buffer stocks. At this point it was believed that all problems regarding work derived from a rigid application of the just-in-time principle, which meant that the whole line stopped whenever and wherever a problem arose. In 1991, when the fourth factory at Tahara was being constructed (Toyota's most automated factory, where the luxury models Celsior, Lexus, Crown-Majesta, and Aristo were produced), the assembly line was divided into eight mini-lines, separated by buffer stocks. Toyota quantified the degree of difficulty at each work station through its TVAL (Toyota verification of assembly line) method. To reduce or eliminate the difficulty of certain tasks, variable height body-carrier platforms and automation were installed. Ordinary conveyors were replaced by large rectangular platforms linked together. Operators were able to perform their tasks standing on the platforms without having to walk. The heights of the car bodies could be varied according to the task to be performed and the height of the operator. The most tedious tasks from an ergonomic standpoint were replaced by automated operations. Lastly, some tasks related to final quality control were transferred to the end of the mini-lines. However, the transformation of the assembly line was not fully realized, nor was a new type of groupwork introduced.

These latter changes were, however, accomplished in construction of the TMK assembly factory. The assembly line was divided into eleven mini-lines in such a way that each mini-line corresponded to a function of the vehicle, which could be made the responsibility of a work group. This line made a number of advances over the one at Tahara (Shimizu 1995). The body-carrier platforms (the same as those used in the fourth Tahara factory) were equipped with a device to automatically adjust their height to accommodate the height of the operator. The rate of automation was lower, though the objective remained to eliminate as many ergonomically arduous tasks as possible through the TVAL system. Each work group was responsible for a mini-line. A quality control post was added to the quality control each operator had to carry out on his own work; the group leader could stop the line by taking advantage of upstream and downstream buffer stocks of three-to-five vehicle bodies. Operators rotated work posts within their group, and learned how to do all the jobs. If they wished, they could be transferred to another group to learn more tasks. This form of organization granted greater autonomy and responsibility to the work groups, yet it was also more efficient than a traditional line, since a stoppage at one point did not paralyse the entire assembly factory. Production without stocks, also known as 'one-by-one' production, had been abandoned, and was now seen as too rigid an application of just-in-time principles to assembly-line production.

Some elements of the TMK experiments were transferred to other factories. Toyota's older factories lacked the space to construct similar assembly lines, and the opportunities to make substantial investments were reduced by economic recession. Nevertheless, the same principle of work and production organization was adopted at the Motomachi factory during the restructuring of the assembly line for the RAV4 model. The assembly line was divided according to vehicle function into five segments. Body-carrier platforms were not used, but conveyor belts were installed at ground level, upon which the operators could stand, a system similar to the one used at Mitsubishi.

Another significant change that had been implemented at TMK was diffused throughout Toyota in 1995. The night shift was eliminated. Instead, alternating shifts worked only during the day; from 6.00 to 14.50 and 15.05 to 23.55 at TMK, and from 6.30 to 15.15 and 16.15 to 1.00 at Toyota. It was therefore no longer possible to lengthen the working day by adding substantial overtime worked between the two shifts.

By humanizing work and modifying its management and wage system, Toyota had most certainly not renounced all that it had created based upon

## A New Toyotaism?

just-in-time principles and autonomization. The *kanban* system continued to be applied, and *kaizen* activities to reduce production costs still took place. What was different was that the rigid application of Ohnoist principles had been rejected. The concept of 'autonomization' had even evolved to include the autonomization of the working group.

Changes were not restricted to the domain of production. In 1992, Toyota created the DUO sales network to sell Volkswagen/Audi cars, and eliminated the clause which prohibited dealers from selling the vehicles of other marques. From 1996, Toyota planned to sell GM cars (the Cavalier model). The volume of imported components was also increased. The internationalization of purchases, like the construction of transplants, was designed to attenuate trade conflicts and to create an image of the company as a 'good citizen' in each of its host countries.

### 3.3.3. Uncertainties over the Future

Yet by the mid-1990s uncertainties about the future of the new Toyotaism began to emerge within the company. Toyota's share of the Japanese domestic market fell to below 40 per cent between January and June 1995. The rate of profit was lower than it had been during the first oil crisis (see Fig. 3.4). Toyota now faced three uncertainties.

The model range had not been adapted rapidly enough to meet the demands of a changing market. The increased demand for minivans and recreational vehicles had encroached on the four-door saloon market, formerly a bastion of profitability for Toyota (see Fig. 3.2). Shifting the product range upmarket had brought with it increased production costs (see Fig. 3.3), and reducing vehicle equipment levels in order to reduce production costs had not enabled the company to recover adequate levels of profitability. Moreover, Toyota was slow to equip its cars with airbags to improve their safety. In 1995, a new managing director, Hiroshi Okuda, was named as replacement for the ailing Tatsuro Toyoda. Cars were now being equipped with airbags and their safety levels increased. It nevertheless seemed that Toyota would be forced to radically revise its commercial strategy if it was to recapture lost market share.

Faced with the American threat to impose a prohibitive tax (100 per cent) on Japanese cars, Toyota was obliged to promise to increase the proportion of its production undertaken in the USA. In 1995 the company decided to increase its production capacity in North America from 735,000 vehicles in 1994 to 1,100,000 in 1998 with the construction of another factory in Indiana. With domestic markets stagnant or in recession, the further relocation of production began to threaten the security of Toyota's employees and suppliers in Japan.

Other uncertainties derived from changes in the TPS itself. Toyota announced the implementation of a new order system in 1996. This would allow dealerships in Japan to transmit orders daily, with vehicles delivered within ten days. Dealers would therefore be able to respond more appropriately to the market and reduce stocks of vehicles at sales points. However this would make the smoothing of production volumes, an essential principle of TPS, more difficult to achieve.

The inappropriateness of the product range in the face of market developments could certainly be overcome. But relocation of production outside Japan and the new order system would probably have structural impacts on the Toyota model. In this respect the future of the 'new Toyotaism' remained uncertain, and further restructuring appeared likely.

## 3.4. CONCLUSION

The Toyota industrial model has been considered to be a post-Fordist model. Its successes are undeniable. In fact, between 1951 and 1995 Toyota never lost money. Yet by the mid-1990s the model had entered a period of restructuring.

It might be thought that this restructuring was only a new adaptation of the TPS, similar to the many changes the company had introduced previously. Each time difficulties were encountered, Toyota was able to overcome them, and even turn them to advantage to improve its system. The company was able to continue in this vein until the end of the 1980s without questioning its management of production costs and personnel, in other words the core of its system. Yet by the mid-1990s reorganization focused on this very form of management, making it possible to speak of a new Toyotaism emerging.

This new Toyotaism could be seen in the design of the new assembly line, the greater autonomy of working groups, and the new industrial relations frameworks (wages system, training, duration of work, working hours, and so on). The goal was to create a production system which was both more successful and more humane.

Yet the new Toyotaism had not broken completely with the old system. Just-in-time principles were still alive, even though their rigid application had been abandoned. The concept of autonomization remained, although it developed towards giving greater autonomy to the working group. The unilateral management of production costs was replaced by autonomous *kaizen* activities focusing on the costs of raw materials and components, and by a management of labour force costs and production efficiency which was more rational and less constraining. The wage system had been radically changed, but the remuneration of productivity, which grew out of Ohnoism, still applied to blue-collar workers. These links to the older form of Toyotaism suggested that both management and the union were acting carefully to preserve the viability and coherence of the TPS. Their prudence partly explained the slow implementation of reforms; further explanation lay in the economic recession which began in 1991.

The differences between Toyota's management of production efficiency and wage system and those at its transplants and its Kyushu subsidiary pose questions about the future of these subsidiaries. In the mid-1990s productivity increases at the subsidiaries were comparable to those at Toyota. Perhaps they were benefiting from know-how acquired from Toyota. But without incentives to raise production efficiency, such as the remuneration of productivity, would they be capable of manifesting the same dynamism in the future? If they could create another form of incentive just as efficient as the Toyota wage system, they would have contributed to the invention of the new Toyotaism.

At this stage in Toyota's history what did appear certain was that the Toyota model glorified by Womack *et al.* (1990) was already out of date, and that Toyota was searching for a new system in an uncertain social and economic environment.

## NOTE

Translated by Sybil Hyacinth Mair.

**STATISTICAL APPENDIX 3: Toyota**

| Year | Production in Japan[a] | | | Export volume | Foreign production | Total workforce[b] | Turnover | Gross operating profit[c] | Financial profit[c] | Pre-tax, profit[c] |
|---|---|---|---|---|---|---|---|---|---|---|
| | Total | Passenger cars | Commercial vehicles | | | | | | | |
| 1970 | 1,516,969 | 978,345 | 538,624 | 481,892 | — | 40,365 | 784,471 | 73,521 | 7,883 | 81,403 |
| 1971 | 1,759,204 | 1,219,647 | 539,557 | 786,287 | — | 41,024 | 923,068 | 68,249 | 8,124 | 76,373 |
| 1972 | 2,021,529 | 1,456,660 | 564,869 | 724,552 | — | 41,447 | 1,083,321 | 86,124 | 10,118 | 96,242 |
| 1973 | 2,308,098 | 1,631,940 | 676,158 | 720,640 | — | 42,927 | 1,325,860 | 100,015 | 10,535 | 110,550 |
| 1974 | 2,165,186 | 1,489,738 | 675,448 | 856,265 | — | 44,228 | 1,473,852 | 46,487 | 19,200 | 65,687 |
| 1975 | 2,421,305 | 1,768,736 | 652,569 | 868,352 | — | 44,584 | 1,809,871 | 57,385 | 25,627 | 93,305 |
| 1976[d] | 2,426,983 | 1,741,342 | 685,641 | 1,177,314 | — | 44,474 | 1,995,742 | 148,561 | 37,509 | 184,276 |
| 1977 | 2,611,291 | 1,813,017 | 798,274 | 1,413,235 | — | 44,798 | 2,288,069 | 167,678 | 44,163 | 210,120 |
| 1978 | 2,865,025 | 1,961,703 | 903,322 | 1,382,174 | — | 45,203 | 2,617,407 | 153,082 | 45,721 | 198,803 |
| 1979 | 2,862,613 | 2,033,003 | 829,610 | 1,383,648 | — | 45,233 | 2,802,469 | 158,289 | 40,045 | 198,334 |
| 1980 | 3,251,290 | 2,270,198 | 981,092 | 1,785,445 | — | 47,064 | 3,310,181 | 233,232 | 58,348 | 291,580 |
| 1981 | 3,251,471 | 2,266,765 | 984,706 | 1,716,486 | — | 48,757 | 3,506,412 | 140,183 | 87,327 | 227,511 |
| 1982 | 3,158,547 | 2,230,157 | 928,390 | 1,665,793 | — | 51,034 | 3,849,544 | 230,513 | 75,670 | 306,183 |
| 1983 | 3,182,718 | 2,305,424 | 877,294 | 1,664,361 | — | 57,846 | 4,892,664 | 304,543 | 94,048 | 398,592 |
| 1984 | 3,376,224 | 2,435,226 | 940,998 | 1,800,923 | — | 59,467 | 5,472,682 | 406,482 | 115,285 | 521,767 |
| 1985 | 3,540,646 | 2,458,671 | 1,081,975 | 1,979,955 | 136,307 | 61,665 | 6,064,420 | 505,891 | 142,118 | 648,009 |
| 1986 | 3,652,211 | 2,624,038 | 1,028,173 | 1,875,763 | 152,524 | 63,890 | 6,304,859 | 329,387 | 158,998 | 488,385 |
| 1987 | 3,599,174 | 2,643,236 | 955,938 | 1,770,937 | 192,260 | 64,797 | 6,024,910 | 248,364 | 149,644 | 398,008 |

| Year | | | | | | | | | |
|---|---|---|---|---|---|---|---|---|---|
| 1988 | 3,854,692 | 2,904,071 | 950,621 | 1,815,721 | 244,371 | 65,926 | 6,691,299 | 369,087 | 152,619 | 521,706 |
| 1989 | 4,006,796 | 3,026,947 | 979,849 | 1,669,130 | 471,581 | 67,814 | 7,190,590 | 400,522 | 169,341 | 569,863 |
| 1990 | 4,028,149 | 3,181,919 | 846,230 | 1,677,127 | 677,055 | 70,841 | 7,998,050 | 538,677 | 195,126 | 733,803 |
| 1991 | 4,118,885 | 3,232,391 | 886,494 | 1,703,589 | 669,912 | 72,900 | 8,564,040 | 338,787 | 235,531 | 574,318 |
| 1992 | 4,033,357 | 3,179,080 | 854,277 | 1,698,236 | 764,466 | 75,266 | 8,940,898 | 124,864 | 250,998 | 375,862 |
| 1993 | 3,856,614 | 3,124,973 | 731,641 | 1,539,005 | 888,715 | 73,046 | 9,030,857 | 103,629 | 182,819 | 286,448 |
| 1994 | 3,446,897 | 2,771,550 | 675,347 | 1,445,161 | 1,051,271 | 71,573 | 8,154,750 | 76,780 | 137,254 | 214,034 |
| 1995[e] | 2,601,675 | 2,060,020 | 541,655 | 1,063,529 | 840,094 | 69,748 | 6,163,885 | 155,126 | 81,079 | 236,205 |
| 1996 | 3,174,300 | 2,566,513 | 607,787 | 1,166,630 | 1,283,123 | 68,641 | 7,957,152 | 235,271 | 105,463 | 340,734 |

*Notes:*

[a] The volume of production is for the whole Toyota group (including subcontracted assembly). Figures for passenger cars include estate/station wagon versions of passenger cars.

[b] The increase of the workforce from 1982 to 1983 is largely due to the merger of Toyota and TMS in 1982.

[c] Million yen.

[d] The accounting period changes in 1976. Prior to 1976, financial results were published twice yearly; the first six months from June to November, the second from December to May. After 1976, annual financial results were published for the period from July to June. Accordingly, the figures (except for workforce) for the 1970–4 period represent the sum of the two six-monthly periods, making the year June to May. The figures for the year 1975 represent the thirteen months from June 1974 to June 1975.

[e] The accounting period changes in 1995. Therefore the figures for the year 1995 represent only nine months from July 1994 to March 1995, and the figures for the year 1996 represent the twelve months from April 1995 to March 1996.

*Sources:* Toyota, financial reports, and *Nikkan Jidosha Shinbun* [Automobile Industry Handbook] (1995).

# BIBLIOGRAPHY

Asanuma, B., 'Jidosha-sangyo ni okeru Buhin-torihiki no Kozo', *Kikan Gendaïkeïzai*, 58 (Tokyo, 1984), 38–48.
Bélis-Bergouignan, M.-C., and Lung, Y., 'Processus de diversification et flexibilité productive dans l'industrie automobile japonaise: Toyota et Nissan', *Actes du GERPISA*, 12 (1994), 13–42.
Boyer, R., and Durand, J.-P., *L'Après-fordisme* (Paris, 1993).
—— and Freyssenet, M., 'The Emergence of New Industrial Models. Hypotheses and Analytical Procedure', *Actes du GERPISA*, 15 (1995).
Clark, K. B., and Fujimoto, T., *Product Development Performance* (Boston, 1991).
Cusumano, M., *The Japanese Automobile Industry: Technology and Management at Nissan and Toyota* (Cambridge, Mass., 1985).
Mishina, K., 'What is the Essence of Toyota's Manufacturing Capability? Self-Manifestation by the Transplant in Kentucky, 1986–94', Proceedings of Third GERPISA International Conference (Paris, 1995).
Monden, Y., *Shin Toyota Sisutemu* (Tokyo,1991*a*).
—— *Toyota no Keïeï Sisutemu* (Tokyo, 1991*b*).
Nomura, M., *Toyotismu* (Kyoto, 1993).
Ohno, T., *Toyota Seïsan Hoshiki* (Tokyo, 1978).
Parker, M., and Slaughter, J., *Choosing Sides: Unions and the Team Concept* (Boston, 1988).
Shimada, H., *Humanware no Keïzaïgaku* (Tokyo, 1988).
Shimizu, K., 'Humanization of the Production System and Work at Toyota Motor Co. and Toyota Motor Kyushu', in Å. Sandberg (ed.), *Enriching Production* (London, 1995), 383–403.
—— and Nomura, M., 'Trajectoire de Toyota: rapport salarial et système de production', *Actes du GERPISA*, 8 (1993), 29–67.
Shimokawa, K., 'Jidohsa', in K. Shimokawa, S. Yonekawa, and H. Yamazaki (eds.), *Sengo Nihon Keïeïshi* (Tokyo, 1990), 67–142.
Suzuki, N., *Amerika shakai no nakano Nikkeï Kigyo* (Tokyo, 1991).
Tanaka, T., 'Toyota Jidosha no Genka-kikaku to Kaïzen-yosan', in T. Tanaka (ed.), *Gendaï no Kanri-kaïkeï Sisutemu* (Tokyo, 1991).
Toyota, *Toyota Jidosha 20 Nen Shi* (Toyota, 1958).
—— *Toyota Jidosha 40 Nen Shi* (Toyota, 1978).
—— *Toyota Jidosha 50 Nen Shi* (Toyota, 1987).
Ueda, H., 'Jidosha Sabgyo no Kigyokaiso Kozo', *The Quarterly Journal of Economic Studies*, 12/3 (Osaka, 1989), 1–29.
Womack, J. P., Roos, D., and Jones, D., *The Machine that Changed the World* (New York, 1990).

# 4

# *Nissan: Restructuring to Regain Competitiveness*

## MASANORI HANADA

In 1960, Nissan still produced only 56,000 passenger cars. There were two models, the Datsun 310, which later became the Bluebird, and an Austin model built under licence, which was later replaced by the Datsun 2000 and then the Cedric. More commercial vehicles were made than passenger cars, and the gap would widen in the years that followed. It was not until 1968 that passenger car production overtook output of commercial vehicles. During the first half of 1960s new factories were built with a view to launching mass production: Oppama in 1961, Zama in 1965, and Murayama, a factory built by Prince, which was taken over in 1966. In 1967, the characteristics and business environment of Nissan were quite distinct from those of Toyota, and the company had not adopted the same profit strategy (Boyer and Freyssenet 1995*b*). The company's production organization and employment relationships lacked the coherence of Toyota's.

## 4.1. THE CONFIGURATION OF PRODUCTION AT NISSAN IN 1967

While Toyota developed its rural and provincial character, Nissan was a 'metropolitan' company. The company was located in the Tokyo region, and its workforce was more educated, more mobile, and more radical. Events at Nissan had a national impact. The Nissan union was cooperative but also extremely influential in terms of decision-making. Its engineers gave the company a reputation for technically excellent cars and modern factories. Nissan was receptive to American management methods. Through the Fuyo group, of which Nissan was an influential member, the company was closely linked to the political and financial establishments (Cusumano 1985: 55). In the 1950s, in accordance with the wishes of the government, Nissan became an assembler under licence to the foreign manufacturer, Austin. The company then deliberately focused on foreign markets, as a response to government requests for manufactured goods to be exported to finance the importation of raw materials and capital goods. Moreover, with the support of its *keiretsu* bank, it later took over Prince, a producer on the verge of bankruptcy, which also manufactured two models admired for their style and technical features.

Nissan was more vertically integrated than Toyota, yet its ties with group suppliers were less trusting. Nissan's supplier network had not been formed in the same way. Toyota's first-tier suppliers either belonged to its group to start with or were formed by the outsourcing of activities. The reverse was the case for

Nissan suppliers, who started off as independent, on the whole small and dispersed, and then became increasingly integrated through technological assistance, medium-term contracts, and partial ownership—at least 25 per cent—by Nissan. As a result, ties were less solid than those between Toyota and its suppliers, despite Nissan's greater financial involvement.

President Kawamata, nominated in 1957, opted for an export strategy and hence for competitiveness in the international market: 'we shall stake out our territory in foreign markets', he had declared. Total Quality Control methods were implemented to simultaneously increase productivity and quality (Nissan Motor Company 1983). At Nissan, this was carried out from the top down, in contrast to Ohno's approach which consisted of inventing a solution to a particular problem locally before its widespread application elsewhere. A special department was created to train about a thousand managers and foremen, who were then charged with passing on their training. On the basis of the results obtained, Nissan won the Deming Prize in 1960 (Udagawa 1995: 69). Production under licence of the Austin 40 and cooperation with this producer also familiarized the company with the demands of foreign markets. The prototype of the Datsun 310, the future Bluebird, was tested in 1958 in North America. In 1960, a distribution company was established in the USA. Nissan's first success in penetrating the American market came with its pick-up trucks, and through the distribution network thus created the company was able to begin selling the Bluebird. Sales subsidiaries were set up in Mexico in 1961, in Canada in 1964, and in Australia in 1966. The import substitution and integrated production policies adopted by numerous developing countries led Nissan to assemble, or arrange assembly of, CKD (completely knocked down) kits, with assembly operations opening in Taiwan in 1957 and in Peru in 1963. Nissan built a factory at Cuernavaca, Mexico in 1966, even though it was only producing 230,000 passenger cars per year in Japan. As early as 1963, the company exported 16.7 per cent of its vehicle production, compared with only 7.7 per cent at Toyota (Cusumano 1985: 394).

Nissan's sales distribution network was created later than that of Toyota. Accordingly, its dealers were less solid financially. Nissan was obliged to own 45 per cent, on average, of their capital, as against 9 per cent at Toyota. This imposed a long-term financial commitment. In 1967, there was still only a single distribution network, which was organized into regions.

Nissan's union, like Toyota's, was cooperative, but it had far more power in negotiations. The long dispute of 1953—three months punctuated by confrontations and the arrest of protesters—was only ended by the creation and actions of a second union by employees who feared that the company would close. Two years later the second union was able to unseat the first, more radical and combative union. It created a federation of company unions at the group, including those at suppliers and distributers. While the second union cooperated with management, it also proved to be autonomous and interventionist, as a result of the conditions under which it was created. Management owed its victory in the social conflicts, which it only just won, to the union. The 'joint committees' established at the company, factory, and work area levels after the 1953 conflict to deal with 'all problems relating to management of the company' (Yamamoto 1981), and to negotiate salaries and working conditions, gave the Nissan union important powers and influence over managerial strategy, to the point that it was possible to speak of a dual management system. The union supported management attempts to increase productivity but only in return for promotions and changes in working hours, overtime, and even the overall strategy of the company. Indeed union acceptance was often a precondition. The union was greatly influenced by the personality of its leader, Ichiro Shioji, and it increasingly exercised power in an authoritarian manner, even over its own members, stifling all criticism and obtaining benefits for individuals on condition that they supported the union and its leader. For instance, one precondition for promotion became that a candidate had previously occupied a union position. The outcome was the gradual development of a powerful 'union clan' within the management structure itself, with all the consequences such a situation might have. For example, when it came to individual salaries, the union had a great deal of influence. Since the individual salary depended on the evaluation of the hierarchical superior (see below), the union was able to influence this via its delegates, who with union support had become team, group, or work area supervisors. The workforce was bound to both management and union; though the interests of both might well converge.

Nissan's wage system, adopted in 1957, was markedly different from that at Toyota. Whereas until 1990 Toyota had a system which directly linked the individual's monthly salary to the effort made by each work group to reduce standard times (Shimizu 1993), salaries at Nissan were mainly based upon classification, age, and evaluation by the hierarchical superior. As at other companies, the sum of monthly salaries only accounted for 70 per cent of the overall annual salary, with bonuses, paid twice yearly, equivalent in total to five or six monthly salaries. Moreover, at retirement, the employee received a payment which could amount to 40 times the monthly salary, depending on seniority. The annual rise in salaries was negotiated each spring. In practice the rise was coordinated with those at other companies. However, what was negotiated was the total sum the company paid its employees, so that the company could use its discretion over how this would be distributed. The monthly salary, excluding overtime (paid at 130 per cent and 140 per cent for Sundays and public holidays), was comprised of the base salary, production bonuses, special bonuses, bonuses based on grade, function, and difficulty of task, and family benefits (Hanada 1994).

The basic salary was the level of salary at which an individual was hired, augmented by successive annual individual increases. The starting salary was based on level of education, sex, and sector (such as production, administration, or design). The annual individual increase depended upon the rate negotiated during the *shunto* period and on the evaluation, carried out by the hierarchical superior, of the individual's skills and accomplishments (Hanada 1994). The evaluation criteria were not known by the employees (Roshi Kankei Chosakai 1965). In 1961, the basic salary accounted for 58 per cent of the average monthly salary. The amount of the production bonus was calculated by multiplying the basic salary by an output coefficient for the work group. The output criterion was the relationship between actual working time and paid working time (Kamii 1994). The production bonus amounted to 16 per cent of the monthly salary, a much lower proportion than at Toyota. A special bonus was introduced in 1956 with the classification of work posts into four major categories, the first two of which were in turn subdivided into three subcategories. The amount of this bonus therefore initially depended upon classification, as well as upon age and once again upon the hierarchical superior's evaluation. This accounted for 10 per cent of the salary. The criterion was indeed the individual's age and not seniority within the company; this was to take account of changes in expenditures over the course of a lifetime. This was one of the benefits won by the social conflicts of the post-war period. In addition to the above components, there was also a bonus related to grade (3 per cent) when the post occupied did not correspond to the hierarchical level the employee had attained, a function-related bonus (1 per cent) when, alternatively, the employee's function was above his hierarchical level, a bonus related to how difficult the post was (1 per cent), and, lastly, a family benefit paid by the company as a function of the number of dependants (11 per cent). The latter sum was independent of family income levels, and the definition of dependants was broad, encompassing the spouse, under-18s and over-60s. State family benefits were not introduced until 1972, for households with modest incomes, and they were added to benefits paid by companies.

Variable wages were paid to employees of the same age, sex, level of education, and classification, depending upon the evaluation made by the hierarchical superior, the output of the work group, the number of dependants, and the particular characteristics of the post occupied. However, it is not known how wide variations have been, since the company has not been obliged to disclose to its employees how the diverse elements of the salary are calculated.

This incoherent configuration of production organization and employment relations at Nissan was viable during a period of expanding demand. However, it began to run into difficulties after the second oil crisis. The solutions the company developed were to prove inadequate when faced with the recessionary period of the early 1990s, when Nissan was to lose money for two years running.

## 4.2. THE START OF DIVERSIFIED MASS PRODUCTION BASED ON DOMESTIC DEMAND AND EXPORT MARKETS, 1967–1973

Output took off in 1967, with a rise in passenger car production from 231,000 to 352,000; it did not stop rising until a peak of 1.49 million vehicles was reached in 1973. Growth was pulled along by domestic demand as well as by exports (Bélis-Bergouignan and Lung 1994b: 21). The proportion of passenger cars exported rose from 22.7 per cent in 1967 to 35 per cent in 1973. Operating profits were over 7 per cent between 1968 and 1972, yet this was three points below Toyota.

By 1967 Nissan had six models of its own, and its takeover of Prince in 1966 had provided another two upper mid-range models, Gloria and Skyline. In 1965, the company had launched a luxury model, the President, and in 1966 a model to compete with Toyota's Corolla, the Sunny (Bélis-Bergouignan and Lung 1994*b*: 38). These new models were the start of a wave which would include four further models: the Datsun 1800 in 1968 (which later became the Laurel), the Fairlady in 1969, the Cherry in 1970 (later to become the Pulsar), and the Violet in 1972 (production of which ceased in 1982). On the eve of the first oil crisis, Nissan offered ten models covering all market segments. New models were first derived from earlier models, although further new models were rapidly introduced. Thus, from 1963 to 1965 the President was a derivative of the Austin produced under licence. Likewise, the Fairlady was preceded by a derivative of the Datsun 310 from 1963 to 1968. The Sunny replaced the Datsun 1000/1200, also a derivative of the Datsun 310, which had been produced from 1966 to 1970. From 1971 to 1973 the Violet was a derivative of the Datsun 1600 which followed the Datsun 410, itself a replacement for the Datsun 310. In the end, all these models had their own platforms with the exception of the Gloria, which was built on the platform of the Datsun 2000 (the future Cedric). In short, platform sharing was very weak: there were nine platforms for ten models. The average volume per platform was 165,000 units, while Toyota's average volume stood at 275,000 units for an equivalent model range (Bélis-Bergouignan and Lung 1994*a*).

Nissan faced two difficulties in satisfying soaring internal demand and developing its exports: a lack of production capacity and a scarcity of workers. In 1968, Nissan's passenger car output exceeded its production capacity. The company was forced to come to an agreement with Fuji Heavy Industries to assemble the Datsun 1000/1200 in one of its factories. Fuji was in fact operating under capacity owing to poor sales of its minicar (360 cc). In the same year, Nissan built a factory at Tochigi, first to install a foundry, and by 1970 to assemble the Cedric, Gloria, and Cherry. Production of commercial vehicles was also climbing, although less rapidly: from 368,000 in 1967 to 543,000 in 1973. Nissan found it difficult to expand operations on the congested urban site it owned in the Tokyo region and to find the labour force it needed. Hence in 1973 the company decided to build a factory in a completely different region, on the island of Kyushu, at Kanda, in order to assemble Datsun pick-ups destined for the American market. In 1982, this factory was converted to make mechanical components and assemble passenger cars.

Nissan's main problem during this period was a shortage of workers; the labour force at the company had increased from 9,433 in 1960 to 20,917 in 1965 and by 1973 it had reached 53,508. The traditional recruitment of younger workers leaving secondary school, with nine years of schooling, was inadequate. As a result, Nissan made extensive use of temporary workers, who accounted for between 9 per cent and 13 per cent of the total labour force. With the hiring of foreign or female workers, in the latter half of the 1960s, Nissan was also forced to begin recruiting workers from among school leavers with twelve years of education.

From 1969 onwards the company signed new contracts with retirement-age personnel to employ them for a further five years after they reached 55. The age of retirement was formally raised from 55 to 60 years in 1973. To retain its labour force, Nissan was also compelled to undertake measures to ameliorate working conditions. In 1971, the company introduced the five-day working week, which had been introduced some years earlier at Honda (Nissan Motor Company 1983). It was decided to halt further expansion of the company in its home region and instead expand in rural regions which still possessed a reserve of workers.

The early automation of operations at Nissan was not unrelated to these problems. Automation could be presented as a means of dealing with the penury of workers, a means of improving working conditions, and as a means of better responding to the expectations of younger, more educated workers. This logic served as an additional justification for methods which were already preferred by the production engineers, who were more powerful at Nissan than at other companies.

Tensions in the labour market had important ramifications for the evolution of wages and for the respective weights of the different elements involved in calculating monthly wages. Absenteeism now figured among the criteria used to determine the individual annual increase in the basic salary and the special bonus; an indication that employees had felt that they could be absent from work with no sanction. The special bonus grew to assume particular significance. From 10 per cent of the average monthly wage in 1967, it rose to 39 per cent in 1969 and 68 per cent in 1973. There were two reasons for this. The salaries at which workers were hired, which later determined their basic salaries, had to be greatly increased in order to attract workers in a tight labour market. There were soon distortions in relative salaries between newly hired younger workers and workers with more seniority. The problem was in part resolved by increasing the part of the wage due to classification and age, through the special bonus (Hanada 1994). The second reason was that the type of post occupied became much more important than other criteria, even if the number of categories of workers remained low, at eight. In this respect, Nissan had moved towards the American model of job classification. A further indication of this tendency was that the production bonus fell from 16 per cent of average monthly salary in 1960 to 11 per cent in 1967 and then to 7 per cent in 1973. It was eliminated in 1978. Hence Nissan progressively abandoned the use of the salary as a direct means of obtaining increased productivity, in contrast to the method used by Toyota.

Besides attracting and retaining a new workforce, in the mid-1960s Nissan also had to re-mobilize existing workers and to motivate new recruits who were better educated than the older workers. Total Quality Control methods reached a ceiling and no longer stimulated much interest. Hence Nissan's management decided in 1966 to introduce quality circles, with the active participation of the union. By 1970, 49 per cent of Nissan's workers were engaged in this type of activity (Maruyama and Fujii 1991).

The distribution network was reorganized in 1969 into channels, that is, by groups of models and not by region as before. By 1970 there were five channels.

Until 1971, dealerships had passed orders along each month, specifying models and versions. They assumed the risk for unsold models, if their forecasts had been inaccurate. During a period when households were furiously purchasing cars and everything could be sold, risk was low and the system greatly helped with the programming of production. However, when growth slowed and forecast sales were not achieved dealers suffered financially. A new system was set up whereby lead-times for orders were shortened to ten days, and part of the risk for the slower selling models was taken on by Nissan (Okamoto 1995). In 1971, the company developed a computerized system for processing orders and data about purchasers and the automobile market, which later became the basis of an integrated system for managing orders and production.

Nissan pursued its export strategy, particularly towards the USA and Europe. While sales in these two regions accounted for 28 per cent of Nissan's total vehicle production in 1964, they accounted for 66 per cent by 1972. The abandonment of fixed exchange rates in 1973 might have compromised Japanese exports, particularly those of Nissan, by reducing the extent of undervaluation of the yen, but in fact the opposite occurred.

## 4.3. EXPORTS COMPENSATE, BUT COMPETITIVENESS IN THE DOMESTIC MARKET DECLINES, 1974–1979

The first oil crisis provoked a decline in demand in the domestic passenger car market of 22 per cent in 1974. Nissan's production fell by 15.6 per cent, the result of an acute drop in domestic sales of 26.9 per cent, compensated by a spectacular 21.3 per cent surge in exports. Exports thus increased their share of total production from 35 per cent to 50.3 per cent. The 1974 recession revealed that the company was not competitive with Toyota in the domestic market. However, Nissan was now very competitive in export markets, owing particularly to the success of its compact cars in the USA in the new economic context created by the aftermath of the oil crisis.

This conclusion was confirmed in the years that followed. By 1979, Nissan had struggled to regain its 1973 levels of domestic sales. It was exports that enabled production to return to its 1973 level as early as 1975 and to reach 1.74 million in 1979. In that year Nissan's domestic market share was 30 per cent. Yet operating profits failed to grow in parallel. In 1973 they fell to 4.54 per cent of turnover, and then to 1.72 per cent in 1974. However profits then recovered, to oscillate between 4 per cent and 6 per cent until 1979. Toyota too was unable to return to the rates it had achieved between 1968 and 1972, but its profits declined to a much lesser extent and rose again more quickly. Nissan's new managing director, Takashi Ishihara, set the company the target of emulating Toyota and increasing productivity by 30 per cent over three years.

In 1976 management launched a campaign to increase productivity. But for the first time since 1953 the union opposed the programme, declaring that the blind pursuit of productivity would lead to the deterioration of working conditions, thus expressing the persistence of a discontent which had not dissipated since the new measures were implemented in 1966. An agreement was finally reached whereby the union collaborated in the P3 programme (participation, productivity, progress) in exchange for greater control over working conditions. For the workers, the agreement implied certain guarantees but also a dual pressure, exerted by both management and union, to participate in increasing productivity.

In 1977 it was decided to extend the application of the Action Plate Method (APM) to all suppliers. Eight supplier companies had been experimenting with the APM since 1974, with the aim of reducing stocks and reducing lead-times through reorganizing work and production equipment, reducing the amount of time needed to change tools, and eliminating faults (Nissan Motor Company 1985:104). This was the *kanban* method under another name. An industrial engineering department was established in the purchasing department to instigate change at supplier companies. But the policy proved a failure and the system was abandoned in 1984 (Monden 1991). Deliveries could not be made regularly and on time, owing to the congested traffic conditions in the Tokyo region where Nissan's factories and suppliers were located. However, Nissan also believed that the *kanban* method had limitations, and this justified the search for a better method. From Nissan's perspective, the *kanban* method was incapable of achieving zero stocks. It required time to manage the labelling system; by contrast, a computerized system worked in real time. Finally, it required a smoothing of production incompatible with the variability of orders and the reduction of delivery lead-times. Nissan preferred to adopt a centralized, computerized system of management, a technical rather than a pragmatic solution.

During this period, Nissan introduced only two new models: the Silvia, first launched on a Sunny platform in 1975 and then on its own platform in 1978, and the Auster in 1977. By 1979 the range comprised twelve models on ten platforms, with average volumes of 148,000 units per model and 174,000 per platform. These were the peak volumes reached between 1946 and 1995. The oil crisis led to postponement of expansion of the range. However, the company began preparations for a new wave of models when its export success was confirmed during the second half of the 1970s. These models would be introduced during the following phase, when overall annual production stabilized between 1.9 million and 2.0 million. Nissan now decided to share some of its platforms, twelve years after Toyota. For the first time the company clearly changed its strategy towards delivering product diversity.

## 4.4. STAGNATING PRODUCTION LEVELS AND DECLINING PROFITABILITY, 1980–1988

The second oil crisis appeared to have the same effect on Nissan as the first: a drop in domestic sales (– 7.0 per cent) and renewed growth in exports (+ 24.4 per cent), to the point that 1980 production surpassed that of 1979, reaching 1.94 million. Worldwide production of passenger cars amounted to 2.01 million.

However, the difference with the preceding phase was that exports did not continue to grow, and they no longer compensated for the gradual decline of domestic sales which lasted until 1988. Nissan's domestic market share by 1988 had fallen to 23 per cent, a decline of one percentage point per year since 1980. Foreign production would gradually compensate, but it only helped to maintain global passenger car production at approximately 2 million. Operating profits declined from 5.61 per cent in 1980 to 1.39 per cent in 1988, and there was even a loss-making year in 1986 (−0.25 per cent).

These problems are all the more remarkable since Nissan launched no fewer than six new models between 1980 and 1985. Three were launched in 1980: the Leopard, also known as the Infiniti M30, the Langley which later became the Presea, and the Laurel Spirit which would later become the NX. Two were launched in 1982: the March (Micra in export markets) and the Prairie. Finally, the Be-one was launched in 1985. Three of these models shared their platforms with others: the Laurel-Spirit with the Sunny, the Be-one with the March, the Langley with the Cherry. Nissan now had a range of eighteen passenger cars on thirteen platforms, thereby confirming its policy of platform sharing (Bélis-Bergouignan and Lung 1994b). Yet diversification moved at a faster pace than platform sharing, and average volumes per model and per platform had decreased to 103,600 and 143,500 respectively in 1985, levels below the global average.

In the early 1980s, Nissan appeared not to comprehend the nature and breadth of the crisis it was about to experience. In 1981 a three-year plan to reinforce sales capacity and reform design capabilities had been launched. The number of sales points and the number of sales personnel were increased, notably by sending workers no longer needed in the factories into the sales network: 2,000 employees by 1985. Each of the network's distribution channels was authorized to sell compact models like the March. In 1987, a structure for coordinating the channels was established for each region, with the aim of avoiding fruitless competition. Moreover, each dealer now had a complete range. The order system was again modified in 1982. Orders were placed every ten days, but the dealer, and thus the client, could modify certain specifications up to seven days before production of the vehicle ordered started, on condition that the engine and body type remained unchanged (Okamoto 1995). A similar system had been established ten years earlier by Toyota. The difference was that Nissan's system relied heavily on computerization. It created a computerized network which linked dealers, suppliers, factories, and after-sales service, to permit the instantaneous transmission of sales and manufacturing information and its continual analysis (Market Analysis and Planning System, or MAP).

As far as its products were concerned, Nissan began to recognize that the technical innovations of its models, which had helped to make its reputation, did not necessarily correspond to the actual desires of clients, and that its range lacked coherence (Itami et al. 1988). However, the measures undertaken to remedy the situation were inadequate given the gravity of the situation. Project leaders were assigned to each new model, responsible for coordinating the various departments involved in design. But the project leaders carried insufficient weight to substantially

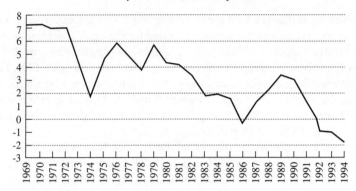

FIG. 4.1. *Evolution of rate of Nissan operating profit*
*Note*: Rate is operating profit/turnover (before tax).
*Source*: Nissan, financial reports.

change the traditional process for designing new vehicles. The labour force continued to grow until 1984, reaching 59,615 employees. In 1986, it was decided to reduce numbers to a level appropriate to output, and employment declined to 51,237 in 1988. Like other producers, Nissan attempted to re-establish its profitability by raising product prices, which the company justified through additional features.

Nissan responded quickly to changes in international trade relationships which threatened its exports: the yen's appreciation in 1979 and increased protectionism in the USA and Europe. The company immediately opted for direct investment in each region. In 1980 it acquired 36 per cent of Motor Iberica of Spain for the production of commercial vehicles, and in 1983 it opened a factory at Smyrna, Tennessee, for the assembly of small pick-up trucks and, by 1986, the Sunny, known locally as the Sentra. Alfa-Romeo began to assemble small Nissan cars. Then in 1986, Nissan commenced operations at its factory in Sunderland in the United Kingdom for assembly of the Bluebird. The opening of this factory had been delayed by opposition from Nissan. The renewed appreciation of the yen in 1985–6 accelerated the substitution of exports by foreign production. Nissan was not satisfied simply with opening factories; the company also opened design offices to adapt its models to the North American and European markets: in 1979 and 1986 in the USA, and in 1988 in Europe. The company also internationalized its financial management. Nissan-Europe was founded in 1989 to coordinate all European activities and Nissan-North America was created in 1990 to coordinate North American activities (Imai and Nira 1992).

The failure of import substitution policies in many developing countries, and its subsequent abandonment, led Nissan, like many other producers, to cease building new transplants in these countries. Instead, the company established a network of factories in South-East Asia which each contributed to the production of models destined for these countries. Given that the market in each country was too small, it was impossible to obtain adequate economies of scale and acceptable production costs despite low wage levels. By creating an international division of labour at the regional level, Nissan was responding both to the requirement for local production and to the need for adequate production volumes (Tabata 1995). Taiwan became the principal assembly location, and in 1985 Nissan purchased a 25 per cent share in the capital of its local assembler, Yue Loong Motor.

Yet Nissan's dynamism abroad did not fully compensate for the deterioration of its position in Japan. It was widely believed that this was owing to the company's industrial relations system, or more precisely its dual, management/union management system, which hindered the implementation of necessary corrective measures. In practice, the tensions multiplied. In 1982, the union decided to pull out of the P3 campaign and organize production line stoppages. This followed a fatal accident and management's unilateral decision to invest in the United Kingdom. A conflict broke out which ended in the dismissal of union leader Ichiro Shioji and a redefinition of the union's role in company affairs. The management had succeeded in mobilizing enough support among employees to obtain Shioji's replacement (Saga and Hanada 1995).

Management immediately adopted a series of reforms to work organization and systems for employee classification, promotion, and wages. Tasks to be carried out at each workstation were precisely defined, clearing up the ambiguities that had resulted from previous compromises, and reducing the opportunities for union control over working conditions. All workers were obliged to participate in small group activities, with each individual assuming a pre-determined role. The organization of participation became the exclusive preserve of management. In each department a manager, at the level of factory area manager, was responsible for promoting Total Gembra Kanri, which consisted of training group leaders to develop plans for improving quality, reducing costs, and improving safety. Total Productive Maintenance (TPM) was integrated into this procedure. Promotion to posts of team leader, group leader, and factory area manager were now to be decided by management, based upon criteria of competence and appropriate level of training, although management continued to promote employees with a union background.

The wage system was modified in 1987. This reform, which did not change the system's basic principles, consisted of little more than making public and precise the elements which made up the special bonus. The special bonus, which had come to account for 75 per cent of average monthly salary, was replaced by the work-related wage (32 per cent), the age-related wage (26 per cent), and the output-related wage (6 per cent), as well as by an increase in the part due to the basic salary from 17 per cent to 27 per cent. The parts due to grade, task difficulty, and family benefit remained unchanged at 4 per cent, 1 per cent, and 6 per cent respectively. The 'work-related wage' corresponded to the 'classification' element in the previous special bonus. However, the subcategories 'entry-level worker' and 'mid-level worker' were eliminated, reducing the number of worker classifications from eight to four. This became a framework for the level of functions fulfilled. There were two objectives: to facilitate mobility, and to make promotion conditional. Employees were not assigned to levels solely on the basis of their capabilities, but also on the basis of the functions they fulfilled. The possibility of demotion was introduced for employees who proved incapable of performing the functions assigned to them. Demotion appeared to be used when operators turned out to be unable to control or supervise automated machinery. The output-related wage was inappropriately named. In fact it was fixed according to employee activities and actual capacities, as evaluated by the hierarchical superior. The evaluation was based on a grid now composed, for the sake of precision, of forty-four rather than twenty-six levels. The age-based wage made explicit the part of the salary due to the age criterion in the special bonus (Hanada 1994).

Following this reform, wages were composed of three major elements: one linked to seniority, comprising the basic wage and the grade-related wage, and which accounted for about 30 per cent of the average monthly salary; a second linked to change in responsibilities during the employee's time with the company, which combined the age-based wage and the family benefit, and accounted for about 30 per cent of the salary; and a final part linked to work, comprised of the work-based wage which was related to classification, the output-related wage related to the hierarchical evaluation of individual capacities, and bonuses linked to the post occupied, which together accounted for about 40 per cent of the salary.

Promotion and hierarchical evaluation therefore remained the only two factors capable of modifying the individual's wage career other than starting wage, age, and household dependants. Hence, 'an employee is motivated to pursue an average path to avoid being penalized over the long-term' (Hanada 1994). Reform had introduced greater transparency and greater flexibility in the assignment of workers, and also made promotion conditional. Whilse it was as objective as possible, hierarchical evaluation still depended upon local level compromises which were themselves a function of circumstances, and was not necessarily directly linked to the improvement of company performance.

In parallel to this transformation of the employment relationship, Nissan once again reorganized and reinforced its distribution networks, which had always been judged to be of only middling efficiency. The reform of the product design departments introduced following the second oil crisis did not yield the anticipated result. The project leaders failed to acquire the authority needed to deal with the various design departments, beyond mere coordination, if new models were to be designed more as a function of customer expectations than of the technical preferences of the engineers. In 1987, Nissan separated the project teams from the design department and placed them under the direct authority of senior management. At this point, a process of simultaneous engineering was adopted, in which the production method departments and the factories became involved early on in the design process, in order to reduce lead-times and costs.

## 4.5 NEW ATTEMPTS TO REGAIN STRUCTURAL COMPETITIVENESS, 1989-1995

For three consecutive years, from 1989 to 1991, the bubble economy in Japan permitted Nissan's output in Japan to return to the peak levels it had experienced in 1980: 1.95 to 2.0 million passenger cars per year. This resurgence was owing solely to the domestic market, since exports were now declining owing to foreign production. Nissan was obliged to hire new employees again, and total employment in Japan rose to 56,873.

The recovery of the domestic market once again stimulated a substantial wave of investments: the construction of a new factory at the Kanda site on Kyushu, and the launch of six new models between 1988 and 1990. The second Kanda factory was promoted as the 'dream factory' before it had even been opened, owing both to flexible processes and novel working conditions. With a capacity of 240,000 vehicles per year in two shifts, this factory was to be capable of producing the entire range of Nissan vehicles, with its high level of flexible automation, and to attract young workers by making working conditions more acceptable, replacing the assembly line with automated guided vehicles. Five of the six new models shared their platforms with another model, accentuating the platform sharing process that had started in the early 1980s. By 1990, Nissan had a range of twenty-four models and fourteen platforms. However, again diversification advanced faster than commonization. Whereas Nissan's output in Japan attained the highest level in the company's history, the average volume per model in Japan fell to 84,200 while the average volume per platform remained stable at 144,300, similar to the trough period of 1985.

And yet Nissan did not truly take advantage of soaring domestic demand. Despite increasing sales, the company's share of the domestic market continued to decline, dipping below 20 per cent in 1991. Operating profits, which had been negative in 1986, only rose to 3.46 per cent in 1989, whereas Toyota recovered its mythical rate of the late 1960s, touching 10 per cent.

Nissan's worldwide output reached a peak in 1991, with 2.4 million passenger cars produced. It dipped again to 2.0 million in 1994, in spite of the solid growth of output abroad, which rose to 659,000 the same year. Production had fallen spectacularly in Japan: down 33.6 per cent between 1990 and 1994, due to declines in both domestic sales (–28.2 per cent) and exports (–35.4 per cent). The growth in production abroad was slightly greater than the decline in exports. In short, Nissan's competitiveness in Japan continued to decline, to an 18.0 per cent market share in 1994, and competitiveness abroad stagnated. In 1992, 1993, and 1994, Nissan made significant losses: –0.86 per cent, –1.02 per cent, and –2.20 per cent respectively. The second factory on Kyushu entered into service in 1992, at the very moment the market was fast contracting. Nissan now had substantial structural overcapacity in Japan, and its recent massive investments brought heavy financial burdens. In 1995, the company was forced to close one of its five assembly factories, the Zama plant in Tokyo, and by March 1994 the labour force had fallen back to 51,398 workers.

Reforms such as the abandonment of the dual 'management-union' management system, the establishment of better cooperation with a union which was now without direct power in the company, new ways to organize work, classifications, wages, and promotions; all proved insufficient to arrest the loss of competitiveness. The company now had to be profitable with output reduced to 1.8 million in Japan. Nevertheless, by the mid-1990s, two axes of reorganization appeared to be emerging which might give Nissan's production system the coherence which had been absent for so long. The first was a policy aimed at increasing volumes through further platform sharing and through finding new and international sources of components. The second was a policy to synchronize flows of materials and information in real time as a means to constrain and channel productivity and quality.

Despite the reticence of the design and marketing departments, the company appeared to have determined to reduce the number of platforms from fourteen to six, or one per market segment. It also began purchasing components from the suppliers of its competitors, in order to lengthen production runs and benefit from their innovations. Hence, from 1995 Nissan bought braking systems from Aisen Seiki and electrical parts from Nippondenso, which were among Toyota's principal suppliers. Moreover, Nissan declared that its own suppliers should no longer rely upon the company as the sole source of orders, given the enduring low levels of production volumes in Japan. One of the rules governing inter-firm relations in Japan had been violated (*Nihon Keizai Shinbunsha* 1995). In the same vein, Nissan no longer wished to be only a producer of complete vehicles, but to become a supplier of automotive components to other producers. The company established a systematic international division of labour in order to expand production runs and thus benefit from lower unit production costs, drawing lessons from the system of dividing activities amongst different South-East Asian countries introduced during the 1980s (Shimokawa 1994). Nissan did not hesitate to close its factory at Clayton in Australia in 1992, which had produced 60,000 passenger cars in 1990, when lower customs tariffs made imported vehicles more competitive (CCFA 1995).

In 1994, Nissan for the first time published a brochure entitled *The Nissan Production System*, and within its pages the company engaged in a certain self-criticism: 'At Nissan, thinking patterns varied from one department to another (Production, Purchasing, Design, Administration, etc.). We lacked sufficient sensitivity to changes taking place in the world about us, and we were incapable of taking them into account . . .' Indeed the difficulties of implementing reforms during the 1980s clearly accentuated the different perspectives of different departments, and the inadequate understanding of new market trends: 'The aim [of the Nissan Production System] is to establish values that are invariant over time . . . and are capable of making the company profitable in all circumstances. To do this, Nissan, along with its foreign subsidiaries and its suppliers, must focus its efforts on the shared goal of improving the efficiency of the whole'. This declaration reveals the concern to develop coherent reforms applicable across company departments which appeared at a number of European and American producers during the late 1980s (Boyer and Freyssenet 1995*a*). It appeared that a coherent and efficient production system was being constructed around the computerized system for managing orders and production that had gradually been implemented since the 1970s. The objective was to move towards zero stocks, which *kanbans* do not in fact permit, by adapting to fluctuations in actual demand and not the forecasts made by dealers. The real-time reception of actual orders in a computer network enables the producer itself to make forecasts of sales without the mediation of dealers. The system for programming production can then function like the seat reservation system employed by airline companies. The manufacturer defines the number of cars per model based on demand forecasts and production capacities. Once the predicted number has been reached by real orders, the next customer to arrive is transferred to the 'following flight', in other words to the next production programme. The programmes are set definitively every ten days, with daily adjustments for certain specifications possible up to six days before production starts. Hence the customer can be informed of the exact date of delivery of his vehicle at the time the order is placed. The integrated computerized system, moreover, enables the customer to be kept abreast of any delays and to be given the reason for them (Takeuchi 1995).

In this system the producer alone assumes the risk for inaccurate predictions, the responsibility for adjusting the planned programme to real orders, and the responsibility for late delivery. In order to respect delivery lead-times, all information concerning production has to be exact, the causes of repeated errors must be eliminated, quality has to be controlled at each stage, and the reassignment of workers according to demand variations must be easy and rapid. The company's guarantee to the customer of delivery lead-times is translated into constraints on worker involvement, flexibility, and improved performance. Nissan's system of computerized and centralized order and production management may, if all these conditions are met, produce a greater reduction of stocks than that obtained by *kanban*: in other words, greater productivity and improved quality. This means that the real-time reception of the operations each individual carries out and the controls this makes possible must be accepted by employees. The future of this socio-productive system thus depends upon its compatibility with the evolution of the aspirations and behaviour of employees at Nissan, its subsidiaries, and its suppliers. Nissan began by establishing this system at its rural factories: Kyushu in the south, and Iwaki in northern Japan.

In contrast to Toyota's system, which is based on the localized, gradual, and additive reduction of standard times by work groups in order to provide customers with quality products at competitive prices and lead-times (Shimizu 1993), the system developed by Nissan compels all employees to work in such a way that the delivery lead-time promised to the customer is respected. Lead-times and their reduction become norms, set in advance, which are to be attained, rather than the results of constant efforts to economize in all aspects of production. Was it that Nissan's management believed that adherence to lead-times given to customers would be more meaningful and acceptable for its employees and its suppliers than a constant call to eliminate waste in time, materials, and energy?

Besides these two major new orientations, Nissan pursued reforms already started in its distribution networks and in its employment relationship. The Prince network had absorbed the Cherry network in order to reduce the number of dealers, although without reducing the number of sales points. The number of salesmen was increased by 2,000, while the Nissan workforce was reduced by between 5,000 and 7,000 employees. Nissan invested capital in its loss-making dealers, and it was possible that this tendency would ultimately lead to Nissan absorbing its distribution networks. From 1996, the part of the salary related to work was increased by ten points while the part related to seniority was reduced, so that 'salaries better reflect the tasks and the efforts made by each employee', according to management. This measure permitted younger employees to increase their salaries a little more quickly if they received good marks from their superiors and were promoted to a higher category. For the first time management published a manual explaining how marks were awarded, and which reminded employees that promotion was conditional. It seemed that the accent placed upon the individualization of the salary was to be accompanied by much more transparent criteria.

More problematic was Nissan's decision to diversify into aerospace and space. Having pursued a policy of very active research and development over many years, Nissan now possessed significant scientific and technical competencies at its research centres, which it had been mobilizing in cooperation with aerospace companies like Martin Marietta of the USA since 1982 and Fuji Heavy Industries more recently. Nissan's turnover in these fields was 38 billion yen in 1991 (Shimokawa 1994).

## 4.6. CONCLUSION

When everything could be sold during the period when households were rapidly equipping themselves with automobiles, Nissan had achieved remarkable financial results, albeit inferior to those of Toyota. The shortage of workers which was particularly marked in the Tokyo region had forced the company to substantially increase starting salaries, to recruit more educated employees, to improve working conditions, and to develop participatory practices in an attempt to make work more attractive. The particular management–union system of management, the outcome of the 1953 conflict, ensured both social peace and significant productivity increases.

When market growth slowed, particularly after the first oil crisis, Nissan's lack of competitiveness in the domestic market became clear, but it was more than compensated by export markets. Profitability declined but remained relatively high. Nissan's export strategy, first developed in the mid-1950s, became vital for the company. Inferior to Toyota in Japan, Nissan was still superior to the American and European producers, in terms of both quality and costs, since it had small models and the yen was undervalued.

The re-evaluation of the yen at the end of the 1970s and the trade conflicts of the early 1980s led Nissan to invest in factories in the USA and Europe.

Production abroad gradually took the place of exports. The rapid growth of demand during the bubble economy period did not prevent Nissan's domestic market share from continuing to decline. Overseas sales no longer compensated for the reduction of sales in Japan. The replacement of union leaders, the collapse of the union's power with respect to promotions, salaries, and working hours, together with other reforms, proved inadequate to reverse this tendency. Nissan appeared to be seeking competitive advantage by adopting a systematic international division of labour in order to increase economies of scale. It attempted to achieve a measure of internal coherence via a computerized system for synchronizing flows which placed constraints and channelled productivity and quality, which it had never before succeeded in doing for any length of time.

Nissan appeared to be a company which had made decisions based upon more or less coherent compromises among the demands of profitability, the traditional power of engineers, the power and the practices of the union, and the successive demands of the government. In this respect, the company was not very different from several European and American producers. The export strategy had led Nissan to emphasize product quality at an early stage, but it had been necessary to regularly re-motivate its employees behind this goal. Salary levels, the undervaluation of the yen, and the organization of suppliers, enabled Nissan to export its products at competitive prices until the 1990s. However, Nissan was unable to realize adequate economies of scale since it did not share platforms between its models and it suffered from excess product diversification. Moreover, substantial expenses in research, design, and capital (the technical path chosen to increase productivity), meant that Nissan was not profitable unless it sold proportionately more vehicles than its competitors.

Nissan was trying to remedy both problems. Longer production runs were being sought through an international division of the different stages of the production process. A computerized system for synchronizing flows introduced real constraints on quality and productivity if the delivery deadlines given to each customer were to be respected. This new trajectory contained a contradiction, and it demanded that certain conditions be met if it were to be viable. The international division of the production process inevitably reintroduced stocks which the synchronization of flows was simultaneously attempting to eliminate. The computerized synchronization of flows presupposed that employees were prepared to accept this constraint even though it increased the prescribed part of their work.

# NOTE

Translated by Sybil Hyacinth Mair.
I would like to thank Michel Freyssenet for his help in reconstituting the analysis of Nissan's trajectory and organizing this chapter.

## STATISTICAL APPENDIX 4: Nissan

| Year | Production | | | Exports | | | Production overseas | | | Employees[d] | Turnover (million/yen)[e] | Profit Operating (million yen)[e] | Rate (%)[e,f] |
|---|---|---|---|---|---|---|---|---|---|---|---|---|---|
| | Total[a] | Passenger cars | Commercial vehicles | Total[a] | Passenger cars | Commercial vehicles | Total[b,c] | Passenger cars | Commercial vehicles | | | | |
| 1969 | 1,209,620 | 741,174 | 468,446 | 301,721 | 196,071 | 105,650 | — | — | — | 45,930 | 669,050 | 48,969 | 7.32 |
| 1970 | 1,421,122 | 949,349 | 471,773 | 461,026 | 324,581 | 136,445 | 22,343 | 11,957 | 10,386 | 47,570 | 799,316 | 58,747 | 7.35 |
| 1971 | 1,666,124 | 1,181,871 | 484,253 | 680,989 | 510,198 | 170,791 | — | — | — | 51,972 | 981,412 | 69,215 | 7.05 |
| 1972 | 1,903,414 | 1,368,667 | 534,747 | 703,024 | 522,975 | 180,049 | — | — | — | 53,508 | 1,176,425 | 83,568 | 7.10 |
| 1973 | 1,996,427 | 1,437,932 | 558,495 | 722,077 | 532,903 | 189,174 | — | — | — | 52,819 | 1,270,833 | 57,666 | 4.54 |
| 1974 | 1,851,271 | 1,309,924 | 541,347 | 863,332 | 623,833 | 239,499 | — | — | — | 51,612 | 1,429,637 | 24,655 | 1.72 |
| 1975 | 2,111,957 | 1,532,858 | 579,099 | 975,297 | 712,967 | 262,330 | 47,334 | 23,727 | 23,607 | 51,454 | 1,770,198 | 81,705 | 4.62 |
| 1976 | 2,301,444 | 1,599,440 | 702,004 | 1,124,528 | 776,000 | 348,528 | — | — | — | 52,577 | 2,024,624 | 120,416 | 5.95 |
| 1977 | 2,353,729 | 1,680,784 | 672,945 | 1,308,416 | 914,605 | 393,811 | — | — | — | 54,411 | 2,246,393 | 104,155 | 4.64 |
| 1978 | 2,374,023 | 1,751,203 | 622,820 | 1,003,598 | 713,289 | 290,309 | 102,538 | 60,051 | 42,487 | 55,747 | 2,306,685 | 88,822 | 3.85 |
| 1979 | 2,412,069 | 1,784,183 | 627,886 | 1,247,683 | 919,941 | 327,742 | 126,335 | 83,934 | 42,401 | 56,702 | 2,738,868 | 153,585 | 5.61 |
| 1980 | 2,648,674 | 1,933,700 | 714,974 | 1,475,453 | 1,035,439 | 440,014 | 115,170 | 72,385 | 42,785 | 56,284 | 3,016,190 | 124,458 | 4.13 |
| 1981 | 2,575,110 | 1,869,737 | 705,373 | 1,409,937 | 965,902 | 444,035 | 134,855 | 89,925 | 44,930 | 57,800 | 3,198,724 | 134,340 | 4.20 |
| 1982 | 2,404,169 | 1,814,946 | 589,223 | 1,341,810 | 969,104 | 372,706 | 172,303 | 91,497 | 36,996 | 58,962 | 3,187,722 | 102,214 | 3.21 |
| 1983 | 2,518,491 | 1,888,857 | 629,634 | 1,383,007 | 974,267 | 408,740 | 192,540 | 76,384 | 51,039 | 59,615 | 3,460,124 | 61,456 | 1.78 |
| 1984 | 2,473,191 | 1,834,042 | 639,149 | 1,375,075 | 984,152 | 390,923 | 333,645 | 83,993 | 132,318 | 58,925 | 3,618,076 | 70,845 | 1.96 |
| 1985 | 2,438,520 | 1,836,182 | 602,338 | 1,408,024 | 1,034,091 | 373,933 | 295,571 | 142,302 | 153,269 | 57,612 | 3,754,172 | 58,999 | 1.57 |
| 1986 | 2,275,098 | 1,813,680 | 461,418 | 1,293,892 | 1,030,757 | 263,135 | 317,613 | 143,480 | 174,133 | 54,573 | 3,429,317 | -8,499 | -0.25 |
| 1987 | 2,159,748 | 1,750,940 | 408,808 | 1,138,069 | 945,081 | 192,988 | 396,105 | 212,945 | 183,160 | 51,237 | 3,418,671 | 47,610 | 1.39 |
| 1988 | 2,239,280 | 1,802,833 | 436,447 | 1,084,746 | 890,273 | 194,473 | 487,528 | 286,246 | 201,282 | 52,808 | 3,580,110 | 92,010 | 2.57 |
| 1989 | 2,371,769 | 1,988,273 | 383,496 | 987,614 | 828,660 | 158,954 | 572,266 | 329,389 | 242,877 | 55,326 | 4,005,550 | 138,664 | 3.46 |
| 1990 | 2,379,634 | 1,900,068 | 479,566 | 959,120 | 806,696 | 152,424 | 585,480 | 328,402 | 257,078 | 56,873 | 4,175,013 | 119,660 | 2.87 |
| 1991 | 2,323,720 | 1,936,566 | 387,154 | 964,073 | 813,410 | 150,663 | 639,033 | 391,849 | 247,184 | 55,566 | 4,270,523 | 33,775 | 0.79 |
| 1992 | 2,036,664 | 1,683,211 | 353,453 | 900,463 | 732,636 | 167,827 | 748,804 | 492,385 | 256,419 | 53,071 | 3,896,888 | -33,639 | -0.86 |
| 1993 | 1,749,814 | 1,484,303 | 265,511 | 629,990 | 514,000 | 115,990 | 900,541 | 676,348 | 224,193 | 51,398 | 3,583,482 | -36,631 | -1.02 |
| 1994 | 1,589,393 | 1,379,099 | 210,294 | 611,215 | 521,741 | 89,474 | 944,294 | 661,151 | 283,144 | 49,177 | 3,407,512 | -74,810 | -2.20 |
| 1995 | 1,713,982 | 1,472,796 | 204,151 | 593,597 | 522,360 | 71,234 | 893,011 | 639,313 | 272,094 | 44,782 | 3,518,153 | 40,027 | 1.14 |
| 1996 | 1,610,542 | 1,458,928 | 203,848 | 598,224 | 518,553 | 79,691 | — | — | — | 41,266 | 3,609,441 | 112,917 | 3.13 |

*Notes:*

[a] The year is the fiscal year (1 April to 31 March). The data are for Nissan Automobile, and do not include overseas activities except for overseas production.

[b] Production, export, and overseas production are by volume. Prior to 1977, CKD (completely knocked down) kits were included in production volumes. Production in other countries is not included since Nissan is only a minority shareholder.

[c] Overseas production is the sum of production at Nissan subsidiaries in the USA, Spain, Australia, and Mexico.

[d] Employment is the figure at the end of the fiscal year.

[e] Turnover, profit rate, and employment include all Nissan's activities (automotive, textile machinery, aeronautical). The textile machinery and aeronautical sectors account for approximately 2 per cent of turnover.

[f] Profit rate = operating profit/turnover (pre-tax).

*Sources:* Nissan Annual Financial Reports.

# BIBLIOGRAPHY

Bélis-Bergouignan, M.-C., and Lung, Y., 'Le Mythe de la variété originelle: L'internationalisation dans la trajectoire du modèle productif japonais', *Annales*, 3 (1994a), 541–67.

—— 'Processus de diversification et flexibilité productive dans l'industrie automobile japonaise: Toyota et Nissan', *Actes du GERPISA*, 12 (1994b), 13–41.

Bosch, G., 'Working Time and Operating Hours in the Japanese Car Industry', in D. Anxo, G. Bosch, and D. Bosworth, *Capital Operating Time: An International Perspective*, (Dordrecht, 1994).

Boyer, R., and Freyssenet, M., 'The Emergence of New Industrial Models: Hypotheses and Analytical Procedure', *Actes du GERPISA*, 15 (1995a).

—— —— 'The World that Changed the Machine', GERPISA (Paris, 1995b).

CCFA (Comité des Constructeurs Français d'Automobile), *Répertoire mondial des activités de production et d'assemblage de véhicules automobiles* (Paris, 1981, 1987, 1995).

Chanaron, J.-J., 'Vers un affaiblissement de l'organisation Kereitsu au Japon?' *Lettre du GERPISA*, 100 (1996).

Cusumano, M., *The Japanese Automobile Industry: Technology and Management at Nissan and Toyota* (Cambridge, 1985).

Hanada, M., 'Modalités de la fixation des salaires au Japon et en France: Etude du bulletin de paie de Nissan et Peugeot', *Japan in Extenso*, 31 (1994).

Imai, K. (NIRA—National Institute for Research Advancement), *21 Seiki Kigyo to Network* [Twenty-First Century Oriented Form of Firm and Networking] (Tokyo, 1992).

Itami, T., Kagono, T., Kobayashi, T., Sakakibara, K., and Ito, M., *Kyôsô to Kakushin* (Tokyo, 1988).

Kamii, Y., *Rôdô Kumiai no Shokuba Kisei* [The Influence of the Enterprise Union at the Shopfloor Level] (Tokyo, 1994).

Maruyama, Y., and Fujii, M., *Toyota to Nissan* (Tokyo, 1991).

Monden, Y., *Shin Toyota Sisutemu* [New Toyota System] (Tokyo, 1991).

*Nihon Keizai Shinbunsha*, 'Nissan ha Yomigaeruka' (Tokyo, 1995).

Nissan Motor Company, *21 Seiki heno Michi: Nissan Jidosha 50 Nenshi* (Tokyo, 1983).

—— *Nissan Jidoshashi 1974–84* (Tokyo, 1985).

Okamoto, H., *Gendai Kigyo no Seihan Togo* (Tokyo, 1995).

Roshi Kankei Chosakai, *Roshi Kankei Jittai Chosa*, iii (Tokyo, 1965).

Saga, I., and Hanada, M., 'Récente évolution des relations professionnelles et de l'organisation du travail', Proceedings of Third GERPISA International Conference (Paris, 1995).

Shimizu, K., 'Trajectoire de Toyota: Rapport salarial et système de production', *Actes du GERPISA*, 9 (1993).

Shimokawa, K., *The Japanese Automobile Industry: A Business History* (London, 1994).

Tabata, T., 'Globalization of Business', in *Japanese Automobile Industry from Challenge to Peaceful Competition* (Tokyo, 1995).

Takeuchi, Y., *Iwaki Kôjô no Chôsen* (Tokyo, 1995).

Udagawa, M., 'Kigyôkan Kyôsô to Hinshitsu Kanri', in *Nihon Kigyô no Hinshitsu Kanri*, Hôsei University, Industriel Information Center (ed.) (Tokyo, 1995).

Yamamoto, K., *Jidôsha Sangyô no Rôshi Kankei* (Tokyo, 1981).

# 5

# The Globalization of Honda's Product-Led Flexible Mass Production System

ANDREW MAIR

Honda became the world's largest motorcycle producer twelve years after it was founded in 1948. The company entered the automobile industry during the early 1960s. It then grew continuously for three decades, overtaking established automobile producers to rank tenth in the world and become one of a Japanese 'Big Three' alongside Toyota and Nissan. During this period Honda developed an image as a 'different' company with an idiosyncratic trajectory led by innovative products (Mair forthcoming *a*). Growth was seriously challenged in the 1990s. Yet by now Honda had deepened and globalized an industrial model of 'flexible mass production'. Accordingly, Honda remained profitable during the 1990s Japanese recession. At the same time the crisis forced a rethink of the product innovation strategy.

This chapter reviews the origins of flexible mass production at Honda. It then examines the globalization phase of the 1970s and 1980s. Lastly, it analyses the crisis in the early 1990s and the implications of Honda's response for its future trajectory. The discussion focuses on Honda in the automobile sector, which accounted for approximately 80 per cent of turnover by the mid-1990s, though key links to motorcycle activities are made as appropriate.

## 5.1. THE HONDA MODEL IN THE MID-1960S

What was the configuration of the Honda model in the late 1960s? Twenty years of motorcycle manufacture had already sown the seeds of a viable model of flexible mass production. Expansion during the 1950s had been based on the twin foundations of Soichiro Honda's innovative engines and product designs and his partner Takeo Fujisawa's novel organizational model with its creative new ways to organize a large company consistent with permanent product innovation.

Soichiro Honda developed very high speed (up to 14,000 revolutions) internal combustion engines to boost their power. His engineers studied combustion and new materials. By the early 1960s the company was regularly winning international motorcycle races. The strategy of inventing new technologies instead of licensing them was succeeding in this competitive arena. Honda's product concepts were an equal factor in its parallel success in the market. The 1958 Super-Cub model was exemplary. Its small but powerful 50cc engine was significant, but its success was rooted in its design concept, a clean and easy-to-use machine for the masses. The SuperCub represented an enduring dual approach to products at Honda: part advanced technology, part new concept.

In the late 1950s Honda turned to the emerging Japanese car market. The company already possessed many of the basic product and process technologies. However MITI (the Japanese Ministry of International Trade and Industry) was trying to prevent new entrants and force industry consolidation to realize internationally competitive economies of scale. Honda remained outside the government–industry policy consensus, arguing that Japan needed entrepreneurial strategies to meet Western companies on their own terms. The company developed a small truck and sports car, launched in 1963, so that it could claim to be an established producer. Both were offered in different variants during the 1960s. However, volumes remained low and there were no mass production facilities. By the time S series (sports car) production ended in 1970, only 26,000 had been made.

Honda was now racing both Formula One and Formula Two cars, adding to its technical-sporty image and letting the engineers develop ambitious projects. To doggedly develop particular technologies (such as high-speed engines) without direct regard to the market was by now an embedded part of the Honda model alongside market-oriented designs. To solidify this strategy, Fujisawa had enforced the spin-off of research and development activities into Honda R&D Co. Ltd, founded in 1960 with Mr Honda as president. The structure permitted a certain independence from demands made by the rest of the company, as well as allowing the product engineers to create their own organizational forms.

Indeed Fujisawa was developing novel organizational features across Honda. In 1953 he devised a pyramidal structure as a stopgap measure. Neither Honda nor Fujisawa really wanted this model, yet they planned to compete in global markets, and recognized that the company would have to grow to be successful. However, by 1960 Fujisawa had created the basis for a different kind of organization which would permit growth without losing the capacity to innovate. He had studied bureaucratic corporations, which he believed wasted most of their human talent. The challenge was how to build a whole company in Mr Honda's image: a company to support his work and utilize his innovations.

The first characteristic of Fujisawaism[1] was its permanent culture or ideology rather than organizational structure. This was a 'warp' onto which 'wefts' of organizational form could be woven and altered with minimal disruption. Secondly, Soichiro Honda, with his individualist antipathy to status and class, became a walking metaphor for appropriate behaviour. Thirdly, an 'expert system' was developed that awarded promotions to technical experts who did not wish to enter management. Fourthly, new ideas were 'owned' by their proposers; thus a suggestion system was started in 1953, and a 'notebook system' was created for employees to record innovations and retain when they moved posts (in part to stop superiors from taking credit). Fifthly, delegation of authority was emphasized. Younger staff were made responsible even when their superiors could solve a problem, since younger staff were likely to find new solutions to accelerate innovation. Sixthly, horizontal and diagonal network linkages were created. By the mid-1950s there was a unified information system comprehensible across the company. Information was graphed and posted. To prevent hoarding of information, there were no drawers on desks. Perhaps the best known feature of Fujisawaism was the 1964 'joint board room' where senior managers were groomed to replace the founders, sharing a single large office that encouraged communication and debate.

The Honda Workers Union was formed in 1953. Major conflicts with leftist unions were avoided, but problems remained. Union pressure forced Fujisawa along certain paths, including pay bargaining. Union mistrust slowed the introduction of the 'expert system'. Union control over overtime work prevented unannounced longer working days. However, by the late 1950s Honda's commercial success permitted bonus demands to be accepted and what were called 'modern' (routinized) union relations to be established. Honda joined other Japanese companies with individual evaluation procedures and salaries, and large bonuses relative to basic wages: a system consistent with Fujisawaism's stress on individual merit.

By the late 1960s Honda had various manufacturing sites in Japan: Wako (1953, north of Tokyo), built for motorcycle production and later switched to car engines; Hamamatsu (1954; central Japan, the original company home), for motorcycles and power products such as power tillers; Suzuka (1960, west of Nagoya), the main motorcycle plant, and Sayama (1964, adjacent to Wako), built for early car production. Wako also housed R&D activities (from 1957), while the headquarters was in Tokyo from 1953. Manufacturing in the 1950s and 1960s does not seem to have been particularly innovative. Initial equipment was purchased in the USA and Germany. The Honda Koki Engineering Factory, forerunner of Honda Engineering, was set up in 1962, and Honda now made much of its own equipment: a progression from product technology. While purchasing generic tools (machine tools, presses), Honda Engineering built competencies in product-specific tools (for assembly and later welding).

While Honda used a simple mass production model, Fujisawa controlled production flow tightly to keep costs down and the rapidly growing company solvent. A shift from factory sections separated by in-process stocks and paperwork to coordinated low-inventory production reduced the inventories/sales ratio from 12.5 per cent in 1954 to 3.2 per cent in 1958. By the late 1950s Fujisawa's strategy for operations was to reduce costs towards global norms and not to take short-term profits during booms in Japan. When the domestic market faltered, competitors that had expanded marginal output to meet demand found Honda undercutting their prices and benefiting from a spiral of cost and price reduction, market share growth, and economies of scale (Abbeglen and Stalk 1985, Sakiya 1987). With 80 per cent of the Japanese motorcycle market by 1960, volumes now permitted internationally competitive prices.

Fujisawa also improved product distribution. Honda had no *keiretsu* linkages for distribution or finance. From 1951 Fujisawa was forcing distributors (wholesale) to pay for popular products in advance, to generate cash flow for investments.

# The Globalization of Honda's System

By 1952 he was circumventing distributors to sell the Cub model directly to bicycle dealers (retail). The goal was to control distribution for Honda as manufacturer, reversing the norm in Japan. Control over customer relations was deemed vital because the product required strong after-sales service. A Honda service training institute was set up in 1957, a dedicated Honda parts centre in 1963, and Honda Credit Service Co. (for loans) in 1966. By 1966 there were sales outlets throughout Japan. The dealer network was now being improved to prepare for mass automobile distribution. Most dealers were unfamiliar with cars, and Honda created new subsidiaries to provide marketing expertise, deal with traded-in cars, and service the vehicles. On the eve of mass automobile production, Honda had built a strong position.

Motorcycle exports started to the Philippines and Taiwan in 1952, and the USA in 1954. An internal task force now recommended that Honda should enter emerging markets in Asia and existing markets in Europe. However Fujisawa took an unorthodox view and decided to focus the meagre foreign currency granted by MITI on North America, which, while wealthy, had only a small motorcycle market. His goal (like Soichiro Honda's racing) was to enter the most difficult and stretching market first to improve the company's competitiveness. American Honda Motor Co. was established in 1959, though Fujisawa had to engage in political lobbying to get even half the foreign currency he sought. Again Honda sought direct control of distribution rather than use the trading companies that dominated Japan's external relations. A small team was sent to California in 1959, with no fixed plan (Pascale 1984). After a slow start when they tried to fit into the traditional distribution network for large motorcycles, they sold SuperCubs in clean bicycle and sports shops (similar to the SuperCub distribution system in Japan). Honda had created a new market sector that boomed in the early 1960s (though it was to collapse in the mid-1960s when the SuperCub was not upgraded). The SuperCub also provided entry into Europe, with the first sales subsidiary in Germany in 1961, and the first Japanese manufacturing investment in an advanced industrial country at Aalst, Belgium, in 1963. Honda struggled to manage this factory, conflicting with trade unions but learning lessons about how to transfer its methods abroad. Other overseas investments included assembly of SuperCub kits in Taiwan (1961), China (1965), and Thailand (1966).

## 5.2. FROM MASS CAR PRODUCTION TO GLOBAL SALES, 1967–1982

There were experiments with Fujisawaism in manufacturing and its systematization in research and development in the fifteen years after mass car production started in 1967. Sales growth began in Japan, but by the mid-1970s there were significant exports, rising from 6 per cent of output in 1970 to peak at 70 per cent in 1977–80 even as volumes tripled. Car sales in the USA rose from 39,000 in 1973 to 375,000 in 1980; by the early 1980s North America accounted for over half of Honda car sales. Total turnover increased fourfold in seven years

(1973–80), and by 1980 income from car sales had overtaken (still-rising) income from motorcycle sales.

### 5.2.1. The Product Range

The car model range (excluding the single small truck) evolved from one model and one platform in 1965 to three models and two platforms in 1968 and to four models and three platforms in 1972 (Jetin 1995), reflecting the development of a range for the Japanese market. The breadth of the range was then stable until 1974, and shrank to two models and two platforms between 1976 and 1978, as the first cars were replaced by a narrow range focused on North America. The range then expanded to six models and four platforms between 1978 and 1981, as new models were spun off this second range for the Japanese market.

The first mass production car was the N360/600, a minicar with an air-cooled engine produced from 1967 to 1973. The N360 was a mass product like the SuperCub, and became the best-selling minicar in Japan. By 1970 Honda had gained an 8.8 per cent market share. More than 1 million units were sold in four years as Honda rode a wave of domestic economic growth to gain important economies of scale (over 200,000 vehicles per year). Saloon, station wagon, and coupé versions were made, as were different engines and transmissions and options like sunroofs. Yet Honda's next, larger car failed. The 1300, a four-door saloon introduced in 1969 (1970 coupé also) was anticipated to move the range upmarket but sold only 115,000 units before it was withdrawn in 1975, despite facelifts and a relaunch. Significantly for the Honda industrial model, the 1300 failed because it was designed as a 'driver's' or 'enthusiast's' car, not a mass car (see below).

Honda had a major crisis in 1970–1. Safety problems with the N360, combined with a new consumer movement in Japan, led to a collapse in sales. Income from motorcycle sales had to compensate for losses due to the 1300 and now the N360 until a new microcar, Life, was introduced in 1971. This would be Honda's last crisis for twenty years. The way senior managers handled it persuaded Mr Fujisawa that both founders should now retire (1973).

The Civic (1972) was the model the 1300 might have been, with a front-wheel drive water-cooled engine and emphasis on comfort in a small car. It evolved with new body versions (two-, three-, four-door, station wagon) and a new model was introduced in 1979. Sales of the Civic fed Honda's rapid growth after 1975 (especially in North America: see below). By 1976 1 million Civics had been sold, by 1979 2 million, and by 1982 3 million. In 1976 the Accord was launched, a car later to take over the lead role from the Civic in North America. Despite the importance of minicars in Japan (Life sold 500,000 units: 44 per cent of Honda's sales in Japan between 1971 and 1975), in 1975 Honda ceased making them to focus on producing the very profitable Civics and Accords for North America. Meanwhile the sporty two-door Prelude (1978) completed a range of three models on three platforms targetted at North America. By 1975 a California design studio was open to help Americanize design. In 1980–1 Honda returned to the Japanese market with the Vigor (a version of the Accord), the Ballade (ditto the Civic), a new Quint/Integra, and a new small City/Jazz. These models were not sold in North America, though some were sold elsewhere. They provided a range for the Verno second dealership chain, launched in 1978.

*5.2.2. Product Technologies*

The key technology achievement of the 1970s was the 1972 CVCC (Compound Vortex Controlled Combustion) engine, which enabled Honda to become the first car maker to meet new American and Japanese emissions standards. The CVCC story became an exemplar of success in Honda. Yet despite its importance, it was not the only reason for Honda's commercial success and was not necessarily representative of Honda. Overall product concepts and durability of components underpinned the Civic's success in North America. Moreover, the CVCC followed a significant failed engine programme. Mr Honda had remained convinced that air-cooled engines were the technology of the future. He banned work on water-cooled engines and enlisted his engineers in the search for a masterpiece air-cooled engine to crown his career. Despite a partial technical success, the commercial price was high. The new engine needed expensive aluminium parts, and the vehicle it equipped—the 1300—had to be heavy to cope with its power. The result was a niche 'sporty' vehicle when Honda needed a mass market product: hence the poor sales. Moreover, demands to reduce pollution were already sounding the death-knell for air-cooled technology. The company turned to water-cooled engines from 1969 following an internal confrontation that left Mr Honda isolated. The stubborn pursuit of a single technical path had lost the company the chance to build a car like the Civic in 1968 instead of 1972.

The CVCC engine represented Honda's ideal self-image. The design solved an apparently intractable technical puzzle (Mair 1996): it was a technological, socially positive response to market change, it was launched ahead of competitors, and it initiated a new high-tech image for Japanese manufacturers. Strategically, the CVCC engine also marked a repositioning of Honda technology in a new social climate, away from speed and sportiness towards environmentalism. After 1968 Honda had withdrawn from Formula One racing; indeed the Formula One engineers had been reassigned to the CVCC project.

*5.2.3. The Research and Product Development Process*

Following the air-cooled engine debacle, Honda R&D began to apply the principles of Fujisawaism more systematically. No longer would it be Mr Honda's fiefdom. Research engineers now selected their own projects, often competing to solve the same problem, under the umbrella of a rigorous, competitive, and transparent project approval procedure involving not just researchers but product developers and marketing staff. A flat organization was formalized, with research engineers lined up horizontally (Mito 1990).

Product development was organized differently. The SED (Sales, Engineering (i.e. manufacturing), Development) system was formally introduced during the mid-1970s for the Accord, a form of simultaneous engineering designed to smooth and accelerate product development. The process was elaborated for the 1981 City/Jazz model, with its very young (average age 28) development team, also consistent with Fujisawaism (Nonaka 1991).

These principles now linked the parent company formally to Honda R&D. They also permitted a strict model replacement cycle, a response to the collapse of SuperCub sales in North America and the later failure of the 1300. A four-year cycle was adopted for most cars, however well they sold. There were minor exceptions as the schedule of new model introductions was re-jigged, but only the small Acty commercial vehicles had a long irregular cycle. This rhythmic supply-led planning of product development activities helped to accelerate processes. The 1979 Civic brought a policy under which major mechanical parts were frequently retained, after having been introduced two years earlier, creating a two-year gap between new models and new engines, each with a four-year cycle. The purpose was to attract customers with regular novelty. Engine compartments and engines had to be carefully designed in advance. Two product 'families' shared key engineering characteristics, one based on the Civic, the other on the Accord.

In 1979, Honda R&D opened a site at Tochigi, north of Tokyo, as a centre for product development. Basic research continued at Wako. In 1980, sensing a market shift in North America back to larger powerful cars, Honda returned to Formula Two racing and then to Formula One. Each model change saw Civics, Accords, and Preludes gain longer wheelbases and larger engines as they became further Americanized.

### 5.2.4. Manufacturing

Automobile production grew rapidly from under 10,000 in 1966 to 280,000 in 1970 (mirroring the SuperCub ten years earlier). Volumes then stagnated owing to the problems of the N360 and 1300 and oil-crisis recession in Japan, recovering to 320,000 in 1975. Production capacity grew from 320,000 in 1975 to 850,000 in 1981 (from 400,000 to 1,000,000 including small trucks) to cope with booming demand in North America, with second assembly lines built at Suzuka and Sayama.

With only two to three models (Civic and Accord, then Prelude) Honda realized huge economies of scale. Despite the company's small size, model- and platform-specific scale economies were second only to Toyota in Japan (Bélis-Bergouignan and Lung 1994; Jetin 1995), both in 1970 (high N360 volumes), and in 1980 (high Civic and Accord volumes), and were far higher than Nissan, Mitsubishi, or Mazda with their broad ranges designed for Japan. Honda's average volumes per model/platform were 138,000/138,000 in 1970, fell to 60,000/75,000 in 1972, peaked at 325,000/350,000 in 1978–9, and then declined to 210,000/280,000 in 1980 as the company returned to the Japanese market.

Total employment at Honda grew from 9,000 in 1966 to 18,000 in 1970, and remained stable between 1970 and 1976 despite the N360, 1300, and recession problems, though in 1972 Honda adopted a five-day week. Employment rose to 25,000 between 1976 and 1981 as automobile output tripled.

Honda was now developing its own production system, learning from the much admired Toyota Production System but also including quite different principles. While a Honda Production System (HPS) was not made explicit, its principles became increasingly clear (Mair 1998). The HPS is a push system with production planning based on market forecasts and product variety kept low (since Honda products were already 'different' in terms of technology and design). Hence production could be simplified and economies of scale generated. Mixed-variant, and by 1980 mixed-model, production balance capacity utilization (large cars at Sayama, small cars at Suzuka). Products are designed to fit existing equipment with only jigs altered. Common 'design philosophies' are adopted (architecture of components invariant across models to facilitate assembly tasks and tool sharing). Variants are produced in batches of sixty or (later) factors of sixty within larger (e.g. 240) batches per model. Perfect component quality is needed for all sixty cars per batch to be completed. Logistical planning is simpler than for Toyota's 'one-piece-flow' system. Line balancing is a different problem. Between models, some work teams are rejigged. For batches of high-specification variants, some supervisors plan to work on the line.

Honda Engineering was formally established in 1970 to provide innovative manufacturing solutions. Factories are planned for structural flexibility, with long-term use of the same people, sites, and production equipment even as models and product types change. The Suzuka factory embodied this 'flexifactory' concept (Mair 1994b), making motorcycles from 1958 and cars from 1967 but stopping motorcycle production by 1992. Thus the HPS is not intended to possess 'short-run' flexibility to produce to 'customer order', but can be 'rejigged' and restarted quickly; this is Honda's flexible mass production system.

Honda claims to have a 'human-centred' approach to work, less rigid in defining standard operations than Toyota. Employees rotate tasks; indeed they must be flexible, changing posts, models, departments, and product types. But the HPS does not rely on unplanned long shifts to make up lost production. When needed, Saturdays are used. Working hours are the lowest in the Japanese automobile industry.

In the mid-1970s there were experiments with new forms of flexibility at a new motorcycle factory at Kumamoto, Kyushu, opened in 1976 (Mito 1990), precursors of what other Japanese companies would do in the same region twenty years later. Following Fujisawaism, the factory was planned by young engineers who tried out ideas that, it was thought, might be needed when Honda built the factories it was already planning in North America. A system was devised to control and reduce stocks of components and plan logistics from central Japan. There was careful negotiation with the local community to avoid disruption. The plant was to be profitable at 50 per cent capacity. Finally, the free-flow assembly line' was introduced, in which products on the line stopped in front of the worker, who pressed a button to move them on when his task was complete. Buffer stocks between workstations allowed workers to control their work pace, giving a sense of control over the system; work could be temporarily speeded up or slowed down. Quality was improved because the product was stationary and workers had time to complete their task. The free-flow line was an explicit attempt to eliminate the negative aspects of machine-paced assembly work, though there was no 'crisis of work' in Japan at the time. It was not in fact copied in North America, though other lessons of Kumamoto were transferred. However, a more sophisticated version with different goals was later built at Suzuka (see below).

### 5.2.5. Suppliers

Honda spun off subsidiary supplier companies such as Keihin Carburetor Co. (1973), Asama Giken (1973), and Press Giken (1975). Most suppliers, however, were drawn from the Toyota and Nissan supply chains. Honda did not set up its own 'suppliers' association', an approach that mirrored the absence of *keiretsu* links, including the *keiretsu* of Honda's bank, Mitsubishi Bank. This did not preclude programmes to teach suppliers new production methods, and the development of close relations, including assignment of 'guest engineers' to Honda for product development projects.

### 5.2.6. Distribution Network

The new Honda automobile dealers in Japan took advantage of the minicar boom with the N360 and Life models. Their sales then declined during the early 1970s (problems of the N360, 1300, recession, and withdrawal from microcar production in 1975). Market share slipped from 8.8 per cent in 1970 to 5.4 per cent in 1980 and unit sales fell from 360,000 to 270,000. However, while production focused on North America, plans were laid to return to the domestic market with the second dealership network and new models (see above).

The N600 was first exported to the USA in 1970. The oil crisis opened the way for imports of small, efficient Japanese cars that later gained a reputation for durable mechanical parts. Honda had the added advantage of publicity for the CVCC engine, which not only met 1975 environmental standards but used leaded or unleaded fuel, whereas catalytic converters required hard-to-find unleaded fuel. Honda did not replicate its SuperCub distribution system (retail shops) for car sales, but adopted the existing American system. Prior to the 1973 Civic CVCC, few dealers would sell Honda cars (in addition to their existing franchise), but by the time the Accord and Prelude arrived, Honda was able to sign up exclusive dealers.

### 5.2.7. Expanding International Activities

A feasibility study for manufacturing in the USA started in 1971. In 1975 a formal project team was set up. By 1977 a site at Marysville, Ohio was chosen to make motorcycles: a low-risk experiment. Production began at Honda of America Manufacturing (HAM) in 1979. By the time protectionist pressures heated up in 1980–2, HAM was building an automobile factory (capacity 150,000 per year) adjacent to the motorcycle factory. Japan adopted 'voluntary' export restraints in 1981, but by 1982 Honda was already producing cars in North America with the benefit of two years' experimentation with logistics, community relations, and human resource management at Marysville.

Exports to Europe expanded slowly. In 1978 Honda Europe was founded to improve the distribution of cars and parts. The relative neglect of Europe did not preclude a 1979 agreement to provide knocked down (KD) kits to the struggling nationalized producer British Leyland. Assembly of the new Ballade began at the Cowley plant in the United Kingdom in 1981 for sale as the Triumph Acclaim. Simultaneously, the internationalization of motorcycle production continued (Mexico 1971, Brazil and Italy 1977, Nigeria 1979). Japanese facilities increasingly prepared KD kits for this network of assembly plants. Small KD car assembly plants were established in Indonesia (1975), New Zealand (1976), Malaysia (1977) and Taiwan (1977) to meet local restrictions.

### 5.2.8. Organization

Honda continued to promote individualism, youth, and a certain equality while Japanese culture emphasized the opposites (groupism, respect for age, status). The constant theme at Honda remained how to overcome the organizational rigidities that Japanese culture was believed to foster ('big business disease'). Yet conditions of employment at Honda remained typical of other large companies in Japan (with the notable exception of fixed working hours in the factories), and were still viewed as consistent with Fujisawaism. The internal hierarchy had seven ranks, including four 'blue-collar' ranks: junior common, common, senior common, and foreman. Company-level bonuses, individual bonuses, and bonuses based on individual evaluations made up a large proportion of the overall salary. The retirement age was 55. The Union membership included assistant managers. Benefits were provided to employees: sports grounds, company apartments and dormitories, vacation cottages, libraries, hobby rooms. A company savings plan offered higher returns than banks. Quality circles (NH or New Honda circles) were unpaid and 'spontaneous' (voluntary), but though 'informal' (not company organized), they did include company competitions and prizes. Employees were viewed as a 'fixed' asset, consistent with the 'flexifactory' operations strategy.

Mr Honda and Mr Fujisawa retired in 1973, the company's twenty-fifth anniversary. Mr Kiyoshi Kawashima became company president, starting a tradition in which the presidency of Honda R&D was the step prior to company presidency. Mr Kawashima sought organizational forms to permit decentralized flexibility alongside strategic and operational integration. He marked his ascendency by establishing three transversal 'expert' committees among senior directors (managers) to develop policy towards personnel, equipment, and finance for the expanding company. The Fujisawaist purpose of the committees was to cut across vertical sectionalism.

Growing internationalization brought further reorganization in 1979, as three 'regional task forces' were set up for Japan, North America, and Europe and Oceania, alongside the expert committees. In a third wave of reform in 1981, the expert committees were abolished and six internal divisions created (not to be confused with Sloanist separate divisions): motorcycles Japan, automobiles Japan, power products Japan, Europe and Oceania, North America, and developing countries. Operations in Japan were developed enough to warrant separate product divisions, but not so those elsewhere. By the early 1980s Honda was about to become a global company, able to focus simultaneously on two major markets, Japan and North America. The HPS was consistent with this. Wide product variety was unnecessary, since Honda products were already sufficiently 'different' in both North America and Japan. The narrow range brought economies of scale. Fujisawaism had diffused and deepened internally to improve the management of research and development. Honda's own flexible mass production model had matured.

## 5.3. EMERGENCE OF THE GLOBAL LOCAL CORPORATION, 1982–1991

The period from 1982 to 1991 started with the opening of the North American automobile plant and the intensification of relations with Honda's European partner following a serious return to compete in the Japanese market.[2] For a decade sales grew in all three regions, yet expansion of manufacturing in North America dominated, posing questions about the transferability of Honda's model and about the multinational form that would replace the 1970s 'export' model. Growth during the 1980s was financed by profits from North American production (as growth during the 1970s had been financed by profits from North American sales). Growth in Japan culminated in the 'bubble economy' of the late 1980s. In Europe, by contrast, Honda's strategy required few resources.

### 5.3.1. 'Localization' in North America

HAM reached full production of Accords after eighteen months in 1984. It was planned that the new plant should break even in 1985 and pay back its

## The Globalization of Honda's System                                121

investment in 1987. In fact it broke even in 1983, before full production was reached, and had paid back its investment by 1984. HAM was now making profits to feed company growth. Several reasons explain this success. The Accord was a high quality simple car with few variants built in a factory at full capacity. Honda took advantage of the price rises that followed export restrictions in 1981 (no Japanese rivals made cars in North America until 1984). The new labour force had low health and retirement costs. The HPS was based on simplicity, with inexpensive (if ingenious) equipment designed by Honda Engineering. Even after a second assembly line had been built at Marysville in 1985–6, the total investment to produce 360,000 cars per year was only $590m. When the value of the yen doubled in 1985–6 in reaction to Japan's continuing trade surplus, Honda already had significant activity in the dollar zone and could shelter behind new price rises forced on Japanese competitors. An engine plant opened at Anna, Ohio, in 1986 (by 1990 making aluminium engines, gearboxes, brakes, and suspension systems), an assembly plant in Canada in 1986, and a fourth assembly line at a new plant at East Liberty, Ohio in 1989, bringing total car and engine capacity to about 600,000 units. Honda's speed in building the new plants was a final factor permitting profits to be reaped.

At first, Honda experimented with North American parts makers. But their quality was poor, they proved unable to meet delivery deadlines, and they generally lacked seriousness in meeting Honda's expectations. 'Local content' in the early 1980s remained at about 30–40 per cent, and only a few Japanese supplier factories were built (with Honda financial aid) to make essential and bulky sub-assemblies. By 1986, with increased volumes and yen revaluation, Honda's Japanese suppliers were both able and obliged to invest in North America. By 1990 over eighty Japanese supplier 'transplants' were making deliveries, raising 'local content' to 75–80 per cent. A division of labour was established between the Japanese and North American suppliers. The former made complex components while the latter supplied generic raw materials and components. Honda had created a parallel automotive industry in North America.

The precise Marysville location was selected for its local labour market, with small industrial and market towns and farms populated largely by Americans of German origin, and distant from the domestic automotive industry so that workers with 'bad habits' could be avoided (Mair *et al.* 1988). A small group of Japanese helped new American managers select workers from the many applicants. Three interviews were held for each worker. Ideal employees were in their late twenties, married with children and a mortgage. A bias against older workers, racial minorities, and women was only corrected after investigations by a Federal government agency.

Human resource management policies new to both Honda and North America were thought necessary for the HPS and Fujisawaism to function. Lengthy discussions focused on new names such as 'associate' (which was adopted for all employees), whether uniforms were 'too Japanese' (they were adopted for 'pragmatic' quality reasons), and so on. Experience at the motorcycle plant allowed many elements already to be in place by 1982. Lessons were also drawn from bitter experience in Belgium, where trade unions and Japanese managers had each tried to impose methods on the other, with little innovation.

The key to the new system was 'single status' employment, in which all employees shared the same uniforms, parking lots, restaurants, and private health-care. All offices were open-plan and many had windows so that they could be viewed from production areas. All production workers were placed in one category (with separate categories only for maintenance workers and team leaders) and received the same wage with no allowance for seniority, job done, or individual merit. This un-Japanese system was consistent with Western principles of equality. Consistent with Fujisawaism, it reduced obstacles to managerial authority, individual initiative, transparency of merit, and open communications. Also consistent with Western ideologies, there was a measure of democracy. Workers voted on the organization of the working day (breaks, leaving hours), holidays, and when to make up for any lost output. A panel of production associates reviewed management dismissals of workers and had the power to reinstate; one in five were reinstated. More challenging than getting American workers to accept this system was to ensure that American managers could work in it. They had to learn how to function with few signs of rank and order, as the tenets of Fujisawaism made up the organizational 'warp' (see Shook 1988).

Monetary reward was central to the new employment relations. Honda stressed to its American workers that they were well paid and that the company accepted that they worked primarily to better themselves. Basic pay rates rose every six months throughout the 1980s. There were two types of bonuses, but neither was individualized or cumulative. A fixed attendance bonus for all workers with perfect attendance rose from about 3 per cent to 6 per cent of the basic wage between 1987 and 1991. A variable annual bonus based on worldwide profits (the same for all workers) was equivalent to about 5 per cent of the basic wage during the late 1980s.

The influence of trade unions had been the greatest fear of Japanese managers. HAM became the first non-union car assembly plant in North America since the 1930s. The United Automobile Workers union (UAW) attempted to organize the workforce in the mid-1980s (under US law, a majority vote would oblige the company to negotiate). Honda then increased wages and benefits to mimic those at Ford, General Motors, and Chrysler, and permitted an anti-union campaign. With new workers continually hired and promotions possible during a phase of expansion, the union was unable to gain sufficient support and called off the vote. Notably, the union campaign had stressed production speed-up, replacement of jobs by robots, and arbitrary managerial decisions: a campaign suited to American Fordism but not meaningful to workers in a company dominated by the HPS and Fujisawaism.

Honda transferred the HPS in its entirety, from the flexifactory principle to the labour process and to supply-chain relationships. The first assembly line at Marysville was a shorter, lower capacity version of a Sayama line, with costs kept down when Honda Engineering designed new flexible welding equipment in which one installation undertook the tasks of several previously: flexibility not for variety but for scale. Unit costs were lower than at Sayama. A large group (300–400) of Japanese staff oversaw all technical aspects of production during the 1980s and gradually trained American engineers. A subsidiary of Honda Engineering was established at Marysville in 1985 for the day-to-day management of production equipment and to develop competencies such as cutting stamping dies for body parts. Each new production line received the latest equipment, sometimes before the Japanese factories did.

The flexifactory concept provided 'structural' flexibility in North America (Mair 1994*b*). In 1985 Marysville became the first car factory on the continent to replace a model without closing. The final 1981–5 Accord was still at the end of the line as the first 1985–9 Accord started production. The changeover took place on one U-shaped assembly line while a second line was being constructed around it. Moreover, during the expansion it was decided to increase capacity from 300,000 to 360,000 to meet high demand in Japan (where export capacity could then be diverted to domestic sales). The flexifactory concept was now being used to balance capacity utilization at the global scale. Moreover, all four North American lines could make both Accords and Civics, a dimension of flexibility not seen in Japan. The North American factories were managed as an integrated production network with models and body variants regularly reassigned during the 1980s to maintain full capacity with the simplest possible mix on each line.

The HPS production planning system based on market forecasts was well suited to demand in North America. A sign of its operation was that stocks of finished cars built up before output was cut at the start of the US recession around 1990. On a daily basis, just-in-time logistics obliged most first-tier suppliers to locate within two hours of Marysville where they prepared batches of sixty parts. Producers of coloured parts also made them just-in-time, and could change the colour at only a few hours' notice (if HAM rejigged schedules to cope with problems). In search of zero-defect deliveries, HAM raised the cost of errors by insisting on a detailed explanation of the least mistake.

In their work, Honda's new American workers had to be disciplined (attend work every day), flexible (willing to be assigned to new jobs), and focus on quality. They could volunteer to participate in innovation (help improve the production process). Single status and monetary incentives were the key to the first three behaviours. Training in what was called 'the Honda Way' was vital for the last. As well as training by Japanese trainers and visits to Japan for prospective specialists, an eight-hour video of the whole Marysville production process gave all employees a wider view. Some Japanese staff became semi-permanent, others were twinned with American managers to teach by example, and still others were temporary, assigned for new projects. Posting abroad became a valued career move for Japanese staff that opened up new ways of thinking. There were parallel management hierarchies, with Americans in line management positions and Japanese engineers in technical positions.

When the manager of the Suzuka plant was promoted to manage HAM in 1984 he recognized that many of the precepts of Fujisawaism were proving difficult to explain to Americans. A programme was launched to make the Honda Way more explicit. Discussions sought culturally appropriate metaphors (related to farming, or motor racing, for instance). The Honda Way became a series of documents and videos, in which Honda staff explained what it meant to them. For some it was a spirit, for others a dream, for others a feeling, for others it was hard to describe (Shook 1988). Fujisawaism as a management philosophy was being transferred in new mediums and a new language to North America. Besides Americanized concepts like teamwork (associated with patriotic nation-building) and equality (associated with democracy), HAM developed other concepts for the local context. One of these was 'pride', which was linked with the pride Americans felt for their high schools or universities and now involved belonging to Honda. 'Pride' explained why employees did not compromise on quality. Managers too were expected to set high standards and stick to them rigidly in search for perfection.

Workers were organized into teams (of up to twenty) under a team leader. All tasks within the team were learned, with regular rotations. There was no reserve pool for absentees. Team leaders were to guide rather than supervise. They arrived early to check reports from the previous shift and plan a short team meeting. They reported every defect they discovered to its source. Team leaders were the breeding ground for future managers, and promotions implied being moved around departments, and from factory to factory, to improve understanding and enhance horizontal communication.

Two schemes encouraged worker innovation. The 'VIP' programme added up points scored by individuals on suggestions schemes, safety schemes, and so on, to award annual winners a prize such as use of a car for a year. According to HAM, by the late 1980s, 59 per cent of suggestions were implemented. A quality circle (NH circle) scheme was set up to solve complex problems, although not until 1985, six years after production had started. By now the HPS and Fujisawaism were thought to be sufficiently embedded. Participation in these schemes marked out employees for promotion to team leader and beyond: there were no individual assessments with an impact on salary. While single status and equal pay delivered the basic behavioural attributes Honda sought, it was uncertain to what extent they limited inter-worker competition and so reduced active participation in innovation.

Honda's American dealers prospered. The products were scarcely advertised, yet so great was demand that during the early 1980s some dealers bribed Honda managers for allocations (the scandal emerged a decade later). By the late 1980s Honda dealers had the lowest inventories, highest unit sales, and highest profits in the USA. In 1986 the 'Acura' dealer network was established to sell the new Legend, the Integra which replaced the Quint, and from 1990 the NS-X sports car: all produced in Japan but sold in North America under this new brand. Consumer surveys consistently rated Acura as having the best quality cars in North

America. Honda now showed that the Japanese could make and sell 'luxury' cars. The Acura models were more profitable and less price-sensitive, and Honda of America switched its imports to these products after the 1985–6 yen revaluation. Over 80 per cent of unit sales, however, came from the three-model Honda range, the Civic and Accord models with their high quality, attractive design and regular updating, and the sporty Prelude. From 1989 to 1992 the Accord was the best-selling car sold by any producer in the USA. While marketing remained low-key, customer service was viewed as essential, a lesson transferred from Japan. Honda focused on repeat business, with customers and their families to be won over one by one for the long term. Acura permitted Honda buyers to stay loyal as their incomes rose. Notably, however, Honda refused to offer the minivans and pick-up trucks that were increasingly popular during the 1980s despite repeated pleas from dealers, so convinced was the company that the growth niches of the 1990s would be luxury and sporty cars.

In 1984 the California design office became an independent subsidiary of Honda R&D. It was responsible for the concept of the sporty 'CRX' Civic introduced in 1983 as well as a two-door Accord introduced at Marysville (not at Sayama) in 1988. With Honda Engineering established at Marysville, the framework was in place for a product development capability including design, equipment-making, and launch. The SED process was being systematically transferred to California and Ohio. During the late 1980s the rear section for an Accord station wagon was developed in North America for launch at Marysville in 1990. HAM took part in this simultaneous engineering operation. Not only were assembly workers able to study prototypes and comment on ease of assembly, but a sloping rear window was designed so that it fitted existing equipment. Yet the station wagon failed to compensate for the absence of a minivan, in part because it was rather small.

By 1990 Honda R&D of America employed 340 and Honda Engineering 170 staff. The growth of design and development was accompanied by organizational changes to permit further regional autonomy, in particular the establishment in 1987 of Honda North America, based in Los Angeles, as a regional umbrella including Canada and Mexico.

## 5.3.2. Consolidation in Japan

Tochigi remained the product development centre, with expanded facilities. By the late 1980s perfection of the SED process permitted new models to be brought from concept to manufacture in two years. The model replacement cycle remained four years, however; speed was used to cut costs and increase flexible response to the market. The six model and four platform range developed during 1979–80 was retained until 1985 (during this period Honda made significant investments in new motorcycles to beat back Yamaha's challenge in Japan: see Abbeglen and Stalk 1985). After 1986 Honda's overall range expanded to thirteen models and ten platforms in 1990 and seventeen models and ten platforms in 1992, as the Acura models were introduced in North America and the range in Japan was broadened during the 'bubble economy'.

Domestic market share sacrificed in the 1970s was regained in the early 1980s. A third Japanese distribution channel opened in 1985, when Honda re-entered the minicar market with Today. By 1989 the Accord had spun off into different 'models' for the three Japanese networks. A small sporty car, Beat, was introduced in 1991. The NS-X limited production supercar, introduced in 1990, had an aluminium body, mid-body three-litre engine, and its own production plant at Tochigi. While model diversification responded to the 'bubble economy', Honda did not offer long lists of options but option 'packages' consistent with HPS batch production.

The sporty image was rebuilt during the 1980s. Honda won six Formula One racing titles in succession from 1986. Sporty cars like CRX, Beat, Prelude, and NS-X were joined by two-door coupé Civics and Accords. Research targeted economical, low-pollution engines that also offered rapid acceleration. The VTEC (variable valve timing and lift technology) engine was announced in 1988. Honda was quick to offer electronic fuel injection and sixteen-valve engines across its range, and by the early 1990s was broadening VTEC sales too. Yet consumers did not always share the views of Honda R&D. Certain that the four-cylinder engine could be made more powerful and smoother, Honda made V6 engines only for the Legend even though many American consumers still believed that better engines had more cylinders. By the early 1990s Accord sales were suffering.

Production capacity in Japan expanded from export boom levels of 850,000 in 1981 to 1.2 million in 1990 (from 1 million to 1.4 million including small trucks). The cramped Sayama plant grew upwards and decanted non-assembly activities to suppliers or logistics centres. Its lines made the Accord and derivatives, and the Prelude and Legend. Suzuka continued to make smaller cars like Today, City, Civic, Concerto, and Integra. A third car assembly line opened at Suzuka in 1989, in time to meet rising domestic demand. Following Kumamoto, this was Honda's second large-scale experiment with new assembly methods. Adjustable height automated guided vehicles (AGVs) carried bodies on part of the line in a new type of free-flow line. Bodies stopped at each workstation to facilitate automation. A target of 30 per cent task automation was announced, but by the early 1990s automation had peaked at 15–20 per cent (estimates vary). The AGVs were already expensive, and robotization added investment costs yet was not cost-effective. Less expensive engineering solutions facilitated manual assembly: on part of the line bodies were turned 90 degrees to ease work on vehicle fronts and rears. Work rationalization remained focused on technology and technical–organizational solutions. The less costly innovations were transferred to the new plants overseas, but the free-flow line with AGVs was not.

Average model and platform volumes globally fell from 210,000 and 280,000 in 1980 to 136,000 and 176,000 in 1990 (still second only to Toyota). In Japan, volumes fell to 95,000 and 125,000 in 1990, as 'niche' model production remained there while the narrow range manufactured in North America produced figures of 300,000 and 300,000. The overall decline in model and platform scale concealed other means of achieving economies of scale, however, including shared production lines and equipment, engine compartments, components and 'design philosophies'. 'Platform' represented by wheelbase was no longer a significant indicator. By the early 1990s batch sizes for some derivatives were being reduced to fifteen and below in Japan as the overall product mix became more complex. However, in North America (and later in Europe) batch sizes remained sixty or thirty. Models unrelated to Accord and Civic were made at their own sites: the NS-X at Tochigi, the ACTY van from 1985 and the Beat (1991) at subsidiary body maker Yachiyo Industry Co., which became a 'mini-producer'. The flexifactory strategy allowed employees to be retained when products changed, and bridged the HPS and a human resource strategy in which long-term employment and familiarity with Fujisawaism were intertwined. The labour shortages of the 'bubble economy' period were partly met through employment of ethnic Japanese labour from Brazil, particularly at supplier companies around Sayama.

Mr Kawashima retired after ten years in 1983 to be replaced by Tadashi Kume, also from Honda R&D. In 1990 Mr Kume was replaced by product engineer Nobuhiku Kawamoto. Both followed the Fujisawaist approach of reinvigorating management with new organizational forms. Mr Kume strove to build 'diagonal' linkages with top managers visiting the front line. A new joint board room was set up at the new Tokyo head office in 1985. Mr Kawamoto changed tack in 1991, increasing individual responsibilities among senior management and allowing individual offices (Mair 1996).

### 5.3.3. Europe and Elsewhere

An assembly plant opened in the United Kingdom in 1992, though Honda had already deepened other activities in Europe. The Ballade built by British Leyland as a KD assembly from 1981 and badged as a Triumph Acclaim was followed by the new Ballade in 1985, badged as a Rover 213/216 with higher local content. During the early 1980s the two companies jointly designed a large car, the first (1985) Honda Legend and Rover 800, which each company also manufactured for the other. From a simple KD arrangement in which Honda surprised British engineers and workers with its simple designs, precise parts quality, quantity and ease of assembly, the companies switched to a complex joint project which suffered from major differences over design and development processes. These, plus the poor quality parts sent from Britain to Japan caused problems at Sayama where the Rover 800 was built in 1986–7. When the Legend was made at the British Cowley site, Honda rejected many on quality grounds. The Legend/Rover 800 project dramatically illustrated the clash of different industrial models (Mair 1994a, see Chapter 15).

Three more joint projects followed. Now Honda led, with the British company (now called Rover Group) spinning off its own cars from Japan-oriented Honda derivatives (Concerto, Ascot/Inspire, Domani). In 1986 the British started to make small numbers of Ballades for Honda. Larger volumes (up to 35,000 per year) of the Honda Concerto were made at the British Longbridge plant between 1989 and 1995 alongside Rover versions. Rover had purchased Honda Engineering automation, and was now capable of making a car that Honda could sell (particularly in still-protected European markets).

The newly founded Honda of the UK Manufacturing (HUM) opened an engine plant at Swindon in England for the Concerto, also supplying engines to Rover. At the instigation of Rover Group's parent British Aerospace, following its understanding of Japanese practice (which was not in fact Honda's), HUM and Rover Group arranged mutual minority ownership (20 per cent swop) in 1989. HUM's own car plant opened in 1992 with production of a new Accord model (Europeanized spin-off of a Japanese car). This plant, with a capacity of 100,000, was modelled on the Canadian factory, which itself had drawn lessons (though not the AGV equipment) from the third line at Suzuka. Meanwhile Honda had used its Rover Group links to study and select European components makers. A subsidiary of Honda R&D was opened near Frankfurt, Germany, in 1985. By 1990 there were 60 employees working on market research, competitor products, and studies of seating and suspension systems for Europe.

The British managers at HUM sought to overcome the British social class system with its myriad social distinctions that were judged to contradict Fujisawaism. New employees were told that divisions and barriers would not be tolerated. Office staff were taught factory tasks, and university engineers were first assigned to production. Uniforms and Honda caps were obligatory. The only exception to this single status model was the allocation of a 'company car' to senior managers (a tax-free benefit common in the United Kingdom). HUM managers did not create a distinct European equivalent to 'Americanization' at HAM. They did avoid trade union representation of employees, making HUM the first non-union car plant in Europe for several decades (Mair forthcoming *b*).

Growth in Europe was slower than in North America. By 1990 Honda sold only 9 per cent of its vehicles in Europe, with a market share of 1.2 per cent, compared to 50 per cent/6.2 per cent in North America, and 35 per cent/8.7 per cent in Japan. Not until 1991 did the European Community agree to open protected European markets to Japanese imports, and Honda had no distribution network in Italy or Spain until the late 1980s. Honda designers also found it difficult to understand consumer tastes in Europe, and at first clung to the idea that cars should be designed for the 'most demanding' market and that sales elsewhere would follow. The USA was the global benchmark, California the US benchmark, and Germany the European benchmark (hence the location of R&D in Europe). By 1990, however, Honda R&D had adopted a 'total car concept' in which specific packages of components were to be fitted together differently in diverse regional markets.

An uneasy model range for Europe was made up of models designed for Japan or North America, combined with Ballades then Concertos made by Rover. The model mix constantly changed (to suit Honda's other markets). While some models were replaced too soon, the Concerto became outdated but continued in production. Honda refused to build or purchase diesel engines despite their growing popularity in Europe, viewing the diesel as 'yesterday's technology', inferior to Honda's own VTEC engines.

### 5.3.4. Towards the 'Global Local' Organization

Motorcycle manufacturing continued to set the pattern for the internationalization of car operations, with production starting at new factories in Mexico in 1986, Indonesia and Pakistan in 1987, and small generators produced in India from 1988. During the late 1980s motorcycle assembly continued to be shifted out of Japan, with Kumamoto converted to supply KD kits for Asia and Latin America. Indeed the Japanese motorcycle market was increasingly supplied from Honda factories in the USA, Italy and Indonesia. Within Europe the focus of motorcycle production shifted from Belgium to Spain and Italy, following markets. The Belgian factory retained its site and labour force but from 1990 added production of car parts.

By the late 1980s this 'global local' pattern in which a unified production system simultaneously responded to local market and political constraints also began to emerge in automobile operations (Mair 1997). The pattern was clearest in the bipolar Japan–North America relationship, which by 1992 accounted for 77 per cent of Honda car sales (27 per cent in Japan, 50 per cent in North America) and 95 per cent of production (62 per cent and 33 per cent, respectively). The markets and production bases of the two regions now operated synchronously, with each region meeting demands (for design skills, for products) in both.

The resulting 'global local corporation' was built on the foundations of the HPS and Fujisawaism, with their simplicity of model range, flexifactories, use of best technologies, and organizational flexibility. In 1987 Honda declared its aim to build a 'self-reliant' company in North America. 'Self-reliance' did not imply operational separation from Japan but a product development capability within a new division of labour between Japan (design and production of niche models, design of mass models) and North America (design and production of niche variants). In 1987 HAM began exporting cars, initially to avoid political barriers to exports from Japan (Accords to Taiwan). Two-door Accords and Accord station wagons were later exported to Japan.

By 1990 a group of American product development engineers had been sent to Japan on five-year assignments to open a new path for the transfer of Fujisawaism and the HPS to North America. As Japanese employees sent to Ohio returned to Japan (replaced by new people appropriate to the developing operations in North America) there was now a substantial group in Japan with overseas experience. Senior managers were already seeking innovations in North America for transfer to Japan. The native individualism of American workers, consistent with Fujisawaism, was viewed as a positive force that Honda could channel productively.

### 5.4. THE HONDA MODEL IN QUESTION, 1992+

The early 1990s saw simultaneous problems for Honda in Japan, North America, and Europe, owing in part to recessionary conditions. In Japan sales declined from 686,000 vehicles in 1990 to 548,000 vehicles in 1994, in the USA from 854,000 in 1990 to 716,000 in 1993, and in Europe from 175,000 in 1992 to 160,000 in 1993. Turnover fell sharply in 1993–4. Declines in unit sales were compounded by a sharp rise in the yen's value that put severe pressure on costs incurred in Japan. Profits declined to 0.6 per cent in 1994. Moreover, in North America Honda was beset by criticism of its cars' local content during the 1992 American presidential elections, and by the public emergence of the early 1980s dealer bribery scandal. Then in 1994 British Aerospace suddenly sold Rover Group to rival car maker BMW. In addition to these short-term problems, many Western competitors had transformed their own industrial models by the early 1990s. The Japanese monopoly over quality combined with low cost disappeared. Further, Toyota had now caught up with Honda in volume of investment in North America. Finally, sales of the niche sports and luxury cars where Honda had staked future growth had declined most significantly. Honda's industrial model was faced with a serious test.

#### 5.4.1. Japan

With a shrinking Japanese market and unprofitable exports (Honda's exports were reduced from 715,000 in 1990 to 434,000 in 1995), the key issue became cutting costs and production capacity. By 1993 a shift at one Sayama line had been cancelled and 'non-production' days instituted. By 1994 a shift at Suzuka had also been cancelled. Some workers were reassigned to make KD motorcycle kits to meet growing demand in Asia. In 1993 it was announced that employment would fall by 3,000 over three years. Yet sustaining the Honda industrial model over the medium term required sales to rise in Japan. There were a series of overoptimistic forecasts of domestic sales during the early 1990s. In 1994 a target was set for sales to rise sharply to 800,000 by 1997. The long-term solution was therefore to be full capacity production of 1 million vehicles per year, with one-fifth of these (low-volume high-value cars) exported. New model launches were to permit the target to be achieved. Meanwhile, cost-cutting led to price cuts; prices for a 1994 Today variant were 20 per cent lower than for its predecessor.

While Japan remained the heart of Honda research, engineering, product development, and strategic decision-making, it was now declining as a manufacturing centre. By 1995 when sales and production abroad recovered somewhat, a milestone was reached that had not been expected until the late 1990s; Honda now made less than half its vehicles in Japan. Moreover, an increasing proportion (over 10 per cent) of sales in Japan were now imports from HAM: Accord four-door, coupé, and estate, and Civic coupé. The new exchange rates made these sales more profitable than vehicles made in Japan.

### 5.4.2. North America

After a decade of growth, by the early 1990s a plateau was reached in North America. The East Liberty plant could not at first reach full output (one shift was not hired). However, by 1994 demand recovered, imports from Japan fell, and East Liberty prepared for full capacity. Further transfers of activity to the low-cost production sites in North America were planned. All three assembly plants would increase their capacities, to total 720,000 by 1997 (up from 610,000). Anna's capacity would rise to 750,000 engines. A new 15,000 unit assembly plant in Mexico (for Latin American markets) was announced. Local content was to be increased with new investments to build high-precision parts still imported from Japan. Local content of engines, for instance, would rise to 90 per cent by 1998. Finally, exports were to rise even higher than the 105,000 cars (20 per cent of output) already sent from Marysville and East Liberty (not including exports to Canada) in 1994, mostly to Japan.

Research and development capabilities in North America continued to deepen. By 1993 a new Acura model was being developed: not merely a body variant but a new model (though most new Honda models had high commonality with other models). By 1994 the absence of recreational vehicles and minivans was finally addressed by a rebadged Isuzu vehicle and launch of Honda's own Odyssey minivan. The expansion of Anna would now permit production of six-cylinder engines (see below on product policy). Yet HAM faced potential problems. Political pressure remained high. Many Marysville workers were now in their mid-forties: older, less enthusiastic, and slowing physically. Yet Honda appeared to believe that its competitive advantage over other automobile producers—and therefore profits—remained greatest in North America.

### 5.4.3. Europe

Experienced employees from the Swindon engine plant were switched into car assembly when the Europeanized Accord (based on the 1992 Japanese Ascot Innova) and five-door Civic (based on the 1992 Japanese Domani) models were introduced from 1992 and 1994. Large runs of each car were made (e.g. half a shift of Accord followed by half a shift of Civic, neither model having body variants). The plant used the HPS small batch system with lots of sixty (e.g. engine variants) broken down into thirty for different colours. The focus continued to be on disciplined manual work, with rotation within teams but not among them. NH circles were introduced in 1996, seven years after manufacturing started.

Only collaboration with Rover Group made HUM, with its 100,000 capacity, economic. Versions of both new Honda models were made by Rover Group, with whom most suppliers were shared. The purchase of Rover Group by BMW therefore posed questions about the viability of HUM. However, plans were made to add a third model at Swindon in 1997 (Civic estate) and increase capacity to 150,000 units (half of projected sales in Europe). By 1995 local content had risen to 90 per cent in response to yen appreciation and the presence of good independent suppliers in Europe (unlike North America where local content remained lower). Any future small decline in European local content would result from components imports from North America, not Japan.

Difficulties with product design for Europe persisted. Few resources were devoted to modifying cars for Europe since spin-offs from Japanese cars were considered adequate. Yet across Europe Honda products were viewed as competent but uninteresting; the bland Accord had to be given a facelift after three years. The range remained inconsistent, with variants of Civic (four-door, three-door, and coupé) and Accord (estate and coupé) imported from Japan and North America where they no longer resembled the variants made at Swindon.

### 5.4.4. Asia and Latin America

Honda sales and production in other regions, particularly the ASEAN countries, accelerated in the early 1990s. Thailand became the hub of a new Honda ASEAN production complex, with a new 100,000 capacity assembly plant (and R&D offices) from 1996 supported by parts production at former KD factories in Thailand, Malaysia, the Philippines, and Indonesia as Honda spread its investments for political reasons while creating a division of labour to permit economies of scale.

New KD assembly plants were opened in Pakistan (1994, initial capacity 6,000) and Turkey (1996, initial capacity 20,000), with plans to build a factory in India too. All these Asian investments were joint ventures with local partners. Honda R&D now designed an 'Asian Car' (a simple, cheap Civic version) for production in Thailand. If the ASEAN production complex were to support the KD plants in Turkey, Pakistan, and India, there would be a fourth Honda 'world region': Asia. Total sales in the Asia region surpassed 170,000 in 1994. Given the pace of growth and Honda's competitive position relative to non-Japanese producers until the East Asian economic crisis, the region seemed likely to become more significant to Honda than Europe. It remained to be seen how the geographical decentralization of supply chains across Asia would affect the HPS. However, Honda had long experience with both KD exports and global transport of components. The HPS push system might prove well suited.

## 5.4.5. Using the Global Local Corporation

It was in part a balanced geographical portfolio that permitted Honda to avoid the worst impacts of recession in Japan. By 1995 two-thirds of sales were outside Japan. Moreover, exports from the low-cost production bases in North America created high unit profits to balance losses elsewhere. Products and production bases in each region could now meet demand elsewhere, both to fill market niches and to counteract currency fluctuations. Europe, for instance, received low volume products from Japan and North America to complement high volume European products.

All this was predicated on the HPS flexifactory principle. An example is revealing. As late as 1993 the 1994 Odyssey minivan was expected to be made in North America where demand for minivans was highest, and the 1993 Accord, whose platform it shared, was built. However, demand for station wagons, minivans, and recreational vehicles rose rapidly in Japan in 1993–4. The Accord station wagon (made at Marysville) sold better in Japan than in North America. Accordingly, the Odyssey was built at Sayama where it used idle capacity. Not until 1998 was a minivan plant opened in Canada.

Mr Kawamoto reformed Honda's structure in steps to manage the global local corporation. In 1991 new product-oriented divisions were created to align product strategy. In 1992 the automobile division was regionalized (four world regions, each with substantial strategic and operational autonomy). In 1994 the motorcycle and power product divisions were regionalized and linked to the four automobile regions: Japan, North America, Europe, and Asia.

## 5.4.6. Product Focus

Notwithstanding Honda's carefully fostered technology image, commercial success had always rested on well-designed and well-made mass-market products: SuperCub (1958), Civic (1972), and Accord (1976) in particular. By 1994 Civic and Accord still accounted for half of Honda's unit sales in Japan and three-quarters in North America and Europe. Yet the company continued to launch innovative automobiles, from the 1300 to the CVCC Civic to the four-wheel-steering Prelude to the aluminium NS-X, each time trying to develop and exploit market niches associated with technology (especially in sporty cars like Prelude, CRX, Beat, and NS-X). Moreover, Honda was quick to introduce new technologies into its mass market products, like transverse-mounted FWD, aluminium engines, sixteen-valve engines, electronic injection, double wishbone suspensions, antilock brakes, always with systems designed by Honda R&D.

The financial crisis brought the relationship between mass products and technology-related products into question since the mass products no longer financially supported advanced research projects and some advanced and niche products were now losing considerable sums of money. The Beat, sold only in

Japan, was a 'bubble economy' over-indulgence (as was the NS-X) and sales collapsed at the end of the boom. Indeed combined Japanese sales of the Legend, NS-X, Beat, and CRX fell from 57,000 in 1992 to 8,000 in 1994. Combined European sales of the same models collapsed from 28,000 to 12,000. Moreover, Honda R&D had steadfastly refused to enter the growth niche for recreational vehicles in North America. Parts of Honda's industrial model, particularly the domination of Honda R&D and product technology, had led to major strategic problems, which the SED process had proved incapable of correcting. Fortunately, other elements of the model, Fujisawaism, the HPS, and the global local corporation, prevented more severe financial problems.

The early 1990s therefore saw an urgent rethinking of Honda's product strategy, with debate about the role played by Honda R&D. When it came, the change was swift. Production of the Beat at parts maker Yachiyo Industry was stopped in 1992. A mid-engined sporty Integra for 1993 was abandoned during development. A relationship to Ferrari was announced in 1993 to recoup some of the NS-X investment through exchange of engineering know-how. There was a rapid entry into the recreational vehicle and minivan niches, particularly when they began to expand quickly in Japan. By 1994 Honda was rebadging two large Isuzu off-road vehicles for sale in Japan (supplying cars to Isuzu, which stopped making them, in return). A Land Rover off-road vehicle was imported to Japan from Rover Group for sale as a Honda model. By 1995 two other Isuzu off-road vehicles built in the USA were sold as Honda products in North America. Honda R&D now developed the Odyssey minivan, based on the 1993 Accord, which soon proved popular in Japan. By 1996 a new small Honda off-road vehicle, the CR-V, was launched, to complement the rebadged Isuzus and Land Rovers. All this flowed from a new so-called 'market in, product out' philosophy meant to ensure that Honda products were driven less by product technologies than by market demand. Honda also started to purchase diesel engines (from Rover Group in 1995) for assembly into cars at Swindon, and debated dropping the long favoured but expensive 'double wishbone' suspension system from some cars.

Simultaneously, new cars for the Japanese market continued to be launched as spin-offs from existing models. By 1994 Honda sold nineteen different models in Japan, compared to eight in Europe and eight in North America. A decline in gross model/platform economies of scale for Honda as a whole from 136,000/176,000 in 1990 to 82,000/123,000 in 1994 was due to the addition of low volume models in Japan (where the respective figures declined from 95,000/125,000 to 60,000/90,000). However, the search for new economies of scale continued. 'Commonization' of components was not new to Honda which had 'spun off' new models from existing ones since the 1981 Ballade. But it reached new heights in vehicles that looked quite different to the consumer. The 1992 Domani shared 60 per cent of parts with other models and the 1993 Today shared 40 per cent of parts with its predecessor. The Odyssey shared 50 per cent of parts with the 1993 Accord despite a different wheelbase and an entirely different body shape. Now even the jigs on automated equipment (previously changed for new models) were designed for reuse. Investment on production equipment for the 1993 Accord was halved over previous levels. When the Accord was changed at Marysville in 1993 the gap between production of old and new models had been reduced to two hours.

On the other hand, Honda continued to invest in new mechanical and electronic technologies. New versions of the VTEC engine were announced in 1993 and 1995, offering better relationships between power and fuel economy and pollution standards that met strict 1997 California ULEV criteria. Honda withdrew from Formula One racing in 1992 and announced that the racing engineers would be assigned to electric vehicle projects. The first electric concept car was built in 1994, as was a car solar panel that boosted engine power by 2 per cent. Innovations in electronic controls flowed: 'drive-by-wire' on NS-X, automatic transmissions that 'hunted' for the correct gear, and a new electronic navigation system.

### 5.4.7. The Model in Crisis?

Did the Honda industrial model go through a real crisis in the early 1990s? The crisis was real, but was due largely to one vital element; other elements in fact helped Honda overcome the crisis. Most significant is that the element in crisis was the one most associated with Mr Honda: the self-image of a company led by product technology and sporty vehicles. Fujisawaism with its organizational creativity and flexibility, the HPS with its flexible mass manufacture of high quality cars, and the global local corporation with competitive advantage from global interaction between local production and marketing bases all triumphed.

Could the Honda model as a whole continue to function dynamically without strong high-technology challenges that engineers could meet in the form of automobiles that succeeded in the market? Was a continued search for 'holy grail' technologies needed to make the Honda model work? Notwithstanding Honda's leading technological position in the world automobile industry, it remained to be seen whether the Honda R&D engineers could at last create a 'breakthrough product' which would translate this leadership into significant competitive advantage; there were now strong echoes of the late 1960s, which produced the CVCC engine. If there was to be no such breakthrough, a key element of the industrial model that had guided Honda's trajectory for fifty years would remain unproven.

# NOTES

1. This organizational model has not previously been characterized as such, but its significance merits this label; see Mair (forthcoming *a*), Mito (1990), Pascale (1990), Sakiya (1987). Within the company it is called Hondaism.
2. This phase of Honda's trajectory is examined in considerable detail in Mair (1994*a*; 1997).

## STATISTICAL APPENDIX 5: Honda

| Year | Worldwide vehicle sales[a] | Vehicle production in Japan[a] | Exports from Japan[a] | Vehicle Production North America Europe[a] | Net sales plus operating revenue[b,c] | Proportion of turnover abroad[c] | Profits[b,c] | Profits/ turnover[c] | Employees Japan[d] | Employees worldwide[d] |
|---|---|---|---|---|---|---|---|---|---|---|
| 1967 | — | 150 | — | 0 | — | — | — | — | — | — |
| 1968 | — | 319 | — | 0 | — | — | — | — | — | — |
| 1969 | — | 365 | — | 0 | — | — | — | — | — | — |
| 1970 | — | 393 | 23 | 0 | — | — | — | — | 17,500 | — |
| 1971 | — | 309 | 34 | 0 | — | — | — | — | 18,100 | — |
| 1972 | — | 331 | 38 | 0 | — | — | — | — | 18,300 | — |
| 1973 | — | 355 | 74 | 0 | — | — | — | — | 18,300 | — |
| 1974 | — | 429 | 120 | 0 | — | — | — | — | 18,500 | — |
| 1975 | — | 414 | 201 | 0 | — | — | — | — | 18,500 | — |
| 1976 | — | 560 | 307 | 0 | — | — | — | — | 19,100 | — |
| 1977 | — | 665 | 445 | 0 | — | — | — | — | 20,000 | — |
| 1978 | — | 743 | 496 | 0 | — | — | — | — | 21,000 | — |
| 1979 | — | 802 | 547 | 0 | — | — | — | — | 20,800 | — |
| 1980 | — | 957 | 659 | 0 | — | — | — | — | 22,900 | — |
| 1981 | — | 1,009 | 668 | 0 | — | — | — | — | 25,000 | — |
| 1982 | — | 1,015 | 613 | — | — | — | — | — | 26,900 | — |
| 1983 | — | 1,031 | 613 | — | — | — | — | — | 28,000 | — |
| 1984 | — | 989 | 633 | — | — | — | — | — | 26,900 | — |
| 1985 | 1,283 | 1,120 | 668 | 145 | 2,740 | 73 | 129 | 4.7 | 28,100 | 55,700 |
| 1986 | 1,365 | 1,236 | 708 | 238 | 3,009 | 70 | 146 | 4.9 | 30,700 | 59,000 |

| Year | | | | | | | | | | |
|---|---|---|---|---|---|---|---|---|---|---|
| 1987 | 1,585 | 1,241 | 695 | 341 | 2,961 | 68 | 84 | 2.8 | 29,700 | 63,000 |
| 1988 | 1,727 | 1,293 | 681 | 416 | 3,229 | 64 | 99 | 3.1 | 30,000 | 65,500 |
| 1989 | 1,903 | 1,363 | 685 | 446 | 3,489 | 63 | 97 | 2.8 | 31,200 | 71,200 |
| 1990 | 1,936 | 1,384 | 715 | 540 | 3,853 | 66 | 82 | 2.1 | 31,600 | 79,200 |
| 1991 | 1,915 | 1,358 | 662 | 550 | 4,302 | 68 | 76 | 1.8 | 31,500 | 85,500 |
| 1992 | 1,961 | 1,200 | 589 | 563 | 4,392 | 67 | 60 | 1.4 | 31,100 | 90,500 |
| 1993 | 1,793 | 1,151 | 571 | 537 | 4,132 | 67 | 37 | 0.9 | 31,000 | 90,900 |
| 1994 | 1,753 | 993 | 510 | 645 | 3,863 | 67 | 24 | 0.6 | 30,600 | 91,300 |
| 1995 | 1,794 | 967 | 434 | 765 | 3,996 | 67 | 62 | 1.6 | — | 92,800 |
| 1996 | 1,887 | 1,092 | 371 | 884 | 4,252 | 64 | 70 | 1.7 | — | 96,800 |
| 1997 | 2,184 | 1,307 | 544 | 921 | 5,293 | 65 | 221 | 4.2 | — | 101,100 |

*Notes*:

[a] 'Vehicle' refers to all passenger cars (including minicars) and commercial vehicles. Production and sales figures are 1,000 units. Production in Japan and exports from Japan figures are on calendar year basis. Production abroad figures are on calendar year basis and refer only to USA, Canada, and UK (excluding production for Honda by Rover).

[b] Net sales, operating revenue, and profits are in billion yen.

[c] Figures for sales by value, operating revenue, turnover abroad, profits, and employees refer to Honda as a whole (including motorcycles and power products). Figures for worldwide sales, revenue, and profits, are for the fiscal year ending that year. All figures for the fiscal year 1987–8 are estimated as follows: 12/13 of seven months to 30 September 1987 plus six months to 31 March 1988.

[d] Figures for employment refer to Honda as a whole. Employment figures in Japan do not appear to include affiliates and sales staffs. Worldwide employment figures may be calculated by Honda on a different basis (i.e. may include sales companies).

*Source*: Japan Automobile Manufacturers Association, Honda Annual Reports, with thanks to Koichi Shimizu. The data are incomplete but allow the best comparisons possible.

# BIBLIOGRAPHY

Abbeglen, J. C., and Stalk, G. Jr., *Kaisha: The Japanese Corporation* (New York, 1985).

Bélis-Bergouignan, M.-C., and Lung, Y., 'Processus de diversification et flexibilité productive dans l'industrie automobile japonaise', *Actes du GERPISA*, 12 (1994).

Jetin, B., 'The Historical Evolution in the Variety of Car Supply: an International Comparison', GERPISA manuscript (1995).

Mair, A., *Honda's Global Local Corporation* (Houndmills, 1994a).

—— 'Honda's Global Flexifactory Network', *International Journal of Operations and Production Management* 14/3 (1994b), 6–23.

—— 'Honda Motors: A Paradoxical Approach to Growth', in C. Baden-Fuller and M. Pitt (eds.), *Strategic Innovation: An International Casebook* (London, 1996).

—— 'Strategic Localization: The Myth of the Post-National Enterprise', in K. R. Cox (ed.), *Putting Space in its Place: Globalization and its Politics* (New York, 1997).

—— 'The Honda Production System', manuscript (1998).

—— 'Learning from Honda', *Journal of Management Studies* (forthcoming a).

—— 'Local Human Resource Management in Global Context at Honda of the U.K. Manufacturing, *Employee Relations* (forthcoming b).

—— Florida, R., and Kenney, M., 'The New Geography of Automobile Production; Japanese Automobile Transplants in North America', *Economic Geography*, 64 (1988) 352–73.

Mito, S., *The Honda Book of Management* (London, 1990).

Nonaka, I., 'The Knowledge-Creating Company', *Harvard Business Review*, November–December (1991), 96–104.

Pascale, R. T., 'Perspectives on Strategy: The Real Story behind Honda's Success', *California Management Review*, 26/3 (1984), 47–72.

—— *Managing on the Edge* (Harmondsworth, 1990).

Sakiya, T., *Honda Motor: The Men, the Management, the Machines* (Tokyo, 1987).

Shook, R. L., *Honda: An American Success Story* (New York, 1988).

# 6

# *The Unique Trajectory of Mitsubishi Motors*

KOICHI SHIMIZU AND KOICHI SHIMOKAWA

The Mitsubishi Motors Corporation (MMC) is a member of the Mitsubishi group (formerly *zaibatsu*), so most of the company's investments have been financially supported by members of the group. However, this means that interest payments have sometimes weighed heavily on MMC's profitability. MMC produces all types of vehicles: passenger cars, commercial vehicles, and industrial vehicles as well as minicars. The company's diversity saved MMC when it ran into difficulties during the transition to mass production of passenger cars. From the latter half of the 1980s, MMC pursued a balanced course of development having at last discovered how to combine technological and commercial innovation in the design of its products so as to meet market demands.

## 6.1. MMC PRIOR TO THE FIRST OIL CRISIS

Mitsubishi Heavy Industries (MHI) had begun building automobiles in 1918, when it made twenty-one units adapted from Fiat's Tipo model. The *zaibatsu*'s companies produced tanks, battleships, bombers, and fighters during World War II. These facts accounted for the dismantling of the Mitsubishi *zaibatsu* following Japan's defeat, with MHI being divided into three companies, each sited in a different region. Two of the companies focused on the automotive industry. Mitsubishi Nippon Heavy Industries, sited in the east, produced trucks and buses under the marque Fuso, while New Mitsubishi Heavy Industries (NMHI), sited in the central region, produced scooters, three-wheelers, automobiles, Willis Jeeps, and commercial vehicles in its two factories at Nagoya and Mizushima, and Mitsubishi engines at its Kyoto factory. In 1964 these three companies merged to form the new Mitsubishi Heavy Industries, which adopted an organizational form of divisions based on product lines: naval, engineering, automotive, aeronautics, and special vehicles.

In the late 1960s, MHI launched a minicar, Minica, the sales of which rose rapidly between 1964 and 1970. However, the first passenger car, the Colt, was not a success. Having learned from this failure, MHI designed a new model, the Colt Galant, with a new engine (Saturn) which met Japan's 1966 anti-pollution regulations. Launched in 1969, this model went into mass production, at 10,000 units per month. Yet at this point MHI remained principally a producer of minicars and commercial vehicles.

During the late 1960s some producers sought to forge links or even to merge, as a means to resist equity participation by foreign companies now that the country's economy would be open to foreign capital. In 1968 MHI decided to merge with Isuzu, in the context of a merger of their respective banks, Mitsubishi Bank and Daiichi Bank. The failure of the bank merger in 1969 led MHI and Isuzu to abandon their own plans. Meanwhile, Chrysler was also proposing future cooperation with MHI. Besides seeking to make headway in the Japanese market, the American company also wanted to produce small automobiles with the aid of MHI. In 1969 the two companies decided to form a joint subsidiary, Mitsubishi Motors Corporation (MMC), and signed a distribution contract for MMC's products in the USA and other markets through Chrysler's sales networks. In the end Chrysler's participation was limited to 15 per cent of MMC's capital, a result of its own deteriorating finances caused by its European investments. Hence, MMC was founded as an American–Japanese joint venture with the capacity to produce 500,000 vehicles per year and 20,095 employees at three factories: the Kawasaki factory which specialized in the production of Fuso commercial vehicles, and the factories at Nagoya and Mizushima which produced passenger cars (Debonair and Colt Galant), commercial vehicles, and minicars.

MMC's production areas had been converted from aeronautical to automotive use. In terms of production equipment, Mitsubishi turned to the most recent methods such as automated guided vehicles to transport vehicle bodies and a computerized system, installed in 1966, to manage the assembly lines (ALC: assembly-line control). However, lack of space at the Nagoya plant, for example, made linkages between the different parts of the assembly line difficult and caused efficiency losses. Work was organized into two shifts: the day shift worked six days per week and the night shift five days. Workers were trained to be polyvalent, though rotation of work posts was not the rule. By the late 1960s, for example, employees worked on several machine tools simultaneously in the mechanical components production area at Mizushima. There had been a *kaizen* system since 1954, though quality circles were not introduced until 1965. As was the case in other Japanese companies using the same methods, improvements only involved equipment that had already been installed, whereas responsibility for the design of larger scale technical innovations was assumed by engineers.

When the company was formed, MMC's union had been separated from that of MHI. Yet until 1979 the two unions cooperated in negotiations with MMC and MHI in order to obtain the same conditions of work. The crisis in the shipyards then made it difficult to pursue identical wage policies. While MMC's union sought to cooperate with management it retained a strong position in negotiations. As a consequence, the reduction of working hours was obtained more quickly here than at other automobile producers. Hence a five-day working week with an official 1,960 hours worked per year was adopted in 1973.

MHI had continued to make strong distinctions between blue-collar workers (paid by the day) and white-collar workers (monthly salary), even though this differentiation had been abolished in most companies as part of the democratization of industrial relations in the period immediately following World War II. Creation of the new MMC brought with it the elimination of this distinction; henceforth all employees had the same status. Instead they were organized into hierarchical functional groups: management, administration, engineering, staff, production, specialized tasks, and medical staff. Within each function, employees were classified hierarchically, by levels which reflected their competencies and seniority; there were five levels for production workers, for instance, and two for supervisors.

The wage system was also revised. From 1971 monthly salaries (not including overtime pay) were comprised of a part that was the individual basic salary (*honkyu*, HK), a part based on attendance (*kinmukyu*, RA), a part based on skill or ability (*shokunokyu*, RCQ) and various allowances. The annual salary increase took place in April at the time of the *shunto* (spring wage negotiations). Individual salary increases were based on two evaluations (the two *satei*): the *satei* for the employee's activities determined individual basic salary increases, attendance-related pay, and the annual bonus; the *satei* for ability determined ability-based pay and prospects for promotion. The *satei* therefore played a vital role in getting employees involved in their tasks: a common thread running through Japanese companies.

MMC differed from other companies in its training policies. From 1970, MMC has employed holders of secondary education diplomas as factory workers. They were trained internally for one year before assignment to a work post. MMC workers therefore possessed advanced technical skills, which was one source of the company's high productivity.

In 1970 MMC suppliers formed an association, *Kashiwa Kai* composed of 227 companies supplying the passenger car division and 124 supplying the commercial vehicle division. Some of them had strong links to the Mizushima factory. In 1962 they had formed a producer cooperative in order to modernize their equipment, raise their level of technology, improve working conditions, and recruit their workforce collectively. In 1966, the cooperative was installed in an industrial park *Sojya Danchi* built close to the Mizushima factory. The factory and the producer cooperative collaborated so closely at the technical level that the former looked upon the latter as one of its own factory areas. By 1968, MHI was already requiring suppliers to control and guarantee the quality of the parts they supplied, so that the company itself would no longer have to control the quality of components received from its suppliers. To do this MHI periodically issued suppliers with certifications, following inspection of their work processes and after assisting them when necessary.

Mitsubishi Motors Sales (MMS) had 114 dealerships, organized into two networks: a network for passenger cars and minicars, and a network, Fuso, for trucks and buses. In 1966, these two networks were merged, but they were separated again two years later, an instability which reflected the commercial difficulties Mitsubishi was encountering. At its inception, MMC sought to establish close

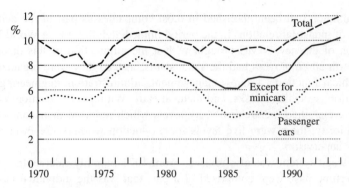

FIG. 6.1. *Mitsubishi Motors Corporation: market share*
Sources: Mitsubishi Motors Corporation (1993) and Nikkan Jidosha Shinbun.

collaboration with MMS, by jointly deciding an annual programme and long-term strategy. Despite these efforts, sales of passenger cars remained lacklustre. The Colt Galant, launched in 1969, was not really able to break into a market dominated by Nissan's Bluebird and Toyota's Corona. Even though new variants were introduced, sales of the Colt Galant/New Galant still did not improve. Nor was the Lancer, introduced in 1973, capable of competing with Nissan's Sunny and Toyota's Corolla. Moreover, the sales of the Minica declined steadily as demand for minicars was displaced into the cheaper passenger cars. While in 1972 MMC had a model for each major market segment—Debonaire (upmarket model), Colt Galant (mid-range), Lancer (mass market), and Minica (mini)—sales of its passenger cars (including minicars) remained stagnant until the first oil crisis (see Fig. 6.1 and Statistical Appendix 6).

MMC's production system seemed to be a combination of Fordism, Japanese techniques for managing work, and more specific company traits (a higher technological level inherited from aeronautical production as well as a highly skilled workforce). Yet these coexisting elements did not result in a coherent and viable model. Inconsistencies between the elements of the MMC socio-productive configuration were reflected in a high degree of instability of top management. The growth of the company slowed in line with the progressive decline of the market for minicars. MMC was temporarily overtaken even by Honda, a late entrant into the automobile industry, in the market for passenger cars, as MMC's market share remained below 6 per cent. The company's indebtedness, owing to its poor capacity for self-financing (around 10 per cent), weighed heavily on its finances. Consequently, the rate of profit after taxes fluctuated around 0 per cent (see Fig. 6.2). The company's viability over the medium term depended in part on its ability to design a passenger car that could be sold in large numbers.

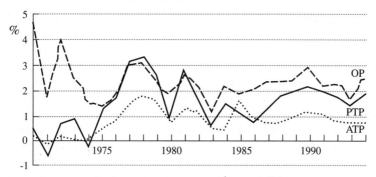

FIG. 6.2. *Profits as a proportion of turnover at Mitsubishi Motors Corporation*
*Notes*: OP: operating profit; PTP: pre-tax profits; ATP: after-tax profits.
*Sources*: Mitsubishi Motors Corporation (1993) and its Annual Reports.

## 6.2. THE END OF A LONG TUNNEL, 1975–1994

It was after the first oil crisis that production of passenger cars took off at MMC, rising from 287,000 units in 1975 to 519,000 in 1978. MMC recorded the highest profits in its history and improved its financial position. Yet production dipped again between 1980 and 1986 because of a reduction in passenger car sales in the Japanese market, although (weak) profitability was maintained by increased sales of commercial vehicles and exports. After 1989, there was more balanced growth of sales of passenger cars, commercial vehicles, and minicars, even though the Japanese economy was in deep recession after the bursting of the bubble economy. MMC became the number three producer in Japan.

Passenger car sales in Japan picked up again in 1975, after the first oil crisis, and then increased rapidly until 1979 owing to the success of two new models. Other producers had difficulties in rapidly designing engines which accorded with anti-pollution norms and were energy efficient. From 1972, MMC had possessed an engine, the improved Saturn, which met the anti-pollution regulations established in 1973. Strengthened by this progress, by 1976 MMC had developed two new engines (G11B Orion and G32B Saturn) which met the 1978 norms. And yet sales of the New Galant and Lancer models continued to be poor because they did not really correspond with customer tastes despite being technically advanced and of high quality. The new Galant Σ/Λ, launched in 1976, and its new version, Eterna, marketed in 1978, did meet with success. Sales of the Galant actually outstripped those of Nissan's Bluebird. During the 1976–9 period 750,000 units were produced, making the Galant the first MMC car to be mass produced. The Galant and the Mirage (the company's first front-wheel-drive model, launched in 1978) enabled MMC to increase its share of the passenger car market in Japan from 5.7 per cent to 8.6 per cent between 1975 and 1979. By the latter half of the 1970s, MMC had truly established itself in the passenger car market.

This improvement was also due to a strengthening of the sales network, increased production capacity, and modernization of the factories. By 1978 MMC had three networks: the Galant chain, the Car Plaza chain for the Mirage, and the SPD (Single Point Dealer) group in places where there had not been Galant chain sales points. A new factory with production capacity of 15,000 passenger cars per month was built at Okazaki. At this time it had the highest level of automation in Japan, with ten robots in the paint shop, seventy-one robots in the body shop (88 per cent automated), and automatic marriage of engine to body. The old factory at Oye was rebuilt in 1979 to install a new type of assembly line; subassembly lines were directly connected to the main assembly line, and operators followed the main line along by standing on a parallel conveyor belt (obviating the need to work while walking along). This all occurred more than ten years before Toyota adopted similar ideas. By 1978 MMC was using *kanbans* to order components, as an economical means to manage mixed production on the assembly lines. That same year the Mizushima factory began assembling the Lancer and Mirage models on the same assembly line (though welding lines continued to specialize in one model). With the construction of the engine factory in 1979 at Shiga, MMC became a flexible mass producer.

The size of the workforce remained remarkably stable despite the rapid growth of production. This was a response to the uncertainties which followed the first oil crisis, and was achieved through longer overtime, automation, and rationalization of the production system.

And yet the early 1980s were very difficult years for MMC. Customers showed less and less interest in the Galant/Eterna and Mirage models. Their sales in the domestic market dwindled from 210,000 units in 1980 to 112,000 in 1986, thus cutting MMC's share of the passenger car market from 7.9 per cent down to 3.7 per cent. As far as the domestic automobile market was concerned, MMC was in crisis.

This reduction was largely compensated by export of passenger cars and the sales of mini commercial vehicles in the domestic market. Indeed exports grew to 77 per cent of passenger car production in 1986 (following a dip in sales in 1982 and 1983, a consequence of the crisis at Chrysler and the self-imposed limitation of exports to the USA). As a result, MMC's annual production averaged 1 million vehicles during the early 1980s.

Yet MMC continued to face problems in designing passenger cars. Accordingly, the company reorganized its design activities in 1985. Teams were formed around each planned new model. Each was headed by a product manager and collaborated with both the design department and the experts in production cost management. To consolidate its financial structure, MMC merged with MMS and terminated its contract with Chrysler, giving 20 per cent of its capital to Chrysler in 1985, and placing its own shares on the Japanese stockmarket in 1988. The company could now attract funds from capital markets to improve its financial position.

MMC's sales in the Japanese market grew in parallel with the economic boom of the bubble economy period during the late 1980s, and continued to grow throughout the long recession which began in 1991. In 1994, MMC recorded the highest operating profits in its history, thanks to the success of its passenger car and commercial vehicle models.

At the lower end of the range, a new Mirage (1300–1800 cc) was introduced in 1987, and the Lancer, which used the same platform as the Mirage, was given its own platform in 1990. In the middle of the range, the new Galant introduced in 1988 was equipped with a double overhead camshaft (DOHC) engine, an active suspension, and four-wheel drive. According to the MMC president, its radical shape truly distinguished the car from its competitors (the Bluebird and Corona). As with the lower end automobiles, the Eterna and Galant models no longer shared the same platform. With this differentiation strategy, the Galant/Eterna series proved very successful: a 'miracle which gave MMC the opportunity to take off again' (Mitsubishi Motors Corporation 1993: 303). At the top of the range, in 1990 MMC launched the Diamante/Sigma models, which also shared their platform. The luxury market was being stimulated at this point by a change in the system of taxation (replacing the tax on vehicles, 23 per cent of sales price, by a 6 per cent consumption tax), and by a reduction of 27.7 per cent in voluntary automobile insurance payments. The 2,500 cc version of the Diamante/Sigma series sold particularly well. Of models categorized as commercial vehicles, the Pajero, launched in 1982, replaced in 1991, and which permitted the driver to select either two-wheel drive or four-wheel drive, and the Delica, a 'box car' equipped with a four-wheel-drive system in 1981, were introduced into the niche 'recreational vehicle' (RV) market which was to develop rapidly during the 1990s.

MMC's excellent performance was the result of a successful marriage between advanced technology and innovative model designs. MMC was responsible for creating the market for RVs and automobiles with 2,500 cc engines. MMC also pursued the flexibilization and automation of its production system. The new body plants installed between 1990 and 1992 were able to weld several different bodies: at the Oye factory the Diamant/Sigma models, at Mizushima the Mirage/Lancer, and at Okazaki the Galant/Eterna. The Mirage/Lancer line, entirely automated with 321 robots, was capable of welding ten different bodies. The automation of assembly had started in 1982 for the installation of seats, windows, and tyres. In 1988, MMC began the automation of the trim line, while by 1995, 32 of 107 tasks had been automated: installation of tyres, headlights, windscreen wipers, gear lever, and so on. To facilitate assembly work, components containers which move along parallel to the assembly line were used from 1980. Doors-off assembly began in 1988. The automation of an operation generally took place after possibilities for improving existing arrangements had been exhausted. MMC itself manufactured most of the robots and equipment used in its factories. The priority given to automation was justified by a desire to eliminate tedious and dirty tasks while at the same time increasing productivity.

In the factories, workers were organized into groups of twenty to forty employees with a group leader (*ko-shi*). The group was divided into two teams, each of which was led by an under-group leader (*fuku-ko-shi*), and it worked in two shifts. The teams worked alternate days (8.10–17.00) and nights (21.15–6.30) with an hour and a half of overtime per team. Task rotation within teams was introduced in 1986 to make employees polyvalent and in order to reduce fatigue. At the Mizushima factory workers rotated posts every two or four hours. There were also *kaizen* groups composed of ten members which made minor improvements but which could work on the assembly line if production had to be increased.

Seventy per cent of parts were delivered by suppliers. The *kanban* system was no longer used except in the body parts stamping department, although the company abided by some of the principles of just-in-time: reducing production costs by reducing stocks, smoothing of production, supply of small and frequent batches. The weekly production plan was established by the head office in Tokyo. Upon receipt of the production plan, the factory ordered parts from suppliers and started the production process, making minor adjustments in the last two days. The parts were then delivered directly and regularly to work posts, or to a warehouse where workers converted them into small batches to be taken to the work posts. Quality was not controlled at this point, having previously been checked by the suppliers.

By the end of the 1980s, social tensions had begun to appear. Strict control over growth of the workforce reduced opportunities for promotion. The average age of employees rose from 30.8 to 38.0 years between 1970 and 1989, yet the number of supervisory and management positions had not increased. Many candidates for these posts were unable to attain them, often remaining at level 5 for workers (just under *fuku-ko-shi*) and level 1 (*huku-ko-shi*) for supervisors. The lack of opportunities for promotion reduced the motivation of employees. To deal with this problem, in 1991 MMC established a new hierarchical organization which separated rank from hierarchical position (see Fig. 6.3). Rank depended upon an employee's competency and experience, whereas hierarchical position remained related to posts available. These two classification systems operated separately, in parallel.

The new organizational form was accompanied by a change in the wage system. The new system included the basic salary (BS), rank-related pay (RRP), pay related to hierarchical position and individual activities (FAP), and allowances (A). The standard salary was made up of BS (45 per cent) + RRP (20 per cent) + FAP (30 per cent) + A (5 per cent). The BS, considered to be a salary based on seniority, increased in line with increases in the cost of living (calculated by age), as well as by rank. The rise in BS was therefore cumulative. The RRP was fixed according to rank. The FAP added together a payment by hierarchical position and a non-cumulative increase determined by the evaluation (*satei*) of the employee's activities, undertaken annually. In this system employees of the same rank and age received the same BS and RRP, and the element influenced by the *satei* was

|  | MANAGEMENT | | (Equivalent) |
|---|---|---|---|
|  | Level/qualif. | Hierarchical position | |
|  | Sanyo | Management | (Huku-Shocho) |
|  | Sanji | | (Bucho) |
|  | Shukan | | (Jicho) |
|  | Shuseki | | (Kacho) |

| MANUFACTURING | | OFFICE AND ENGINEERING | | |
|---|---|---|---|---|
| Level/qualif. | Positions | Level/qualif. | Positions | |
| Shunin | Instructor | Shunin | Instructor | (Kakaricho) |
| Ko-shi | Supervisor | Huku-Shunin | | |
| Huku-Ko-shi | or expert | 4 | Shu-Tanto | |
| 5 | Superior | 3 | Tanto | |
| 4 | production worker | 2 | Employees and engineers | |
| 3 | Production worker | 1 | | |
| 2 | | | | |
| 1 | | | | |

FIG 6.3. *Mitsubishi Motors Corporation: organizational hierarchy*

only one part of the FAP. If they worked better and received an improved evaluation, it was therefore possible for employees to make up for any handicap in the following year. The new wage system was deemed more transparent and equitable while at the same time preserving individual incentives. Above all, it permitted wage increases through increased rank even if hierarchical posts were all filled.

It could be said that by the end of the 1980s MMC had adopted a coherent and viable production system. However, its financial structure remained fragile, and its rate of self-financing remained low, still under 30 per cent, even after its shares were placed on the stockmarket. Even though operating profits for 1994 were the highest in the company's history, the net rate of profit was only 0.7 per cent; financial structure continued to be the company's Achilles' heel.

## 6.3. TOWARDS GLOBALIZATION

Despite this weakness, financially aided by the trading firm Mitsubishi Shoji, and with the support of Chrysler in distribution, MMC developed a strategy of internationalization, followed by globalization. Chrysler's sales networks permitted MMC to increase its exports, especially to the USA, during the 1970s. The crisis at Chrysler at the end of the 1970s had held exports back, but also gave MMC the opportunity to create its own distribution networks in markets heretofore dominated by Chrysler. In the USA MMC founded a sales company, Mitsubishi Motor Sales of America, in 1981, and went on to do the same in Europe following Chrysler's withdrawal from that market. Europe became MMC's best market overseas, with 224,600 vehicles sold in 1991.

In terms of the globalization of manufacturing, there were three types of production unit. The first was joint venture subsidiaries already formed by MHI and local capital, in South-East Asia and Portugal. The second was companies whose control had passed from Chrysler or joint ventures to MMC: MMAL in Australia, CARCO in the Philippines, and DSM (Diamond Star Motors) in the USA. The third type consisted of companies in which MMC participated at the request of a national government or local company, such as Proton in Malaysia and Nedcar in the Netherlands. Of these various ventures, the manufacturing plants that could be viewed as true MMC transplants were Proton, DSM, and Nedcar.

MHI had exported automobiles and assembled them in Malaysia since 1958. In 1981, the Malay government launched a national project for automobile production. Proton was formed by Heavy Industries Corporation of Malaysia (70 per cent), Mitsubishi Shoji (15 per cent), and MMC (15 per cent). MMC was responsible for building the factory. Before installation, the functioning of all equipment was first verified by MMC at its Mizushima factory, and 322 Proton employees were trained there between 1983 and 1986. Proton proceeded to launch its first national car, the Saga, with 70 per cent local content. In 1988 the Malaysian government decided to hand over the company's management to MMC because of its deteriorating financial situation. With reductions in production costs, easing of financial costs and modification of product prices, Proton was revived. In 1989 Proton began exporting Sagas to the United Kingdom, and its output now increased steadily.

In 1985, Chrysler and MMC founded DSM in the USA, to supply automobiles to both Chrysler and MMSA. MMC planned the production equipment for the DSM factory at its Okazaki factory. The factory was highly automated with 477 robots; the body shop was more than 90 per cent automated, while assembly was about 20 per cent automated, with 116 robots. The factory was managed by an ALC system with 100 terminals. All section-head posts, except for that of accountant, were occupied by MMC employees sent from Japan. One hundred and seventy American employees, the majority of whom had no experience of working in the automobile industry, had been sent to the Oye factory for training, particularly in equipment maintenance. In 1988 the DSM factory began production of two models on the same platform, MMC's Eclipse and Chrysler's Plymouth Laser/Eagle Talon. In 1990 the plant began production of the MMC Mirage model. While DSM was itself successful, Chrysler's own crisis led it to sell all its shares to MMC, so that DSM became a full subsidiary of MMC in 1991.

As far as Europe was concerned, MMC signed an agreement in 1990 with Volvo and the Dutch government to manage the existing Nedcar company; MMC contributed to the capital and assumed management of the company. The new equipment for the factory had been developed at the Mizushima (assembly) factory and the Kyoto (engine) factory to be installed in 1994. Dutch employees were sent to both these factories to learn about the equipment, while a further 600 employees received on-the-spot training in Japan. Nedcar produced cars for both Volvo and MMC, becoming the first Japanese automobile transplant on the continent of Western Europe (Dankbaar 1995).

Among the problems faced at the MMC transplants were the training of maintenance staff and the implementation of teamwork. In fact, MMC particularly needed workers trained to maintain automated equipment, given the high priority it gave to automation, whereas other Japanese producers prioritized the training of team leaders and supervisors so that they could develop teamwork and *kaizen* activities. DSM adopted the same form of organization as used by MMC, although the former had a formal distinction between direct workers who were unionized and paid by the hour, and salaried, non-unionized workers who were paid by the month (Suzuki 1991). Teamwork at Nedcar was a hybridization of the Scandinavian tradition (socio-technical design) and MMC organization (Dankbaar 1995). Whatever form was taken, MMC sought to use teamwork to improve quality, productivity, safety, and motivation. Moreover, it sought cooperative relations, without conflict, with trade unions. At DSM, the United Auto Workers union accepted, as it had at the General Motors–Toyota joint venture NUMMI in 1984, a reduction in the number of blue-collar categories to three, rotation of work posts, *kaizen*, teamwork, not to mention increased productivity and improved quality, in exchange for job security. The UAW also accepted to resolve problems through discussions between management and union without the need to resort to strikes (Suzuki 1991; Abo *et al.* 1991).

As well as exports and overseas production, MMC's international activities included collaboration with Korean automobile maker Hyundai and German automobile maker Mercedes-Benz. Collaboration with Hyundai began with technical assistance for the design of the Pony in 1973. MMC and Mitsubishi Shoji took 5 per cent of the Korean company's capital in 1982, rising to 6.3 per cent in 1991. In contrast, collaboration with Mercedes-Benz did not proceed far, although MMC sold the German manufacturer's products in Japan.

By 1991, annual production at MMC's overseas manufacturing units exceeded 500,000 vehicles, equivalent to one-third of MMC's production in Japan (or one-quarter of worldwide production). Furthermore, these units too began to export their products: MMAL in Australia to New Zealand and Japan, MSC in Thailand to Canada, Proton to New Zealand and the United Kingdom, DSM to Japan. The mutual supply of components, known as BBC (brand to brand complementation) was being developed among facilities in the ASEAN countries. Starting with three countries, this international division of labour progressed to involve nine countries, including Japan and Korea (Hyundai) during the first half of the 1990s. MMC thus seemed to be moving towards developing a network of manufacturing units worldwide in order to realize economies of scale at the global level.

## 6.4. CONCLUSION: THE INSTABILITY OF THE MITSUBISHI PRODUCTION SYSTEM

Mitsubishi's production system was purported to be 'lean' like that of other Japanese producers. Yet it differed from Toyota's system on a number of vital points.

First, there was the company's strategy, which after numerous trials and errors succeeded in making MMC a successful automotive producer in a country where there were already several producers, focusing on niches such as minicars, minivans, four-wheel-drive and recreational vehicles, and by designing technically advanced products. A further differentiating feature was the significance of automation in increasing productivity. *Kaizen* activities were not organized as systematically as Toyota and there was no link between salary and increased productivity at the team or factory level. Other than through job security, the involvement of workers was achieved through emphasis on technical skills in training and promotion, and on individual evaluation. The flexibility of production, particularly necessary with a range of highly differentiated vehicles and with variable sales, was obtained by turning to flexible automation and through the polyvalency of a workforce which changed posts systematically every two or four hours, through a system of two shifts per day, which permitted each shift to work one and a half hours of overtime per day when necessary. The *kanban* method was not used. Production was planned on a weekly basis for both MMC factories and suppliers. Lastly, MMC adopted a strategy of internationalization through collaboration and transplantation to support the position it gradually and with some difficulty secured in the Japanese market.

Neither Fordist, nor Toyotaist, MMC's production system resembles instead the 'innovation-flexibility' model (Boyer and Freyssenet forthcoming), which reduces the risks of a strategy focusing on technical and commercial innovation by means of greater flexibility in production and a low break-even point. Yet even though the growth of MMC accelerated during the 1990s, the company continued to suffer from poor profitability. Automation involved burdens on its capital, and the company's low rate of self-financing left it with high debt repayments. Hence by the mid-1990s MMC was seeking to reduce the length of time required to design new models to twenty months, to import components from countries where production costs were lower, and to have complete subassemblies manufactured by suppliers. These measures were all the more necessary as Mitsubishi faced increased competition from other producers in the recreational vehicle segment. The company had also announced, in 1996, that it would be cutting its workforce by 2,000 employees over the next few years. Nevertheless, the company has not abandoned its goal of a 5 per cent share of the global market by 'globalizing' its production.

# NOTE

Translated by Sybil Hyacinth Mair.

## STATISTICAL APPENDIX 6: Mitsubishi

| Year | Production | | | Exports[b] | Employment[c] | Gross sales[d] | Production cost[d] | Management and sales cost[d] | Profits | | |
|---|---|---|---|---|---|---|---|---|---|---|---|
| | Total[a] | Passenger Car | Minicar | | | | | | Operating[d] | Gross[d] | Net[d] |
| 1970 | 468,607 | 112,091 | 220,329 | 31,588 | 20,095 | 221,581 | 210,606 | | 10,450 | 936 | 426 |
| 1971 | 471,365 | 159,465 | 169,768 | 97,955 | 22,436 | 266,318 | 261,797 | | 4,520 | -1,626 | -282 |
| 1972 | 470,088 | 175,804 | 132,597 | 77,323 | 22,688 | 288,125 | 276,662 | | 11,462 | 2,077 | 397 |
| 1973 | 560,094 | 229,392 | 124,561 | 111,783 | 22,227 | 371,529 | 362,316 | | 9,212 | 3,126 | 0 |
| 1974 | 485,278 | 199,085 | 100,638 | 177,266 | 22,912 | 417,187 | 392,171 | 18,809 | 6,206 | -1,156 | 0 |
| 1975 | 533,041 | 287,018 | 64,056 | 187,240 | 22,775 | 446,675 | 396,650 | 43,850 | 6,174 | 5,603 | 2,068 |
| 1976 | 685,240 | 380,229 | 121,263 | 286,726 | 22,508 | 590,503 | 515,831 | 64,271 | 10,400 | 10,065 | 4,666 |
| 1977 | 836,615 | 504,417 | 129,456 | 380,197 | 22,654 | 741,200 | 642,397 | 76,543 | 22,259 | 23,325 | 10,457 |
| 1978 | 919,207 | 519,126 | 137,554 | 380,456 | 23,338 | 874,529 | 754,382 | 92,642 | 27,505 | 29,157 | 15,666 |
| 1979 | 970,708 | 510,274 | 155,233 | 410,383 | 23,303 | 903,035 | 769,322 | 113,083 | 20,629 | 23,380 | 14,088 |
| 1980 | 1,124,213 | 623,908 | 178,737 | 598,389 | 23,460 | 1,107,945 | 965,573 | 122,437 | 19,934 | 10,037 | 8,334 |
| 1981 | 1,079,693 | 564,046 | 212,142 | 557,086 | 24,421 | 1,082,193 | 914,074 | 139,572 | 28,546 | 30,614 | 13,190 |
| 1982 | 971,269 | 550,531 | 189,342 | 494,654 | 24,268 | 1,061,375 | 880,536 | 157,930 | 22,908 | 18,064 | 12,666 |
| 1983 | 978,074 | 480,400 | 205,676 | 474,200 | 24,084 | 1,173,631 | 1,002,760 | 158,556 | 12,316 | 7,055 | 5,555 |
| 1984 | 1,110,362 | 496,627 | 287,102 | 565,265 | 23,742 | 1,408,307 | 1,183,237 | 194,016 | 31,053 | 20,737 | 6,637 |
| 1985 | 1,188,636 | 544,811 | 285,152 | 664,900 | 24,737 | 1,578,823 | 1,323,296 | 226,313 | 29,212 | 17,473 | 25,332 |
| 1986 | 1,177,413 | 527,628 | 260,620 | 663,541 | 25,292 | 1,558,670 | 1,347,356 | 179,355 | 31,958 | 10,725 | 14,493 |
| 1987 | 1,257,016 | 585,302 | 262,911 | 667,251 | 25,793 | 1,752,697 | 1,509,641 | 202,420 | 40,635 | 20,808 | 11,016 |

STATISTICAL APPENDIX 6 (cont'd)

| Year | Production Total[a] | Passenger Car | Minicar | Exports[b] | Employment[c] | Gross sales[d] | Production cost[d] | Management and sales cost[d] | Profits Operating[d] | Gross[d] | Net[d] |
|---|---|---|---|---|---|---|---|---|---|---|---|
| 1988 | 1,253,360 | 609,686 | 283,449 | 637,649 | 25,770 | 1,898,828 | 1,632,829 | 221,927 | 44,072 | 33,716 | 12,837 |
| 1989 | 1,253,676 | 649,222 | 270,926 | 592,056 | 26,047 | 2,025,715 | 1,717,422 | 259,517 | 48,774 | 41,419 | 20,242 |
| 1990 | 1,366,508 | 715,695 | 303,429 | 617,019 | 26,228 | 2,313,636 | 1,965,114 | 282,699 | 65,822 | 50,214 | 25,208 |
| 1991 | 1,407,801 | 788,587 | 267,869 | 640,385 | 26,970 | 2,554,055 | 2,221,494 | 276,374 | 56,186 | 50,540 | 27,023 |
| 1992 | 1,406,525 | 855,082 | 243,135 | 665,325 | 27,603 | 2,615,959 | 2,288,869 | 269,597 | 57,493 | 46,567 | 20,232 |
| 1993 | 1,325,407 | 773,228 | 269,251 | 592,814 | 28,678 | 2,455,928 | 2,155,512 | 260,330 | 40,085 | 35,354 | 15,952 |
| 1994 | 1,349,636 | 767,590 | 269,589 | 554,560 | 28,742 | 2,652,517 | 2,317,312 | 267,459 | 67,745 | 48,046 | 18,826 |
| 1995 | 1,284,343 | 671,069 | 321,584 | 489,653 | 28,383 | 2,522,559 | 2,192,014 | 268,184 | 62,359 | 55,393 | 20,468 |
| 1996 | 1,221,171 | 640,866 | 283,904 | 481,169 | 27,827 | 2,585,940 | 2,203,247 | 325,544 | 57,148 | 58,035 | 15,067 |

*Notes:*
[a] Total production includes passenger cars, commercial vehicles, and minicars (the latter also includes both passenger cars and commercial vehicles). It does not include overseas production, although CKD kits are included up to 1977.
[b] Exports are of passenger cars and commercial vehicles.
[c] Employment does not include temporary employees, but does include employees sent to work at an overseas subsidiary.
[d] Unit of account: million of yen.

*Sources:* Mitsubishi Motors Corporation (1993), and Annual Reports.

# BIBLIOGRAPHY

Abo, T., Itagaki, H., Kamiyama, K., Kawamura, T., and Kumon, H., *Amerika ni Okeru Nihon-teki Seisan Sisutemu* [Japanese Production System in USA] (Tokyo, 1991).

Boyer, R., and Freyssenet, M., *The World that Changed the Machine* (forthcoming).

Dankbaar, B., 'Dutch Socio-Technical Design and Lean Production', Proceedings of Third GERPISA International Conference (Paris, 1995).

Mitsubishi Motors Corporation, *Mitsubishi Jidosha Kogyo Kabushiki-gaisha Shi* [History of Mitsubishi Motors Corporation] (Tokyo, 1993).

—— *Annual Reports* (various years).

Nikkan Jidosha Shinbun, *Handbook on the Automobile Industry* (Tokyo, various years).

Shimokawa, K., *The Japanese Automobile Industry: A Business History* (London, 1994).

Suzuki, N., *Amerika-shakaï no nakano Nikkei Kigyo* [Japanese Firms in American Society] (Tokyo, 1991).

Usine Mizushima de MMC, *Mizushima Jidosha Seisakujo 50 Nen Shi* [The History of the First Fifty Years of the Mizushima Factory] (Okayama, 1993).

# 7

# Hyundai Tries Two Industrial Models to Penetrate Global Markets

## MYEONG-KEE CHUNG

Korea emerged as a significant world producer during the 1980s. The growth of the Korean automobile industry has been remarkable. Since 1973, under the 'automobile promotion plan', a combination of active corporate development, joint ventures with multinational corporations, low wages, a protected home market, export promotion policies and government subsidies have supported the creation of this dynamic new industry.

Despite the tremendous successes achieved by the Korean automobile industry since the mid-1980s, by the mid-1990s the industry was still undergoing fundamental changes. These changes included a shift of strategy from expansion of the domestic market to penetration of the international market, a gradual increase in global competitiveness, a restructuring of the production system, a maturing of the domestic market, and declining profitability.

During the first half of the 1990s the Korean automobile industry underwent a series of major internal changes, focusing on reforms to managerial strategies and industrial relations. Significant features of these changes included cost reduction, improved quality, and increased flexibility in the production process. The result was the emergence of new managerial strategies, which were combined with rising research and development (R&D) expenditures, rapidly changing market conditions, and greater availability of information technology.

With this broad context in mind, the historical trajectory of Hyundai is analysed in this chapter at three levels: production and work organization, industrial relations, and inter-company relationships. These factors appear to be the central issues shaping Hyundai's future trajectory. The chapter first discusses the development of the company up to the crisis of knocked down (KD) production and the export expansion of the mid-1990s. It then describes in detail the company's production system and industrial relations, which were based on the Taylorist–Fordist model. Finally, the chapter examines the shift towards the establishment of a new production system at Hyundai, which was designed to bring about a fundamental restructuring of the 'Fordist syndrome' at the company. The early stages of a step by step implementation of a Toyotaist organizational logic and the changes in industrial work associated with it are identified.

## 7.1. FROM KNOCKED DOWN PRODUCTION TO INDEPENDENT DEVELOPMENT, 1967–1984

### 7.1.1. Knocked Down Production and the Relationship with Ford

Hyundai Motor is part of the Hyundai industry group, founded by Ju-Young Chung in 1967. At first, Hyundai assembled knocked down (KD) vehicles (Ford Cortina and Taunus) with the technical assistance of Ford, although output was extremely small. Forty-eight Ford engineers helped Hyundai. Several of these men stayed for one or two years, to help resolve the various problems which arose. The new equipment and components imported from Ford and other American firms arrived late, and Korean engineers had no experience of operating the specialized American machine tools. Another difficulty was the lack of standardization in specifications and quality between parts that Hyundai made in-house and those which were purchased locally. The proportion of locally produced parts purchased for the Cortina reached about 21 per cent in 1968 (Hyundai Motor Co. 1992: 315).

The Ulsan factory, which opened in 1968, employed 2,456 workers and had a capacity of 12,000 units per year. In 1968 the plant produced only 614 units, while importing 1,200 units from Ford of Britain. After one year, total output was still only 6,242. In 1969, Hyundai began production of a second model, the Ford Taunus. By 1971, Hyundai had produced 8,887 units of this model. The company's aggressive growth programme developed and achieved virtually complete localization of passenger car production (that is, imports of whole cars were nearly eliminated) by 1972. The proportion of locally produced parts purchased for cars built in Korea, however, only reached 68 per cent in 1974, as many components were still imported (Hyundai Motor Co. 1992: 321).

In the KD production phase, Hyundai imported both equipment and ideas associated with the Taylorist–Fordist industrial model which was used by Ford. This manufacturing system was characterized by job fragmentation and the transfer of skills to dedicated machinery. At Hyundai, output was erratic and capacity utilization low because the Fordist system was ineffective. Although Hyundai had used a moving assembly line since 1967, production was often interrupted owing to poor quality parts and poor work standardization. Hyundai simply could not achieve the volume required to fully exploit mass production. Although the company concentrated on improving work organization, many elements of craft production remained up until the early 1980s. However, they were gradually replaced by standardization of work and product-focused factory layout.

The market conditions of the 1970s were not suited to the expansion of demand in the domestic market. It is clear in retrospect that the early 1970s initiated the phase of stagnation that followed. The growth rate of GNP declined rapidly and interest rates increased in the wake of the first oil crisis. To protect the balance of trade there was high taxation on passenger cars, which led to a freeze in domestic demand. Approximately 70 per cent of the sales price of passenger cars was composed of government tax. Consequently, the major customers for passenger cars were taxi companies, although these too faced a crisis owing to the growth of real wages and a squeeze on their profits. Accordingly, the demand for passenger cars declined sharply, and with it capacity utilization, which dropped from 52.0 per cent in 1969 to 19.6 per cent in 1970 and 21.8 per cent in 1972. Production volume declined from 6,242 units in 1969 to 2,615 units in 1972. As a result of this crisis, Hyundai laid off 1,300 employees in 1970, a third of the workforce (Hyundai Motor Co. 1992: 346).

Management accepted loans, reduced its workforce, and took other measures to avoid bankruptcy while sales were low, including attempting to negotiate a joint venture with Ford. Hyundai was experiencing problems with access to export markets and production of major components such as engines, and the major objective of negotiation with Ford was to establish its own engine plant and to export its cars. Ford's goal was to retain its market in the USA free from imports from Hyundai. Ford also hoped to be able to treat the factories in Korea as its own subsidiaries and to have them focus on the domestic Korean market. These different intentions brought negotiations to a halt in 1973 and stimulated Hyundai management into a new strategy of self-reliance.

*7.1.2. Production of the First Indigenous Model, and Government Policy*

After seven years of KD and semi-knocked down (SKD) production, Hyundai succeeded in developing Korea's first independently designed and manufactured model, the Pony, in accordance with the government's long-term Automobile Promotion Plan of 1973. The objective of the plan was production of a people's car at a rate of 50,000 units per year. This plan included the development of a parts industry and the establishment of vertical integration all the way through to final assembly. This significant government policy selected which companies would be the final assemblers, and applied taxes and provided subsidies to help realize the objectives. The government also helped control the wages of workers in 'strategic industries' and restricted labour union activities. For this project Hyundai invested $71,253,000 during 1973–5, and several foreign banks agreed to loans of $52,116,000.

Simultaneously, Hyundai began technical cooperation with Mitsubishi, obtaining chassis components and other parts which were difficult to manufacture, such as gearboxes and engines, directly from Mitsubishi, while fabricating the cylinder head and blocks, housings, and transmission cases in-house. Hyundai paid Mitsubishi 13 million yen for the technical licences and 8,500 yen per car as royalty. Ital Design of Italy began work on the car body design in 1973. Hyundai became very active in training the necessary technical manpower: for instance, dispatching 97 technicians and engineers to advanced countries and investing 538 man-months in training during development of the Pony. The company established the first Korean R&D institute for automobile manufacture in 1978. By

1975, Hyundai had also expanded its annual capacity to 55,000 vehicles (Hyundai Motor Co. 1992: 328–78).

To manufacture the Pony, Hyundai built a modern plant for body stamping, welding, painting, and assembly work, with new equipment and layout. The factory had conveyors, transfer machinery, and automated high-speed equipment. Skid and automated roller conveyors were installed, starting a process of flow production. Taylorist standardization of work expanded during the 1970s and 1980s. Hyundai increasingly oriented its organization, production, and work process towards the principles of the Fordist manufacturing model in order to achieve economies of scale and profits. The associated Taylorist work standardization process tended to mean a freezing of standard operations, combined with a vertical separation between workers and industrial engineers at the shopfloor level.

The Hyundai Pony was the first indigenous Korean model, and went into production in 1976. It was equipped with a four-cylinder, 1,238 cc Mitsubishi Saturn engine, and the local content ratio of the cars reached 85 per cent at the start of the phase. Production growth was striking; it soared to 11,692 units in 1976, 24,766 units in 1977, and 64,886 units in 1979. These volumes allowed the introduction of technical innovations and organizational solutions consistent with mass production.

During the 1970s Hyundai basically worked through this particular organizational trajectory, becoming an efficient mass producer of small cars mainly targeted at the domestic market. Alongside organizational reform, Hyundai developed close technical collaboration in the field of production technology. Between 1976 and 1979 there were twenty technical collaborations with foreign companies. Furthermore, since the growth of production volume required a stable and adequate supply of materials and components, Hyundai increased in-house capacity for essential subassemblies and bodies and also made efforts to organize independent suppliers. There is evidence of close financial linkages between Hyundai and parts companies.

### 7.1.3. *The Second Oil Crisis, the Mitsubishi Tie-up, and Factory Modernization*

The early 1980s represented a period of profound crisis for Hyundai, as social, market, and internal forces contributed to significant production overcapacity and poor financial performance. The second oil crisis, which began to affect the Korean economy in late 1979, acted as a substantial brake on the continued growth of the automobile industry. From 1979 to 1980, motor vehicle output declined rapidly from 64,886 to 43,875 units. The rate of capacity utilization dropped to 38 per cent in 1980. Hyundai's management announced the lay-off of 3,311 workers (Hyundai Motor Co. 1992: 492–500).

In the general atmosphere of crisis, in 1981 the Korean government sought to merge Saehan Motors (the former name of Daewoo Motors), which was 50 per cent owned by General Motors (GM), and Hyundai, with the goal of obtaining further economies of scale. Hyundai now found itself embroiled in a contest with GM over the equity ownership ratio, free access to export markets, and the development of new models. In the end, the negotiations proposed by the government to merge all the country's passenger car makers ended in failure, because GM also insisted on managerial control as part of its world car strategy. The government decided that only two car makers, Hyundai and Saehan, were to produce passenger cars (Kia Motors was not permitted to produce passenger cars until 1987).

In 1982, Hyundai set up a cost centre to analyse and control production costs for every fifteen-day period at every division. During the 1980s, a national sales network was organized, supported by Hyundai's direct investment into a sales company. Hyundai offered instalment plans with small down-payments to the domestic middle classes, with 60–70 per cent of the total sales price paid in instalments. In 1982, the company planned to restructure its management organization. However, higher profits were realized from sales promotions, rather than from the restructuring efforts that involved worker lay-offs.

Shortly after the failure of negotiations with GM, Hyundai made the major decision to build a new plant with a potential capacity to produce 300,000 vehicles a year. This new plant was to be the first front-wheel-drive car plant in Korea. Hyundai needed to make a huge investment despite the tight credit restrictions in force at the time. The company therefore negotiated a formal tie-up with Mitsubishi for its capital and technology, and then invested $519 million of international credit and $474 million won from domestic sources in the new plant (Hyundai Motor Co. 1992: 522). Mitsubishi now owned 12 per cent of Hyundai (1982). Hyundai's engineers were sent to Japan, where they visited Mitsubishi operations. Their study of Japanese automobile factories was intensive, and lasted for three months, so that they could understand the advanced transfer machines that were fundamental to the engine machining process. Lastly, Hyundai acquired the blueprints for the transfer machinery used to make bearing caps and differential cases from Mitsubishi Heavy Industry. Hyundai developed this machinery itself, and it was installed in 1983.

The gradual recovery of domestic demand, combined with an aggressive export strategy, helped Hyundai's Fordist production system to mature. Output requirements in the second half of the 1980s had changed dramatically. Hyundai now had to manufacture in much larger volumes, with its potential capacity of 350,000 per year, and it had to start exporting its products. The company focused on improving flow on the assembly line and on purchasing new equipment for the new plant. Hyundai began to rely heavily on automation and robotics; in 1984, fixed-sequence robots were installed to automate spot and arc welding.

In the early 1980s the company introduced its Assembly Line Control (ALC) system, with the help of an IBM 3630 Plant Communication System, to develop synchronized production (Hyundai Motor Co. 1992: 534). The production process become more flexible, to permit a wider range of options (different body colours, bumpers, and wheels, for instance). This system was later integrated into the SPS (Sequence Production System), which aimed to reduce inventory costs and non-availability of parts. The heart of the ALC system is located in the Central Control Room which controls the flow of production and the line speeds in each area of the factory. Assembly takes place on a machine-paced line. Simultaneously, microprocessor-based technology began to influence the factory floor. Computer-aided design (CAD), computer numerically controlled (CNC) machine tools, and computer-aided manufacturing (CAM) were fast becoming important tools in the manufacturing process. Hyundai now had to rely heavily on the importation of equipment such as robots and CNC machines from Mitsubishi and other Japanese companies.

The introduction of the new advanced technology and the achievement of economies of scale made it possible to reduce production costs. For example, the adoption of CAD/CAM reduced expenditures on new model development by around 20 per cent by cutting work in progress. Hyundai was able to reduce the price of the Pony II model by 4.9 per cent in 1983 (Hyundai Motor Co. 1992: 527–50).

After having achieved economies of scale, Hyundai began transforming its relationships with components manufacturers. The company established a supplier's association in 1985 in order to facilitate communications with the parts makers as well as among the parts makers themselves. Hyundai stimulated the growth of its subcontractors through capital aid and managerial assistance. Nevertheless, until the late 1980s the subcontracting system functioned primarily as a shock absorber and a source of cheap labour. Hence the primary factor in components purchasing was price, with quality and delivery times secondary.

To summarize, during the 1980s Hyundai pursued a technology strategy based on Taylorist–Fordist rationalization, which was successful in terms of quantities produced. However, it did not imply the adoption of modified Japanese production concepts even though some Japanese ideas, such as JIT and TQC, were partially introduced. Work organization at Hyundai was still characterized by a sharp hierarchical and functional division of labour combined with tight organization of the flow of materials and work and close management of time. Under this organizational form, workers did not participate actively in company suggestion programmes, quality circles, or *kaizen*. Ultimately, this highly standardized production strategy confronted Hyundai with a number of difficulties.

## 7.2. CRISIS OF THE FORDIST MANUFACTURING SYSTEM, 1985–1990

### 7.2.1. Maturity of the Domestic Market

An important development in the car market was a shift from price competition to increased emphasis on model variety and quality. In this market context Hyundai

TABLE 7.1. *Hyundai product variety*

| Product | 1984 | 1984 | 1985 | 1986 | 1987 | 1988 | 1989 | 1990 | 1991 | 1992 |
|---|---|---|---|---|---|---|---|---|---|---|
| Platform | 4 | 3 | 6 | 8 | 7 | 7 | 7 | 6 | 6 | 7 |
| Engine families | 7 | 5 | 10 | 13 | 13 | 13 | 13 | 14 | 14 | 13 |
| Index of body and engine | 19 | 14 | 20 | 24 | 24 | 22 | 14 | 29 | 27 | 31 |

*Source*: Hyundai Motor Co. (1997a).

introduced a new production strategy involving a broader product mix and variation within each model. Until the mid-1980s all of Hyundai's cars were subcompacts with limited options and colours. The company decided to build a second plant specializing in compact cars (Sonata model), which was completed in 1988. The production capacity of this integrated assembly plant was 300,000. A third plant, specializing in subcompact cars (Lantra model), was completed in 1990 with a capacity of 240,000 units. This was the most automated plant in Korea. These were also the first assembly plants in Korea that could match the standard capacities of American factories. They also permitted Hyundai to offer a full range of products.

All phases of process control were computerized during the second half of the 1980s. However, while Hyundai massively increased the level of automation on the shopfloor during the 1980s, the production system was still based on single purpose machine tools, a detailed division of labour and large-batch production with little product variety. This management strategy was aimed at high-capacity-utilization mass production to reduce costs.

Hyundai pursued diversity, high quality, and high price through model changes and the development of new cars, seeking to expand sales volumes and market share through these strategies. There was a remarkable rush to develop new models. In 1983, Hyundai had four models including two KD models, but by 1992 seven passenger car platforms were being manufactured (see Table 7.1). Demand for compact cars began to exceed that of subcompact cars.

The Korean domestic market declined into slow growth after 1989. Market saturation now made it difficult for the domestic market to expand at an annual rate of more than 10 per cent despite average annual growth of domestic sales of 36.8 per cent in the period between 1985 and 1988. Moreover, after 1988, Kia emerged as a significant passenger car producer linked with Mazda and Ford, and competition heated up among car producers. Hyundai's domestic market share declined from 58.5 per cent in 1989 to 42.8 per cent in 1993 (Table 7.2). Although sales volumes continued to rise, Hyundai suffered from falling profitability: from 2.1 per cent in 1987 to 0.81 per cent in 1993 (see Statistical Appendix 7). The primary reason for deteriorating profitability was a rise in labour and materials costs while car prices were virtually frozen for four years. To maintain its profit

TABLE 7.2. *Evolution of Korean producers' market share and registration*

| Year | Hyundai* | Kia* | Daewoo* | Registration* |
|---|---|---|---|---|
| 1983 | 73.2 | — | 25.4 | 104,290 |
| 1984 | 68.5 | — | 30.0 | 107,897 |
| 1985 | 73.0 | — | 25.7 | 130,212 |
| 1986 | 69.5 | — | 29.2 | 155,831 |
| 1987 | 55.6 | 12.2 | 31.2 | 248,108 |
| 1988 | 54.7 | 18.9 | 24.7 | 321,824 |
| 1989 | 58.5 | 18.1 | 20.5 | 514,289 |
| 1990 | 53.1 | 22.4 | 21.0 | 626,559 |
| 1991 | 50.1 | 24.3 | 17.9 | 774,256 |
| 1992 | 46.7 | 25.8 | 15.4 | 877,331 |
| 1993 | 42.8 | 27.1 | 19.3 | 1,037,600 |

*Note*: *Percentage of the registrations.
*Source*: Hyundai Motor Co. (1993), Kia Research Institute (1994).

TABLE 7.3. *Export of Hyundai by region*

| Region | 1986 | 1987 | 1988 | 1989 | 1990 | 1991 |
|---|---|---|---|---|---|---|
| North America | 268,153 | 340,609 | 358,546 | 152,289 | 144,103 | 124,359 |
| Europe | 11,493 | 26,796 | 20,097 | 20,963 | 28,168 | 67,510 |
| Middle-East Asia | 9,091 | 11,392 | 6,997 | 3,208 | 4,246 | 6,012 |
| South-East Asia | 463 | 1,073 | 1,269 | 1,269 | 1,698 | 2,729 |
| East Asia | 6,823 | 12,851 | 16,815 | 16,815 | 15,841 | 20,127 |
| South America | 5,478 | 9,794 | 5,037 | 5,037 | 5,342 | 7,580 |

*Source*: Hyundai Motor Co. (1992: 835).

rate Hyundai sold fixed assets and marketable securities (thus, for instance, remaining profitable in 1989 despite a fall in output).

### 7.2.2. Rise and Fall of Overseas Markets

Hyundai launched the Excel in 1985 and successfully entered the US market in 1986. After mass production of this new model began, Hyundai's growth strategy became focused on exports to North America. Until 1988 Hyundai benefited from a well-educated, disciplined workforce that cooperated closely with management and accepted low wages and/or poor working conditions. Total compensation per hour in the Korean automobile industry was estimated to be 12 per cent of that in the USA and 43 per cent of that in Japan (Kia Research Institute 1991: 14). The volume of automobile exports reached a peak of 404,881 units in 1988. In this year, 586,658 cars were produced (see Statistical Appendix 7, Table 7.3).

Yet Hyundai's motor vehicle exports decreased substantially after 1989, from 404,881 in 1988 to 213,639 in 1989, a fall of almost 50 per cent. This failure to expand automobile exports was connected with the overvaluation of the Korean won[1] and increased wages[2] which made Korean exports comparatively more expensive abroad. After 1987 labour disputes grew year by year, leading to lockouts and the loss of opportunities to reduce costs. The price discrepancy between the Excel and the Ford Escort was gradually reduced following labour disputes, shrinking from $1,118 in 1986 to $757 in 1990 (Hyundai Motor Co. 1992: 718). Reliance on price alone did not guarantee long-term success in developed markets such as the USA. Moreover, after a promising start in the mid-1980s, Hyundai sales also became sluggish in the USA owing to problems of after-sales service, poor quality, and design.

Despite the need to improve quality in order to remain competitive with overseas producers, Hyundai acquired a reputation for poor quality in its largest export markets. Growing distrust between workers and management was a major factor behind the deterioration of quality. According to the J. D. Power new car initial quality survey, the assembly defects found in Hyundai models rose from 178 in 1989 to 230 in 1990 and 235 in 1991. Consequently, customer dissatisfaction increased and Hyundai's CSI ranking in the J. D. Power survey fell from fourteen in 1989 to thirty-five in 1991 (Korea Institute for Industrial Economics and Trade (KIET) 1994: 255–7). Hyundai's US market share declined as a result. The deteriorating quality was a potential disaster. Even with internationally competitive prices, the poor quality reduced market share in overseas markets. Improved quality was clearly a prerequisite for survival in world markets.

During the phase of export expansion, Hyundai decided to establish an automobile manufacturing plant at Bromont in Canada, with 100,000 annual capacity. Over $300m. was invested in the factory, and production began in 1988. This project was the first investment in an advanced country by an automobile producer from a newly industrializing country. The production system used at the plant was also characteristically Fordist. With a workforce of 1,200, total production in 1990 and 1991 was 27,409 and 28,201, respectively, which was far below capacity and expectations. When Canadian and US sales failed to rise, Hyundai scaled back production targets. Moreover, labour disputes in the main Korean factories disrupted the supply of components to Bromont, creating an additional obstacle to increased productivity and capacity utilization. As a result, the plant was closed in 1991. By the mid-1990s Hyundai was planning to move the assembly-line equipment from the Bromont plant to India.

Beyond the general decline of sales in the USA, the future growth of export markets would be determined by the evolution of both non-price and price competitiveness. The first implication was that Hyundai needed to maintain its access to the US market in order to capitalize on recent investments in advanced consumer and intermediate products. To become more active in the US market, Hyundai established Hyundai Motor Finance Company in 1990. The second implication was that Hyundai would need to diversify and direct its marketing towards European countries and Asian and African markets. The company decided to boost its presence in Malaysia. An assembly factory was built with a capacity of 12,000 vehicles per year. Hyundai also exported CKD vehicles as follows: Egypt (20,000 units per year), Botswana (10,000 units per year), Zimbabwe (10,000 units), Thailand (10,000 units), Philippines (12,000 units). The network of overseas dealers was to be expanded from 2,968 to 3,200 between 1994 and 1995, with the after-service network growing to 3,700 worldwide. However, this internationalization strategy faced the stiff competition of Japanese car markers in South Asia, for example.

### 7.2.3. Managerial Unilateralism, Labour Disputes, and the Crisis of the Fordist Model

Labour relations became an issue of major confrontation between the senior management and the trade union at Hyundai. Workers set up a labour union after the Korean government's declaration of 29 June 1987 that called for a relative relaxation of the historic tight control over labour union activities. The struggle between labour and management began in 1987, with both sides suffering severe damage.

The origins of the struggle can be traced back to the big labour dispute in July 1987 which was essentially a struggle to construct a democratic union. After the government declaration, top management had supported the establishment of a labour union to promote better relationships between company and union leaders. However, most workers rejected this union and set up a new union themselves. The acceptance of the new union required a twenty-day strike.

The union succeeded in establishing the union-shop form of organization, in which blue-collar workers automatically become union members when they commence employment. By 1994, 72.8 per cent of all employees were members of the union. The big dispute at Hyundai in 1987 directly influenced a broader diffusion of labour problems into subcontracting companies. A vicious circle of labour disputes with vehicle producers and consequently with suppliers emerged as an annual event after 1987. During the 1987–92 period there were an average of twenty days of strikes and lockouts per year at Hyundai and seventeen days at subcontracting companies.

The strikes can be attributed to a readiness for conflict on the part of the union which had developed during the long phase of economic expansion. The major issues at Hyundai were wage determination and improved working conditions. The wage system at Hyundai was complicated, characterized by a combination of a traditional seniority wage system with a productivity wage system. According to the Hyundai labour union, wages were composed of basic wages (57 per cent) and extra wages (43 per cent). Basic pay and efficiency pay were the most important components. In the formula to calculate efficiency that the union had to accept, only the number of vehicles actually produced was taken into consideration. The sharing of productivity gains continued to be the primary focus of conflict in collective bargaining up until the mid-1990s.

The second major issue was the widespread demand by workers to improve working conditions on the shopfloor. The reduction of working hours and line speeds was a key issue in the union's activities. While the standard working time was 44 hours, real working time per employee was 50.6 hours in 1993. The annual working time for production workers with shift work was 2,630 hours, 13.3 per cent more than the average Japanese assembler.

The annual labour disputes prevented full capacity utilization of the highly automated production facilities and delayed the improvement of product quality. Hence capacity utilization was reduced after 1988, falling from 90.2 per cent in 1988 to 59.4 per cent in 1990 before returning to 85.7 per cent in 1993 despite the use of overtime and shift work. This became the most important obstacle to cost reduction and increased productivity. Moreover, the traditional authoritarian and paternalistic control of work by supervisors on the factory floor began to erode rapidly as a result of management's loss of hegemony.

Hyundai thus faced the problem of how to manufacture a rising volume of cars while maintaining low wages and traditional authoritarian and paternalistic control over work. Managerial unilateralism led to a crisis of the mass production model at Hyundai. After growing strongly throughout the 1970s and early 1980s, the rate of increase in labour productivity (measured in terms of output per person) slowed significantly after 1987. Production volumes grew at an annual rate of 66 per cent in the period 1985–7 but declined to 7.6 per cent in 1988. As a result of the violent labour disputes, in 1989 Hyundai registered negative production growth for the first time since 1980, with the 1988 peak level not attained again until 1991. Higher productivity resulted from work intensification under Tayloristic forms of work organization but this led to resistance from workers on the production line. One consequence of the slowdown in productivity growth was the rapid increase in the part of unit costs due to labour disputes after 1987, which helped bring about worsening profitability (see Table 7.4, Statistical Appendix 7).

TABLE 7.4. *Annual percentage change in labour productivity, unit labour cost, and unit labour output of Hyundai*

| Year | Output | No. of Workers | Labour Compensation | Output per Person |
|---|---|---|---|---|
| 1986 | 82.3 | 33.4 | 6.3 | 21.5 |
| 1987 | 32.3 | 11.2 | 32.0 | 19.6 |
| 1988 | 7.6 | 16.2 | 38.2 | –7.6 |
| 1989 | –13.5 | 5.6 | 29.7 | –17.8 |
| 1990 | 4.1 | 8.4 | 15.4 | –4.1 |
| 1991 | 11.7 | 5.7 | 22.2 | 5.8 |
| 1992 | 16.5 | 1.6 | 13.0 | 15.1 |
| 1993 | 10.6 | 0.6 | 12.5 | 10.2 |

*Source*: Hyundai Motors Workers Union (1994), Hyundai Motor Co. (1994).

## 7.3. THE STRUGGLE TO ESTABLISH A NEW INDUSTRIAL MODEL, 1991–1994

Hyundai adopted a two-step approach to increasing its international competitiveness. The first step was the adoption of Japanese management concepts, and the second step was the adoption of a high technology approach to rationalize production and increase efficiency. This strategy required the development of human resources, which in turn required innovative industrial relations. The new employment relationship presupposed a shift towards a Japanese type of organizational strategy.

### 7.3.1. The Hybridization of Japanese Management Concepts and the New Organization of Production

Hyundai attempted to launch a massive adoption of Japanese management concepts after 1990, focusing on quality circles (QCs) and a suggestion system. Management set up competitions between QCs using prizes and monetary rewards. Hyundai places greater emphasis on team activities. In 1991, 85 per cent of workers participated in QCs, a participation rate which increased to 95 per cent in 1993. The key functions of QCs were primarily to increase worker involvement in company operations and to boost morale by allowing employees to work in groups to solve problems. Hyundai revived its individual suggestion system in 1990 and launched a new promotion campaign to obtain more proposals from the QCs. The result was an average of eighty suggestions per worker in 1993, compared to only nine suggestions in 1990. In 1993, the proportion of suggestions actually implemented reached 95 per cent. Of the proposals made, 25.1 per cent dealt with quality improvement, 23.4 per cent with improved production procedures, and 13.7 per cent with work safety (Hyundai Motor Co. 1994: 150).

Hyundai also began experimenting with devolved forms of quality control where workers were responsible for their own work, and with devolved maintenance techniques under a TPM (Total Productive Maintenance) programme.[3] Quality controllers and maintenance workers accounted for 21.6 per cent of total production workers in 1994, with maintenance workers accounting for 3.4 per cent (see Table 7.5). One goal of the new programmes was to increase the capacity to manage emergency situations where a line stops or a product on the line has a fault. They were closely linked to job rotation and worker polyvalency.

By the early 1990s job rotation to train polyvalent workers was a prominent theme in work organization. Hyundai had no special policy for job rotation because workers still viewed it negatively as a cause of work intensification. Nevertheless rotation took place frequently on the factory floor, essentially organized by work groups, and not designed to create polyvalent workers so much as to compensate for work-related stress and strain among group members. According to a 1994 union survey, 35.5 per cent of respondents (sample size: 2,056) practised

TABLE 7.5. *Structure of Hyundai workforce 1988–1994 (per cent)*

| Status | 1988 | 1994 |
| --- | --- | --- |
| Blue collar | 62.5 | 60.3 |
| Direct | n/a | 70.3 |
| Indirect | n/a | 29.6 |
| Maintenance | n/a | 3.4 |
| Quality control | n/a | 18.2 |
| Production management | n/a | 8.0 |
| White collar | 37.5 | 39.7 |
| Office | 27.4 | 71.2 |
| Sales | 10.1 | 28.2 |
| Total | 100.0 | 100.0 |

*Source*: Hyundai Motor Co. (1994).

job rotation. Only one-third of them were satisfied. Workers rotated most frequently within the same work group (76.8 per cent), and so job rotation did little to help train the worker for multiple tasks (Hyundai Motors Workers Union 1994, ii: 84–5).

Production costs were reduced by increasing the flexibility of production systems. The main factors behind increased flexibility were the introduction of microprocessor-controlled robotics, other computer numerically controlled machine tools, and automated process control. The automation strategy was an essential feature of the Hyundai trajectory after the crisis of the late 1980s. The first goal was to eliminate the most dangerous and laborious jobs which were a major cause of labour disputes. The second goal was to improve quality and productivity.

By the early 1990s efforts to automate production were concentrated in press shops, body shops, and the trim line.[4] Ninety-five per cent of operations in the press and body shops were automated. The second phase of automation focused on the final assembly line, which retained a conventional organization. Doors, seats, and front and rear windows were now installed by robot systems, with the level of automation in final assembly at the third plant rising to 5 per cent. By 1990 Hyundai had installed 952 robots and planned to raise the level of automation further, with 157 new robots to be introduced by 1995 (Hyundai Motor Co. 1994: 350).

The stamping area in the first plant included a line of seven 400–1,500 tonne hydraulic ram presses. At each, a worker stood alone and stamped parts. These were automatically transferred between presses on a bed of rollers. The panels were then loaded onto a robot carrier and conveyed to automated framing stations where the body was assembled. Only a few direct workers were required. The transfer of body panels from the press shop to the body shop took five minutes. The stamping plant was now oriented towards small batch production, with batch sizes now reduced to 1,500 panels. Dies were being changed five to six times per shift with set-up times of around twenty minutes. The buffer stock amounted to just ten units. In the final line at the body shop a robot able to recognize the body's shape ground down imperfections and applied the finishing touches to welded body parts, a task previously done by skilled workers using grinders and considered an important factor in a car's final quality.

A new trend towards outsourcing of component modules was also a preliminary step towards a lean production strategy. An increasing proportion of work, such as the making of rear and front panels and side frames, was bought in. This technology strategy, based on the progressive automation of all manufacturing processes, continued with the implementation of a new version of the ALC system. The new ALC system included a partial automation of the assembly process with manufacture undertaken by computer-based automated processes.

Automation also permitted technological and organizational flexibility to be increased by decoupling worker tasks from the production process. One result was that employment was affected. The proportion of white-collar workers at Hyundai increased from 37.5 per cent in 1988 to 38.7 per cent in 1994 while that of blue-collar workers dropped correspondingly from 62.5 per cent to 60.3 per cent. The change resulted mainly from a decline in direct blue-collar workers, whereas the number of indirect blue-collar workers (who accounted for 29.6 per cent of blue-collar workers in 1994: see Table 7.5) increased gradually. The company published estimates of the redundancies expected to result from automation in the period 1992–3. The actual redundancies were 160 workers in 1992 and 364 workers in 1993 (Hyundai Motor Co. 1994: 350). Overall employment at Hyundai only grew by 1.9 per cent during the period 1991–3 while production volumes increased 16.5 per cent over the same period.

The utilization of part-time or temporary workers was also expanded. In 1994 Hyundai employed around 2 per cent of production workers on a part-time basis. Their working conditions were generally inferior to those of full-time workers. At the same time, around 5,000 workers on the Hyundai shopfloor were employees of subcontractors and undertook subassembly jobs such as door assembly or fuel tank assembly. These flexible personnel policies had a twofold effect. They helped reduce the physical strains that can contribute to conflictual labour management relations and they potentially reduced labour costs. However, these personnel policies caused other problems in the production process. Most of these workers had no sense of obligation to their work, and their higher absenteeism caused production line imbalances.

Work organization was based on clear divisions of labour according to functions, tasks, and hierarchical level. The divisions between worker and manager or engineer, blue collar and white collar, direct and indirect worker on the shop floor were crystal clear. There were five hierarchical levels above production workers: *chochang* (group leader), *banchang* (foreman), *kisa* (senior foreman), *juim* (subsection director), *kichang* (section director). The jobs of *kisa*, *juim*, and *kichang* were held by the university graduates. The key part of the organizational hierarchy for the workers was therefore the *cho* and the *ban*. According to company data, a *chochang* directed eight to ten general workers on average while a *banchang* had around thirty subordinates including two to three *chochang*. Group leader and foreman had managerial responsibility for immediate production activities and played crucial roles in the organization, design, and allocation of work on a daily basis. Their job was to supervise, or 'boss', shopfloor workers. They were selected by management, who had the right to personally assess the workers. This dense managerial structure played a critical role both in mobilizing workers to undertake rationalization activities and in controlling workers on the shopfloor.

This organizational hierarchy remained inflexible, which in turn generally inhibited cooperation and horizontal communication. The most fundamental issue in restructuring Hyundai was therefore the reform of work and production organization linked with human resource management. Flexible production systems required flexible work practices and worker commitment to quality control and maintenance. There was considerable pressure to change the traditional model of work organization.

### 7.3.2. Human Resource Management

Further elements of the Japanese production system were adopted by Hyundai during the 1990s. The new strategy required that human resource management focus on a long-term commitment between employee and employer. One result was an 'exclusive' workers' recruitment policy. Hyundai became eager to recruit new production workers who were recommended by existing workers. The proportion of new workers supplied in this way rose to 80 per cent. This system aided management by increasing loyalty and creating cooperative attitudes. As a result of the policy, many employees were closely related. The hierarchy within the company played an important role in human relations in private life. Hyundai attempted to create 'familyism' as a precondition for productive industrial relations. Recruits who were graduates of technical high schools started their jobs after five days' vocational training on the shopfloor. Those who were less skilled had to undertake six months' vocational training in the factory before being assigned to their jobs. To sum up, the acquisition of skills to work on the line required about one week.

Accordingly, the new trends in labour management relations in the 1990s resulted from conscious innovation. Before 1987 important decisions on industrial relations were made by top management. Middle managers had relatively few rights and duties concerning industrial relations. Union leaders ignored their role and communicated directly with top management. Middle managers often complained that workers could not be disciplined. In order to establish constructive relationships Hyundai is attempting to decrease employee distrust by concentrating on improving welfare facilities and solving personnel problems.

The central thrust of the new industrial relations policy was a reform of wages and organizational structures. The wage system for production workers at Hyundai in 1994 was as described in Table 7.6. Basic pay and the overtime

TABLE 7.6. *The wage structure of Hyundai in 1994*

| Wage structures | Won | per cent | per cent |
|---|---|---|---|
| Standard Wage | 690,841 | 57.7 | |
| basic pay | 581,098 | | |
| position allowance | 4,978 | | |
| family allowance | 12,000 | | |
| welfare allowance | 15,000 | | |
| efficiency pay | 19,934 | | |
| long-service allowance | 47,778 | | |
| others | 10,053 | | |
| Extra wage | 505,739 | 42.3 | |
| overtime allowance | 356,157 | | |
| shift allowance | 93,961 | | |
| others | 55,621 | | |
| Monthly Wage | 1,196,580 | 100.0 | 73.3 |
| Bonus (650% per Year) | 374,206 | | 23.0 |
| Long-service Bonus | 53,574 | | 3.7 |
| Average Monthly Wage | 1,624,360 | | 100.0 |

*Note*: Bonus = $\dfrac{\text{Standard Monthly Wage} \times 650\%}{12 \text{ months}}$

*Source*: Hyundai Motors Workers Union (1994).

allowance were the most important categories in the wage system. The extra wage was in fact performance-based pay dependent on the production volume. Hence the wages of production workers were unstable. Basic pay was calculated according to education level and age. A position allowance was paid to supervisors (*chochang, banchang, kisa, juim, kichang*) according to their position. A family allowance was paid to each employee and depended on the number of family members. A bonus allowance was one of the major components of the total wage. The individual wage thus varied depending on the employee's position. The wage of a *chochang* was on average 24.3 per cent higher than that of a production worker, while his length of service was 4.2 years longer. The wage of a *banchang* was one-third higher than that of a production worker. The higher position allowance was a reward for the personal control of labour on the shopfloor. This wage system was also one of the obstacles to improved industrial relations at Hyundai.

The wage system had now to be reformed. The new wage system differed from the old system in the way 'age pay' and 'ability pay' were calculated. 'Age pay' was based on the seniority principle. 'Ability pay' was based on qualifications and personal assessments. The ratio of age pay to ability pay was 70:30. The new wage system also reduced the role played by productivity pay. After 1987 the wages of individual workers did not vary depending on their performance. The system for evaluating the performance of production workers was abolished at the demand of the union. The bonus allowances now basically depended on the company's business performance. The strength of the labour union was an important factor in the determination of the bonus allowances.

The reformed organizational structure was characterized by the integration of white-collar and blue-collar workers. The existing status system was a primary cause of blue-collar dissatisfaction. Organizational innovations were designed to increase the motivation of production workers, in part by solving problems related to promotion bottlenecks. The union and management started to discuss and coordinate the new organizational form and wage system in 1994, although initially without reaching agreement on a method for assessing personnel and the proportion of ability pay in the total wage.

Parallel with these reforms to the wage system and organizational structures, management stressed the upgrading of skills. This was seen as a precondition for automation on the shopfloor. The basic approach to acquisition of skills by shopfloor workers was on-the-job training. Hyundai offered several training programmes to upgrade the skills of workers, with five different courses: industrial relations, job training, new recruit training, skills development, and foreign language training. In particular, management emphasized 'work morale education' (learning the Confucian class ideology, based on harmony, respect, and sacrifice) in order to stabilize industrial relations; this alone accounted for 52 per cent of all training programme costs in 1990.

With more automated machines introduced into the production process, it became necessary to train workers to operate them. Hence the company established its own technical centre, which also offered training to employees of suppliers. Training costs therefore increased rapidly, with Hyundai investing 0.136 per cent of total sales value in training and education in 1991 compared to 0.058 per cent in 1986 (Lee 1994: 165–8). At the same time, the upgrading of skills depended more on relationships with senior workers on the shopfloor. According to a union survey, 39.6 per cent of respondents preferred to learn skills directly from senior workers, while only 18.5 per cent wanted to acquire their skills through company programmes (Hyundai Motors Workers Union 1994, ii: 84).

### 7.3.3. Product Strategy, Just-in-Time and Inter-Industry Organization

By the mid-1990s Hyundai company strategy was to expand production capacity in order to improve market segmentation. Hyundai dominated the medium to deluxe segments of the domestic passenger car market with the top three best-selling models and planned the timely introduction of two new models in 1995, a subcompact model and a medium deluxe model, maintaining the dominance already achieved in these segments. As a result of its aggressive export strategy, Hyundai had been able to produce and market a full range of passenger cars, from small to deluxe models.

Hyundai's major product investment programme had included the so-called Y-2 project (2,000 cc deluxe model), X-2 project (subcompact car), and J-car project (medium deluxe model) and was completed between 1988 and 1990. For the Y-2 project Hyundai had spent 1,852m. won ($2m.). The lead-time had been

3.2 years. The X-2 project cost 7,358m. won and had a three-year lead-time. The J-car project cost 32,456m. won and had a 2.6-year lead-time. In total, Hyundai invested around $48m. in new model design and development projects between 1986 and 1992 (Hyundai Motor Co. 1992: 642–95).

For the design and development of new models Hyundai relied on technology transfer and styling concepts from Mitsubishi and other foreign companies. Hyundai also centralized its own design and development facilities which directly employed 4,100 professionals, engineers, and workers and adopted a CAD/CAM system for new model development that helped to reduce the time required to develop new models. For instance, the lead-time for the Y-2 project was thirty-eight months, but the J-car project required only thirty months. Reduced model change costs were in part due to the use of common parts and components across model generations, although the rate of commonality was only 16 per cent.

Hyundai also began changing its relationships with subcontractors. An important outcome of the shift in subcontracting from unilateral to bilateral relationships was recognition of the strategic role played by subcontractors in determining the quality and cost of cars. Subcontractor participation in the design process helped reduce costs and lead-times. The bilateral relationship between customer and supplier led to formalization of the 'black box' design concept. Hyundai shared its laboratory facilities to permit bilateral design and development of new components between the company and its subcontractors. Around fifty engineers from fifty subcontractors were resident in these facilities and another fifty engineers from other suppliers cooperated with Hyundai's engineers on a less regular basis. However, the proportion of 'black box' components development was only 29 per cent, in sharp contrast to the situation at Japanese automobile companies.

In the early 1990s Hyundai also started to rationalize and reorganize its logistics through the use of modern information and communication technologies such as the 'value-added network'. A rationalized logistics system was set up in 1994 with 329 suppliers. With the development of this system Hyundai could completely control the logistics chain, which permitted greater flexibility in logistical planning. With the adoption of just-in-time deliveries, the unit of time used for delivery scheduling was radically reduced, from months to days to hours. By 1993 approximately 76.4 per cent of all components bought (by volume) was delivered daily, 17.1 per cent weekly, and 6.5 per cent hourly (KIET 1994: 184). The average inventory at Hyundai was 0.6 days of production. While this delivery system was similar to the Japanese just-in-time system, Hyundai also installed warehouses at its plants to collect and distribute parts supplied by subcontractors. Many of these component manufacturers were so small and their technical abilities so poor that they simply could not afford small lot production.

Hyundai could not, however, eliminate many suppliers and set up a single sourcing system. The labour unrest left Hyundai little choice but to build a dual- and/or multiple-sourcing system in order to avoid work interruptions. Each component was supplied by an average of 2.3 subcontractors in 1992, while in 1985 each component was supplied by only 1.3 firms (KIET 1994: 174). This new trend towards a multiple-sourcing system had a twofold effect. On the one hand, it was difficult to achieve significant economies of scale and improved productivity or quality. On the other hand, the competition between suppliers resulted in lower costs. In practice, this purchasing policy placed strong pressure on suppliers.[5]

## 7.4. LIMITS TO ESTABLISHING A NEW MODEL

Hyundai's future survival and enhanced position in the world car market now depended more on high value-added products than ever before. It was natural, therefore, for Hyundai to invest in human capital formation and automation of production facilities as a competitive foundation for the company. Hyundai's future depended on how management faced these critical challenges. The company would have to learn fast and adapt to the new situation quickly. Workers had already begun to realize that even Hyundai's miracle depended on increased productivity and improved quality, as was well illustrated by the experience of the Japanese car makers during the early 1990s.

To build up a new production system required the motivation of employees. The hybridization of the Japanese management concept into a mass production paradigm required the stimulation of voluntaristic forms of behaviour. However, Hyundai found this very difficult. Workers did not report problems, repair their own machines, or take the initiative to spot and correct faults. Middle management complained that workers were only interested in wages and improving working conditions but not in productivity nor in participating in work design. Manufacturing managers needed to establish a system which would ensure that workers took part in TPM activities.

This situation caused tensions between company and workers. Despite the positive results in industrial relations, the personnel policies of management had not resulted in a sense of shared commitment between workers and management. 'Self-quality control' and 'self-maintenance' were only theoretical on the shopfloor. Most workers were only superficially involved in TPM activities. Workers often argued that this production method was designed to intensify work. While the time cycle was unchanged, workers did extra jobs like quality control and maintenance. Vertical communication and hierarchical coordination based on a strict division between workers and engineers/managers in Korea were rooted in the cultural characteristics of Confucian class ideology. These generally inhibited information flow and created tensions. Under these circumstances human resource management focused on imparting a work ethic that might prevent labour disputes. It remained to be seen how Hyundai would maximize worker skills and development training since the main focus of human resource management remained stable labour management relations, and skills upgrading was still considered secondary.

The reform of the wage system also raises doubts about the extent to which workers can become involved and participate in the company's strategy. The union must try to increase wages and improve working conditions in the face of the significant gains in productivity which are being made through the application of new production concepts, so that the attitudes of workers gradually change, resulting in the establishment of new and productive worker–management relations.

In sum, by the mid-1990s market conditions for Hyundai had worsened. Two new car makers were planning to enter the market. Samsung would be entering automotive production with the technical cooperation of Nissan in 1998 and with a production capacity raised from 250,000 units in the first year of production in 1998 to 500,000 units in 2000. Ssangyong would begin the production of passenger cars with the technical cooperation of Daimler-Benz in 1998 with a capacity of 700,000 units per year. In 1998, when Samsung and Ssangyong launch their passenger cars, the combined production capacity of the Korean automobile industry will reach 5 million units. Competition will be severe and Hyundai will have to tackle many problems if it is to retain its leading position. Huge capital expenditures will be required to increase production capacities. If Hyundai also loses international competitiveness, it could face an overcapacity problem.

## 7.5. CONCLUSION

By the mid-1990s Hyundai had reached a critically important crossroads. The previous strategy was based on the conviction that an increasing degree of product standardization was inevitable, and that the key competitive strategy would remain one of lowering unit production costs through increased economies of scale and low wages. This mass production strategy was rapidly eroded by increased labour costs and poor quality. It also caused labour disputes. Immediate solutions were found in the reduction of unit costs through increased volume, automation, and elimination of waste. The efficient design and manufacture of vehicles of high quality in higher market segments was also an appropriate strategic choice.

The Japanese production methods that had therefore been adopted were bringing about a hybridization of lean production into mass production. By the mid-1990s it remained unclear to what extent this would be embodied in the production process in Hyundai. The key question remained that of human resource management. Yet management did not appear to be pursuing a human-oriented manufacturing system, but rather a man-machine-oriented production system even though mutual trust between management and workers had become an important factor in improving the company. This tendency was rooted in a basic mistrust of management by workers and a lack of interest in the humanization of work by workers and the union. Hyundai's transition, based on flexible automation and Japanese organizational models, was having to take into account the specific Korean cultural conditions and industrial relations practices. Hence Hyundai's hybrid model would be very different from Toyotaism or the lean production system. Hyundai's future trajectory would be governed especially by developments in industrial relations. However, the road ahead remained unclear.

## NOTES

1. One US dollar was worth 861 won in December 1986 and 716 won in December 1990.
2. Unit labour costs increased from 6.7 per cent in 1986 to 9.3 per cent in 1990 and 11.3 per cent in 1993.
3. Under this programme, from 1989, Hyundai implemented a 100 PPM (parts per million) campaign that implied a 0.01 per cent rejection rate in assembly work.
4. Hyundai began heavy investments in advanced and labour-saving manufacturing technologies: 30b. won was invested in automation in 1992, or 0.51 per cent of total turnover. Investment was doubled in 1993.
5. Hyundai required suppliers to reduce unit costs by about 2–3 per cent every year.

**STATISTICAL APPENDIX 7: Hyundai**

| Year | Employment | Production[a] | Export[a] | Gross sales[b] | Profit[c] | Profit ratio |
|---|---|---|---|---|---|---|
| 1970 | 1,957 | 2,356 | — | n/a | n/a | n/a |
| 1971 | 1,705 | 2,398 | — | n/a | n/a | n/a |
| 1972 | 1,608 | 2,615 | — | n/a | n/a | n/a |
| 1978 | 11,679 | 57,054 | 12,195 | n/a | n/a | n/a |
| 1979 | 12,149 | 71,744 | 14,493 | n/a | n/a | n/a |
| 1980 | 9,845 | 43,875 | 12,357 | n/a | n/a | n/a |
| 1981 | 8,838 | 55,992 | 15,199 | 296,758 | −16,447 | −5.54 |
| 1982 | 9,129 | 78,071 | 13,573 | 430,149 | 11,288 | 2.62 |
| 1983 | 10,636 | 93,015 | 16,052 | 577,415 | 25,694 | 4.45 |
| 1984 | 12,372 | 123,110 | 48,286 | 669,252 | 18,362 | 2.74 |
| 1985 | 17,324 | 225,945 | 118,583 | 1,047,037 | 28,781 | 2.75 |
| 1986 | 26,008 | 411,985 | 297,964 | 1,906,429 | 38,333 | 2.01 |
| 1987 | 28,918 | 545,110 | 403,419 | 2,840,211 | 59,698 | 2.10 |
| 1988 | 33,606 | 586,658 | 404,881 | 3,411,146 | 42,421 | 1.24 |
| 1989 | 35,494 | 507,626 | 213,639 | 3,806,510 | 45,207 | 1.19 |
| 1990 | 38,463 | 528,343 | 225,263 | 4,643,171 | 67,511 | 1.45 |
| 1991 | 40,649 | 589,904 | 254,108 | 5,605,244 | 53,762 | 0.96 |
| 1992 | 41,195 | 687,434 | 281,966 | 6,079,027 | 41,629 | 0.68 |
| 1993 | 41,450 | 762,439 | 337,363 | 7,181,184 | 58,233 | 0.81 |
| 1994 | 44,083 | 882,952 | 354,643 | 9,052,254 | 136,788 | 1.51 |
| 1995 | 44,024 | 983,833 | 432,948 | 10,339,100 | 156,600 | 1.51 |
| 1996 | 45,840 | 1,052,886 | 493,971 | 11,489,835 | 86,803 | 0.75 |

*Notes*:
[a] The volume of passenger cars.
[b] Including sales of commercial vehicles (unit: million won).
[c] Profit is calculated by deducting the provision for income tax from income before income taxes.

*Source*: Hyundai Motor Co.

# BIBLIOGRAPHY

Cho, H.-C., *The Strategic Choice of Korean Automobile Industry*, in Korean, (Seoul, 1993).
Chung, M.-K., 'Transformation of the Subcontracting System in the Automobile Industry: A Case Study in Korea', International Motor Vehicle Program (1993).
—— 'Transforming the Subcontracting System and Changes of Industrial Organization in the Korean Automobile Industry', Proceedings of Second GERPISA International Conference (Paris, 1994*a*).
—— 'Production System in Korean Automobile Industry', in Korean, *Economic Studies*, 3 (1994*b*), 127–39.
Hyundai Motor Co., *The History of Hyundai*, in Korean, (Seoul, 1992).
—— *Hyundai Internal Reports*, in Korean, (Seoul, 1994).
—— *Korean Automobile Industry*, in Korean, (Seoul, 1997*a*).
—— *Hyundai Internal Reports*, in Korean, (Seoul, 1997*b*).
Hyundai Motors Workers Union, *Report of Activities*, i and ii, in Korean, (Ulsan, 1994).
—— *Report of Activities 1996*, in Korean, (Ulsan, 1997).
Kia Research Institute, *Korean Automobile Industry*, in Korean, (Seoul, 1994).
Kim, A.-K., 'Market Conditions of Imported Cars in Korea', in Korean, *Journal of Automobile Industry*, 6 (1994), 15–9.
Korean Institute for Industrial Economics and Trade, *Development Tendency towards Twenty-First Century of Korean Automobile Industry*, in Korean, (Seoul, 1994).
Lee, Y.-H., *Fordism and Post-Fordism: Hyundai, Toyota, Volvo*, in Korean, (Seoul, 1994).

# PART II

# THREE DISTINCT TRAJECTORIES AT NORTH AMERICA'S BIG THREE

# 8

# The General Motors Trajectory: Strategic Shift or Tactical Drift?

MICHAEL S. FLYNN

If the automotive industry is the prime example of mass production, General Motors (GM) is surely its epiphany. With a less centralized Sloanist organizational structure and marketing strategy blended with its Fordist production model, GM became and remains the world's largest vehicle assembler (see Chapter 2). The advantages of this model and structure were clear: for four decades after World War II, GM was the model of corporate success, a bulwark of economic stability, and indeed a motor of economic growth. However, the GM variant shared a number of weaknesses with the traditional Fordist model: it constrained the workers' role to tending machines, emphasized standardization and repeatability, and fostered risk-averse strategies better suited to stable than to contested markets. GM, like other automotive companies, eventually became overly bureaucratic, unresponsive to its markets and customers, and tinged by an arrogance tolerated by society only because of its size and financial success.

GM's trajectory since World War II falls naturally into three distinct phases. The first, GM's quarter-century, lasted until the end of 1973, when GM reached the pinnacle of its success, dominating the North American market and worldwide industry. The second period, the oil shock phase, began in 1974 as the first oil shock pummelled the economy and ended in 1984 after the industry recovered from its worst recession in the post-war period. GM seemed to weather this phase substantially intact, and if its 1980 loss was its first in sixty years, it still averaged double-digit returns over the period. The third phase, GM's downward spiral, began as market share fell in 1985, and continued as of 1995, as GM implemented a strategy of revived Sloanism to restore its dominance.

The rise of Japanese competitors certainly challenged the superiority of the GM model and approach, especially during the third phase. Operating with a profoundly different model and structure, they, along with Ford and Chrysler, loosened GM's control of its domestic markets, a dominance which began well before World War II. If Toyota became the paradigm of the superior new model, so GM came to represent the obsolescence of the old, perhaps because it fell so far. Yet the unremitting focus on the competitive prowess of Japanese automobile producers, and the uniformly unfavourable comparisons of the GM model to 'Toyotaism' and 'lean production' seem often to ignore a basic fact: in the mid-1990s, GM remained by far the world's largest producer of motor vehicles. Indeed, GM's survival raises the possibility that the announced demise of the Fordist–Sloanist system, suitably amended, might be premature.

Understanding GM's trajectory is critical to answering questions about the capacity of Fordism and Sloanism either to resist the encroachment of these new industrial models or to adapt to the new competitive environment. Three questions about GM are especially important. First, were GM's difficulties in changing and adapting to new competitive environments rooted in its modified Fordist model, or did they reflect its particular experiences and circumstances? Secondly, did changes at GM and its evolution through 1995 affirm or reject its traditional model? Thirdly, did the new efforts the company launched during the 1980s enhance its competitiveness, or were they irrelevant, perhaps even damaging distractions? The analysis in this chapter suggests that GM's significant success during the first phase solidified strategies and views that were less effective in its changing circumstances and environment in later phases. The legacy of its ingrained and successful hybrid Fordist/Sloanist model was very difficult to shed, and still shaped GM's efforts and trajectory through 1995. Five of GM's six major efforts in the 1980s, as initially conceived, would arguably have made GM even more Fordist, albeit moving towards a newer, higher technology version. These included its technology investments, the creation of Saturn, the Nummi venture, its corporate reorganization, and the acquisition of Electronic Data Systems. GM in the mid-1990s continued to struggle to reverse its trajectory and regain its dominance in the North American market and worldwide industry. GM survived as the world's largest car maker partly owing to modifications of its traditional model, and partly owing to the revival of lapsed elements of that model.

## 8.1. GM'S QUARTER-CENTURY, 1948–1973

The growth of the post-war North American automotive market was fuelled by a flourishing economy, an expanding and dispersing population, and an increasing need for work- and home-related transportation by women.[1] New technologies such as automatic transmissions and power-assisted steering and brakes augmented product offerings and enlarged profits. The industry enjoyed a golden age, producing its 100 millionth US passenger car in 1952, and its 200 millionth just fifteen years later. The Big Three (Chrysler, Ford, and GM) and American Motors (AMC) constituted the domestic industry (or 'Detroit') after Studebaker-Packard ceased operations in 1966. However, as this phase progressed, the industry's Fordist environment became less stable and certain, especially in regard to government demands and market composition.

### 8.1.1. GM's Situation and World-View

GM dominated the North American market and industry throughout this initial post-war period, when its financial strength and performance was legendary. A favourite with investors seeking both income and capital security, in 1955 GM became the first American corporation to net more than $1 bn. in profit. Capital was not a major problem; when GM needed capital, it turned to equity financing or commanded the best available rates from lenders. The corporation pursued a world governed by the basic Fordist model: a market driven by falling costs and oligopolistic competition, ready access to capital, work tightly tied to an increasingly automated assembly line, and production employees collectively bargaining for gains based on productivity. GM, like other North American assemblers, emphasized product design and technology, paying less attention to process techniques and manufacturing technology, and even less to workers, who were viewed more as obstacles to smooth production than as resources, ideally to be replaced by hard automation. The production rule was eminently Fordist: 'get the iron out the door'.

GM was more insular than Ford, where international experience had some value. European countries had forced the US assemblers to establish local assembly operations, and they so accepted the philosophy of 'building where you sell' that they often forgot this had originally been a political rather than a business decision, and pursued it in other markets as well. These overseas operations largely functioned as independent, parallel industries, rather than as integrated divisions of the US parent. The one exception was Canada. A 1965 treaty permitted duty-free movement of vehicles and parts across the border. This 'Autopact' assured Canada a minimum production share (set at Canada's share of the combined markets, typically about 11 per cent) and allowed the US manufacturers to meet an important Fordist goal: production rationalized to serve the larger, combined North American market. Towards the end of this phase, imports became a more important element of Big Three strategy. Chrysler sourced small cars from its affiliate, Mitsubishi Motors Corporation (MMC), as early as 1971, while Ford sourced compact pick-up trucks from Mazda and GM from Isuzu, in which GM had purchased an equity stake in 1971.[2] However, exports continued to play virtually no role in Detroit's strategic thinking. Strong domestic market growth weakened the argument for exporting, as did parallel overseas production, such as GM's Opel subsidiary in Germany: why compete with yourself?

GM did assemble its vehicles worldwide, producing nearly 20 per cent of its vehicles outside North America by 1973, but it remained anchored in the US mid-west, with its executive offices, extensive research laboratories, and many of its assembly and parts plants located in southern Michigan, especially around the city of Detroit. Its North American endowment of over a hundred plants had gradually spread from this base, and, by 1973, the company assembled cars in two dozen plants and light trucks in over a dozen, reaching from Massachusetts through Ontario to California (Ward's 1974).

*8.1.2. The North American Vehicle Market Becomes Progressively more Differentiated*

But for most of this first period the market still approximated the Fordist ideal: numerous nearly identical replicates spilling from the factory to customer. In 1955, the US industry produced nearly 8 million passenger cars, and GM accounted for just over half of the total. Moreover, Chevrolet's 1955 model, which had been introduced in time to be the corporation's 50 millionth vehicle in late 1954, itself accounted for nearly 23 per cent of the total, and the Buick (10 per cent) and Pontiac models (8 per cent) each accounted for substantial production shares as well (Ward's 1956). Economies of scale were critical, and the other manufacturers were loathe to challenge GM's substantial volume advantages, fearing price competition which they could only lose. They generally avoided distinctive styling, allowing GM to test the market, then producing similar versions themselves a few years later. If outsiders tended to view all these companies as strong, only GM viewed itself that way. The others saw their survival as precarious in the looming presence of GM. But GM, apprehensive about government antitrust action to force its break-up, restrained its efforts to capture further market share.

This climate provided little incentive for competition, and the offerings of all the manufacturers largely converged on dimensions such as price and quality, fostering an extremely stable competitive environment. Competition for the marginal percentage or so of market share was based on styling variations, comfort, and performance (especially power and acceleration). In this environment, the number and location of retail and service outlets were crucial. Here, too, GM had a distinct competitive edge: it had the most dealers, the widest distribution, and its dealers were generally viewed as the best. These dealers were independent franchisees, but throughout this first phase, their business interests roughly coincided with the assemblers: sell more units, and the bigger and more option-rich, the better (Cray 1980).

The car companies reacted more to each other's competitive moves and decisions than to customer preferences and desires. They talked to their customers, but less often listened to them, and came to believe that customers should simply buy whatever the firms produced. For example, imports had small market shares concentrated in the sports car and luxury segments. Detroit viewed the small volumes of sports cars as unattractive, since they did not fit the Fordist economy of scale paradigm, and had little concern over the small import share of the growing luxury car segment. However, the US success of Volkswagen (VW) soon signalled a challenge in the mass market entry-level segment. In 1955, the VW Beetle quadrupled German imports, vaulting Germany past the United Kingdom as the main source of imports into the USA. The Big Three initially responded reluctantly, introducing smaller compact cars, which they viewed as unexciting, difficult to sell at a profit, and likely to cannibalize the sales of more profitable vehicles. However, smaller, often imported, cars continued to gain share in the 1960s. The Big Three then tried to shift their customers' preferences to the new specialty vehicle, typically more powerful, stylish, and fun to drive than standard American cars. This segment emerged in GM's 1962 mid-year offerings of its compact Corvair, but fully flowered in Ford's 1964 Mustang, with first-year sales of over 400,000. This demonstrated Detroit's ability to define its market, but diverted its attention and effort from the growing small car segment.

The Beetle remained the best-selling US import throughout the 1960s, capturing just over 50 per cent of import sales in 1969, when import sales first reached 1 million vehicles, at 11 per cent of the market, and Toyota became the first Japanese company to reach 100,000 sales.[3]

### 8.1.3. GM's Organization

In 1955, GM's Chevrolet Division rivalled Ford in size, complexity, and the scope of its internal activities, operating twelve assembly plants and eleven plants manufacturing engines, transmissions, bodies, and major components. Chevrolet developed its basic 1955 model, including a new transmission and engine, in roughly eighteen months. This vehicle alone accounted for the overwhelming share of the division's production volumes, revenues, and profits, so it received most of the divisional Chief Engineer's attention and time. The car's volumes required as many as seven plants, making manufacturing a high priority because of the need to ensure standardization across so many facilities. In effect, Chevrolet in the mid-1950s operated with closely integrated and coordinated product and manufacturing engineering. In the early portion of this phase, GM's product strategy truly was a model of Fordist focus and efficiency.

However, if GM's production operations and world-view were decidedly Fordist, GM's chairman, Alfred Sloan, had by the 1930s created a management structure and marketing strategy that substantially departed from that model. First, he decentralized most decision-making to the divisional level while maintaining centralized, rigorous financial controls. Secondly, he implemented a market strategy based on divisional product differentiation to attract all potential customers (a car for every purse and purpose), but with adjacent divisions overlapping at the price boundaries to spur internal competition, improve efficiency, and accelerate performance (Cray 1980). The GM divisions—Chevrolet, Pontiac, Oldsmobile, Buick, and Cadillac—spanned the entire price range, provided distinctive market images, while offering the most expensive Chevrolet in the same price range as the least expensive Pontiac, and so forth up the line. GM also maintained two light truck divisions, Chevrolet Truck and the more upmarket GMC.

A strong element of GM culture, its unusually consensus-oriented management style, also departed from the Fordist model. The corporation was managed by committee, vastly leveraging the company's talent pool. Moreover, GM attracted and retained top talent, with management careers that were less 'up or out' than the other assemblers. GM had strong and talented management throughout its many divisions, functions, and levels.[4]

There was some erosion of the Sloanist elements in GM's blended Fordism later in this trajectory phase. For example, GM created the Assembly Division (GMAD) in 1965 to assemble almost all of the corporation's vehicles, centralizing what had been a divisional responsibility. This move was prompted by three circumstances. First, the car divisions had perhaps become too independent, resulting in duplicate effort, reduced scale economies, and divisional product preferences which sometimes blocked efficient production. While Buick and Chevrolet manufactured virtually identical engines, each insisted on minor distinguishing differences and separate manufacturing plants. Secondly, divisional decisions could blur product lines and strategies. GM Styling—a centralized function—developed a new car for Cadillac in 1962. Cadillac rejected it, but Buick then adopted it, reducing corporate strategy to the sum of divisional decisions. Thirdly, it made it more difficult for the US government to attempt to break up GM, since the divisions were less viable as independent, full-scale companies.[5]

*8.1.4. GM's Labour Relations throughout this Period are Uneven*

There were sporadic bitter strikes, including a United Automobile Workers (UAW) work stoppage of 67 days in 1970 resulting in a 6 percentage-point light vehicle market share loss to 38 per cent.[6] The next year, 1971, saw an extremely bitter confrontation between GMAD and the UAW at Lordstown Assembly, directly resulting from GMAD's Fordist attempts to improve productivity by replacing workers with robots, then markedly increasing the pace of the workers who served those robots. GMAD had managed to turn factory automation simultaneously into a job threat and worker speed-up tactic. When the strike was settled, the image of GM as the hard-hearted profiteer at the expense of workers lingered, and, for the next twenty-five years, Ford and Chrysler enjoyed an important edge in labour relations. Wage rates were competitively neutral in theory, since the UAW pursued 'pattern bargaining', targeting one company for negotiations and then requiring all other companies to accept this initial settlement as the model for their own contracts. However, pattern contracts indirectly benefited GM because of its scale economies. UAW strategy targeted GM when pay increases were critical and Ford when new elements of the social wage were at issue (Lacey 1986). The UAW did well for its membership, but higher labour rates were of minor concern in a growing and oligopolistic market.

*8.1.5. Supplier Relations*

GM sought high levels of vertical integration, ensuring control of product design and cost, while offering profit opportunities in the face of constrained vehicle shares. GM purchased numerous supplier plants and integrated them into its operations, becoming, and remaining as of 1995, the most vertically integrated of the world's major vehicle producers. Its purchasing strategy emphasized price and reliability as selection factors. GM demanded—and typically received—large discounts. Nevertheless, GM offered suppliers substantial volume advantages over its rivals, important because most suppliers were themselves wedded to Fordist economies of scale. Most suppliers considered GM their premier customer, providing GM first refusal on new technology and products. However, GM's high vertical integration created some tensions. Independent supplier companies felt that they were often forced to compete against artificially low bids by internal (or captive) suppliers, and worried that their own proprietary technology and business information would be exposed to these competitors.

## 8.1.6. Government Relations

GM was not reluctant to influence government, and successfully did so on numerous occasions throughout this phase. However, when government efforts to influence GM and the other assemblers became more common in the 1960s, GM resisted this 'interference' in product decisions about vehicle safety and emissions.

Cars grew more powerful and faster, heightening safety concerns as people reacted to the growing absolute number of traffic accidents and fatalities, rather than the falling rates. After twenty-four US states required seat belts, they became standard equipment for the 1964 model year. Major legislation and a Cabinet-level Department of Transportation came in 1966, and the National Highway Safety Bureau issued twenty vehicle safety standards for implementation in 1968. The consumer advocate Ralph Nader published a blistering attack on GM's Chevrolet Corvair in 1966, *Unsafe At Any Speed*, portraying GM as a callused company that knowingly sold vehicles that posed deadly risks for operators and passengers. GM's unsavoury efforts to undercut Nader's credibility accelerated consumer safety concerns and suspicions, further exacerbating a major customer and public relations problem, and, in settling Nader's civil suit, ended up funding a group that has monitored and criticized the industry's behaviour since. California required emission control equipment in 1960, and, beginning in 1963, Congress passed legislation establishing a range of air quality standards, culminating in 1970's Clean Air Act, which further tightened emission standards. This inaugurated a long period of contention over vehicle emissions. By 1973, so many different government authorities and agencies regulated various aspects of its activity that the industry's relationship with government became extremely complex and often hostile. Different agencies took actions with little apparent coordination or regard for their combined effects, while the industry resisted government initiatives, labelled them technically impossible, and waited for a roll-back in regulation that had not developed as of 1995.[7]

The Big Three and GM ended this period in dramatic fashion. In 1973, the USA and Canada still accounted for 36 per cent of all motor vehicles produced in the world. The Big Three set North American production and sales records of nearly 13.5 million and over 15 million units respectively. GM worldwide assembled 7 million cars, added another 1.6 million light trucks, employed over 800,000 workers, achieved a return of 21 per cent on assets, and reaped record corporate profits. However, there were clouds on the horizon, and the successful strategies and initiatives of the past were already beginning to reveal future competitive weaknesses.

## 8.2. THE OIL SHOCK PHASE, 1974–1984

Two external events fundamentally altered the form of North American automotive competition during the second phase of GM's post-war trajectory. First, in October 1973, OPEC cut off supplies of Arab oil, and gasoline prices rose sharply, quickly escalating concerns about fuel availability. Secondly, in 1979 Iran led a second cut-off of oil exports. The short-term effect of these twin oil shocks was dramatic. Vehicle sales plunged by 20 per cent in 1974 and 1979 from their previous annual levels, as each oil embargo in turn threw the North American economy and industry into serious recession. The North American consumer suddenly cared about fuel economy, and turned to fuel-efficient imports, a preference that receded as fuel concerns and prices quickly abated after the first shock. After the second oil shock, consumers again demanded fuel efficiency, and this time demand persisted longer than it had after the first shock. The long-term effects were equally dramatic, as two more elements of the Fordist environment collapsed: the Big Three's oligopolistic market control began to crumble, and job security began to rival wage rises as labour's paramount bargaining concern.

### 8.2.1. *GM's Situation and World-View*

The first embargo hit GM especially hard, and it lost 2 percentage points of market share in 1974, primarily to Chrysler and Ford. This reflected the relative concentration of GM in the full size or family passenger car segment, the segment hardest hit by the oil shocks. However, the industry and GM recovered rather quickly from the first oil shock, and 1978 set another new record for light vehicle production and sales. GM introduced its small, fuel-efficient Chevette in 1975 and an entire line-up of smaller, front-wheel-drive, fuel-efficient cars by early 1979, so the second oil shock had little effect on its share.[8] GM, with only one less assembly plant than it had in 1973, averaged a share of over 43 per cent for this entire phase, about 1 percentage point below its level in the 1960s.

In the early 1980s, GM began a string of efforts that would reinforce the Fordist elements of its approach, while further retreating from Sloan's modifications. These efforts may have been elements of a grand and coherent strategy, but appeared less and less related as time passed. Rather than charting a strategic shift, they formed a tactical drift at GM that became even worse in the next phase. Moreover, undertakings that were successful failed to diffuse throughout the corporation, restricting them to limited experiments at best.

GM considered the Japanese important during this period, but viewed them as less serious competitive threats than did Ford or Chrysler. Nor had GM yet recognized that the Japanese assemblers and suppliers offered a new and very different model of vehicle manufacturing. Throughout this phase GM's performance continued to be impressive, and that very success trapped it in its Fordist assumptions and views.

GM explained Japanese car makers' success in ways that de-emphasized its own competitive weaknesses. The company viewed Japanese sales as largely based on fuel economy, an advantage destined to be eroded as the oil situation stabilized and American preferences for larger cars reawakened. GM would be ready, as it introduced a series of mid-sized platform cars planned for high volume sales beginning in 1981, and in 1980 it had already begun work on its subsequent generation mid-sized cars, the GM-10 programme. Few Japanese assemblers achieved the scale economies for Fordist low cost, so GM initially believed their lower manufacturing costs must be rooted in its own recalcitrant and high wage labour rather than any fundamental process disadvantage.

Many GM managers dismissed early Japanese quality advantages as unimportant, reflecting appearance items and consumer misunderstanding of real (i.e. engineering) quality. If younger buyers were more likely than older to prefer Japanese cars, GM believed that, as they aged, these buyers would purchase the larger vehicles GM thought more suitable to their changing family and financial circumstances. These explanations insulated management from blame and also blocked effective competitive responses.

## 8.2.2. Market Change Accelerates

The share of total vehicle sales taken by light trucks grew from roughly 20 per cent in 1974 to 25 per cent by 1984. Explanations for this shift ranged from the less powerful and smaller passenger cars available in the market to the masculine image and relative value of the pick-up truck. Chrysler introduced minivans in 1983, contributing to GM's 1984 light truck share loss of nearly 4 percentage points. Passenger car volumes dropped, and the record sales year of 1978 saw the top five selling cars average under 450,000 copies, down from nearly 1 million in 1955. Import share increased, primarily at the expense of Ford and Chrysler, which lost 8 points and 2 points, respectively, while GM managed a 1 point gain. The source of imports changed, as the Japanese assemblers moved from 5 to 18 points, gaining share at the expense of European importers as well as the Big Three, and, importantly, expanding that share as concerns for fuel economy waned between the oil shocks (Fig. 8.1). Japanese assemblers greatly expanded their distribution networks, typically by 'dualling' with established Big Three dealers whose franchisees wanted smaller, less expensive, and more fuel-efficient vehicles to fill out their customer offerings. The production base became internationalized as well, as VW began production of the Rabbit in Pennsylvania in 1978, and Renault took effective control of AMC in 1979. Honda first established motorcycle operations, then added passenger cars in 1982, the year Nissan opened a light truck assembly operation in Tennessee. However, Fiat left the US market after 1983.

Consumer memories of the oil embargo had faded, and Pontiac, Oldsmobile, and Buick all found demand for V-8 engines exceeding supply. In most

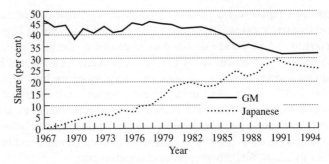

FIG. 8.1. *Selected North American light vehicle market shares, 1967–1994*

*Note*: Data from 1967 to 1973 inclusive are displayed to provide a sense of the first phase. The last year in which Japanese assembler share fell below 1 per cent, 1967, was also the year GM produced its 75 millionth vehicle in the USA.

instances Chevrolet V-8s were functionally equivalent, so they were installed in these other divisions' cars. Lawsuits, based on customer beliefs that these divisions, and therefore their engines, differed and had varying value forced GM into a costly rebate and extended warranty settlement of claims. More importantly, many consumers felt that GM had somehow deluded customers into buying inferior products, shaking consumer trust in GM.

In September of 1980 Roger Smith, the Chairman-designate, indicated that he anticipated no major or substantial changes in GM's business philosophy or operating principles. Indeed, GM confidence in the early 1980s must have been high. Under Smith, GM would eventually invest over $50 bn. to open eight new plants and refurbish nineteen others with the latest in automated technologies (Keller 1989). If it opened plants in Europe that would later require it to close plants in North America, this only reflected GM's historic parallelism and continuing blindness to North American exporting as an important part of a successful worldwide strategy. But it also showed the value of this worldwide activity: European operations provided the corporation with important revenues during a period of weak markets in North America in the early 1980s.

GM still pursued a Fordist product model, and its GM-10 programme was to provide mid-sized cars, based on one platform, to all divisions except Cadillac, reaping enormous economies of scale. GM soon recognized that Japanese cost advantages might be based on more than low wages, but remained wedded to the Fordist belief that any Japanese productivity edge resided in superior factory floor automation, not in the fundamental balancing of product, process, and people. By the autumn of 1981, GM reported it had 1,200 robots installed or on order. In fact, in 1983 GM entered a joint undertaking with Fujitsu-Fanuc of Japan to provide itself and the rest of industrial America with the latest in Japanese robotics (Ward's 1982).

In early 1983, GM formalized an agreement to undertake small car production jointly with Toyota at a closed GM plant in California, securing a competitive small car and an opportunity to learn Toyota technology for its own new small car project, Saturn, announced at the end of 1983.[9] Saturn, a new division, was to build an affordable and profitable small car, suited to the standards and tastes of import owners. Saturn was also to be a new model for high technology production at GM, reflecting Japanese technology and lessons to be learned from GM's own new Detroit area high technology plant scheduled to open in mid-decade. Saturn would also depart from the Fordist model of labour relations, and build on effective relationships between management and labour.

In 1984, GM purchased Electronic Data Systems (EDS), a data processing company, both to diversify its business and acquire the capabilities of a paperless, high-tech manufacturer. GM also announced that it was reorganizing its car divisions into two groups, thereby reducing the duplication costs of the old system and restoring scale economies. CPC (for Chevrolet, Pontiac, and GM-Canada) specialized in small cars, and BOC (for Buick, Oldsmobile, and Cadillac) focused on large cars.[10] Numerous glitches and extreme confusion plagued the reorganization and, as GM started to settle into its new pattern, further, smaller reorganizations suggested the old structure was reviving. For example, a metal stamping plant was soon identified as part of the Cadillac division.

### 8.2.3. A Time of Increasingly Strained Labour Relations

The UAW's relationships with the different companies began to diverge. GM, wedded to Fordist economies of scale, expanded capacity, opening plants in Oklahoma (1978) and Louisiana (1981), far from its historic homeland. This Southern Strategy was widely viewed as anti-union. GM elected to open these plants without union representation, but it is unlikely that anti-union sentiment was the primary driving force. The population was moving south and west, so this put assembly plants close to the market, and management also hoped to attract workers with a stronger work ethic. This attitude further strained relationships with the UAW, which in any case soon organized these plants. As Japanese competition stiffened and the North American economies worsened in 1982, acrimony between the UAW and the companies heightened over who was to blame for the Japanese cost advantage, widely believed to be of the order of $2,000 for a small car, and what, if any changes would be required to close this gap. GM stressed the role of high wage rates, which aggravated the UAW, while Ford targeted lower productivity, and Chrysler identified both these factors, plus Japanese tax advantages.[11]

Ford was first to succeed in establishing a better relationship with the UAW (see Chapter 9). The 1982 contract called for profit-sharing and joint efforts to improve competitiveness, especially in the areas of productivity and quality. GM experienced selective strikes before reaching its own agreement, with a less generous profit-sharing formula which became a source of morale problems, as did management's continual threats of lay-offs and criticisms of high UAW wages while GM paid record executive bonuses. It is difficult to assess how much of GM's poor relationships during this period were due to actual differences from Chrysler and Ford, and how much to a management arrogance and indifference to labour that sparked focused resentment of GM.

### 8.2.4. GM's Supplier Relations Change

As the industry entered the 1980s, each of the Big Three pressured their suppliers to lower part prices, improve quality, and make technical investments. Near the end of this phase they initiated more structured programmes to help suppliers improve. GM lagged behind Ford in its efforts to support its suppliers' acquisition and implementation of various manufacturing techniques and routines, such as Statistical Process Control (Flynn and Cole 1986).

GM sought to reduce the number of its suppliers, because of the costs of maintaining so many relationships. The Japanese assemblers' lower levels of vertical integration[12] appeared to be a source of competitive advantage, so GM especially considered more purchasing and less in-house manufacturing of parts and components, reversing another element of its traditional model. However, if supplier reductions and outsourcing seemed to promise more business to survivors, GM also turned its purchasing strategies outwards from North America, making it clear that it would now consider suppliers worldwide eligible to bid on contracts and provide parts, a move which threatened survivors' business levels. Corporate purchasing still seemed weak, and most such strategies to reap economies of scale across divisions and platforms seemed to fail.

### 8.2.5. Government Action Affecting the Automotive Industry Moves into the Trade Arena

Political pressures began to mount as the US balance of trade turned sharply negative, sparking concerns about the large bilateral deficit with Japan, and its single largest component, automotive trade. Pressure for government action increased sharply as falling sales and increased Japanese share yielded a near-depression economy for the traditional industry beginning in 1980, putting Chrysler through a near-death experience (see Chapter 10). In 1981 the US and Japanese governments agreed to limit Japanese passenger car exports to the USA. The Voluntary Restraint Agreement, or VRA, gradually raised these limits over three years, and was renewed for 1984 with a higher quota.[13] In 1982, US courts held that a 25 per cent tariff on light trucks, originally aimed at European vehicles, also applied to Japanese imports.[14]

If government in the first phase had often frustrated the automotive industry by its failure to speak with one voice and to coordinate its many demands, it was the industry in this second phase that often frustrated government on issues of Japanese competition. Chrysler, Ford, and the UAW frequently pursued governmental support and relief, but GM performed well throughout this phase, and opposed government action on ideological grounds. This lack of agreement often stymied government action and drained its sympathy for the industry's situation.

Congress, in response to the first oil embargo in 1975, passed the Energy Policy and Conservation Act establishing Corporate Average Fuel Economy (CAFE) standards to take effect in 1978 and rising thereafter, while specifying penalties for failure to achieve the standards. The 1978 Energy Tax Act enacted a special tax for vehicles with fuel economy performance well below CAFE standards.[15] CAFE became a dominant industry concern for the next decade, affecting the companies in a variety of ways. First, CAFE required the industry to reach high levels of fuel economy rather quickly, demanding resources and often steering product design decisions to unpopular compromises. Secondly, the legislation defined two fleets, based on the North American content of the vehicle, which each had to meet the CAFE standard separately, complicating product planning and sourcing decisions.[16] Thirdly, CAFE did nothing to encourage consumers to value fuel efficiency, forcing manufacturers to control not just their range of product offerings, but their mix of actual sales as well. CAFE dominated GM's product development programmes even more than Chrysler's and Ford's. This partially reflected GM's greater reliance on large cars, as well as the fact that the industry in general relied so much on GM laboratories for a successful response to regulatory demands.

GM ended the period in good shape with a 42 per cent market share, having set five annual profit records, and confident that the strategy of high-technology diversification and operation was in place and would soon bear fruit. Indeed, the Japanese share had fallen 1.5 points since 1982. GM largely stayed the course through this middle phase, and that seemed to be a wise and successful strategy. And yet 1984's record profits were lower in constant dollar terms than they had been for six separate years during the 1970s, and GM had fallen substantially behind Ford in labour productivity, remaining only marginally ahead of Chrysler. In fact, GM was adrift in uncharted waters without a strategy suited to its new and developing challenges.

## 8.3 GM'S DOWNWARD SPIRAL, 1985–1994

The entrance of so many Japanese and European competitors destabilized competition in the North American market by the early 1980s, and the market remained fragmented and intensely competitive as of 1995. By 1985, the Japanese challenge dominated Ford's and Chrysler's agendas, and became a growing concern at GM. Thus Japanese competition dominated the strategic thinking of the Big Three throughout the second half of the 1980s even more than regulatory demands drove their strategies in the 1970s. Moreover, the competitive relationships among the Big Three shifted, as Ford and Chrysler were no longer reluctant to challenge GM directly as its dominance faded and it became more the first among equals. Still, if GM's survival was ever seriously threatened during this phase, it was certainly more in terms of its past dominant performance levels than its actual existence.

### 8.3.1. GM's Situation and Strategy

GM's performance and fortunes began to decline sharply in 1985, while Chrysler and Ford improved their competitive performance throughout much of this third phase. Indeed, Chrysler and Ford market share gains came largely from GM, with fewer adverse effects on the Japanese manufacturers until 1992. By 1994, GM operated fifteen car and twelve light truck assembly plants, down from twenty-three and twelve plants, respectively, ten years earlier.

GM continued to acquire other companies, purchasing Hughes Aircraft in 1985, with a view to enhancing its engineering and electronics capabilities, including access to technical personnel at Hughes. Whatever the ultimate value of this purchase, it created some image problems for GM. It followed so closely on the purchase of EDS that some observers questioned how committed GM was to the core business of auto production. It also sparked confrontations between Ross Perot, the founder of EDS, who had joined the GM board, and Roger Smith which would eventually result in the buyout of Perot's stock and his resignation from the Board. Many among the public saw Perot as the savvy entrepreneur challenging the wrong-headed managers of an arrogant and oversized corporation.

If GM remained committed to the automotive business, it clearly was still seeking a Fordist-style technology solution to its competitive challenges and quality and productivity problems, albeit a distinctly high-technology version. GM had planned a new Cadillac assembly plant in the early 1980s, and this high-technology marvel opened in 1985, a showcase for the latest in automated production technology. From the first day, little went right, as workers were unsure of their jobs and robots unprogrammed to do theirs, and it took years for the plant to achieve smooth operation. Hence Hamtramck Assembly, meant to be a stepping-stone to Saturn as the fully modern plant, became an important lesson, especially in contrast to GM's low-technology venture with Toyota. More managers questioned the wisdom of GM's extraordinary reliance on technology, although that questioning remained muted for some time. GM continued to invest heavily in traditional automation, a mistaken choice that fortune had denied to cash-poor Ford and Chrysler.

GM was slow to modify its traditional Fordist production model in the direction of the Toyota production paradigm in spite of the collaboration at Nummi. GM first viewed Japanese productivity as the result of work pace and automation, and was attracted to Toyota's use of the hard technologies, like robotics. It only recognized the importance of the soft technologies like just-in-time later (Flynn and Andrea 1994). Most of the North American industry had already begun to modify its Fordist production techniques by the early 1980s, largely due to

Japanese competitive success. Ford, for example, increased plant productivity by some 20 per cent between 1984 and 1989, while it remained virtually flat at GM (Harbour 1995). Indeed, when Hamtramck Assembly became fully operational, the factory had virtually no unit productivity advantage over Ford's Wixom plant, and less flexibility, despite a much higher investment level. Each of the Big Three learned in the context of its own Fordist paradigm and experience, and GM, mired in the success trap set by its earlier accomplishments, changed more slowly than its two rivals.

The insular outlook of GM in North America created an ironic comparison during this trajectory phase. Opel's competitive position had seriously weakened in the second phase and GM began to pour capital into its major European subsidiary in the late 1970s. With new products and factories, Opel reversed its decline and became a strong European competitor and a star in GM's worldwide operations by the late 1980s. Opel's new plant in Eisenach in Germany, with substantially lower capital investment and very heavy reliance on Japanese manufacturing strategy compared to Hamtramck, became GM's most productive plant by 1994. The particular irony is that, even though much of this turnaround involved key managers from North America, GM in North America did not find it any easier to implement the lessons learned at Opel than it did those from Nummi or Saturn.

*8.3.2. Competition Becomes much more Complex*

GM suffered large share losses. In 1985, Ford's new Taurus/Sable model hit GM's core segment, the mid-sized family car, and GM's car share slid 3 points. Ford's continued success, combined with market share gains by Chrysler and the Japanese, gave GM another 4 point loss in 1986. By 1992, GM's share of the light vehicle market fell to 32 per cent, and remained at that level throughout 1994, some 8 points below 1985's share.

Light truck sales continued to grow throughout this third phase, virtually unaffected by the brief fuel price rise in 1991 at the time of the Gulf War, and they accounted for roughly 40 per cent of light vehicle sales in 1994. Minivans, which served suburbanites and families well, played a major role in the expansion of light truck share over the first half of this phase, while the multiple-use SUVs (Sports-Utility-Vehicles) were major drivers during the 1990s. These vehicles, like passenger cars, spanned the price range from entry-level to luxury segment. Ford and Chrysler captured light truck share from GM and the imports, while passenger car sales pressured GM's reliance on Fordist scale economies. The top five cars in 1994 sales averaged about 280,000 copies, or roughly the output of one efficient assembly plant each. The four late-to-market W-cars, planned to be GM's high volume models for the mid-1980s, peaked in 1990, at a total of 580,000 copies. Figure 8.2. shows that, since 1967, GM's mix of light truck and car sales had closely tracked the overall market ratio, departing by more than 3 points only

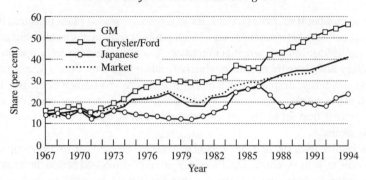

FIG. 8.2. *Light trucks as a proportion of North American total sales, 1967–1994*

in 1984, when its light truck share suffered a sharp loss to Chrysler's new minivan. After 1973, Ford's proportion of light trucks sales exceeded the market ratio, and Chrysler's followed after 1986, a record sales year. The Japanese assembler mix was skewed to passenger cars, apart from the years 1984 to 1986. For GM, years of market dominance suggested that its close market tracking represented excellent strategic positioning, allowing it to follow market shifts, while its competitors faced serious risks should the market fluctuate sharply. In reality, GM faced strong, more specialized competitors which forced it to compete on two fronts, perhaps explaining its eroding market share.

Japanese makers' shares totalled 29 per cent in 1990, but receded to just under 26 per cent in 1994, by which time almost 60 per cent was sourced from a North American production capacity in excess of 3.3 million units. The rapid strengthening of the yen from 240 to 160 to the dollar between 1985 and 1986 increased the dollar cost of goods exported from Japan by 50 per cent. but lowered the yen cost of direct investment by one-third, providing economic pressure for a second wave of Japanese automotive investment in North America. By 1989, all the Japanese passenger car makers except Daihatsu had established assembly plants in North America, transferring many of their competitive advantages to these new facilities. Other producers experienced mixed outcomes in North America. Renault abandoned the US market after 1986, while Hyundai became the first Korean manufacturer to enter the North American markets: Canada in 1985, then the USA in 1986. Based on its initial sales success, Hyundai opened a plant in Canada, but closed it in 1993 as sales fell. Renault sold its equity in AMC to Chrysler in 1987, VW sold its last US-produced cars in 1988, and Peugeot left the US market in 1991. BMW and Mercedes established US production facilities to assemble a sports car and an upmarket SUV, respectively. European and South Korean makers combined captured about 4 per cent of the 1994 light vehicle market, but the presence of so many competitors complicated GM's challenge.

The Japanese assemblers changed the automotive market's competitive dynamics. As fuel economy receded as a purchasing criterion, many buyers chose Japanese vehicles because of their higher quality, whether measured by vehicle defects or reliability. The 1981 quality gap of some five defects per car had declined 95 per cent by 1991, but this probably slowed rather than reversed Big Three share loss. Styling became more a contested dimension, especially as GM stumbled in the mid-1980s with its 'look-alike' A-cars; some Japanese vehicles had become style leaders by the decade's end.

As was true of GM in the first phase, Japanese assemblers may have been reluctant to exploit their cost advantages in direct price competition for fear that increased market share would create political problems. Rather, their lower costs in phase two funded product development and strategic undertakings, such as entry into new segments. This greatly expanded the range of choice available in the market and disrupted the traditional relationships among vehicle size, comfort, and available options (Office for the Study of Automotive Transportation (OSAT) 1992). The move to larger and more expensive cars met their customers' needs as these changed with age or higher income. Larger vehicles, like the Honda Accord and Toyota Camry, and upmarket models like the Nissan Infiniti, competed successfully in the mid-sized range, the heart of GM production volumes, and in the luxury segment, the core of its profits, and a segment already contested by upmarket European producers. The yen again appreciated in the early 1990s, and Japanese vehicle prices rose above comparable Big Three offerings. By then, Japanese assemblers were less concentrated in price-sensitive segments and any price difference perhaps constituted a Japanese premium rather than a Big Three advantage.

GM's deterioration through the first half of this phase took on the characteristics of a downward spiral, as each problem seemed to accelerate its momentum and move it to a new and lower level. This pattern was characteristic across numerous performance dimensions, including market share, plant performance, or necessary capacity reduction. GM had tremendous difficulty adapting its past practices to a more differentiated and competitive marketplace.

### 8.3.3. GM Organization and Structure

By the beginning of this period, GM's swift and effective product development execution of twenty years earlier slowed and stumbled. The W-cars illustrate this problem. In September 1987, the Buick Regal, the first of the W-cars from the GM-10 programme initiated in 1980, reached the market some two years late. The last of the four cars, the highest volume Chevrolet Lumina, would not become available until May 1989. The eighteen months to develop the 1955 Chevrolet was extremely rapid by the standards of the day, and vehicles had become more sophisticated and complex in the intervening years. Yet these factors alone are insufficient to account for the protracted development of the W-cars. Indeed, GM lengthened and complicated this programme by changing engine, transmission, and body-style requirements a number of times. This also moved its key decisions farther from the market. For example, the W-cars' production timetable was set early, when two-door sedans enjoyed a brief sales surge, so the once-more popular four-door models were unavailable at introduction in 1987.

Throughout the 1980s, Japanese car makers generally had a significant edge in product development, bringing new vehicles to market with substantially lower total investments and in much less time. A number of factors have been offered to explain this advantage, most notably differences in organization (platform versus functional, integrated versus separated product and manufacturing engineering), the role of suppliers (level and timing of involvement), and differences in general efficiency between the two industries.

Since 1955, GM had experienced a fundamental evolution in its product development process and the organizational structure that supported it. These transformations largely occurred in the day-to-day practice of organizational routines, and were typically not heralded by major announcements of changes in the organizational chart. Hence, they may appear less important than they in fact have been in the erosion of GM's product development performance. Over time, the number of models grew and their average volumes decreased. The Chief Engineer became a vice-president with responsibility spread more evenly across a number of vehicles, with new concerns for the status and role of engineering *vis-à-vis* other functions like finance and marketing. Engineering careers became more dependent on the engineering function, rather than the car division. The number of plants required to produce a car decreased, and 'one vehicle, one plant' became the rule rather than the exception, and a binding rule, as in the case of the W-car. In effect, GM gradually lost its divisional platform focus and integrated effort, and adopted the functional organizational form with all its risks of more fragmented focus and effort.

Changes in the market, combined with the internal complexity of the corporation's task, drove a change in product development processes and structure that eventually adversely affected performance. To be sure, this was not the first time that such a sequence had unfolded in the automobile industry. Ford, in its early days, pursued manufacturing strategies similar to the Japanese strategies that it adopted in the mid-1980s, while Sloan emphasized the critical role of quality and supplier selection and development, again not so very different from the 1980s and 1990s.

GM still planned for a Fordist market when the market was already differentiating vehicles to an extent that prevented the achievement of traditional volumes. The rule of 'one vehicle, many plants' in the first phase had shifted to 'one vehicle, one plant' in the second phase, and was becoming 'many vehicles, one plant' in this third period. For example, GM originally expected its W-cars to fill at least seven plants and assigned them to four, but they reached volumes sufficient for only two. In 1992, the programme's four assembly plants captured just 16,000 more sales than the two Ford plants producing the Taurus and its platform-mate, the Mercury Sable. However, GM was unable to consolidate production because of its inflexible capacity, reflecting its Fordist emphasis on dedicated equipment. This lack of flexibility meant that in 1994 GM ran above capacity in light truck plants and well below capacity in car plants (Harbour 1995).

The 1984 reorganization illustrates a pattern which became more frequent as time passed and more detrimental as GM's resources became tighter. Besides eliminating duplication, GM top management had other motives for reorganizing in 1984. They were concerned about the concentration of power in two divisions, Fisher Body and GMAD, their resistance to change, and inefficient performance. The solution was to form two new groups which would absorb elements and functions from each of these divisions. This reorganization, presented as a return to Sloan's decentralization, actually seemed to carry GM further along the road to Fordist centralization, as it combined functions of Fisher Body, GMAD, and the five car divisions into two groups. The car divisions were to be strictly marketing organizations, with other functions ceded to the two main groups. This further subordinated the marketing strategy to Fordist demands for production scale.

GM often adopted quick fixes rather than carefully considered organizational solutions, although the former frequently had wide-ranging effects. The 1984 reorganization damaged morale, as many long-time GM managers found their networks disrupted and their traditional divisional loyalties undercut (presumably intentionally) but with no communication structure or sense of membership to replace them (presumably unintentional). GM's change efforts often seem to have been predicated on mis-estimates of the benefits of the old forms—an effective informal network and loyal workforce in this instance—and of the adverse consequences of the new policy of undercutting worker loyalty and initiating an extended period of corporate turmoil. GM increasingly stumbled, undertaking costly, low-yield efforts to transform itself. Even when it overcame the success trap and recognized the need for change, it often fell into the quick-fix snare, experiencing great difficulty in developing and implementing effective changes.

GM suffered from a paradox. If it sometimes seemed careless and quick in its decisions, it could also ask an important question in the wrong way, and move at a glacial pace in answering it. By 1980, for instance, GM was considering whether to shed a division, particularly focusing on whether Buick and Oldsmobile any longer met distinct market needs. Both of these divisions had production shares of about 6 per cent from 1967 to 1971 inclusive, but then began to diverge, with Oldsmobile as much as 2 points above Buick by the late 1970s. GM looked at Buick, asking if the division had lost its rationale as the market moved away from the full-size, near-luxury cars that were its mainstays. However, by 1982, at the bottom of the downturn, Buick's share reached 9 per cent. Because buyers of expensive cars have more stable incomes, upmarket cars often improve their share in a recession. GM concluded that Buick still met a market need and should remain a division. Oldsmobile's performance peaked in 1981 to 1983 at 11 per cent, but had plummeted to 6 per cent by 1987. Although Oldsmobile continued to deteriorate for the rest of this phase, falling to 3 per cent by 1992 to 1994, GM decided that it was also worth preserving. Two factors seem to have been critical. First, numerous GM executives were emotionally committed to the Oldsmobile name, and argued that the division was a victim of corporate problems, a 'no fault' defence. Secondly, a nameplate that had accounted for over 1 million units of production as recently as 1986 was too valuable to eliminate.

GM seemed unable to ask the more general question of whether it needed to shed a division, and, if so, select which one was best to eliminate at the time, regardless of fault or prospects. Rather, its reviews of specific divisions seemed destined to conclude that no division should close. The performance of any division rose and fell because of market fluctuations and its own product introduction cycle, so sufficiently lengthy reviews virtually assured no closures. Further, GM seemed unable to separate analytically the value of a nameplate as a market identifier and as a producing division. Consequently, it seemed to ignore the option of folding divisional nameplates into the same production unit, much as Ford did with its Mercury nameplate. Debate about the appropriate number and structure of divisions continued at GM up until the mid-1990s, as the corporation wrestled with the question of whether and how Saturn should expand.

More time will probably have to pass before analysts can draw a balanced picture of GM's trajectory under Roger Smith. Rarely has a top manager been so personally identified with corporate success and failure. His admirers saw him as a true visionary, perhaps too far out in front of the industry. They pointed to the better than expected benefits to GM of the EDS purchase, even if those benefits were relatively long and costly in developing. When Saturn finally appeared in 1990, it was a successful vehicle in terms of sales and customer reaction, offering GM an alternative model of production and marketing. Smith's critics saw him as an autocratic manager, undercutting the Sloan-inspired committee management of GM just when it was most needed, following peripatetic and ad hoc whims rather than a coherent strategic response to the critical challenges facing GM, and committed to Fordist, if high technology, solutions to complex problems. Whatever the extent of Smith's responsibility, he finished his Chairmanship with GM substantially weaker than it had been at the beginning of his tenure. Market share was falling, capacity problems were massive, and the low-cost producer had become a high-cost producer. The ship seemed to be foundering, and the noble experiments of the 1980s seemed isolated lifeboats, with little corporate-wide benefit.

In 1990, GM's Board selected a new Chairman, Robert Stempel, an engineer who was expected to emphasize getting GM's product programmes back on course. Widely respected within GM, Stempel was popular for his low-key, participatory management style, a welcome contrast with Smith's high energy, directive approach. However, in October of that year, GM announced it would close 8 plants. The week before Christmas 1991, it slated another 21 of its 125 plants, and 70,000 jobs for elimination by mid-decade. GM did not identify all the plants in some cases simply listing groups from which some would be selected for closure. For example, GM had introduced a new version of its long-serving model

Chevrolet Caprice at the end of 1989, then added a more upmarket version, a famous name from GM's past, the Buick Roadmaster, and a station wagon. Based on optimistic sales forecasts, GM assigned these vehicles to two plants, Willow Run in Michigan and Arlington in Texas. It quickly became clear that one plant was more than adequate, and, in 1991, these plants were paired, with one to close and one to survive.

GM went through a protracted and fairly public process to decide the fate of these two plants. Contrary to the expectations of most analysts, GM announced in early 1992 that it would close Willow Run, immediately provoking a spate of questions and criticisms. Whatever the merits of GM's decision about which plant would close, even more fundamentally, GM seemed to have again asked the wrong question. Neither plant was a poor performer, and, if Willow Run had an edge in cost, productivity, and quality, Arlington was still among GM's best. GM chose between two of its better plants, neither of which should have been anywhere near the top of the list for closure. The vagaries of GM product allocation and customer preference had narrowed the choice to these plants, at least partly because GM's plants were inflexible, and converting them from one vehicle to another was relatively expensive. The final irony came a few months later, when GM considered cancelling the Caprice. GM had closed an effective assembly plant, and it drew nearer to abandoning the large, rear-wheel-drive segment it once dominated.

After two years of mounting losses, the Board removed Stempel, feeling progress was too slow, and that Stempel was too reluctant to take the decisive steps necessary to restore GM to health. GM's crisis had arrived. The Board appointed John F. Smith, Jr. as President to make operational decisions, while the Board Chairman, John G. Smale, the former CEO of Procter & Gamble, took on many of the duties of the Chairman. The new President was generally considered a sound financial manager, and while less well known throughout the corporation, shared Stempel's reputation as a consensus manager who was quite comfortable in delegating responsibility to subordinates. He moved quickly to stem GM's losses. GM share stabilized in 1992, and it shifted away from less profitable fleet passenger car sales and vehicles sourced from other makers. Since the early 1980s, GM had questioned its level of vertical integration, viewing UAW wages in its internal supplier plants as a major competitive handicap. Smith began serious efforts to shed some of these plants.

The brief tenure of Stempel and the appointment of Smith seemed to have restored an important element of the Sloan variant of the Fordist model: management by committee and consensus. GM was reinventing its past decision-making practices. If authority was still more centralized than in its earlier days, GM was again in a better position to exploit its considerable talent pool, and management morale seems to have improved. While GM's decision-making still resided at a higher organizational level than was the case at Ford or Chrysler, that decision-making now seemed better informed by information from lower levels of the corporate hierarchy.

### 8.3.4. GM's Labour Relations

GM often suffered from the effects of decisions made at earlier times and in different circumstances. For example, decisions made in the second half of the prior period strongly influenced GM's labour relations in the first half of this period. GM's market share had been relatively unaffected in the 1982 recession, and GM reduced its hourly workforce by about a third, much less than Ford's near-50 per cent, and Chrysler's over 50 per cent. As GM recovered, it called more workers back than did either Chrysler or Ford, returning to 85 per cent of its 1978 level by 1986.[17] Yet these efforts did not yield a better workforce relationship at GM, as its severe vehicle share loss in the second half of the 1980s meant further facility and workforce reductions.

Chrysler and Ford accomplished major and enduring workforce reductions during the 1982 downturn, when the situations they faced justified such actions to their employees and the UAW. GM missed this opportunity, and GM workers faced a higher risk of redundancy during the better times that followed. In 1986, GM announced it would close eleven major plants by 1990, lay off 29,000 workers, and reduce its salaried workforce by 25 per cent. Workers had difficulty understanding why GM was able to weather the downturn at the beginning of the decade yet took such drastic action in a record sales year, and then followed these with the reductions in 1990 and 1991 discussed above. Many managers found it difficult to seek effective productivity improvements and workforce reductions while wondering what their own futures might hold.

By 1994, GM had reached the same staffing levels relative to 1978 as had Ford and Chrysler. But its blue-collar job reductions still exceeded its salaried reductions, and thus violated equality of sacrifice, a principle important to the UAW in defending reductions to its membership. This led to resentment and suspicion on the plant floor. The morale of hourly paid workers at GM also suffered with the adoption of profit sharing in 1982, since the workers often received little or no bonus. In 1988, GM hourly bonuses were only about 10 per cent of those received by Ford workers, and paled in comparison to its own executive bonuses.

GM's inflexible capacity and its difficulty in converting capacity from one vehicle to another often meant that market success and the vagaries of allocation of product to plant determined the ultimate fate of a plant, without regard to the local workforce's efforts and commitment. The ill-fated Pontiac Fiero offers an even clearer example of this problem than the situation of Willow Run discussed above. The Fiero was an innovative, mid-engine, low-volume two-seat sports car, built with new processes at a dedicated, specialty plant because of GM's continued difficulty in building multiple models in one plant. The car was underpowered, and after initial sales success from late 1983 up to and including 1985, demand faltered. Moreover, GM had discovered that it was still unable to build cars profitably at smaller volumes. In 1988, GM closed the plant in spite of its workforce's cooperation with management and high levels of productivity and quality. GM workers increasingly felt that their efforts were irrelevant, and that they were victims of the market and management whims.

Meanwhile, Saturn offered GM a very different picture of future labour relations. When it finally took shape, Saturn possessed a very non-traditional approach to labour relations, including more flexible work rules, more worker training and security, and part of worker base pay tied to profitability. The UAW and GM management faced the challenge of changing their historic relationship—coolness punctuated by periods of antagonism—and developed a relationship that defined areas of common interest and effort, while respecting the appropriate arenas of conflict while moderating its degree.

### 8.3.5. GM's Supplier Relations

GM planned to increase outside suppliers' share of production, but it moved somewhat more slowly than Chrysler or Ford in reshaping the supplier relationship. It too developed supplier award programmes, vowed to reduce its numbers of direct suppliers, and planned to increase its reliance on supplier expertise. To some degree, these efforts always seemed to suffer from GM's sheer size and complexity. For example, in the mid-1980s GM decided to shift major engineering responsibility to suppliers and outside specialists. But GM did not immediately shed its own idled engineers, and the supply of qualified engineers available to outside firms was insufficient to meet GM demands. This experience left GM sceptical about the wisdom of such efforts, and also ensured that when GM eventually let a smaller number of idled engineers go, there were no longer jobs for them. This lack of coordination seemed to be all too typical of GM near the end of the 1980s.

The Big Three, including GM, continued to re-evaluate their criteria for selecting suppliers, the traditional allocation of responsibilities to suppliers, and the nature of their transactions with their suppliers. The rhetoric called for moving from price-based, make-to-print, antagonistic purchasing relationships to ones based on multiple performance dimensions, expanded supplier contribution, and a partner-like, trusting dimension. However, GM's relationships with suppliers soon experienced particular strains. In 1990, GM unilaterally announced that it would cut supplier prices by 3 per cent that year and 2 per cent in each of the next two years. Suppliers were enraged, and felt that this proved that talk of new relationships was simply rhetoric. In 1992, GM installed a purchasing manager from Europe who immediately adopted a two-pronged effort to slash purchasing costs. He took major steps to centralize and coordinate corporate purchasing to leverage GM's potential economies of scale, but he blended this eminently Fordist method to achieve a Fordist goal with a Toyota goal, cost reduction throughout the production chain, to be achieved with a Toyota approach, working together to eliminate waste. While this approach was quite similar to Toyota's in theory, in practice it included actions which suppliers felt violated their confidence and even broke contracts. This controversial manager moved to VW in 1993, and by the mid-1990s it remained unclear how well GM would maintain the effective and fair elements of his approach, while eliminating or reducing its more problematic elements.

By 1990, the rapid expansion of Japanese assembly capacity drew in over 300 Japanese automotive suppliers intent on capturing sales from the Big Three, as well as serving their traditional customers. North American suppliers had primarily faced indirect Japanese competition, as their customers lost share; beginning in the late 1980s, they faced direct competitive threats to their traditional business. But if new suppliers increased GM's options in selecting its supply base, the circumstance that drew them to North America—the presence of many Japanese car maker facilities—also meant that suppliers had more options in selecting customers. For many suppliers, Chrysler, Ford, Honda, and Toyota replaced GM as the most preferred customer, especially as suppliers began to look for customers specializing in light trucks or in passenger cars.

The US government unilaterally and abruptly terminated the VRA programme at the end of its 1984 term.[18] Apprehensive about the politically explosive trade deficit, the Japanese government imposed its own export limits and shares upon Japanese assemblers. These rising limits were never reached, and the programme was soon abandoned. The strengthening yen in the mid-1980s and early 1990s spurred hopes for lower trade deficits, but reduced numbers of Japanese vehicle imports were offset by their higher value, and their North American operations sourced many parts and components from Japan. The bilateral trade deficit grew, and automotive parts became a major share of it in the late 1980s. Both the Bush and Clinton administrations targeted autos and auto parts for negotiations, arguing that Japan's automotive market was relatively closed, reflecting years of concerted public and private efforts to restrict imports. The Japanese argued that low shares in the Japanese market reflected the weak efforts of US companies, and that private competitive efforts would suffice, without Japanese government actions.[19]

The manufacturers and suppliers generally welcomed the Free Trade Agreement (FTA) signed with Canada in 1988. For the automotive industry, the major provisions of the Autopact carried over, with additional features that promised to make cross-border movements of automotive goods and investments easier. Then, in 1993, Canada, Mexico, and the USA negotiated a trilateral agreement, the North American Free Trade Agreement, or NAFTA. The Big Three and their suppliers largely welcomed NAFTA, but the UAW opposed it. The companies believed it would promote the development of the Mexican market, remove Mexico's protectionist production rules, and foster the integration of all three markets, as the Autopact and then the FTA did for the Canadian and US markets. The UAW feared low-wage competition for jobs.

Industry–government cooperation on regulatory issues improved by 1994. Joint efforts addressed governmental concerns such as safety, emissions, and fuel economy, and industry concerns, such as manufacturing technology. However, contention continued over California's strict emission standards which effectively required major car makers to market electric and/or hybrid vehicles. The mid-1990s political debate about the devolution of federal power to the states raised the spectre of a balkanized regulatory regime, a particularly troubling scenario to the automotive industry, since it would effectively reduce product volumes.

Throughout this third phase, GM scaled back its definition of success, as management rhetoric changed from 'regain market share profitably' to 'profitability at lower market share'. Smith heeded the Board's instructions, and placed profitability ahead of share. If market share only stabilized, there were other indications of improved performance, ranging from higher quality and productivity to higher profits. GM ended 1994 looking as if it was on the road to recovery, and analysts increasingly felt that if its next major round of new vehicles, due out in late 1996, succeeded, the company would have returned from the brink.

## 8.4. GM'S SITUATION AND CHALLENGES

As of the mid-1990s, GM, if still the front runner, was no longer the odds-on favourite in the car maker dominance sweepstakes for the twenty-first century (Keller 1993). It retained important strengths, including its volumes, workforce, and worldwide presence, but had yet to demonstrate that it could rapidly shift its assumptions and strategies to meet unanticipated market developments and competitive challenges. Moreover, it still trailed its major rivals in the efficiency and quality of its company-wide execution of the automotive basics: designing, building, and marketing vehicles that customers want and can afford. These challenges were not trivial, but nor were they insurmountable.

GM's fall from such dominance to first among equals raises the question of how and why its position changed so substantially. A number of factors must be taken into account to reach a satisfactory explanation. First, GM's environment began to change radically, and that change accelerated after 1973. Demand shifted to new product segments, new competitors entered the marketplace, established competitive relationships were destabilized, government regulatory demands mounted, and both labour and supplier relationships altered. Secondly, GM's remarkable success and market dominance over so long a period reinforced practices and strategies which later undercut its performance and position. Some believe GM's market dominance bred an arrogance that prevented GM from recognizing important changes in its market and competitive environment. There is merit to this view, although it can be as oversold as can the notion of a culture-bound 'Japanese' production model. Moreover, the notion of 'GM arrogance' simply personalizes corporate behaviour, is too reductionist in nature, and often ignores substantial differences in corporate behaviour and circumstances over time.

Although arrogance was not the key culprit, GM's dominance early in its trajectory did contribute to its competitive problems later on. GM fell into the success trap, and simply continued previously successful routines in the face of altered circumstances, failing to change with its environment. The fact that GM was so dominant simply made it more difficult to abandon its Fordist roots and past practices to adopt new ones. GM continued to plan for huge product volumes long after it seemed clear that the market would not support them. It also invested in Fordist technology solutions to quality and efficiency challenges, again well beyond the point when these solutions seemed optimal. GM maintained and reinstated employment levels in the mid-1980s that were appropriate to its past performance, but not supportable in the new market and competitive climate.

GM also endured extensive organizational drift, as the effects of changes, whether planned or unplanned, gradually multiplied and combined into major functional changes. Changes in GM's product development process induced by the development of centralized functions, the growing power of divisions like Fisher Body and GMAD, and the erosion of the 1984 reorganization all provide examples of this. The elimination of divisions and plants, perhaps delayed and resisted because of the success trap, when implemented, often seemed more the result of drift than clear decisions. The decline of GM's competitive performance and efficiency was more a drift than a plunge until the late 1980s, and GM continued too long to measure its effectiveness against timid and dated competition. GM's history of success, and its maintenance of that success during the mid-1980s, inoculated it against recognizing environmental shift and internal drift.

GM's earlier competitive dominance had often permitted sharp reversals of misfortune, such as its rapid recovery from the strike-induced 1970 share loss, and this supported a belief in quick fixes. It is difficult to consider GM's hopes for Nummi, Saturn, and the acquisition of EDS as anything but the wish for quick, simple, and targeted solutions to amazingly enduring, complex, and diffuse problems of company performance. However complicated these efforts actually became, GM initially approached them as relatively simple and automatic panaceas.

The success trap, organizational drift, and a quick-fix culture provide more persuasive and satisfactory explanations of GM's difficulties than does simple arrogance. To a certain extent, then, GM's difficulties in changing and adapting to its new competitive environment were rooted in its modified Fordist model, but they also reflected its particular experiences and circumstances. GM's evolution through 1995 represented a rejection, even if unconscious, of its traditional Sloanist model at the same time as it reflected a reaffirmation of elements of the earlier Fordist model. Many of the efforts GM launched during the 1980s to enhance its competitiveness were initially irrelevant, perhaps even damaging, distractions. Over time, most of them have yielded clear financial and operational benefits to GM, but they were costly and took a long time to develop. GM's tragedy may be that it had all the internal resources it required, but was simply unable to marshal and deploy them as fully and as speedily as necessary in a rapidly changing competitive world.

GM's challenges over the balance of the 1990s centred on resolving issues of its scale, culture, and capabilities. First, it had to complete its capacity alignment to a reasonable market share target, permitting it to leverage still very high volumes and return to a low cost position. GM needed flexible plants, so that it would not continue to be capacity-constrained in popular vehicles while forced to operate plants with excess capacity for less popular models. Secondly, GM still had to reconcile and coordinate its needs for product differentiation across its vehicles and divisions with its process commonality requirements for functional and scale efficiencies. Determining the proper mix of common and distinctive efforts across engineering, design, and manufacturing remained a significant challenge. Thirdly, GM still required further change in its corporate culture, moving from the rather insular and self-satisfied 'Generous Motors' of old to a much leaner, more focused, and more effective production climate, without alienating its employees and in fact using them as effective resources. GM had to develop the capability to respond flexibly and rapidly to changing circumstances across its whole activity spectrum, from the designer's board to the dealer's outlet. Fourthly, GM needed to define its organizational structure, including how to coordinate its increasingly global activities and respond to its North American markets. GM had to decide whether it would truly integrate its international activities, allow them to continue as complete and parallel industries, or achieve some balance of complementary independence. The market shift to light trucks, the addition of Saturn, and GM's declining production volumes since the late 1970s, all raised the possibility that GM should eliminate, combine, or realign some of its divisions.

Fifthly, GM had to identify its core capabilities or competencies, align its activities accordingly, define the appropriate level of vertical integration, and settle on a strategy for reaching it, including a coherent and consistent approach to managing its supply base. Finally, GM needed to identify important lessons from its many pockets of excellence and diffuse them throughout the corporation, a task as daunting as any of the others.

## 8.5. INDUSTRIAL MODELS

Is GM in the early stages of the emergence of a new industrial model that will become the prevalent organizational form and the basis for company success and dominance? Analysis of GM's trajectory suggests that this is not likely to happen. First, the drivers for this development are weaker than they may appear. The competitive success of some of the Japanese car makers has been dramatic, but that competitive success has been more in terms of their own past levels than comparative dominance over their rivals. Toyota, the leading Japanese producer, had not yet matched GM or Ford worldwide sales levels, achieving 55 and 70 per cent of those levels, respectively, in 1994. Nor had Toyota approached the extent to which Ford early and GM later in this century dominated their worldwide rivals. Secondly, pressures countering the development and diffusion of such a model are stronger than they may appear. Automotive companies such as GM are large and complex, do not readily change, and often reflect the previous rather than the current competitive dynamics of their industry and marketplace. 'New' models are more often variants of earlier models. The Fordist model developed and became dominant in a new, not an established, industry. Ford's competitive dominance early in the century effectively defined the automotive production paradigm; GM's later dominance altered rather than replaced it.

Environmental and competitive circumstances do change, and there is no one perfect organizational form that best suits all situations and corporate challenges. GM's past four decades suggests that the periods of environmental stability that might allow a new model to develop have been brief indeed. Market and competitive instability places a premium on organizational flexibility and rapid response. These may be the key characteristics of emerging industrial models, but they may also be competitive demands that prevent the routinization and time required for a true model to develop. The trajectory of GM suggests that any candidates for a new and dominant automotive industrial model will be likely to differ from the Fordist model on some key dimensions, but probably not all. Such a model will need to foster structures that respond rapidly and flexibly to a changing automotive market. Its organization of production will have to wear a more human face in order to attract and hold the flexible, skilled workforce needed to participate in improving performance. The model will also have to support the integration of effort across traditional company and role boundaries. Finally, it will have to support response to a regulatory environment that will become ever more demanding across a wide range of vehicle performance dimensions.

# NOTES

1. This chapter treats the industries of Canada and the USA as one combined North American industry because of a history since the mid-1960s of integrated decisions and special trade arrangements. The Canadian market is similar, if not identical, to the US market in its segmentation and import penetration.
2. These vehicles were typically called 'captive imports', since they were imported by the domestic assembler and sold under its label, e.g., the Mazdas as 'Ford Couriers' and the Isuzus as 'Chevy Luvs'. They counted in the seller's market share, but the maker's production, and were imports in assessing trade balances. At the time, Mazda was still called Toyo-Kogyo, but Mazda is used throughout this chapter to avoid confusion.
3. i.e. non-Canadian imports. Because of the Autopact's integration of the market and production base, vehicle trade between the USA and Canada is excluded from estimates of exports and imports. These vehicles do not have the same competitive implications as other exports and imports.
4. Henry Ford II's attempts to replicate GM's management strength by hiring the occasional top GM executive probably failed because Ford lacked both the GM culture and a strong middle management cadre.
5. Some saw the creation of GMAD as an anti-union move, on the theory that GM was creating a centralized, powerful unit that could speak with one voice and discipline

the United Automobile Workers union. However, anti-union sentiment probably played less of a role in this development than did these other motives.
6. Light vehicles include traditional passenger cars and light trucks, which in turn include traditional pick-ups, vans, and sports-utilities. In North America, light trucks have increasingly become substitutes for cars, rather than commercial or special-use vehicles. Unless otherwise attributed, all market shares reported in this paper are for North America, and are calculated from a database constructed from information reported in various annual series: *Ward's Automotive Yearbooks*, Motor Vehicle Manufacturers Association (MVMA) Facts and Figures, and Automotive News' *Market Data Books*.
7. Canadian and US automotive regulations are nearly identical, rarely creating joint compliance problems.
8. However, the Japanese assemblers gained 5 points in 1980 from Chrysler and Ford.
9. The establishment of a technical information centre near Nummi to collate information and arrange access reflects the learning from the Toyota goal of the venture. Interviews with Nummi graduates confirm its intended use in Saturn. That it was in fact used so little is probably the result as much of the 'go it alone and different' mentality at Saturn as any GM lack of interest or even GM's original focus on hard technology. See Keller (1989: 129–36), or Flynn and Andrea (1994, *passim*).
10. On the day of the announcement, people at GM began referring to the groups as 'Cheap Plastic Cars', and 'Big Old Clunkers' to outsiders! Presumably, divisional loyalties spurred this sarcasm.
11. The UAW was also under considerable stress, and faced at least four new challenges itself. First, strikes became a more costly tactic, as lost sales might go to non-union companies. Secondly, Japanese operations were and remained in the mid-1990s major organizing challenges to the UAW. Thirdly, the UAW experienced a tremendous decline in membership, and thus political influence. Fourthly, the Canadian Automobile Workers separated itself from the UAW and this complicated bargaining strategies and tactics.
12. This was as much a result of government regulations as was the US industry's preference for parallel offshore industries (see Cole and Yakushiji 1984).
13. Since the US car market plummeted in 1982, the agreement did not limit Japanese share gains, nor, in all probability, sales. Canada's quota was based on market share, so it did constrain Japanese sales.
14. Originally developed in retaliation for European limits on American poultry imports, this 'Chicken Tariff' was particularly galling to the Japanese assemblers, because they imported so many more trucks than did the Europeans.
15. CAFE had no effect on the majority of importers, whose vehicles were in general already fuel-efficient, reflecting the much higher price of fuel in their home markets. However, the special tax for heavy consumption was imposed on a number of luxury imports.
16. This provision called for a 'domestic' fleet of vehicles with at least 75 per cent US and Canadian content and an import fleet of vehicles below that level. Sponsored by the UAW, it was designed to prevent the Big Three from meeting CAFE by sourcing all small, 'CAFE-positive' vehicles from Europe or Japan, while manufacturing reduced volumes of CAFE-neutral or negative vehicles in North America.
17. Calculated from GM Annual Reports.
18. In March, 1985, the close of the Japanese fiscal year.
19. Canada departed from the US trade position, taking no part in these agreements after VRA.

## STATISTICAL APPENDIX 8: General Motors

| Year | Light truck Production[a] | Passenger car Production[b] | Labour[c] | ROA (%)[d] | US export ratio (%)[e] | Turnover[f] in millions | Passenger car Production[g] | Light truck Production[h] | Labour[i] |
|---|---|---|---|---|---|---|---|---|---|
| 1967 | 90,810 | 5,318,397 | 728,000 | 22.6 | 1.0 | 20,026 | 4,430,085 | 753,850 | 600,211 |
| 1968 | 1,122,488 | 6,021,729 | 757,000 | 25.0 | 0.8 | 22,755 | 4,930,093 | 915,266 | 608,965 |
| 1969 | 1,149,601 | 5,928,290 | 794,000 | 29.0 | 0.8 | 24,295 | 4,812,563 | 940,335 | 617,345 |
| 1970 | 882,788 | 4,335,771 | 696,000 | 4.6 | 1.0 | 18,752 | 3,201,863 | 676,057 | 548,088 |
| 1971 | 1,245,695 | 6,490,099 | 773,000 | 20.5 | 0.7 | 28,264 | 5,259,135 | 1,012,455 | 621,770 |
| 1972 | 1,271,611 | 6,406,527 | 760,000 | 22.7 | 0.6 | 30,435 | 5,129,875 | 1,071,067 | 610,694 |
| 1973 | 1,638,800 | 7,004,892 | 811,000 | 21.5 | 0.8 | 35,798 | 5,696,542 | 1,398,276 | 639,091 |
| 1974 | 1,472,111 | 5,093,953 | 734,000 | 7.6 | 1.4 | 31,550 | 4,036,017 | 1,223,537 | 555,578 |
| 1975 | 1,364,836 | 5,125,202 | 681,000 | 11.1 | 2.0 | 35,725 | 4,086,587 | 1,151,033 | 548,336 |
| 1976 | 1,850,000 | 6,718,000 | 748,000 | 22.1 | 1.6 | 47,181 | 5,422,978 | 1,526,666 | 603,760 |
| 1977 | 1,977,000 | 7,091,000 | 797,000 | 23.1 | 1.5 | 54,961 | 5,784,483 | 1,700,129 | 654,711 |
| 1978 | 2,209,000 | 7,273,000 | 839,000 | 21.7 | 2.3 | 63,221 | 5,857,383 | 1,874,137 | 680,399 |
| 1979 | 1,992,000 | 7,001,000 | 853,000 | 14.9 | 2.6 | 66,311 | 5,651,019 | 1,641,261 | 618,365 |
| 1980 | 1,321,000 | 5,780,000 | 746,000 | -3.4 | 2.7 | 57,729 | 4,576,876 | 943,380 | 516,706 |
| 1981 | 1,170,360 | 5,497,052 | 741,000 | 1.0 | 2.1 | 62,699 | 4,382,472 | 1,001,289 | 521,933 |
| 1982 | 1,280,436 | 4,869,672 | 657,000 | 2.3 | 1.5 | 60,026 | 3,507,791 | 1,124,217 | 441,019 |
| 1983 | 1,539,085 | 6,098,880 | 691,000 | 12.1 | 0.7 | 74,582 | 4,513,930 | 1,392,513 | 463,331 |
| 1984 | 1,755,382 | 6,315,265 | 748,000 | 9.1 | 0.4 | 83,890 | 4,891,062 | 1,604,760 | 559,500 |
| 1985 | 2,029,003 | 7,090,145 | 811,000 | 7.1 | 0.4 | 96,372 | 5,385,461 | 2,394,604 | 568,900 |
| 1986 | 1,852,577 | 6,523,163 | 877,000 | 2.7 | 0.4 | 102,814 | 4,653,717 | 1,702,445 | 539,500 |
| 1987 | 1,891,762 | 5,605,301 | 813,400 | 3.6 | 0.9 | 101,782 | 3,799,401 | 1,755,214 | 472,500 |

| Year | | | | | | | | | |
|---|---|---|---|---|---|---|---|---|---|
| 1988 | 2,080,575 | 5,662,843 | 765,700 | 5.6 | 1.8 | 110,229 | 3,763,518 | 1,970,403 | 443,500 |
| 1989 | 2,064,822 | 5,547,285 | 775,100 | 5.5 | 2.2 | 112,533 | 3,405,644 | 1,927,257 | 418,800 |
| 1990 | 1,936,335 | 5,208,221 | 761,000 | 0.3 | 3.3 | 110,797 | 2,905,629 | 1,725,046 | 355,100 |
| 1991 | 1,669,546 | 4,994,682 | 756,000 | -3.1 | 3.7 | 109,157 | 2,619,221 | 1,474,308 | 423,200 |
| 1992 | 1,875,890 | 4,989,938 | 750,000 | -2.3 | 4.5 | 118,572 | 2,557,755 | 1,658,209 | 396,000 |
| 1993 | 2,031,762 | 4,852,168 | 711,000 | 1.9 | 4.0 | 125,252 | 2,816,648 | 1,853,649 | 361,800 |
| 1994 | 2,240,143 | 5,183,238 | 693,000 | 5.5 | 4.4 | 141,575 | 2,833,629 | 2,063,984 | 347,200 |
| 1995 | 2,842,217 | 5,777,366 | 709,000 | 6.2 | 3.5 | 143,666 | 3,159,507 | 2,197,090 | 434,000 |
| 1996 | 2,856,000 | 5,544,000 | 712,000 | 9.0 | n/a | 145,341 | 2,755,733 | 2,200,271 | 424,000 |

*Notes and Sources*:

[a] GM worldwide light truck production under 10,000 lbs. GVW (Ward's Automotive, AAMA (American Automobile Manufacturers Association)).

[b] GM worldwide passenger car production (Ward's Automotive, AAMA).

[c] GM total worldwide employment, including subsidiaries (GM annual reports 1967–95, GM Fact Book, 1993).

[d] ROA = operating profit/total assets.

[e] Export Ratio = US light vehicle exports/production. Figures are for US manufacturers combined and do not include exports to Canada (Motor Vehicle Manufacturers Association, Ward's Automotive).

[f] Turnover = sales of manufactured products.

[g] GM North American passenger car production (Ward's Automotive, AAMA).

[h] GM North American light truck production under 10,000 lbs. GVW (Ward's Automotive, AAMA).

[i] GM total North American employment, excluding subsidiaries (1984–94). US employment, including subsidiaries (1967–83) (GM Annual Reports 1967–95; GM Fact Book, 1995).

# BIBLIOGRAPHY

American Automobile Manufacturers Association, *Motor Vehicle Facts and Figures* (Detroit, various years).
—— *World Motor Vehicle Data* (Detroit, 1996).
Automotive News, *Market Data Book* (Detroit, various years).
Cole, R. E., and Yakushiji, T. (eds.), *The American and Japanese Auto Industries in Transition: Report of the Joint US-Japan Automotive Study* (Ann Arbor, 1984).
Cray, E., *Chrome Colossus: General Motors and its Times* (New York, 1980).
Fine, C. H., and St Clair, R., 'Meeting the Challenge: U.S. Industry Faces the 21st Century. The U.S. Automobile Assembly Industry', The International Motor Vehicle Program, IMVP Draft Report (Cambridge, Mass: 1995).
Flynn, M. S., and Andrea, D. J., 'Corporate Learning from Japan: Partnering, People and Process Technology', *Proceedings of the ESD IPC'94, Conference and Exposition* (1994), 19–30.
—— and Cole, R. E., *Automotive Suppliers: Customer Relationships, Technology, and Competition* (Ann Arbor, 1986).
Halberstam, D., *The Reckoning* (New York, 1986).
Harbour, J., *The Harbour Report 1995* (1995).
Keller, M., *Rude Awakening: The Rise, Fall, and Struggle for Recovery of General Motors* (New York, 1989).
—— *Collision: GM, Toyota and Volkswagen and the Race to Own the 21st Century* (New York, 1993).
Lacey, R., *Ford: The Men and the Machine* (Boston, 1986).
Office for the Study of Automotive Transportation, 'Competitive Survival: Private Initiatives, Public Policy, and the North American Automotive Industry' (Ann Arbor, 1992).
Ward's Communications, *Ward's Automotive Yearbook* (Southfield, various annual editions).
Womack, J. P., Jones, D. T., and Roos, D., *The Machine That Changed the World* (New York, 1990).

# 9

## *Globalization at the Heart of Organizational Change: Crisis and Recovery at the Ford Motor Company*

### GÉRARD BORDENAVE

Ford has retained its position as the second largest automobile producer in the world ever since the company lost its status as leading producer to General Motors in the 1920s. The Ford name has long been associated with the idea of 'mass production', a form of manufacturing the company developed and then helped to diffuse throughout the world (Wilkins and Hill 1964). While every company has its own particular characteristics, and Ford has certainly not remained unchanged over the decades, Ford's original industrial model (Bordenave 1992; Boyer and Durand 1993) gradually evolved into a set of organizational canons for the automobile industry. Developments and variations over time have tended to enrich rather than distort the basic Ford industrial model.

The penetration of Japanese manufacturers into previously oligopolistic markets across the world has had an enormous destabilizing effect, raising questions about the source of the Japanese competitive advantage. Several analysts have concluded that the Japanese have a new industrial model which is distinct from the traditional Western Fordist model, and which some have labelled 'lean production' (Womack *et al.* 1990).

Western vehicle producers like Ford face a twin challenge, to both their competitiveness and their industrial models. While the imitation and incorporation of certain Japanese practices which have been shown to produce better results for the Japanese is one possible response, it has remained unclear whether this represents the only possible future. The accumulated organizational know-how of companies like Ford, crystallized in their routine functions, has created 'reserves' of possible problem-solutions which cannot be underestimated. The old principles have not necessarily lost all their operational force. Moreover, their very existence may create significant resistance to the introduction of new principles. A further possibility is experimentation with completely original solutions that differ from both the Japanese model and the traditional Western model.

These ideas suggest that the analyst of a company like Ford would do better to thoroughly investigate the diverse range of real transformations as they have unfolded in practice, rather than positing a fixed and 'necessary' new model a priori. Indeed rather than 'one best way', it is quite possible that analysis of the automobile industry will reveal multiple future trajectories (Boyer and Freyssenet 1995).

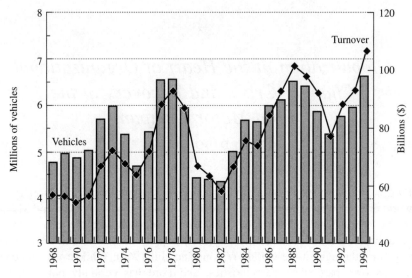

FIG. 9.1. *Vehicle factory sales and turnover of Ford*
*Note*: Monetary data are in 1994 prices.

The case of Ford is particularly instructive. The company's North American operations experienced a notable financial recovery during the late 1980s. Proponents of the 'lean production' thesis were quick to argue that Ford was the most advanced Western producer as far as the adoption of lean production was concerned, and that this was what explained the company's recovery. Yet a more thorough analysis is merited. We cannot simply rely on evidence of improved performance to arrive at conclusions as to its causes. This chapter therefore describes the development of Ford's trajectory over the 1968–95 period, to set the changes of the 1985–95 period—and the claims that have been made about them—in a longer and fuller context. The analysis reveals the multiple transformations that played a role in the company's recovery, transformations that cannot be reduced to the adoption of lean production principles. In particular, the chapter reveals the important role played by Ford's strategy of globalization, which is not one of the distinctive characteristics of lean production.

Figure 9.1 summarizes Ford's performance from the late 1960s to the mid-1990s and presents an overview of the company's trajectory.[1] The first conclusion to be drawn from an examination of this data series is that long-term growth, measured by the number of vehicles sold or by turnover, was weak, particularly in terms of vehicles sold. The company was maintaining its position, but was making little progress. Within the difficult context of the emergence of new rivals in its markets this was not a particularly poor performance, but it does emphasize the extent to which competitive conditions between the late 1960s and the mid-1990s differed from the strong growth characteristic of earlier decades. The second point is that performance was very strongly cyclical. The impacts of three recessions—1974–5, 1979–82, and 1990–2—is clear. The third point is that the 1979–82 recession was more acute for Ford, in duration and intensity, than the other recessions.

This initial overview leads to a periodization of Ford's trajectory in which three phases can be identified. In the first phase, from the end of the 1960s to the end of the 1970s, the traditional Fordist model was losing its momentum. In the second phase, during the early 1980s, Ford was managing a difficult crisis and experimenting with new principles of organization in an emergency situation and under the pressure of events. The third phase, from the mid-1980s to the mid-1990s, was characterized by strategic consolidation and the coalescence of a partially restructured industrial model, in the context of an apparent recovery in the company's performance. While serious problems were still being faced, changes during this period were less driven by events.

## 9.1. THE 1960S: REAPPRAISAL OF THE FORDIST MODEL

In 1978 Ford established record sales, reaching a level not surpassed again until 1994. The company's factories delivered 6,557 million vehicles that year. However, the overwhelming recovery in sales and the steady increase in profits since the recession linked to the first oil crisis in 1974–5 could not entirely mask the slowing momentum of the Fordist organizational model in the relatively stable form it then assumed. What were the significant features of this model, and what were the signs of imminent crisis?

### 9.1.1. Ford's Own Version of a Stable Fordist Model

The Fordist industrial model had experienced remarkable development since the company established the original mass production model. Despite these changes the company remained loyal to several of the model's founding principles. At the same time, while Ford shared a number of characteristics in common with the majority of Western automobile producers since its industrial practices and modes of organization were imitated and some of them acquired a strong normative value, the company also clearly displayed its own specific practices, the results of its own particular trajectory. Among the elements peculiar to Ford, attention should be drawn to Ford's practices in the management of international space.

Ford's industrial model during the 1970s can be characterized as Fordist in the sense that it still relied on the systematic application of most of the organizational principles adopted by Henry Ford I at the beginning of the century. Analysts frequently consider work undertaken on a moving assembly line, parcelled into short cycles with the time given for each task strictly controlled, to be the ideal image of Fordism. However, rather than considering the presence or absence of this type of work as the unique criterion for identification of Fordism, it is more useful to understand the development of an overall logic that spread from the factory to the company as a whole. From this viewpoint the guiding principles of Fordism include: the standardization of products and modes of operations; the division of work into fragmented and highly specialized activities; the imposition of a coherent framework upon activities through external control of persons executing tasks, which is exerted hierarchically by means of procedures which are predefined or integrated into production equipment; the general separation between design and production which imposes a division between skilled and unskilled workers; and remuneration schemes for employees based on position held and working time. These principles have in turn led to characteristic practices. First, the system is driven from upstream and in sequence: design, produce, sell. This approach implies a reduced ability to respond to market changes, and the maintenance of high stocks. Secondly, there is strong vertical integration, and other suppliers are often technically dependent, so that the production process can be strictly controlled. Lastly, there is a high level of turnover among unskilled sections of the workforce as well as recourse to 'hire and fire' practices—external flexibility—in order to accommodate the fluctuating pressures of the economic context.

Other Western automobile producers have shared these general characteristics. In Ford's case, however, they are greatly accentuated. The company has remained highly centralized, with its headquarters controlling facilities and subsidiaries very tightly. When Henry Ford II assumed control of operations in the 1940s the new management team considerably reinforced financial control procedures. Management structures at Ford were characterized by a preponderance of the finance function combined with significant hierarchical control, rigidly compartmentalized functions, and interpersonal relations notable for aggression and lack of cooperation. Moreover, the company's efforts to improve the production process were dominated by technical approaches, in particular automation, with little emphasis placed on human resources. Not until 1979 did Ford in the USA sign an employee involvement contract with the United Automobile Workers union, whereas General Motors had launched its 'quality of work life programme' six years earlier.

As with the industrial models of other automobile producers, Ford's precise organizational model differed from the original Fordism in various ways. The first amendment had been the mass production of a differentiated product range. Henry Ford I's industrial model was quite clearly based on a single product with a very limited number of variants. This philosophy proved to be extremely damaging to Ford in competition with General Motors, which had been reorganized by Alfred Sloan during the 1920s, and which had developed into the world's largest automobile producer owing to its superior performance. General Motors' organizational form, incorporating separate divisions with differentiated product ranges, combined with shared components, continuous revision of the product range, and the development of structures, procedures, and know-how necessary to manage a complex industrial group, became the standard for the automobile sector. After considerable hesitation and procrastination, Ford eventually adopted these principles itself. It might be noted in passing that increased product variety might therefore be considered a long-term tendency—as opposed to a recent phenomenon—in the automobile industry.[2]

A second modification of the original Fordism was the reduced emphasis placed on the vertical integration advocated by Henry Ford I. Determined to gain total mastery of the production process, he had sought control right back to the very sources of the raw materials, from purchasing rubber plantations and possessing an armed navy across to producing his own energy and steel at Ford's gigantic industrial complex at River Rouge. However, when the general development of the American industrial fabric permitted Ford to turn to specialized external suppliers offering cost advantages, intricate forms of vertical integration no longer appeared to be a necessary condition for Fordism. Accordingly, Ford became even less integrated than General Motors, which had developed its own integrated components sector.[3]

A third modification of the initial model stemmed from the changing environment in terms of employment relations, and resulted in the establishment of more regular and codified relationships with employees. Henry Ford I's paternalistic vision was vigorously hostile towards any collective action by workers. In contrast to his main rivals, he led a fierce rearguard action against the establishment of the UAW in his company during the late 1930s. The issue was not finally resolved until the war years and the arrival of his grandson Henry Ford II to head the firm. And yet the practices of collective bargaining and, more broadly, post-war social legislation, would later serve as an institutional framework for what became know as the 'Fordist compromise', in the terms of the French regulation school (Boyer 1986). In exchange for planned increases in remuneration combined with fixed bargaining procedures for workers, the organization of production and productivity increases were placed clearly under managerial control.

Once established, this compromise enabled conflict to be channelled. However, it was established and maintained unevenly across Ford's worldwide operations. Particular difficulties appeared during the late 1960s and early 1970s, with differences emerging according to the country. Conflict became endemic in the United Kingdom compared to the USA or other European countries where Ford operations were located (Tolliday 1991). Indeed variable social conditions in different countries led to considerable diversity in the company's employment relationships. This became particularly important when Ford began to take such variations into account in the management of its geographical structure at the international level.

At the start of the 1990s Ford was advanced in the management of international space. It produced 63 per cent of its vehicles in North America, 29 per cent in Europe, and 8 per cent in the rest of the world. The proportions fluctuated annually as a function of varying economic conditions in the different world regions. However, over the long term the pattern had been relatively stable. Respective proportions at the beginning of the 1970s were 70 per cent, 24 per cent, and 7 per cent. Ford was therefore relatively developed in comparison to its competitors in terms of production outside its country of origin. Comparison of production of automobiles alone (not counting trucks) in 1992 reveals that the proportion of production outside the country of origin was 52 per cent for Ford, 42 per cent for General Motors, 38 per cent for Honda, and 23 per cent for Fiat (Bélis-Bergouignan et al. 1994). Ford was the most internationalized of the producers and had held this position longer than any other producer, even General Motors.

During the 1970s Ford's deliberate management of its 'production space' focused largely on the integration of the European regional market, mirroring the thoroughly successful process of integration in North America which had taken place during the 1960s. The founding of Ford of Europe in 1967, as a managerial and coordinating structure, was the organizational point of departure for a unification of product ranges which had until then been different at the British and German subsidiaries. The aim was to fully profit from economies of scale in design, production, and distribution by organizing operations at the continental scale. Ford would thus move from a multi-domestic spatial configuration to a multi-regional configuration which would correspond to the institutionalization of free trade within the European Community. The movement towards continental integration involved the reorganization of facilities over space through competition among existing and new facilities for the allocation of production activities. In effect, Ford relocated the centre of gravity of its European industrial apparatus towards Southern Europe, with considerable investments made in Spain to produce a new small automobile for the European market. The overall impact of restructuring during the 1970s was to redistribute final assembly activities, with a net reduction in the United Kingdom share, but maintenance of the German share.[4] Faced with a crisis of employment relations in the United Kingdom, Ford's ability to manage its spatial organization now permitted the firm to threaten relocation away from the trouble spots (Bordenave and Lung 1988).

*9.1.2. The Signs of Latent Crisis in the Fordist Model*

It was during the 1970s, and as a result of the difficulties experienced by the company in responding to new conditions in its economic and social environment, that the Fordist industrial model was first questioned. In comparison with the previous decade there was now increasing uncertainty about product markets and the organization of work. Yet Ford's reaction appeared inadequate and did not arrest the erosion of its relative performance. The difficulties encountered by the company during the 1970s, combined with an appearance of a certain immobilization, are clearly reflected in comparisons with the 1980s. Between the two periods the indicators reveal a turning-point representing a real break in the company's mode of functioning.

Ford's competitive position crumbles in a less certain environment. During the 1970s three disruptive influences unsettled the world's automobile markets. First, there were the oil crises of 1973 and 1979, which not only provoked global fluctuations in macroeconomic demand but also destabilized market structures, with a shift towards smaller models (or diesels) when fuel prices increased and reversion to larger models as prices relaxed. Secondly, there was a net increase in public regulation of safety, environmental protection (primarily emissions of pollutants), and energy-saving measures which required costly modifications to vehicles. Thirdly, the decade witnessed the growing strength of Japanese exports. The initial comparative studies of Western and Japanese companies underlined the net competitive advantage of the latter in terms of production cost and productivity as well as product quality and features. The 1970s were also marked by increasing uncertainty in employment relationships. So much resistance emerged towards the deepening or even maintenance of the dominant form of work organization that it is legitimate to speak of a crisis of work. Further automation was frequently presented as the appropriate response. Yet this typically Fordist approach itself met with a number of technical, economic, and social difficulties.

Ford, which was already oriented towards small cars, might have benefited more than General Motors from the increase in energy prices. In reality Ford's share of the US market was whittled away throughout the 1970s. By the start of the 1980s the company controlled 17 per cent of the automobile market, compared with 25 per cent a decade earlier, whereas General Motors had maintained its position throughout the period. Ford joined Chrysler and American Motors Corporation in losing market share to the Japanese producers. Its luxury range, which essentially dated back to the beginning of the 1970s, already seemed old and outmoded technically. Moreover, Ford adapted its range to the new market conditions more slowly and in a less definite way than General Motors, which deployed considerably more resources. The company's new small car, the Ford Escort, was not launched until 1980, and then in the midst of an economic recession. Ford also suffered from a poor image in terms of product quality: comparative studies often ranked Ford in bottom place among the Big Three American producers.

The causes of Ford's relative immobilization during the 1970s are difficult to identify. One factor must, however, be mentioned. An important issue arose over who would succeed Henry Ford II as head of the firm, with the posts of Chief Executive Officer and Chairman of the Board of Directors to be filled respectively in 1979 and 1980. It may well be that the question of succession interfered with the management of the company and paralysed its decision-making processes.[5] More structural reasons probably also played a part. Ford tended to practise a more radical and rigid type of Fordism than other firms. Accordingly, it was hardly surprising that Ford suffered from the inability of its traditional model to permit more profound adaptations to a new environment. Hence Ford had a more compelling need to develop its methods of organization and functional procedures than did other firms. Behind the difficulties experienced during the 1970s lay the very crisis of the model itself, as the serious impact of the 1979–82 recession would confirm.

TABLE 9.1. *Comparison of average annual growth rates in selected indicators for 1968–1979 and 1982–1994 at Ford*

| Indicators | 1968–79 | 1982–94 |
| --- | --- | --- |
| Vehicle factory sales (physical units) | + 2.4 | + 1.9 |
| Turnover in 1994 $ | + 4.8 | + 3.6 |
| Employment (for the whole group) | + 1.5 | – 1.5 |
| Costs per vehicle in 1994 $ | + 2.8 | + 2.0 |
| Number of vehicles per worker | ns | + 3.4 |
| Turnover per worker in 1994 $ | + 3.3 | + 5.1 |
| Labour costs per worker in 1994 $ | + 3.7 | + 2.5 |
| Gross investment in 1994 $ | ns | + 5.4 |

*Note*: ns: average annual growth rate not significantly different from zero.

Table 9.1 indicates average changes in selected variables during the periods 1968–79 and 1982–94. The figures deliberately exclude Ford's deep downturn between 1979 and 1982. They permit a comparison between two long periods, each of which included a recession (see also Fig. 9.1). The clear difference in performance between the periods supports the idea that Ford's mode of operation changed significantly either side of the deep recession of 1979–82, and that this period therefore marks a break in the company's mode of functioning: and therefore in the evolution of its industrial model. The first two lines of the table reveal that the second period was actually more difficult in terms of market conditions, with lower annual rates of growth. The shift from growth to decline in employment appears to be of major significance in comparing the two periods. After 1979 the reduction of employment was continually a component of management decision-making. Accordingly it played an important role in the general dynamic of transformation, a dynamic which linked rationalization of operations to substantial organizational changes.

Yet growth in activities and employment during the 1970s did not prevent other indicators from revealing the crisis of Fordism. Comparison between indicators of productivity (number of vehicles per worker or turnover per worker) and labour costs per worker reveal an apparent breakdown in the 'Fordist compromise', in which increased productivity ought to have matched direct and indirect payments to employees. The first indicator of productivity, the number of vehicles per worker, barely changed during the first period. A year by year examination reveals poor growth up to the recession period of 1974–5, followed by a dip during the recession (which might be expected), and then by hardly any growth during the short recovery from 1976 to 1979. The second indicator of productivity entirely confirms this general pattern. Per capita labour costs grew faster at the beginning of the 1970s (1968–73) than at the end (1976–9), which suggests an early rupture (before the first oil crisis) with the Fordist compromise.

Investment by its very nature is prone to fluctuate, and given that it is influenced by the rate at which automobile products are replaced, the 1970s revealed no clear trend. By 1982, however, a trend could be observed. There was a particularly acute dip in investment in 1975 and 1976, which could be interpreted as a 'wait-and-see' policy in regard to the poor state of the market. This contrasted sharply with major investments made during the 1990–2 recession. Although interpretations are difficult, the lower rate of capital investment during the 1970s may be closely related to the crisis of Fordism which characterized the period. By contrast, examination of the key variables characterizing the dynamics of accumulation during the 1980s and the early 1990s reveals an apparent period of recovery. Indices of productivity grew more rapidly than during the earlier period. Per capita labour costs were much better contained, as a result of lower rises in average wages. Investment rose in a more stable pattern, representing a continuous and sustained pattern of accumulation.

Hence Ford's performance was much stronger following the major crisis at the start of the 1980s. While the company was certainly not shielded from all problems, as we shall see, a set of positive dynamic interactions had been set in motion. Modifications to the Fordist industrial model helped the company regain its strength. The transformation had begun during the recession, under the pressure of events.

## 9.2. THE EARLY 1980S: CRISIS MANAGEMENT AND URGENT EXPERIMENTATION WITH NEW PRINCIPLES OF ORGANIZATION

During the second half of 1979, following two record years for vehicle production, Ford slipped into an economic crevasse as a result of the second oil crisis in combination with the interventionist anti-inflationary measures which then guided the macroeconomic policies of most Western industrial countries. The company entered a major recessionary decline which lasted until 1982, the particular intensity of which revealed the latent crisis of the Fordist organizational model, and which led to demands for far-reaching emergency measures. The emergency measures adopted had the initial dual objective of reducing costs and improving quality. Significantly, however, they then led to a process whereby Ford systematically compared its methods and performance, both internally and externally. This process was buttressed by research into innovative forms of labour relations and cooperation in production.

### 9.2.1. The Context of the Early 1980s and an Exceptionally Deep Recession for Ford

The data in Figure 9.1 illustrate the three recessionary periods Ford has traversed. The first lasted two years (1974–5), the second four years (1979–82) and the third three years (1990–2). All the indicators confirm that the second recession was unparalleled for Ford in the post-war period. That the recession was worse for Ford than its competitors was confirmed by the decline of Ford's market share in the USA between 1979 and 1982, by 3.8 points for cars and 2 points for utility vehicles. The 1974–5 recession had been shorter and less profound in all respects, and Ford continued to make profits. Comparison with the 1990–2 recession is less clear-cut. In 1990–2, profits from financial services reduced the impact of automotive sector losses at group level, which had not been the case in the early 1980s. Moreover, in 1991–2 Ford lost money in both Europe and North America whereas in 1979–82 losses were due entirely to North American operations.[6] Indeed an important aspect of the 1979–82 recession for Ford was its geographical concentration at the firm's centre of operations, which heightened its intensity. Many insiders agreed that the early 1980s was a 'dark period' for Ford. Managers and workers at all hierarchical levels believed that the very survival of the company was at stake. This undoubtedly accounted for their receptive attitude to innovation, along with their new-found willingness to question established practices.

### 9.2.2. A Vigorous Programme to Reduce Costs and Improve Quality

Gaining control of costs was not a new preoccupation for Ford, which had traditionally allotted a central role to cost control in its management framework. The same could be said for quality, although in this case the novelty of the new programmes was greater because the quality imperative was so strongly emphasized.

The first programme was the least novel. Immediate action to reduce the workforce commensurate with the lower level of business activity was combined with an active and ongoing rationalization of the productive apparatus. The outcome was that the workforce fell by 25 per cent over four years and nine factories, including three assembly plants, were permanently closed. That Ford resorted to reducing its workforce was not so unusual. The novelty lay not only in the size of the reductions, but also in the fact that recovery of vehicle output was not accompanied by a significant rebound in employment.[7] The implication was that marked increases in productivity had been achieved owing to a transformation of the labour process. The substitution of capital for labour through the introduction of new production equipment and new technologies figured prominently in this transformation.[8] Indeed a wave of flexible automation combined with a computerization of production had a strong impact on certain segments of the production process. Body construction, together with certain operations in the mechanical components factories, were the most radically transformed segments. By contrast, only limited changes were made to assembly tasks. Ford was following the general trend in the automobile industry rather than staking out an extreme position. Although it did not always succeed, the company attempted to curtail recourse to overly technical solutions with their excessive investments which might also pose problems if they could not be mastered. The company had, however, seen in the new production technologies the potential for improving quality and obtaining greater flexibility. This desire to utilize new production technologies did not in itself represent a dramatic shift from the guiding spirit of Fordism. However, the application of these new technologies might well have propelled the company towards new forms of work organization.

The second path Ford followed to reduce costs and improve product quality did represent a notable evolution in its model of industrial organization. Across its worldwide operations, Ford launched an initiative to significantly reorganize its factories by transferring responsibilities from indirect labour to direct labour. In other words, Ford embarked upon a certain recomposition of tasks which called into question the ultra-Taylorist procedures of traditional Fordism. The focus was above all on quality control, relying upon a single new tool, statistical process control (SPC), which direct workers were trained to use. To a lesser extent, certain indirect maintenance tasks or even management tasks were also transferred to operators. The outcome was a significant reduction in indirect labour, particularly in quality control. This tendency was undoubtedly facilitated by the more autonomous functioning of automated machinery, which often shifted the role of operators towards machine minding. And yet the fact that Ford remained equally attached to non-automated or partially automated tasks indicated that the company was not being led purely by technical change. Also relevant was the introduction of modular assembly methods during the first half of the 1980s, the purpose of which was to separate some subassembly work from the main flow of production. This trend was not uncommon in the automobile sector (Jürgens *et al.* 1989), but Ford made an early and strong commitment to it. As early as 1981 the company's annual report highlighted Ford's efforts in terms of quality, statistical control techniques, and consultation on these subjects with W. Edwards Demming, who was depicted as a major contributor to the improvement of product quality and productivity in Japan.

It is important to recognize the limitations to task recomposition during this period. While certain indirect functions were reintegrated into direct work, the principle that work should be divided into narrow and strictly defined tasks by the work study department remained the basis for organization. The uneven distribution of automation meant that short cycle times were by no means abolished even if improvements had been made in terms of ergonomics. In short, assembly line work continued to be widespread, not only during this period but into the 1990s too. Hence during the early 1980s there were significant changes but not a complete revolution in work organization, a point that will re-emerge below with respect to Ford's approach to innovation involving cooperative forms of production.

The third programme Ford initiated to reduce costs and increase quality focused on relationships with suppliers, and, more broadly speaking, the company's management of internal and external flows of intermediate products. No statistics are available on the proportion of components Ford purchases from external suppliers. Undoubtedly they represent a major proportion of overall purchases, perhaps 60 per cent of turnover in the early 1970s, and more than 70 per cent in the early 1990s.[9] By virtue of its importance, therefore, the purchasing function was seen as a vital factor in cost reduction. The same was true for quality, since quite often disruptions to production or defects in the finished product were due to defective components or supplies.

The first action taken with suppliers was to negotiate lower prices. Ford also sought to have suppliers focus on long-term quality by transferring quality control responsibilities to them and obliging them to adopt a long-term commitment to continuous improvement. Ford created the Q1 (Quality First) label, which was only awarded to suppliers which satisfied criteria measuring both post-production quality and such various internal operations as SPC or procedures to deal with production malfunctions. Award of the Q1 label, which would later be extended to Ford's own components factories, was a 'carrot and stick' approach, reflecting a desire to categorize suppliers so as to reduce their number and ally the company only with selected firms in a new type of relationship based on long-term exclusive contracts. To this action on quality was added a new prominence for measures to reduce stocks, based primarily on a profound revision of the principles of logistics and inspired by Japanese 'just-in-time' delivery techniques. Stock reduction became a tangible element in the transformation of Ford. Overall stock levels declined from 13 per cent of turnover in the late 1970s to 8.5 per cent in the mid-1980s and to 6 per cent by the early 1990s. Stock reductions both reduced costs directly and helped improve quality, since the latter was both a condition and a consequence of operating with reduced stocks.

### 9.2.3. The Major Role Played by Internal and External Comparisons of Methods and Performance

A significant role was played by comparisons of performance between Ford and its competitors and among the company's own factories and suppliers. This activity fulfilled several functions. It served as a means to formulate targets in terms of gaps to be made up, with the situation of the best-placed company or factory becoming the point of reference, and thus encouraged a spirit of 'healthy' competition. At the same time it became a means to apply direct pressure, which might turn into a threat of closure, relocation, or withdrawal of business from a factory or supplier which did not meet the standard. More importantly, an approach based on systematic comparisons helped to keep Ford alert to technological and organizational developments so as to encourage new ideas for reinvigorating both products and management methods.

The practice of comparing and confronting management experiences has been recognized and codified as 'benchmarking'. While this term has recently surfaced in Ford's vocabulary, it was already a long-standing practice at the company to study the competition and single out the 'best in class' among them. Ford could then establish its own differentiating characteristics by comparing, emulating, or surpassing the best technical solutions while bearing in mind Ford's own strategy in terms of quality/price trade-offs. Similarly, Ford had long employed comparative techniques in the fields of industrial organization and productivity, cost and quality performance. This was most evident when Ford organized competition among its factories for the allocation of production activity, or competition among the potential locations of new investments. Ford relied on comparisons of productivity between its United Kingdom and continental European assembly plants to obtain concessions from the British workforce in terms of reduced levels of conflict, involvement of workers, or acceptance of work process reorganization. The practice of making comparisons was also linked to threats of relocation, particularly towards low wage countries. Investments in southern Europe fuelled and sustained these threats. The integration of Mexico into Ford's North American industrial structure, when decisions were made during the early 1980s to establish new production facilities there, fulfilled the same function.[10] While the actual extent of relocation remained limited, more important was the symbolic impact on the effort to mobilize the workforce. Comparative activities leading to the adoption of norms practised by competitors were spontaneously integrated into the Fordist model.

Many of Ford's cost control and quality actions were inspired by similar practices observed in Japan and by a special teacher in Mazda. The process of observation and selection of Japanese practices was made systematic by Ford during this period. In the 1970s many Ford managers and even trade unionists made study trips to Japan, and returned with comparative data measuring cost and productivity gaps. Following the wave of educational visits, a campaign to sensitize managers to the development of new organizational processes was launched, in connection with the establishment of targets to reduce differences. While other companies, in and outside the automobile industry, were studied, it was Mazda that provided Ford with a special window onto Japan. Beyond the simple mutual transmission of know-how, Ford and Mazda benefited from a fully fledged industrial partnership. Cooperation between the two companies dated from the early 1970s, but it deepened considerably after Ford purchased 25 per cent of Mazda's capital in 1979, whereupon the links between the two companies became the automobile industry's most intense and significant cooperation between producers which remained independent. The purchase of Mazda capital by Ford allowed the American company to assume a powerful role within the relationship. But as a minority shareholder Ford was not in a position to dictate conditions to Mazda, with which it had to establish a cooperative relationship based on mutual advantage. The relationship reached its full potential during the late 1980s.

### 9.2.4. Experimenting with Innovative Forms of Cooperation in Workplace Relations

Transformations in the organization of work at Ford factories focused largely on the systematic integration of quality control operations into the responsibilities of direct workers. In fact, other innovative forms of work organization also emerged, but these were applied less systematically, often dependent on local circumstances, and taking the form of relatively isolated experiments. Some of these concerned the activities of production workers while others were more closely related to the activities of managers.

Worker involvement was complementary to organizational change aimed at recomposing tasks and creating work teams. The initial experiments with participation, involving invitations to workers to put forward suggestions to improve the work process, were small scale and isolated, starting in 1977. Based on examples from American and Japanese companies, some managers at Ford promoted the idea of re-instilling in workers the ability to initiate productivity and quality improvements while also making work more attractive. The wider diffusion of the initial experiments did not start until 1979 when Ford's top management signed an agreement with the UAW. The agreement set out procedures, which were to be jointly managed by management and trades union members. Meetings were to be held at factory level, with participants then choosing from among various suggestions which emanated from the basic groups and offering to support them. Ford sought to extend this procedure to all its operations, under the name 'quality circles'. The rather uneven distribution of these circles was due sometimes to the hostility of unions and sometimes to the fact that they seemed redundant given existing co-participative management practices. In the main, workers raised immediate and limited issues. It would be incorrect to interpret these activities as a crucial factor in productivity and quality improvements. Nevertheless, they did sensitize employees to team working, a practice into which the quality circles tended to merge and diffuse.

The theme of worker involvement well illustrates the implications of geographically uneven social relations at Ford. In North America the UAW considerably aided Ford's recovery, in contrast to the more conflictual situations Ford encountered elsewhere. Wage concessions resulting from the 1982 UAW agreement, alongside concessions on work rules in local negotiations, opened up new possibilities for the company (Katz 1985). In return, Ford agreed to discontinue the factory closure policy it had just launched, as well as to partially maintain (under certain conditions) the incomes of laid-off workers. It also introduced a profit-sharing scheme.[11] Lastly, the agreement led to the launch of a joint Ford–UAW training programme.

The creation of work teams with collective responsibility for the management of a group of machines or a specific segment of the work process was based on a managerial initiative. However, the acceptance, at least implicitly, of the unions was required because this form of organization might clash with rules of demarcation between skills when it was proposed to increase the polyvalency and rotation of workers or to transfer skilled maintenance workers to production activities. The adoption of this form of work therefore depended upon the local context of work relationships and also upon the willingness of local managers to promote it. In certain factories it was limited to particular areas, those in which lines of integrated machines made it more appropriate.

While Ford, conscious of variations in local conditions, left the different factories with wide scope for decision-making in the management of their workforces and the organization of their work processes, the company generally encouraged the spread of teamwork, both vertically across the hierarchy and horizontally across disciplines. Training programmes included sessions for learning about teamwork. Whether or not this mode of organizing work was adopted locally, the firm was obliged to carefully negotiate a reduction in the number of job classifications and demarcations between skills which undermined the flexible utilization of the workforce. While these developments, which remained slow, should not be exaggerated, Ford clearly adopted the objective of abandoning the strong functional categorization traditionally associated with Fordism, so as to give more initiative to employees and to encourage more efficient forms of cooperation among them. Moreover, schemes to promote participatory forms of management activity could be observed at various levels of the organizational structure.

The traditional Fordist organizational model was characterized by strict divisions between areas of competency. These included both micro-scale divisions into specialized and assigned tasks within the factories and offices and the overall division of the company into grand, watertight functions (design, manufacturing, sales, and so on) which tended to become autonomous citadels protective of their prerogatives and communicating with each other only poorly and via hierarchical links. In light of Japanese practice in particular, where companies tended to function more organically, the dysfunctions and conflicts caused by excessive barriers and lack of transversal communication were recognized as major faults of the traditional Fordist organizational structure. This diagnosis was arrived at within Ford by a number of working parties organized at managerial level under the banner of 'participative management', which had been called upon to formulate proposals for reorganization in response to the major crisis of the early 1980s. The need to reduce the overarching power of the finance function, to reduce the excessive weight of formal procedures, to counter aggressive styles of management, and to begin the process of breaking down functional barriers had been recognized. The implementation of these organizational changes was not, however, immediate, given the deep roots of the old practices and indeed the scepticism of some about such novel ideas (Pascale 1990; Starkey and McKinley 1993).

The influence of participative management became clear in the organizational innovations associated with the design of the Taurus model, which was launched onto the North American market towards the end of 1985 and which has been claimed to be 'the car that saved Ford' (Shook 1990; Taub 1991). The Taurus team functioned as a project group, transversally organized across the different functional departments of the company. Its members remained attached to their original departments. They were, however, given individual titles and held answerable to the director of the project. Decisions made in the team were recognized to override the interests of the functional departments. The principal functions and the suppliers were involved from an early stage, demonstrating a desire to develop the product simultaneously rather than sequentially. Hence suggestions by workers, suppliers, dealers, and even customer panels which had been consulted over the initial prototypes could be incorporated into the vehicle. Assembly factories were selected four years before production started. Most suppliers were designated two years in advance, the first of them chosen three and half years in advance. The great commercial success of Taurus,[12] combined with its process and product innovations, contributed to its design methods becoming an exemplary point of reference within the company.

The theme of participative management was primarily addressed to the management staff. It drew them into the development of new structures, methods, and attitudes regarding their work. In this it was similar to the attempts to instigate reform in the factories. While there were problems in making concrete these changes at each and every level, there at least appeared to be an intention to favour more collective activities and to obtain greater adherence to the firm's objectives from employees: thus reinforcing their feelings of belonging to the company. Ford's desire to increase its organic cohesion was expressed in the 1985 publication of the charter *Missions, Values and Guiding Principles*. Besides products and profits, the company emphasized personnel, and in particular involvement and teamwork, as its three basic values.[13] These steps towards a valorization of human resources did not emerge by coincidence, and would not replace long entrenched and antithetical work habits over night. There were therefore incongruities between Ford's expressed intentions and what actually occurred.

## 9.3. FROM THE MID-1980S ONWARDS: RECOVERY, STRATEGIC CONSOLIDATION, AND THE COALESCENCE OF A REVIVED INDUSTRIAL MODEL

The transformations of the early 1980s assumed a dual character. On the one hand, they were broadly focused on production activities, with particular emphasis on factory productivity and quality. On the other hand, rapid and tangible improvements were necessary in these areas, and actions were taken under the pressure of crisis. From the mid-1980s, however, the focus of transformation began to shift elsewhere. Towards the apex of the organizational structure, there was further change in terms of work and managerial methods. Change also concerned 'grand strategy'. There was both a redefinition of the boundaries of the firm, that is to say, the nature of its relationships with the exterior, and a revision of its internal procedures and structures. Beyond these two tendencies the strategy of globalization played a vital role in the dynamics of change.

### 9.3.1. *A Context of Clear Recovery, but New Problems at the Start of the 1990s*

Ford's recovery can be traced to the expansion of its market share in North America. Having declined from over 23 per cent in 1978 to under 17 per cent in 1981–2,

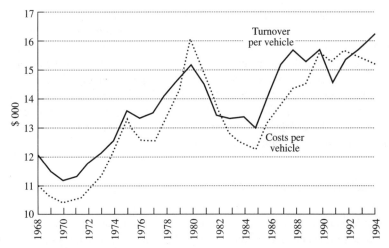

FIG. 9.2. *Turnover and costs per Ford vehicle*
*Note*: Monetary data are in 1994 prices.

by 1987 the company's share of the US automobile market had once again risen above 20 per cent, where it remained into the mid-1990s. For utility vehicles, Ford's share soared to 30 per cent after reaching a low point in the mid-1980s. Ford thus attained a 25 per cent share of the combined domestic market in 1993. The rapid growth of turnover and especially the spectacular growth of profits between 1985 and 1988 confirmed this recovery. During 1987 and 1988 profits exceeded US$6 bn., more than 8 per cent of turnover: rates not seen since 1972 or the 1960s. This rosy period was also a result of the weakening of its rival General Motors, a reversal of the relative situation of the two firms during the preceding period. The long tradition of competitive confrontation between the two leading global producers meant that this development helped establish an atmosphere of calm at Ford after the period of doubt in the early 1980s. However, the recovery proved more fragile than Ford's growth during the second half of the 1980s had appeared to indicate. Comparing the two recessionary periods of the early 1980s and the early 1990s, the economic downturn of 1990–2 hit profits just as hard. The company's capacity to maintain costs seemed to be at issue, since sales and turnover had fallen less than a decade earlier. Figure 9.2 reveals a narrowing of the gap between turnover per unit sold and cost per unit sold.

The return to profit in 1993 and especially in 1994 demonstrated that the situation remained under control and that corrective measures had been taken.[14] However, in the USA, Ford's profitability suffered by comparison with Chrysler, and given the success of Opel, the local subsidiary of General Motors in Europe, Ford's results seem particularly mediocre. Indeed the inversion of the relative performance of Ford and General Motors between the 1980s and 1990s was also accompanied by a complete inversion of the performances of the two companies by geographical region. In the early 1980s Ford's results in Europe were better than those of General Motors, in contrast to the situation in the USA. In the early 1990s the inverse was true. The instability of competitive positions over time, as well as the fragile nature of Ford's economic recovery, was thus confirmed. What is important, however, was that attempts to promote change since the mid-1980s derived from an anticipative rather than a reactive stance, since the pressure of events at this point was less than during the preceding period. Nevertheless the continuous improvements that had been inspired by Japanese practice remained difficult to implement.

*9.3.2 Redefinition of the Ford's Boundaries*

Redefinition of the company's boundaries involved the diversification of its portfolio of activities, changes in the extent and nature of horizontal cooperation with other producers, and changes in its relationships with suppliers. There were implications for the organizational model both in terms of strategy and relational modes.

The boundaries of the Ford Group were substantially modified during the second half of the 1980s. The interplay of acquisitions and sales suggested a dual tendency towards a refocusing of industrial activity on the automobile sector (narrowly defined) and a diversification into the financial services sector. In terms of withdrawal from industrial activities, the 1989 sale of the Rouge steelworks, the last vestige of a large, integrated industrial complex created by the firm's founder, was symbolic. This sale was preceded by the sales of the glass and chemicals businesses. With the sale of Ford Aerospace in 1990, Ford also abandoned diversification towards the aerospace industry, since the expected synergies had not materialized. The sale of Ford New Holland to Fiat Geotech in 1991 reversed Henry Ford's strategy of producing agricultural tractors. As early as 1986, Ford and Fiat had formed the joint company Ford-Iveco Trucks Ltd., through which Ford exited from the heavy commercial vehicle sector in Europe where it was too weak to attain critical mass. The sole acquisition of significance was that of Jaguar, the British specialist car producer, in 1989.[15] Ford was seeking to reinforce its position in the American luxury car market. Purchase of Jaguar effectively created a new division within Ford, since it was necessary for Jaguar to preserve its brand name and dealer network autonomy, even as production systems and management methods were partly integrated. The net outcome of these industrial acquisitions and sales of businesses was a refocusing on Ford's core competencies in automobile manufacturing.

The parallel diversification towards financial services was very significant. Since the 1959 creation of Ford Motor Credit, Ford had had a subsidiary specialized in automobile loans and the financing of its distribution network. The 1985 acquisition of the First Nationwide Financial Corporation marked the start of a series of purchases within the financial services sector. However, the difficulties experienced by the Savings and Loans sector in the USA led to the resale of this bank in 1994. A more permanent acquisition came in 1987 with the purchase of US Leasing International Inc., an institution specialized in industrial leasing (not restricted to the automotive sector), and then in 1989 with the purchase of The Associates Corporation of North America, the most expensive (US$3.4 bn.) of the company's various acquisitions. This particular business had networks in Japan and the United Kingdom. Its various activities focused on individual purchasers, including a network of credit cards. Still other financial institutions were purchased, among them the car rental company Hertz. Ford was first a minority shareholder and then in 1994 acquired and consolidated the whole company. By 1992 Ford's financial sector had 27,000 employees, with assets of US$123 bn. One goal of diversification into the finance sector was to compensate for cyclical fluctuations in the automotive sector, a role that was played during the 1990–2 recession.

The development of cooperative relationships between producers was a particular characteristic of the 1980s. Such cooperation appeared to be a valuable means of responding to the imperatives for model range diversification through the sharing of risk and know-how. Ford made considerable use of cooperation, beginning with its special relationship to Mazda in the Asia-Pacific region and North America.

In the Asia-Pacific region Ford and Mazda created an industrial network in which the design and manufacture of products were partly shared. The principal nodes of the network were Mazda's factories in Japan, Ford's subsidiaries in Taiwan, New Zealand, and Australia, the Korean producer Kia (in which Ford and Mazda both held minority shares), and various enterprises involved in components or automobile assembly.[16] Ford's penetration of the Japanese market was partly based on the distribution network owned in common with Mazda since 1989, the name of which was later changed from Autorama to Ford. In North America Mazda was actively involved in defining the production methods and product (the Mercury Tracer) of the new assembly factory that Ford constructed in northern Mexico, which was considered one of the most efficient sites in the Ford network (Carrillo 1995). In the USA the collaboration between Ford and Mazda was first concretized in the joint venture AutoAlliance, established at Flat Rock, Michigan, which manufactured a model shared by the two producers.[17] Mazda's active participation in the replacement of the Escort and the Mercury Tracer in 1990 became a significant opportunity for Ford to learn Japanese methods of product design and development. However, the fact that from 1991 Mazda began to sell a modified version of a light utility vehicle designed and manufactured by Ford in the USA indicated a shift in the balance of power and perhaps a maturation of the relationship. A similar interpretation may be placed upon the oft-deferred announcement, finally made in 1994, that the two firms would collaborate in Europe, which had previously been kept separate from the other joint projects.[18] Mazda's own difficulties in the mid-1990s, the increased presence of Ford staff at its headquarters, the shifting balance of participation between the two firms in the joint ventures Autorama and AutoAlliance, which had initially favoured Mazda, led, in 1996, to what may be considered a virtual takeover of Mazda by Ford. Ford increased its shareholding in Mazda from 25 per cent to 33.4 per cent, and a Ford manager became the senior manager of Mazda. The Japanese producer remained a separate company, but a reinforcement of industrial and sales cooperation between the two companies seemed bound to follow the new arrangement.

Ford's cooperation with other producers was more limited and localized than its broad relationship with Mazda. There were two preoccupations. The first was to deal with difficult local situations by sharing risks and resources. This was the case with the joint company Autolatina, formed in 1987 with Volkswagen to integrate the activities of the two producers in Brazil and Argentina (Salerno 1994). However, this relationship was short-lived. Its dissolution was announced in 1994, following the improvement of the general economic climate in these countries, coupled with the commercial need to preserve brand names, and the difficulties of developing shared local products capable of matching those of competitors whose products were transferred directly from Europe.

The second preoccupation was to enter narrow or uncertain market niches, where questions of scale and critical mass were of particular importance. Entry into new niches was often approached in this way. This was the case for some of the products of the overall alliance of Mazda, and also for specific products developed with Nissan in the USA and Europe as well as with Volkswagen in Europe. With Nissan, the products in question were a minivan designed by Nissan and manufactured by Ford in the USA, and a compact sports-utility vehicle designed and manufactured by Nissan in Spain. Both vehicles were launched in 1993. With Volkswagen, the joint venture AutoEuropa was created in 1991 for the production of a minivan in Portugal. The product, a joint design based on a VW platform, appeared on the market in 1995. A certain fidelity in Ford's alliances, with a limited number of producers, may be noted in passing: Mazda, Nissan, and Volkswagen. At the same time these alliances with special partners did not exclude relationships with other producers in terms of components supply or more limited objectives. An example was the collaboration among the American Big Three, Ford, General Motors, and Chrysler, in shared research undertaken by various consortia receiving government aid. The issues studied included electric batteries, pollution emissions, new materials, vehicle safety, computerized design and production systems.

Relationships with suppliers or dealers evolved in parallel with those with other automobile producers: towards reinforcement of cooperation and partnership. Suppliers began to be directly linked to the realization of Ford's management priorities (cost control, quality), to the extent that a tendency towards disintegration prevailed. Suppliers were increasingly required to supply complete subsystems in the context of stable contracts. A new award, the TQE (Total Quality Excellence), now complemented Q1 certification. This award recognized more than did Q1 new requirements for organizational qualities and the ability of the supplier to take technological responsibility for its products. Active partnership with suppliers in the design process tended to become systematic. With the emergence of a 'global sourcing strategy', selection of suppliers tended to favour larger and more internationalized companies. The end result was a reorganization of the hierarchical structure in the supplier network similar to that in Japan, with Ford only preserving direct relationships with first-tier suppliers, of which there were many fewer. In North America the number of components suppliers was reduced from 2,200 in 1980 to 1,500 in the early 1990s. In Europe their number fell to below 1,000. By 1995 there were expected to be 1,600 direct component suppliers at 5,500 different locations in North America and Europe as a whole, the largest of them obviously active in both regions.

These developments at the boundaries of the company led to greater permeability and openness towards the exterior. Ford began to integrate the knowledge and experience of other companies into its own basic competencies. The company therefore began to constitute itself into a network so as to exploit the dynamics available from an accumulation of diverse competencies. This network of relationships to other companies constituted Ford's external network, an extension of the company's network of relationships between its own facilities. Moreover, Ford's multinational activities had great future potential in terms of the creation of network forms of dynamics. The globalization strategy aimed to exploit these resources.

### 9.3.3. Fundamental Revision of Internal Procedures

Revision of internal procedures paralleled redefinition of external boundaries. Product design and development activities were the focus of particular attention, as were other management procedures oriented towards customer satisfaction and cost reduction. Characteristic of all these changes was their focus on white-collar work. The improvement of management methods and the efficiency of managerial work became important preoccupations for the company.

It was important to systematize the lessons of the Taurus team as well as Ford's observations of the most innovative research and development practices at competitors, notably the Japanese producers. The objective was to simultaneously improve the lead-time, cost, and quality of the process so as to respond better to market change and consumer expectations. Priority was given to controlling the considerable rise in product design costs, and also to recognizing, upstream and in a preventative way, problems that might become manifest when the vehicle was manufactured and sold. A fundamental principle behind the new procedures was simultaneous engineering, the parallel development of the design work for different parts of the product as well as manufacturing methods. Similarly, it was recognized that the choice of production equipment and the designation and active involvement of suppliers should come very early in order to shift as much design work as possible upstream. The need to adapt fully to commercial requirements also presupposed that relevant management tools should be applied at the stage of product definition. At the beginning of the 1990s, these various requirements were unified into a new codification of the various stages of the product design process. Actions to be undertaken and managerial objectives to be attained at each stage were spelled out in detail, including a general temporal framework for the processes. This new codification of the processes involved in product development was labelled 'World Class Timing' in North America and 'Concept to Customer' in Europe. However, the rules were not fixed. On the contrary, they were expected to be continuously revised as a function of acquired experience and the development of management tools. For instance, Ford's 1992 annual report announced that a future objective was to reduce the then current basic product development cycle from forty-eight to thirty-six months.

Cost reduction had become a permanent preoccupation and strategic objective. There was a notable attempt to reduce levels of complexity.[19] This might mean reducing the number of versions of certain models or variations for different markets, or regrouping options, all of which assumed trade-offs with customer demands for differentiation. However, a considerable part of the reduction in complexity was invisible to the client: for instance the design of a modular vehicle and the simplification of its design, or the utilization of the same components or mechanical parts for separate (simultaneous or successive) vehicles. There was also a reduction in the number of production sites and supply sources for the same vehicle, and a reduction in the number of suppliers, which then supplied complete modules. Reduction of complexity was only one means of reducing costs. However, in its focus on scale requirements and the geographical dimension this could assume in practice, the company revealed a characteristically Fordist approach.

The second important type of action on costs was led by cross-functional teams which concentrated on particular aspects of the process. In some cases such groups included suppliers. They often appeared to be ad hoc, because they were formed on the basis of a particular objective or around a particular methodology. Hence their aim was not to develop into permanent structures. Their actions could later be renewed from a different angle. Significantly, their cross-functional approach was meant to overcome traditional Fordist segmentation. Yet the initiative to form them was broadly devolved from a managerial hierarchy which set overall goals and periodically relaunched their activities. This strong hierarchical momentum to guide relatively decentralized and transversal activities was a characteristic Ford approach.

### 9.3.4. Confirmation of a Globalization Strategy

Ford's international activities permitted the company to adopt an explicit global strategy. A long-standing concern had been to erase local specificities in order to maximize economies of scale by selling quasi-identical products in as many markets as possible. A principal role here was to be played by the 'world car', a key theme in the major reorganization launched in 1994 under the name 'Ford 2000'.

The origins of the world car concept could be traced back to the beginnings of the company, the Ford Model-T car, and the obstinate insistence of Henry Ford I on the imposition of the same product on every market. And yet the history of the company's internationalization was built upon the solid resistance which market realities in turn imposed on the universalist project of the initial Fordism, as well as natural or artificial barriers to free trade (Bordenave 1992). The progressive convergence of lifestyles, the unifying pressures of emerging energy and environmental concerns which favoured small cars, the growth of global trade (particularly Japanese) in automobiles, all these factors brought about a resurgence of interest in the world car in the late 1970s. Ford declared that the Escort, its new small car launched simultaneously on both sides of the Atlantic in 1980, was a world car. This was in fact merely a marketing strategy, since the two vehicles had almost nothing in common beyond their name and the fact that they belonged to the same class of product. With such different market requirements, their design was different, their manufacture different, and they shared practically no components in common.

In 1987 Ford resuscitated the concept of the world car through its 'centres of responsibility' strategy. The objective was to distribute responsibility for developing major products among the development teams in the three global poles according to their competencies. Mazda was the recognized centre of responsibility for small cars, Ford of Europe for medium-sized cars, four-cylinder engines, and manual transmissions, and the American headquarters for large cars, V6 and V8 engines, automatic transmissions, air-conditioning, and electronic systems. These were the highest-volume product lines. The strategy required an internal telecommunications system, together with a unified system for indexing and processing engineering data. The World Engineering Release System (WERS) was created in 1986, to be used by the 20,000 individuals designing Ford products throughout the world. Ford later placed its various worldwide design centres under the authority of the American headquarters.[20] Telematic links created a 'virtual' proximity, to permit the coordinated functioning of geographically dispersed facilities.

The implementation of the ambitious 'centres of responsibility' strategy was not without difficulty. Replacement of the Escort model in 1990 took place separately either side of the Atlantic. The advanced state of the programmes, the separation of the Europeans from Mazda, and the different market conditions perhaps constituted insurmountable obstacles. The CDW27 project in the upper-medium segment was therefore the first to succeed, with the 1993 launch of the Mondeo in Europe (the centre of responsibility for the project), followed by the vehicle's 1994 introduction in the North American market and in the Asia-Pacific region (European exports to Japan). This vehicle could legitimately be described as the first world car for Ford, which expected annual sales of 800,000 units spread over seventy countries. The vehicle was manufactured at three sites, one in Europe, at Genk (Belgium), and two in North America, at Kansas City and Cuautitlan (Mexico). Various components manufactured in the USA were utilized in European production, and vice-versa. A significant organizational innovation was the mobilization of a 140-person launch team at the Genk site eighteen months prior to launch, drawn from the British and German R&D centres, so that any problems of introducing the model could be resolved at the production site itself. Part of this team was then moved to Kansas City in order to transfer its experience to the American launch. The overall cost of the project reached US$6 bn., which many observers judged to be very high, but this did include the development of new engines and new transmissions. However, there were problems adapting the product to the North American market.[21]

The 'centres of responsibility' strategy served as a trial run for the broader and more structured vision which emerged in 1994 under the name 'Ford 2000'. As part of a plan to eliminate duplications in the major mechanical components, engines, and transmissions between the North American and European regions, the Ford 2000 strategy appeared to represent a profound geographical reorganization of the company. It effectively fused Ford's North American and European activities, leading to the disappearance of these world regions as the principal organizational level and replacing them with a directly global form of coordination based on product types. From being multi-regional, Ford's organization clearly became trans-regional, which implied a greater degree of hierarchical control in the management of globalization (Bélis-Bergouignan et al. 1994).

A single entity, 'Ford Automotive Operations', with functional divisions for product development, manufacturing, marketing and sales, and purchasing, now covered Europe and North America. Under this structure Ford of Europe was no more than a functional entity without hierarchical competencies and essentially limited to public relations. The Asia-Pacific and Latin America regions were omitted from this organizational structure. Placed under the responsibility of 'International Automotive Operations' they were nevertheless expected to join it when modalities for doing so could be found: in particular, ways to integrate the cooperative activities with Mazda. Within the division responsible for product development, five 'vehicle centres' were created, each competent in a particular product type. Four were based in the USA and one in Europe, reflecting the recognized expertise in these regions,[22] but the use of telematic communications meant that human resources did not have to be centralized in one place.

The strong tendency towards centralization that this reorganization appeared to represent was tempered by the adoption of a matrix organizational structure. The five vehicle centres shared functions so that they constituted five decentralized poles of competency, thus creating a framework for cross-functional work. Each individual would have two allegiances: his functional department and his centre. He was to work with individuals who belonged to other functions within the centre. This structure reduced the number of hierarchical levels to seven and was expected to reduce bureaucracy and staffing levels. The precise definition of the new arrangement was to be based on the work of 'internal study teams' involving relevant top-level managers. The process of adopting Ford 2000 was progressive, with the transition phase expected to continue beyond the official starting date of 1 January 1995. Before it was launched, 1,700 managers met in 1994 to plan the diffusion of the new principles to their subordinates.

The most obvious objective of these changes was to reduce the geographical duplication of products and components, and therefore to reduce research and development costs as well as realize economies of scale in manufacturing.[23] Suppliers found themselves drawn into these tendencies because the Ford supply base was now conceived of at the global scale. A redistribution and reduction of components sources was expected. The second objective was to improve the efficiency of white-collar work, principally product design and engineering work, by emphasizing cross-functional work and decentralizing responsibilities to units that prioritized the product. That these concerns were not entirely new at Ford does not diminish their importance. The future position of the firm in world competition would greatly depend upon the success or failure of the 'Ford 2000' reorganization.

## 9.4. CONCLUSION: A NEO-FORDIST ORGANIZATIONAL MODEL

The multiplicity of transformations of Ford's management and organization reveals a broad diversity of orientations. To assess the contribution of each would be impossible, and it would be risky to grant one or another of them the main or perhaps exclusive role in the company's economic recovery. It now seems quite arbitrary to focus on the—doubtless numerous—elements which made Ford appear to resemble the canonical lean production model so as to be able to claim that Ford's improved performance was due exclusively to their adoption. To do this would be to neglect the presence and efficacy of the many elements which formed part of a Fordist continuity.

In terms of a shift towards the lean production model, the most significant changes were the transformations that occurred in the factories, notably the recomposition of tasks and integration of some indirect work into direct work, the desire for worker involvement, and the partial organization into work teams. The same wish to surpass functional segmentation and improve more organic forms of cooperation was seen in the development of white-collar work. Also in this category were the changes in the place and role of suppliers in the production system with the emergence of partnership-based relationships. Management of the production flow according to just-in-time principles, the focus on quality, capacity to react quickly, and customer satisfaction led in the same direction. It would therefore be wrong to assume that nothing had changed. Indeed in comparison with other Western producers Ford tended to be at the forefront in terms of the identification and adoption of new concepts.

Yet despite all these transformations, the general principles of Fordism had by no means lost their relevance as fundamental elements in Ford's organization.

Standardization of the product and modes of operation, together with segmentation of work into fragmented and specialized tasks, remained the keystones of the industrial model in the factories as much as in the overall organization of the company. Management of employment levels still generally depended on external flexibility. Remuneration systems had not been fundamentally restructured. Hierarchical coordination to control processes and structures remained strongly embedded. Indeed Ford relied primarily on the improvement of managerial work, which it had been studying, reorganizing, and for which it had been selecting sound methods: in other words more upon the upper levels than the base of the hierarchy. After its close observation of Japanese organizational principles, the company had not sought to adopt an holistic approach which would have been impossible to transpose in its original form. Instead, it had followed a more selective, instrumental approach and adopted a more limited set of particular methods of Japanese origin.

One last point that ought to be emphasized in this review of developments in the Fordist model of organization was the privileged role played by globalization, in other words the practices of spatial management. Ford's international activities represented a specific resource for the company. Moreover, Ford's long-term strategy to manoeuvre locations and industrial sites into competitive positions was still accompanied by a desire to disseminate its methods and products universally. The management of space was in fact built precisely on the tension between differentiation and homogenization. At root, the large-scale restructuring that Ford was undergoing during the 1990s to unify its activities on either side of the Atlantic represented the resumption on a broader spatial scale of the integration of operations within each continent. In this way Ford in the 1990s was returning to one of the earlier ambitions of Fordism, in which the active management of space was an important dimension.

Thus it was through a combination of old and new solutions that Ford attempted to adapt itself to difficult economic situations. Rather than a complete revolution in its industrial model, there was an evolutionary process towards a much amended Fordism: a neo-Fordism. All things considered, this was hardly a novel development for the Ford Motor Company.

# NOTES

Translated by Sybil Hyacinth Mair.
1. All the data derive from Ford Motor Company annual reports. Factory sales of vehicles include cars, vans, and agricultural tractors, in physical units. Agricultural tractors represented less than 2 per cent and disappeared in 1991. Turnover is expressed in constant (1994) dollars (deflator: consumer price index for the United States; calculation by author). Turnover is industrial turnover, excluding Ford Group's

activities in financial services. Even at their most influential, Ford's non-automotive industrial activities never represented more than 10 per cent of overall turnover.
2. For GERPISA, Jetin (1995) has measured the long-term evolution of product variety for the major world producers. According to his calculations the variety of models offered by Ford in the North American market reached its apex in the 1970s and slowly diminished during the 1980s.
3. The most recent data on levels of integration suggest 50 per cent for Ford and 70 per cent for General Motors (but only 30 per cent for Chrysler and Japanese producers).
4. In the early part of this period the two countries shared Ford's European vehicle production equally (48 per cent in the United Kingdom, 52 per cent in Germany). After the late 1970s the division of production was fixed at approximately 28 per cent in the United Kingdom, 55 per cent in Germany, and 17 per cent in Spain. The United Kingdom fell back both in relative terms and in absolute numbers (150,000 fewer vehicles were manufactured per year). However, it should be borne in mind that the United Kingdom retained a strong capacity in components (notably engines).
5. Phillip Caldwell was Henry Ford's first successor. He emerged as a company insider to assume leadership of the company, as did Donald Petersen (in 1985), Harold Poling (in 1990), and Alexander Trotman (in 1993). Turnover among company directors has been high. Directors must include Ford family members, who remain influential. Ingrassia and White (1994) give detailed information on individuals and their activities.
6. From 1979–82 accumulated losses in the USA represented 5.5 per cent of turnover in that region, whereas Europe during the same period experienced an exceptionally high level of profits which somewhat reduced the impact of the massive losses suffered in the North America. In contrast, 40 per cent of losses in 1991–2 were due to European operations.
7. A simple regression analysis shows that variations in employment in the 1970s were linked with variations in production, a link that was not significant after 1982 (Bordenave 1994).
8. Various indicators reveal the growing capital intensity of work. For example, when comparing the periods 1983–92 and 1976–82, the ratio of average value of capital invested in production equipment and machinery to average turnover increased by 18 per cent, the ratio of the former to average number of workers increased by 65 per cent, and the ratio of average gross investment to average turnover increased by 9 per cent.
9. Gross added value can be calculated by adding the cost of labour, depreciation, and profits, and then relating this added value to turnover. This rate decreased from over 40 per cent to less than 30 per cent between the early 1970s and mid-1990s, revealing a clear tendency towards disintegration.
10. By the early 1990s Ford employed more than 20,000 workers in Mexico. In 1983 an engine factory in Chihuaha and in 1986 an assembly plant in Hermosillo were added to the existing factory at Cuautitlan near Mexico City. The company possessed (or participated in) several component factories located in the 'maquiladoras' of the United States–Mexico border.
11. This mechanism led to substantial payments by the late 1980s. In 1988 they reached $3,700 per employee and in 1995 $4,000. Ford thus increased wage cost flexibility.
12. The introduction of front-wheel drive for a car of this size was a risk for Ford. The rounded bodywork to create a more aerodynamic form was also a design innovation (following the European Sierra in 1982). Taurus was also the first Ford automobile for which computer-aided technology was used on a large scale in the design and

production stages. Manufacture made use of high levels of flexible automation in body welding, with the application of robotics and modular assembly techniques.
13. This publication coincided with the arrival at the head of Ford of D. Petersen, generally considered to be a major promoter of participatory management.
14. Profits made in 1994 were US$5.8 bn., or 5.4 per cent of turnover, as against a mere 1.6 per cent the previous year. In 1994 Ford's European operations returned to profit after three consecutive years of losses.
15. In 1987 Ford acquired the craft producer Aston Martin Lagonda. The purchase of Jaguar came after fruitless attempts to acquire Rover and Alfa Romeo. Jaguar had an annual output in the order of 40–50,000 cars at the end of the 1980s and exported more than 40 per cent of its output to the USA. Ford had competed with General Motors to purchase Jaguar, and ended up paying a very high price for Jaguar, which burdened its balance sheet by more than US$2 bn.
16. Ford owned 10 per cent and Mazda 8 per cent of Kia. Kia produced a small car, a derivative of the smallest Mazda, which was exported to the entire Asia-Pacific region and North America. Examples of joint operations were the production of components in Japan and Malaysia and the announced shared assembly plant in Thailand. Mazda became a minority shareholder in Ford's New Zealand-based operations when they were restructured in 1987.
17. When the joint venture was launched in 1984, Mazda alone was responsible for management and production methods. Ford assumed greater responsibility for the factory's operations after increasing its participation to 50 per cent in 1992 (Babson 1995).
18. Ford supplied Mazda with a version of its small European car (the Fiesta), assembled at Ford's Dagenham, United Kingdom, factory. Planned output was 25,000 vehicles per year.
19. In Europe in the mid-1980s, the number of feasible build combinations had risen to 36 million. It had been reduced to well under 100,000 by the 1990s. For body-in-white alone the number of variants diminished from 27,000 to about 500. Jetin's (1995) results for North America reveal that in the late 1980s an increase in the number of platforms was associated with a decrease in the number of models.
20. This organizational change was announced in 1993. Ford's design centres were situated in Dearborn in Michigan, Valencia in California, Dunton in the United Kingdom, Cologne in Germany, Turin in Italy, Hiroshima in Japan, and Melbourne in Australia.
21. In comparison to previous models, the market position of the new vehicles, named Ford Contour and Mercury Mystique in North America, was somewhat too high, and certain features, such as a lack of leg room for back-seat passengers, were criticized as too European.
22. The centres based in the USA were for large front-wheel-drive vehicles, large rear-wheel-drive cars, light trucks, and commercial trucks. The European centre was for small and medium vehicles. Logically speaking, Mazda ought to have worked with the European centre. Jaguar remained separate but was to coordinate its operations with the centre for large rear-wheel-drive cars.
23. The product plan adopted in 1995 after this reorganization forecast an increase in the variety of models offered to the consumer but a decrease of 30 per cent in basic platforms for vehicles, motors, and transmissions. Moreover, the goal of attaining economies of scale did not imply that the vehicles had to be the same everywhere. Adaptation of basic models to local tastes remained a preoccupation.

STATISTICAL APPENDIX 9: Ford

| Year | Sales—worldwide | Employees—worldwide | Turnover—worldwide | Post-tax profit—worldwide | Pre-tax income | Gross investment—worldwide | Sales—USA | Employees—USA | Turnover—USA | Post-tax profit—USA |
|---|---|---|---|---|---|---|---|---|---|---|
| 1968 | 4,653 | 415 | 14,075 | 627 | 1,283 | 879 | 3,167 | 245 | — | — |
| 1969 | 4,849 | 436 | 14,756 | 546 | 1,118 | 957 | 3,111 | 245 | — | — |
| 1970 | 4,770 | 432 | 14,980 | 516 | 1,023 | 1,047 | 2,985 | 229 | — | — |
| 1971 | 4,933 | 433 | 16,433 | 657 | 1,280 | 1,039 | 3,106 | 225 | — | — |
| 1972 | 5,593 | 443 | 20,194 | 870 | 1,626 | 1,154 | 3,563 | 233 | — | — |
| 1973 | 5,871 | 474 | 23,015 | 906 | 1,569 | 1,486 | 3,772 | 250 | — | — |
| 1974 | 5,259 | 465 | 23,621 | 327 | 605 | 1,451 | 3,328 | 235 | — | — |
| 1975 | 4,578 | 416 | 24,009 | 323 | 436 | 956 | 2,677 | 204 | — | — |
| 1976 | 5,304 | 449 | 28,840 | 983 | 1,587 | 1,055 | 3,215 | 224 | 18,199 | 429 |
| 1977 | 6,422 | 485 | 37,841 | 1,673 | 2,746 | 1,762 | 3,971 | 244 | 24,769 | 942 |
| 1978 | 6,462 | 512 | 42,784 | 1,589 | 2,358 | 2,542 | 4,090 | 261 | 27,799 | 809 |
| 1979 | 5,810 | 500 | 43,514 | 1,169 | 917 | 3,440 | 3,227 | 244 | 24,408 | −199 |
| 1980 | 4,328 | 433 | 37,086 | −1,543 | −2,278 | 2,769 | 2,151 | 185 | 18,429 | −2,018 |
| 1981 | 4,313 | 411 | 38,247 | −1,060 | −1,256 | 2,227 | 2,102 | 176 | 19,739 | −1,195 |
| 1982 | 4,268 | 385 | 37,067 | −658 | −487 | 2,968 | 2,074 | 161 | 20,541 | −1,118 |
| 1983 | 4,934 | 386 | 44,455 | 1,867 | 1,804 | 2,333 | 2,666 | 169 | 28,375 | 1,516 |
| 1984 | 5,585 | 390 | 52,366 | 2,907 | 3,422 | 3,515 | 3,287 | 179 | 36,788 | 2,391 |
| 1985 | 5,551 | 369 | 52,774 | 2,515 | 2,730 | 3,737 | 3,201 | 172 | 36,779 | 1,988 |
| 1986 | 5,916 | 382 | 62,868 | 3,285 | 4,142 | 3,409 | 3,498 | 181 | 42,891 | 2,459 |
| 1987 | 6,051 | 352 | 71,797 | 4,625 | 6,256 | 3,674 | 3,653 | 181 | 47,798 | 3,441 |
| 1988 | 6,441 | 359 | 82,193 | 5,300 | 6,612 | 4,712 | 3,918 | 186 | 52,992 | 3,015 |
| 1989 | 6,336 | 367 | 82,879 | 3,835 | 4,252 | 6,695 | 3,710 | 188 | 52,850 | 1,627 |

## STATISTICAL APPENDIX 9: Ford (cont'd)

| Year | Sales—worldwide | Employees—worldwide | Turnover—worldwide | Post-tax profit—worldwide | Pre-tax income | Gross investment—worldwide | Sales—USA | Employees—USA | Turnover—USA | Post-tax profit—USA |
|------|-----------------|---------------------|--------------------|---------------------------|----------------|----------------------------|-----------|---------------|--------------|---------------------|
| 1990 | 5,805 | 369 | 81,844  | 860    | 316    | 7,163 | 3,275 | 179 | 48,761 | 628    |
| 1991 | 5,346 | 332 | 72,051  | -2,258 | -3,769 | 5,723 | 2,866 | 156 | 40,627 | -1,447 |
| 1992 | 5,764 | 325 | 84,407  | -502   | -1,775 | 5,697 | 3,361 | 158 | 51,918 | 514    |
| 1993 | 5,964 | 322 | 91,568  | 2,529  | 1,432  | 6,714 | 3,826 | 167 | 61,559 | 2,822  |
| 1994 | 6,639 | 338 | 107,137 | 5,308  | 5,826  | 8,310 | 4,276 | 180 | 73,008 | 4,159  |

*Notes*:

Sales—worldwide Vehicle factory sales, cars and trucks, worldwide: units in thousands. Includes units manufactured by other companies and sold by Ford.
Employees—worldwide Average number of employees, worldwide, for the whole Group in thousands. Employees in the financial services sector are included from 1976.
Turnover—worldwide Turnover, worldwide, for automotive sector (excluding financial services) in millions of current US dollars. Includes non-automotive industrial activities for a small amount.
Post-tax profit—worldwide Post-tax profit (net income), worldwide, for the whole group in millions of current US dollars. An exceptional provision of $6.9 bn. made in 1992 for post-retirement health-care benefits is not included.
Pre-tax income Pre-tax operating income, worldwide, for automotive sector in millions of current US dollars.
Gross investment—worldwide Gross investment, worldwide, for automotive sector in millions of current US dollars.
Sales—USA Same as Sales—worldwide but for USA only. Includes exports from Canada, Mexico, and Australia.
Employees—USA Same as Employees—worldwide but for USA only.
Turnover—USA Same as Turnover—worldwide but for USA only.
Post-tax profit—USA Same as Post-tax profit—worldwide but for USA only.

*Sources*: All the data derive from Ford Motor Company Annual Reports.

# BIBLIOGRAPHY

Babson, S., 'Mazda and Ford at Flat Rock: Transfer and Hybridization of the Japanese Model', Proceedings of Third GERPISA International Conference (Paris, 1995).
Bélis-Bergouignan, M.-C., Bordenave, G., and Lung, Y., 'Hiérarchie et Multinationalisation, une application à l'industrie automobile', *Revue d'Economie Politique*, 104 (1994), 739–62.
Bordenave, G., 'Le Modèle fordien et son espace', *Actes du GERPISA*, 5 (1992).
—— 'Les Voies multiformes du redressement de Ford Motor Company', *Actes du GERPISA*, 10 (1994).
—— and Lung, Y., 'Ford en Europe, crises locales, crise globale du fordisme', *Cahiers de recherche du GIP Mutations Industrielles*, 17 (1988).
Boyer, R., *La Théorie de la régulation: une analyse critique* (Paris, 1986).
—— and Durand, J.-P., *L'Après-fordisme* (Paris, 1993).
—— and Freyssenet, M., 'The Emergence of New Industrial Models: Hypotheses and Analytical Procedure', *Actes du GERPISA*, 15 (1995).
Carrillo, J., 'Flexible Production in the Auto Sector: Industrial Reorganization at Ford-Mexico', *World Development*, 23 (1995), 162–77.
Ingrassia, P., and White, J. B., *Comeback: The Fall and Rise of the American Automobile Industry* (New York, 1994).
Jetin, B., 'Économies d'échelle et variété de l'offre: une analyse comparative internationale', Proceedings of Third GERPISA International Conference (Paris, 1995).
Jürgens, U., Malsch T., and Dohse, K., *Moderne Zeiten in der Automobilfabrik: Strategien der Produktions Modernisierung im Länder und Konzernvergleich*, (Berlin, 1989). English Translation: *Breaking from Taylorism, Changing Forms of Work in the Automobile Industry* (Cambridge, 1993).
Katz, H. C., *Shifting Gears: Changing Labor Relations in the U.S. Automobile Industry* (Cambridge, Mass., 1985).
Pascale, R. T., *Managing on the Edge* (New York, 1990).
Salerno, M., 'The Historical Trajectory and Future Perspectives of Autolatina's Development in Brazil', *Actes du GERPISA*, 10 (1994).
Shook, R. L., *Turnaround: The new Ford Motor Company* (New York, 1990).
Starkey, K., and McKinley, A., *Strategy and the Human Resource: Ford and the Search for Competitive Advantage* (Oxford, 1993).
Taub, E., *The Making of the Car that Saved Ford* (New York, 1991).
Tolliday, S., 'Ford and "Fordism" in Postwar Britain', in S. Tolliday and J. Zeitlin (eds.), *The Power to Manage? Employers and Industrials Relations in Comparative Historical Perspective* (London, 1991), 81–114.
Wilkins, M., and Hill, F. E., *American Business Abroad: Ford on Six Continents* (Detroit, 1964).
Womack, J. P., Jones, D. T., and Roos, D., *The Machine that Changed the World* (New York, 1990).

# 10

## Reinventing Chrysler

### BRUCE M. BELZOWSKI

From the perspective of the mid-1990s, it was difficult not to be impressed by the strength of Chrysler in product and process innovation, a strength that had led to a complete rejuvenation in sales and profits. The trajectory of this company, which in the previous fifteen years had barely escaped bankruptcy and more recently had lacked money for new model development, was a remarkable story of poor management, luck, resiliency, and the power of personality. It was also a remarkable story in terms of the industrial models that lay behind the public face of restructuring at Chrysler.

Chrysler, perhaps more than any other American vehicle maker, embodied much of what had gone wrong with American heavy industry since the 1960s. The company's near-fatal flaws included placing production and sales above product quality and forward-looking product development, building a global automotive network without mastering the manufacturing process in its home country, buying companies in industries where its management had little expertise, poor management of the workforce, and failure to save sufficient money for product development during industry downturns when cash flow could not cover these expenses. Moreover, Chrysler had also frequently reacted to the strategic actions of General Motors (GM) and Ford, rather than set its own course and unique trajectory. Only in the 1990s did Chrysler begin to march to the beat of its own drummer. Indeed the scale of Chrysler's production and sales in the USA compared to its Big Three competitors must be understood to place Chrysler within the context of both US and world automotive production and sales. On average, between 1960 and 1995, Ford tended to produce twice as many vehicles as Chrysler, with GM producing three times as many.

On a global scale, Chrysler tended to rank between fifth and tenth place among the world's vehicle producers. As a large manufacturer, Chrysler enjoyed certain economies of scale, but not the economies of scope available to other much larger vehicle manufacturers. Thus while it might be tempting to compare the resources of Chrysler with those of General Motors and Ford because of their membership in the North American 'Big Three', such comparisons may not accurately reflect the reality of the past, present, and most likely, the future because by the mid-1990s Chrysler still did not possess the resources to set up and maintain manufacturing operations and supply networks throughout the world. The company was now succeeding in exporting vehicles from North America, but many countries still demanded that assembly take place in their country in order for them to benefit from job creation and technology transfer. Hence the company's different possibilities in terms of multinational operations represented a significant difference between Chrysler on the one hand and Ford and GM on the other.

However, one comparison that can be made among the three manufacturers is that they were all strongly affected by US economic conditions. Each time economic conditions worsened and interest rates increased, production decreased. These swings in the US economy over the past thirty-five years, combined with external events such as the oil shocks of the 1970s, forced the Big Three to respond in order to compete with imports and new entrants into US automotive manufacturing. Of the Big Three, it was Chrysler whose very survival depended on the response of its management and the trajectory they chose to follow. In a broad sense, Chrysler's leaders reinvented the company each time economic conditions forced change. In the 1960s, Lynn Townsend helped rebuild Chrysler's reputation after poor sales and a conflict of interest scandal tarnished the company's name in 1960. Lee Iacocca steered the company through its near bankruptcy in the late 1970s and early 1980s, reinventing the company to become a leader in light truck sales through the introduction of the minivan, hence providing the space for the company to reinvent its product development and manufacturing processes in the late 1980s and early 1990s. Robert Eaton, Chrysler's chief executive officer (CEO) during the mid-1990s, was able to maintain and extend Chrysler's reinvention through improved relations with suppliers, an ability not to repeat the mistakes of the past and to withstand potential hostile takeovers of his now successful company.

Underlying these processes of reinvention, Chrysler can be seen as a case of an industrial model which by the 1990s was evolving from a traditional Fordist model into a hybrid model that attempted to borrow techniques or strategies learned from Japanese competitors and apply them to the Fordist model. Within manufacturing, Chrysler was putting into place what was now called the 'Chrysler Operating System', an effort to formalize the integration of practices which had been experimented with since the mid-1980s. In other areas, such as purchasing and product development, Chrysler was now in the process of formalizing practices which had already made the company a world-class manufacturer in the eyes of many competitors. In particular, in the area of manufacturer–supplier relations, Chrysler was now the furthest along of the Big Three in developing a fundamentally new model of the supplier value chain. In product development, Chrysler's 'time to market' for new vehicles was a marked improvement over past performance, and its product development process was now being closely studied by Japanese competitors. Whether this emerging new model would in fact be identified in the future as a model in its own right remained to be seen, but the turnaround from laggard to cutting-edge innovator in these areas was none the less a remarkable fact.

Chrysler did not, of course, reach this level of success without a struggle. Looking back to the 1960s, Chrysler is best interpreted as an example of how a company fell prey to many of the pitfalls of the Fordist industrial model, with management developing strategies based more on the model than on the realities facing the company, realities which probably made the model inappropriate for Chrysler. The analysis of Chrysler that follows in this chapter is divided into separate decades for ease of presentation, but also because, in general, the major challenges that Chrysler has faced since the 1960s evolved in approximately ten-year increments.

## 10.1. CHRYSLER AND THE FORDIST MODEL IN THE 1960S

### 10.1.1. The Market

Chrysler's modern-day history begins in 1961 when Lynn Townsend, at the age of 42, assumed leadership of the company. The decade of the 1960s is important in understanding the trajectory of Chrysler's industrial model in later decades because the company's actions during this decade laid the groundwork for many of its future problems. Townsend, whose career began in finance, led Chrysler to new highs in production and market share in the expanding US market, yet these highs did not match Chrysler's overall success in the previous decade. During the 1950s, Chrysler had a combined US market share of 16.8 per cent. By the end of the 1960s, Chrysler's combined share for the decade as a whole had dropped to 14.1 per cent. Townsend inherited a situation, in the early 1960s, in which Chrysler was heading into its worst production and market-share years, to be matched only by its near bankruptcy in the late 1970s and early 1980s. But by 1965 he had doubled domestic production, and by 1970 he had increased market share to a company high.

The 1960s were automotive boom years for the US automotive industry and for Chrysler, where competition from imports was just beginning to develop. Even Volkswagen, which dominated the US import market, generated only a 4.3 per cent market share for the decade. Americans, in general, continued their love of full-sized, powerful cars, but mid-sized cars like the Ford Falcon and the Dodge Dart/Plymouth Valiant began to enjoy increased acceptance among the car-buying population. The Falcon was important because it coincided with the first appearance on the automotive scene of Lee Iacocca, who would lead Chrysler in the late 1970s and throughout the 1980s. At Ford, Iacocca, along with product development expert Hal Sperlich, used the existing Falcon platform to develop the Mustang, creating an entirely new segment: the affordable, mid-sized sports car.

Hence, compared to the 1950s, Chrysler lost market share in the 1960s, but like the rest of the Big Three, survived by selling larger volumes of vehicles. Yet further increases in vehicle sales were based more on the continued growth of the US middle class than on the attributes of the vehicles themselves. From 1960 to 1970 the USA moved from 0.85 to 1.0 registered vehicles for every licensed driver, putting over 34 million more vehicles on the nation's roads. The potential for growth was there, and Townsend began building new factories, expanding capacity to take advantage of the growing market. Drawing on his finance background, he also created a credit subsidiary to finance buyers' vehicles purchases through Chrysler itself. Chrysler's product mix consisted of mostly large cars, alongside its low-cost, mid-sized Dodge Dart and Plymouth Valiant models. Chrysler was unable to tap into the mid-sized sports car segment the Mustang had created, and had no product development plans for a US-made small car (a segment in which GM and Ford each sold hundreds of thousands of cars at the beginning of the 1970s).

The Statistical Appendix shows Chrysler's profit as measured by its return on assets (ROA: operating profit divided by total assets) based on worldwide sales. Using 10 per cent ROA as an approximate measure of financial success, Chrysler, in general, was profitable for most of the 1960s. Part of the reason for the drop at the end of the decade, and one of the reasons for Chrysler's future failure in the 1970s, was a system of accounting developed by Townsend that in the end would constrain Chrysler. Known as the sales bank, this procedure made sales look better than they really were. As a car was driven off the assembly line, it was deemed to have been sold to a dealership when in fact it was often sent to a parking lot where it was stored until a dealer requested it. In early 1966 there were about 60,000 cars 'banked', and by February of 1969 there were over 400,000, about a quarter of annual sales.

Once dealers and customers realized that they need only wait until Chrysler resorted to discounts to move vehicles out of the bank lots, the profit margin dwindled. This method of distribution also added costs of refurbishing these new cars, which in many cases endured serious environmental damage from sun and snow. Early in the 1960s this method did not affect the ROA because vehicles were just beginning to accumulate. As the decade proceeded, however, more and more vehicles were banked or sold at discount, and the balance sheet began to look grim. Townsend's excessive 'banking' of vehicles revealed what the Fordist model's 'push system' of production organization could lead to if taken to the extreme, as well as showing how slow management was in making the necessary changes to allow the company to respond to market drivers such as the move to smaller vehicles.

## 10.1.2. Globalization

Because during the 1960s US manufacturers tended to build factories in the countries where they intended to sell their products instead of exporting vehicles from the USA, the export of vehicles (see Statistical Appendix 10) was not generally considered a significant dimension of US vehicle makers' global strategies (the increase of exports from the USA in the mid to late 1980s was a function of new Japanese transplants exporting vehicles back to Japan).

GM and Ford had invested in Europe long before Chrysler, during the early days of the development of the automobile industry prior to World War II, and had formed alliances or created completely new companies. Under the direction of Townsend, Chrysler was to follow GM and Ford's lead. In 1963 the company increased its holdings in the French automobile producer Simca from 25 per cent to 64 per cent, and in 1965 it acquired a large stake in the British firm, Rootes, which claimed nearly 10 per cent of the British market. The Spanish truck maker Barreiros was acquired, and other acquisitions were made in Latin America. While in 1961 about 4 per cent of Chrysler's vehicles were made outside North America, by 1967 the figure had risen to about 24 per cent. By the end of the decade, Chrysler had plants in eighteen countries.

Moreover, by the end of the 1960s, Chrysler managers understood the need to have small cars in its product mix. Designing and building small cars themselves was deemed unprofitable, so they developed what would become a thirty-year relationship with the Japanese car maker, Mitsubishi, to supply small cars and, eventually, engines for Chrysler. Mitsubishi was just beginning to build its automotive division, so its relationship to Chrysler was strategic because it gave the Japanese company broad technical assistance, financial support, and potential market access to the USA (Flynn and Andrea 1994: 13).

Townsend no doubt believed that his company could take advantage of market opportunities in foreign countries (as did GM and Ford), but he allowed the company to lose its focus on its US customers and he failed to invest sufficiently in new products. This failure, combined with a lack of cash reserves owing to capital spending related to the costs of globalization, put Chrysler in the precarious position of continually needing successful product launches and a strong economy in order to maintain its operations. Thus when US vehicle buyers met the 1969 Chrysler models with a distinct lack of enthusiasm, Chrysler found itself without the cash to support its global operations.

One global agreement which proved profitable for Chrysler was the 1965 Auto Pact between the USA and Canada which enabled cross-national integration of automobile production and trade. Following the Auto Pact, Chrysler quickly adjusted its production structure to take advantage of the relatively low Canadian wage rates compared to US wages, and by the end of the decade, had nearly doubled the proportion (to approximately 15 per cent) of North American production performed in Canada.

### 10.1.3. Finance

Owing to the cyclicality of the US automotive industry, Chrysler, like other companies, needed to stockpile cash assets during an upturn in order to support the company through a downturn or a poor product run. Selling more shares might have been another way to raise needed cash, but during a downturn a company's stock price usually fell, hence selling shares during this period would generate less cash and oblige the company to pay more in dividends when the cycle turned up and the company started to make money. Like other companies, Chrysler also wanted to buy back some of its stock when it had excess cash assets, usually during an upturn, in order to reduce dividend payments, increase the value of the stock, and create a hedge against the next downturn; unfortunately, precisely the time when the price was high. Failure to save for the future, and thus avoid such costs, was an error Chrysler was to repeat between the 1960s and 1980s. Expanding the company became an alternative strategy for saving money, based on the reasoning that the company would make more money by extending itself if it was managed correctly, and market growth conditions continued. Unfortunately for Chrysler, the management skills and economic conditions necessary for such a strategy failed to develop. Chrysler therefore found itself following the Fordist strategy adopted by its competitors, in the hope that its past history of successfully following the industry leaders would continue.

*10.1.4. Labour and Work Organization*

In many ways, the history of labour relations between the major US assemblers and the United Automobile Workers serves as a template for how *not* to run a company, though the paradigm had worked for so long in the US market that it became an extremely painful process to change. Having been formed during the 1930s as a response to terrible labour/management relations, the UAW was generally looked upon by the workers as a means of exacting compensation for their contribution to the assembly process. 'Pattern bargaining', which forced all three of the major manufacturers to accept the same wage scale for its hourly workers, gave the workers wages and benefits that allowed relatively uneducated people to become part of the burgeoning US middle class. Chrysler, being the smallest of the Big Three, was affected most by pattern bargaining, but had no option but to pay the wages the union demanded.

Respect for the worker's contribution to the manufacturing process did not follow rising compensation. Throughout the 1950s and 1960s (and even until the mid-1980s), Chrysler management saw no role for company workers in the production process beyond 'doing the job'. This lack of respect for the value of the individual worker bred an adversarial relationship between management and labour that alienated each group and reduced the quality of the product being produced. In general, Chrysler's labour relations were the worst of the Big Three in the 1950s and 1960s. There was more lost time due to strikes at Chrysler than at any of the other companies, and this situation worsened in the 1970s, in part contributing to the company's near bankruptcy in 1979–80.

One effect of increased production volumes at Chrysler in the 1960s, as far as workers and the UAW were concerned, was an increase in employment (and consequently UAW members) during this time. In classic Fordist tradition, Chrysler built new plants with the idea of never being caught short when a particular car or truck line was in high demand. The company also purchased outside suppliers or developed internal suppliers to support their products, a vertical integration philosophy that had been in effect since Henry Ford's original River Rouge Plant where raw metal ore entered one side of the complex and finished vehicles exited the other side. Chrysler's further vertical integration also increased employment, and it doubled both hourly (blue-collar) and salaried (white-collar) employment in the 1960s.

Though many of the external suppliers of the 1960s were not capable of supporting new product development and/or delivering the quantities of parts and materials the vehicle makers needed, eventually, the costs involved in maintaining the internal suppliers also became a constraining factor for Chrysler. The costs involved in pursuing this vertical integration strategy left the company with a permanently poor debt to equity ratio. Banks therefore saw loaning money to Chrysler as a higher risk and consequently loaned money to Chrysler at higher rates.

Another way to look at Chrysler's productivity is to examine the ratio of vehicles produced to the number of employees in the company. Throughout the 1960s and 1970s Chrysler averaged about twelve vehicles per employee; it would not be until the early 1980s and 1990s that the company was able to raise the average number of vehicles per employee to about fifteen.

Chrysler's geographical shift towards Canada also affected profitability. Because wage rates were approximately 30 per cent lower in Canada than in the USA, more labour-intensive work such as final assembly and the production of more labour-intensive parts were performed in Canada, while the production of high-value body stampings, engines, and drivetrain components remained concentrated in the USA. However, the growth of employment in Canada also increased the membership of the Canadian UAW and made it a much larger and more powerful organization than it had been before the Auto Pact of 1965. Consequently, by 1970, the UAW had negotiated nominal wage parity between Canadian and American vehicle workers.

At the end of the 1960s, Chrysler found itself with high fixed costs owing to the construction of new plants and the pursuit of globalization, poor labour relations, high labour costs in both the USA and Canada, high vertical integration costs, high costs of borrowing money, and 400,000 vehicles sitting in fields waiting to be bought. By following the Fordist push system and utilizing a labour process that relied on unskilled Taylorized labour, Chrysler entered the 1970s unprepared to deal with the crises of market and labour that would nearly bankrupt the company.

## 10.2. CHRYSLER VEERS TOWARDS BANKRUPTCY IN THE 1970S

### 10.2.1. The Market

Chrysler opened the 1970s in considerable financial trouble (with a debt, at $791 m., larger than GM's) and ended it in dire financial distress. Chrysler's production began falling in 1973, temporarily rebounded in 1975, and then entered a steady decline from 1976 until it bottomed out in 1980, necessitating government intervention to prevent bankruptcy. The company's US market share for the decade declined from 14.1 per cent over the 1960s to 12.3 per cent over the 1970s.

As the decade commenced, Chrysler's weakness was basically one of not having enough money to develop new products. The solution was still seen to be increased production. After a poor 1969 model year, Chrysler needed a credit extension to support its financial business. Both salaried and hourly workers were laid off, and construction of new factories was suspended. In order to generate cash, Chrysler began to consider, but did not yet sell, its share in its foreign alliances (though potential buyers probably would not have paid high prices since the companies themselves were losing money). Luckily, the economy improved and Chrysler began to set records for production and employment during the period of growth that lasted until the first oil shock in late 1973.

The external force of the oil shock, combined with the internal force of increasing government regulation, were major factors contributing to Chrysler's woes in the 1970s. The 1973 oil shock meant that fuel prices began to rise and suddenly the cost and availability of gasoline became a national political focus. Long queues at gasoline stations became common, and more Americans began to consider fuel-efficient cars when buying a new vehicle. At this point in the decade, interest rates were climbing yet were not high enough to restrict buyers (as they would later in the decade with the second oil crisis). As potential car buyers scanned the market for a new, fuel-efficient vehicle they found vehicles from Japan and Europe more attractive than had been the case in the past. These cars were exclusively imports, and many were in fact at rather early stages of development with poor reliability, indistinct design, and low engine power.

The German Volkswagen Beetle, by far the most successful small import in the 1960s and early 1970s, was a mature vehicle (with its replacement, the Rabbit, making its debut in 1975), but the Japanese imports quickly moved up the development scale, showing improved reliability, better design, lower prices, and good fuel economy. The Japanese were also beginning to expand their distribution networks. Many US-only vehicle dealers decided to become dual dealerships, pairing their domestic vehicle offerings with import offerings to give themselves the fuel-efficient products the US manufacturers could not offer. By the end of the decade, it was the Japanese models rather than the European models that were winning and holding market share.

Chrysler was caught off guard by this sudden switch in customer preferences. The company had not adequately developed a small car strategy because the profit to be made on small cars was lower than for larger cars, and because the market for these cars had been limited. Chrysler's smaller cars offered the buyer a lower purchase price rather than higher fuel economy. Accordingly, when the first oil shock occurred, Chrysler was at a particular disadvantage. The only car it offered in this segment was an import from Mitsubishi, the Dodge Colt. As the economy reeled and customers tried to adjust to higher fuel prices, sales of the Japanese models rose and sales of US models declined. Chrysler's financial problems of the 1970s began to appear as early as 1971 when it attempted to purchase a 35 per cent share in Mitsubishi Motors Corporation, and realized that it did not have sufficient cash for the purchase, having to purchase 20 per cent then, and 15 per cent later. Nevertheless, the Chrysler/Mitsubishi partnership became a forerunner of the joint ventures that many of the major vehicle manufacturers would develop in the 1980s. Chrysler's management decision in the early 1970s to buy into this competitor instead of developing a competency for building a particular type of vehicle itself stands in stark and revealing contrast with the decision in the late 1980s to build Chrysler's own small car, the Neon, based on a perceived need to prove competency in this segment.

The domestic force that contributed to Chrysler's financial slide in the 1970s was increased government vehicle regulation. During the 1960s, the US automobile industry had been slow to realize the popular support government had in instituting regulations for safety and cleaner air. The industry often seemed to spend more time and effort trying to avoid regulations (probably because it saw only increased cost and/or lower profits as the result) than developing the technology that would be needed to meet them. The National Traffic and Motor Vehicle Safety Act of 1966 had been the first major government regulation that affected the industry by requiring redesign and retesting of hundreds of component parts that might cause death or injury in the event of an accident. These regulations took effect in 1970, and the Clean Air Act of the same year demanded that the industry begin to reduce automobile emissions by 1975.

Though safety and clean air were the goals of these regulations, the government made it particularly difficult for the companies through its decision about how these goals were to be met. It insisted on each company developing its own technology to meet the proposed standards, yet only one would then be picked as the approved method. This tactic led each company to expend resources from which in the end it might not receive any return. Though GM's method of air pollution reduction, the catalytic converter, later became the industry's effective method of reducing emissions, each company faced billions of dollars in new expenditures for research, testing, design changes, and tooling expenses. Chrysler, in particular, was disadvantaged because the high fixed cost of safety and air pollution research, and parts manufacture had to be spread over a smaller number of vehicles sold than was the case at Ford or GM. In effect, the government forced the industry to comply without affecting the consumer, but in so doing, set the industry in conflict with its customers as industry-instituted price increases to pay for new features which were not highly visible to the customer. This form of regulation permitted the government to avoid a political fallout from consumer (i.e. voter) dissatisfaction with the absence of clean air and safety equipment by transferring responsibility for these improvements to industry alone.

The US government also reacted to the first oil shock by devising the 'corporate average fuel economy' (CAFE) formula in 1975. The purpose of this formula was to reduce dependency on foreign oil by forcing vehicle makers to develop more fuel-efficient vehicles. Government assigned a fleet-level CAFE score based upon each company's overall vehicle fuel-efficiency and sales. The easiest way for vehicle makers to increase their CAFE performance was to sell more small cars, but the Big Three lacked attractive small cars, and the public was not interested in buying the ones they were offered. Again, Chrysler was worst hit by this particular regulatory approach when it realized the cost of complying with the new regulation: making vehicles lighter, replacing rear-wheel-drive with front-wheel-drive vehicles, and total redesign of the interior and exterior spaces. Researchers at the federal government's Transportation Systems Center think tank estimated that to meet the CAFE requirements over time would cost the Big Three $60 to $80 bn., more than all of the capital invested within the Big Three. Again, as with pollution standards, the government forced industry to respond instead of responding itself (for instance, by raising fuel taxes to deter consumption). Hence while it might have been argued that a two-pronged attack on fuel shortages via fuel-efficient cars combined with higher taxes to deter consumption would be more effective, the government refused to increase fuel taxes for political reasons, preferring to force industry to take the financial risks while the government avoided the political risks.

The costs of these regulations, combined with the costs Chrysler was incurring because of globalization, new plant construction, and vertical integration, damaged the company's ability to develop new products, let alone engage in technological innovation. In some ways the regulations did force the development of new technologies, and they advanced the thinking of the industry in areas they had not previously addressed, but for Chrysler, these regulations effectively added yet another long-term cost which the company could not bear.

*10.2.2. Labour and Work Organization*

Like the rest of the Big Three, Chrysler suffered during 1974 and 1975 and had to lay off massive numbers of employees. During this period Townsend retired and was replaced by his chosen successor, John Riccardo. Though the economy recovered slightly and the industry improved in 1976 and 1977, Riccardo knew the company was in dire straits. There was too little cash to operate it and develop good products.

An obvious example of what was going wrong from both a design and labour standpoint was Chrysler's problems with quality. One case came with the introduction of the Dodge Aspen and the Plymouth Volare. The pair won the Motor Trend Car of the Year award in 1976; however, in 1977 the cars received the Lemon of the Year award from the Center for Auto Safety. During 1977 Riccardo did hire Hal Sperlich, who had just been fired by Henry Ford II, as Vice-President of Product Planning. Sperlich's impression of what was happening at Chrysler was sobering:

Morale was terrible, many of the company's best people had already left, and the plants were in bad shape (factories were filthy and often unheated with broken windows in winter, racial tensions ran high, drugs were used openly in some plants, and quality was a joke). Chrysler had utterly lost its sense of accountability. (Halberstam 1986: 557)

Productivity in the 1970s, as measured by the number of vehicles per employee, remained the same as in the 1960s even with major lay-offs, and Chrysler continued to increase Canadian production because of the lower value of the Canadian dollar against the US dollar and the lower cost to the employer of employee benefits, particularly medical insurance premiums.

The year 1977 became a turning-point for Chrysler because it was the first time the company did not benefit from any market-wide improvement in vehicle sales. Throughout its recent history Chrysler had been able to ride the wave (though a much smaller wave than GM or Ford) of increased sales each time the economy grew out of recession. But in 1977 Chrysler did not recover as GM and Ford did, and in 1978 the company's sales again did not recover while GM and Ford set records for sales and production. When in 1979 the economy began to fall into what would be a long recession, Chrysler had no money banked from good sales years to sustain it.

### 10.2.3. Globalization

Some of the cash that was needed to survive the lean sales years was tied up in overseas ventures. Unfortunately for Chrysler, in its international strategy the company had not picked or developed winners in the global vehicle game. The relationship with Rootes in the United Kingdom was a failure, and that company lost money in six out of nine years. In 1978 Riccardo sold off operations in Australia, Brazil, Britain, France, and Spain. Chrysler was in dire need of cash so these transactions were probably not as profitable as they might have been. Indeed if Townsend had sold these companies earlier in the decade, he might have had more money for product development, which in turn might have saved the company its embarrassment in 1977.

One bright spot for Chrysler's overseas ventures during the 1970s, however, was its relationship with Mitsubishi. Chrysler recognized, before GM and Ford, that the Japanese were replacing the Europeans as the producers of low cost, high quality small cars. Through Mitsubishi, Chrysler not only had a vehicle, the Dodge Colt, that appealed to buyers based on its fuel economy, but during the 1970s Chrysler's engineers were also able to develop front-wheel-drive technology which would in part save the company when applied to the minivan in the 1980s. Mitsubishi for its part picked up production technology for automatic transmissions from Chrysler (Toga 1995). But the relationship that was developing between Chrysler and Mitsubishi would yield even greater benefits for Chrysler during the 1980s when Chrysler began to apply Japanese-style management and labour techniques to its own management processes.

### 10.2.4. Crisis and Rescue

Within a month of his firing by Henry Ford II in 1978, Lee Iacocca had been convinced by Riccardo and Sperlich to accept the challenge of reviving Chrysler, and within five years he had brought the company back to being a respectable player in the US vehicle industry. Chrysler was on a severe downward slope in terms of sales in the late 1970s, as Japanese imports gained market share primarily at Chrysler's expense. Throughout 1978 and 1979 Chrysler reported millions of dollars of losses for each quarter, a pattern that would continue until 1982. Not only was Iacocca faced with domestic and import competition, but Chrysler was so cash poor that he had to spend much of his time convincing bankers and investors of the viability of the company in order to receive extensions and new loans. He also brought to the ailing company a cadre of competent Ford managers willing to take the career risk and the challenge of such a move.

Sperlich immediately began stressing the need, in the next product development cycle, to switch to front-wheel-drive vehicles. He argued that the CAFE requirements were not going to go away, and might even worsen in the wake of the second oil crisis in 1979. But his development staff was small, and he needed time to develop what would become the K-car platform. In 1978 Riccardo initiated the first of many talks with the government about securing loans for Chrysler or a temporary relaxation of CAFE and air quality regulations. By 1979, the government was convinced that a loan guarantee was a possible solution for Chrysler, but there had to be a consensus about the willingness to make concessions by all the stakeholders involved, including labour, suppliers, bankers, dealers, and Chrysler itself. Iacocca led a parade of witnesses before Congress in an attempt to show the value to the American public of saving Chrysler. Iacocca's main arguments included themes of external forces over which Chrysler had no control: energy prices, high interest rates, government regulation, and growth of vehicle imports.

Labour for its part agreed to revise its contract with the company, with the UAW essentially abandoning its 'pattern' agreement to make Chrysler an exception. Suppliers agreed to accept delayed payments, and bankers who were already owed money agreed to support the loan, though many did so reluctantly. Chrysler offered the retirement of Riccardo and the promotion of Iacocca, who would be paid $1 a year, to chairman and chief executive as a symbolic admission of past management mistakes, and it also agreed to a detailed plan to restructure the company. In December 1979 the Chrysler Loan Guarantee Act was signed into law, giving the company access to $1.5 bn., although it took until June 1980 to convince all the banks, including some Japanese banks, to accept the restructuring of Chrysler's debt. Many arguments were made about the need for government to support ailing 'critical' industries. These included keeping jobs that would be lost in the case of bankruptcy, and avoiding lost tax revenues. But some deeper issues about the role of the government were also controversial. High on the list of issues Congress debated was a fear that supporting Chrysler would set a dangerous precedent. Chrysler's rescue could be a signal for managers at other companies to take reckless chances because government support would always be available as long as the company was a sufficiently large employer.

However, the one main issue that worked in Chrysler's favour was Congress's previous experience with the support of failing major organizations. They had rescued the Penn Central railway and Lockheed aircraft in 1970, and New York City in 1975. They had learned their lessons well and set up procedures that helped ensure that the government's money was secure. A loan board was created that had statutory operating control over Chrysler. If Chrysler failed, the government had first claim on most of the vehicle maker's assets. Also, Chrysler was required to submit a yearly operating plan, to ask the board every time it wanted to spend more than $10 m., and to give monthly reports on progress. In the final analysis, Chrysler did survive, and the government received $300 m. of interest Chrysler paid on the loan. Iacocca was able to save the company by leveraging his name and reputation, UAW support during a Democratic administration, his ability to unite Chrysler's cause with the nation's cause during the recession, and his ability to attract Ford talent to a weak company like Chrysler. But the losers were the Chrysler employees and companies that were subsidiaries of Chrysler. Between 1978 and 1982 Chrysler laid off about 60,000 hourly and over 15,000 salaried employees, while closing assembly plants and selling off over twenty supplier plants that supported production. Would these employees have been laid off if Chrysler had gone through normal bankruptcy proceedings? Iacocca argued a conventional bankruptcy would destroy confidence in the vehicle maker and also force dealers into bankruptcy from which there would be no recovery (Levin 1995: 147).

Chrysler ended the 1970s in much worse financial shape than it had ended the 1960s, yet the status of near insolvency was now forcing the company to make some of the decisions in the 1980s that, if made at the beginning of the 1970s, might have prevented the company from needing government aid. Saving money by becoming less vertically integrated and less global, while focusing on innovating through improved product design, might have better served Chrysler a decade earlier. In terms of its industrial model, Chrysler in the 1970s served as an example of how a company needed, but failed, to differentiate itself from other larger competitors by either product or process. If the US market had consisted only of the Big Three, Chrysler probably could have survived as a smaller member of that group. But when faced with multiple strong competitors, a lack of money available for product development, and a series of regulations or external events which drastically alter how a company must do business, only the companies with the financial resources to withstand the shock until they can respond to the challenge can survive.

## 10.3. RECOVERY AND GROWTH DURING THE 1980S

### 10.3.1. The Market

In the 1970s it was the Japanese manufacturers (seven of them were importing cars into the USA by the end of the decade) that benefited most from government regulations which focused on fuel economy and from the impacts of the twin oil shocks. Their vehicle fleets easily met the CAFE standards, and their relatively low prices made them attractive to buyers. By the end of the decade, the Japanese share of the US market grew while Chrysler's declined, and by 1980 the Japanese manufacturers together were selling almost as many vehicles as Ford. As a result, taken as a whole, during the 1980s Chrysler's US market share continued to decline: in the 1970s, Chrysler's combined market share totalled 12.3 per cent, while the combined market share for the 1980s was 10.5 per cent. Yet despite a production performance that continued a three-decade slide in US market share, Chrysler in fact turned the corner during the 1980s, with a recovery based squarely on two products, the front-wheel-drive K car and the minivan.

When trying to understand and dissect the process that led to the success of the Japanese manufacturers in the US market, US manufacturers discovered that the processes used in automotive manufacturing in Japan had evolved considerably. The system that most fully represented this evolution was the Toyota Production System, and the product differentiator that came out of this system was quality. The Toyota Production System was interpreted as one that pursued the goal of increasing vehicle quality through a process of continuous improvement in small increments that began in the executive offices and reached all the way to the shopfloor. Though some of the Japanese manufacturers had reached higher levels of quality than others by the early 1980s, they all had better quality than the domestic brands. Accordingly, the Japanese now began winning over large numbers of new buyers, especially younger buyers.

In this context, early in the decade, Chrysler and the other US manufacturers believed that they had few options in response to increased imports other than to appeal to the government for help. They received it in 1981 through the Voluntary Restraint Agreement (VRA) in which the Japanese vehicle makers agreed to limit their exports to the USA. The agreement lasted until 1984, and the Japanese voluntarily extended it until 1985. In 1981, the limit was set at 1.68 million vehicles per year, although by 1985 it had been extended to 2.4 million.

Chrysler struggled through the year 1980 using $800 m. of the $1.5 bn. the government had loaned. The K cars—the twin Plymouth Reliant and Dodge Aries—were launched, and initial sales were encouraging because the cars were utilitarian and relatively inexpensive. Yet Chrysler needed more successes if the company was to survive in the 1980s. The K car allowed the company to hold on until the economy recovered, though at times it seemed the company was surviving only on the basis of the government loan and the force of Iacocca's personality. Iacocca, indeed, began appearing in television advertisements promoting the virtues of the K car, and trying to link the plight of the nation as it tried to recover from recession to Chrysler's bid for recovery. As discussed above, Chrysler was laying off 50,000 workers and closing twenty-one plants that provided parts and components. The company became less vertically integrated, increasingly relying on independent suppliers to provide its components. Despite relatively weak sales, these drastic measures allowed Chrysler's return on assets to improve in 1981 and 1982. From the perspective of Chrysler's industrial model, this retrenchment from vertical integration, though it had a serious effect on the lives of the people who lost their jobs because of it, was the first step in making Chrysler competitive in the 1980s and the 1990s, because it lowered the company's fixed costs and consequently reduced the break-even point.

As the economy improved in 1983, Chrysler sales improved as well. The forced downsizing of the company, in combination with Iacocca's message that the company was an underdog doing all it could to recover from near financial disaster was beginning to reap financial rewards. By mid-1983 Chrysler was earning profits and decided to repay $400 m. of the $1.2 bn. owed on its loan. Iacocca now announced that Chrysler would pay off the rest of the loan by the end of the year, seven years earlier than was required.

### 10.3.2. The Minivan

In 1977, Hal Sperlich had brought with him from Ford the idea of developing a family/utility vehicle that had the room of a van and drove like a car. Previous Ford research had shown that such a vehicle would have enormous popularity, yet Henry Ford rejected it because of his reluctance to pay for the development of a front-wheel-drive platform (Halberstam 1986: 573). Chrysler's new K cars meant that the company was already developing such a platform. Iacocca and Sperlich were about to create another new segment in much the same way that they created the low-priced sports segment with the Mustang, where they developed a market segment innovation based on an existing platform. They were, in effect, reinventing Chrysler by changing the company's product focus away from one with many more cars than light trucks (the category that includes minivans in the USA) to an equal balance of cars and light trucks. The minivan completely changed the way that American consumers viewed light trucks, and led Chrysler to completely change the way it defined its vehicle market mix.

Following a very successful sales year in 1983, Chrysler launched the minivan in early 1984. The product appealed to customers from its initial launch. Classified as a truck, it sidestepped a number of safety regulations which helped keep prices low, and it helped meet Chrysler's CAFE requirements for trucks because its front-wheel-drive platform produced better fuel efficiency. Each vehicle yielded as much as $5,000 in gross profit, and in the first quarter of 1984 Chrysler earned more than it had earned the entire previous year and the most in the company's entire sixty-year history. By the year's end, Chrysler's profits had reached $2.4 bn., and the return on assets was the highest in the company's modern history (see Statistical Appendix 10).

Perhaps the most surprising aspect of the minivan success was the slow response of GM and Ford. Both companies offered small vans, but both failed to see that Chrysler's minivan offered something their vans did not. The result was that for five years Chrysler had the minivan segment almost entirely to itself,

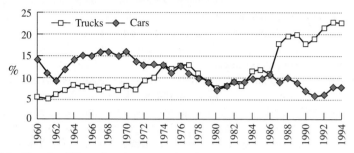

FIG. 10.1. *Chrysler US market share: cars/trucks*

selling twice as many minivans as GM's small vans and almost three times as many as Ford's small vans. Chrysler's dominance of the minivan segment increased its share of the US truck market, and with the acquisition of American Motors Corporation (AMC) in 1987 and its Jeep sport-utility vehicles, Chrysler became perfectly positioned to meet the needs of consumers who by the late 1980s and early 1990s were increasingly seeking the combined benefits of trucks that drove like cars. Consequently, Chrysler's share of the US truck market grew rapidly in the late 1980s even as its share of the car market declined (Fig. 10.1).

In effect, the minivan supported Chrysler financially during the rest of the 1980s and the early 1990s, as it heralded the move towards light trucks by American consumers. Excellent returns on assets in 1985 revealed that Chrysler was back on the automotive cycle in which it benefited from overall economic growth, in particular matching GM's increases and decreases for the rest of the decade.

It was during these profitable years that Chrysler might have learned from the past and begun saving for the next downturn in the automotive cycle, focusing on product development and starting to develop the replacement for the K-car platform. Yet Chrysler did not follow this path, instead adopting strategies that eventually put the company in a precarious financial position once more.

### 10.3.3. Finance

Chrysler now followed GM's lead in investing in non-automotive companies, in the hope of using the profits of these companies to balance losses on the vehicle side during a downturn. In 1985, Chrysler purchased Gulfstream Aerospace, and in 1987 Electrospace Systems. The company also purchased a number of financial services companies in order to have non-vehicle loan activities within its portfolio of companies. From 1984 until 1988, Chrysler also repurchased $1.2 bn. worth of stock, in the belief that the company would be able to generate cash by reselling it when needed. Unfortunately, the stock was repurchased when its price was high, and in 1991, when the company needed to sell stock to support daily operations, the stock had to be sold when its price was low.

Chrysler also purchased part of or all of a number of automobile industry companies. In 1984, 15.6 per cent of the equity of the Italian company Maserati was bought, with the hope of building a car which would raise Chrysler's prestige as a sports car producer, and in 1987 another Italian sports producer, Lamborghini, was purchased. However, the most successful purchase was of the troubled fourth largest US vehicle producer, American Motors Corporation (AMC), in 1987 for $1.5 bn. American Motors' most attractive assets were its Jeep models and franchises which appealed to younger buyers, its joint venture in China, and its new car assembly plant in Bramalea, Canada. What Chrysler did not know at the time but discovered after the acquisition was that American Motors had a group of fine systems engineers who would help Chrysler's later transition to a new way of organizing the product development process, based on team-oriented vehicle platforms. On the other hand, AMC's liabilities included 20,000 employees, two outmoded assembly plants, $800 m. of American Motors debt, and billions of dollars in pension liabilities. Chrysler was also forced through the agreement to market AMC's Premier automobile for a few years. The result was that Chrysler now had seven distinct car lines (only one fewer than GM), with weak product development and marketing to support them, as well as further huge fixed costs to maintain assembly plants.

### 10.3.4. Competition

Throughout the 1980s, Chrysler's product development success rested only on the minivan and, to a lesser extent, the introduction of the front-wheel-drive, compact Dodge Shadow/Plymouth Sundance in 1988. Other product development activities focused primarily on finding new ways to wring as many variants from the K-car platform as possible. The same platform was used for mid-sized cars, the minivan, luxury cars, and even an ill-fated attempt to marry it with a Maserati sports car. Styling was emphasized over platform improvements such as drivetrain, suspension, and chassis. This strategy had worked in the 1950s and 1960s when Chrysler basically followed the lead of GM and Ford, but the US market in the 1980s had become extremely competitive, especially in cars.

The major driving force in this increased competitiveness was the arrival of the Japanese manufacturers. In the early 1980s, Chrysler, through Iacocca, was probably the most vociferous of the Big Three in its attacks on Japanese manufacturers. Iacocca's most common complaints included charges of 'dumping' vehicles (selling vehicles at lower prices in the USA than in the home market), non-competitive labour rates, and unfair trade practices which kept US manufacturers from selling vehicles in Japan. Yet as he railed against the Japanese, Iacocca continued to source small cars and thousands of engines from Mitsubishi, even entering into a joint venture, Diamond-Star Motors, in 1985 to build cars in Illinois. Ford and GM followed similar paths to Chrysler for sourcing small cars, but they were much less vocal in their opposition to the Japanese, letting Iaccoca do most of their arguing. Iacocca argued that the political rules

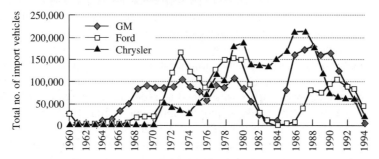

FIG. 10.2. *US vehicle import sales*

forced him to make business decisions such as sourcing from Japan, and that he would gladly not source from Japan if the rules were changed. In fact Figure 10.2 shows that, until the Diamond-Star plant opened in 1988, Chrysler was the largest importer of vehicles, all of which were Japanese.

The Japanese manufacturers responded by doing exactly what Iacocca (and the rest of the Big Three and the UAW) told them to do: 'build where you sell'. They built 'transplant' facilities in both the USA and Canada, and formed joint ventures with both GM (Toyota, Suzuki) and Ford (Mazda, Nissan). What the US firms learned from these ventures was that workers alone did not make the difference in improving the efficiency and, by extension, the quality of vehicles. What was significant was the *process* of designing and assembling the vehicles which relied on involving the workers in decisions and empowering them to make suggestions (which in turn would be implemented) about improving the manufacturing process. Gradually, Chrysler began to understand that its whole industrial model was at stake.

### 10.3.5. Labour and Work Organization

All of the Big Three were forced at one time or another during the 1980s to drastically reduce their number of employees, both hourly and salaried. As a consequence, when the UAW renegotiated their contracts, they increasingly focused on job security rather than on pay increases. In the past, the companies would lay off workers when sales were slow, but new contracts forced the vehicle makers to pay hefty unemployment benefits to laid-off workers. The union was, in effect, forced to tie itself more closely to the fate of each company by forcing management to view it as a fixed rather than a variable cost. Unfortunately, though a contract determined the size of the workforce for the life of the contract, when the new contract was negotiated, workers who had not yet been recalled were not considered part of the potential returning pool of workers.

In the mid-1980s Chrysler and the UAW were able to agree on what was called a 'modern operating agreement' for the new Jefferson Avenue plant which had been scheduled to close in 1988. The City of Detroit offered land and tax breaks, and Chrysler agreed to erect a new $1 bn. state of the art factory with a more team-oriented management style in the plant. Moreover, the UAW agreed to drastically altered work rules, job classifications, and the management of the plant. The number of job classifications dropped from ninety-eight to ten; the union agreed that workers could be organized into teams and that the number of union stewards and company supervisors would be reduced. Instead of 4,500 to 5,000 workers, the new plant would employ no more than 3,500. This new agreement with labour applied only to the Jefferson Avenue plant, and it would take until the 1990s to reach this type of agreement with the other forty-six plants. Nevertheless, the first deep reform of manufacturing and work organization had begun.

Chrysler workers had made considerable financial concessions in the early 1980s as part of the package that saved the company from bankruptcy, and the UAW's Douglas Frazier was allowed to sit on the Chrysler board of directors. But what the workers did not expect were the huge lay-offs that occurred in the early 1980s. It was no surprise in the 1988 negotiations that the workers threatened to strike if Chrysler sold off their parts subsidiary, Acustar, which accounted for about 30,000 jobs. Owen Bieber, who succeeded Frazier, negotiated a settlement that kept Chrysler from selling off Acustar as well as extracting a promise that Chrysler executives would not be eligible for bonuses unless workers were paid a profit-sharing bonus. The union for its part now agreed to cooperate company-wide in implementing the type of modern operating agreement that was in place in the Jefferson Avenue plant.

Reorganizing the workplace was just the first of many changes which were necessary to raise the level of quality of vehicles. For example, Japanese workers had refined continuous improvement techniques over two decades. The American workforce was not going to embrace these new ways of working immediately. Both labour and management had to adapt the ideas of continuous improvement to their particular circumstances, and, by definition, this process of adaption continued well into the 1990s. Chrysler workers were no exception in their uncertainty about using new methods, but they also knew that the company needed to improve so they would not face another set of permanent lay-offs.

The Canadian vehicle workers, on the other hand, were able to negotiate to their advantage during the Chrysler loan negotiations. As part of the loan package, Chrysler had to secure loans from Canada and wage concessions from the Canadian workers. The Canadian UAW made it clear that they would not support the loan without guarantees of jobs. After intense political negotiations, the Ontario and Canadian governments granted Chrysler a loan in exchange for guarantees of investment in Canada, and the investments ultimately meant more jobs. As massive lay-offs began to occur in the USA, the American UAW began to lose support from its members because they had lost wages and jobs. On the other hand, the Chrysler loan negotiations strengthened the Canadian UAW because the union there was seen as protecting its members' jobs. These issues began to create tensions between the two factions and ended with the Canadian UAW breaking off from the American UAW and forming the Canadian Automobile

Workers (CAW) in 1984 when it became clear in the contract negotiations with GM that the American UAW leaders were willing to cooperate with management to force Canadian workers to conform to US collective bargaining patterns.

The role played by labour relations in the development of Chrysler's industrial model during the 1980s reveals an extreme case of the importance of having a clear vision of what is needed for a company to succeed. The Chrysler loan negotiations were quite straightforward because of the company's impending default, but the mid–1980s negotiations were masterful give and take negotiations involving both labour and management. It was eventually understood that, given the need to improve the quality of vehicles and thereby make the company more competitive, Chrysler was better off focusing on reorganizing its workplaces and beginning to institute continuous improvement methods than selling off subsidiary parts suppliers.

### 10.3.6. Product Development

Chrysler also reorganized the way it developed vehicles in the late 1980s. The years 1987 and 1988 were peak years of production prior to a market downturn. They were important to Chrysler because it was during this period that the company undertook a major study of Honda, purchased AMC, and replaced Hal Sperlich with Robert Lutz, who had a very different vision of product development.

The team studying Honda proposed that Chrysler should make two major changes. First, Chrysler should stop concentrating on tactics to increase short-term profit and turn instead to improving quality and customer satisfaction. Secondly, the company should reorganize into cross-functional teams where all the main functions involved in producing vehicles for a platform would work together concurrently on development. These platform teams, made up of 700–800 people, would include representatives from engineering, manufacturing, finance, and marketing who would be part of the team through all phases of development, including design, assembly, and sales of a vehicle (Levin 1995: 207).

The Honda study team found sympathetic ears among former AMC engineers, in particular François Castaing, who would become Chrysler's vice-president of engineering. He had orchestrated the switch to cross-functional teams while at AMC, where they were able to complete multiple projects with little money. The AMC engineers had multiple skills and were good systems engineers, which was something that the US industry often lacked. The Honda study also had support from groups within Chrysler who were working with Mitsubishi in their Diamond-Star joint venture in Illinois. The Japanese-style product and manufacturing engineering process was taking hold in Diamond-Star and winning American converts, influencing and expanding Chrysler's understanding of Mitsubishi's system, and showing management that American workers were capable of working within this system.

Lutz and Castaing realized the need to speed up development of the L/H platform which was to replace the ageing K-car platform when the economy began to enter a mild recession in 1989 and Big Three sales began to decline. (The L/H designation is an arbitrary naming device which in this case represented Chrysler's mid-size platform.) Chrysler was only able to sell its cars by offering large discounts, because the vehicles themselves were no match for the Honda Accord, Toyota Camry, or Ford Taurus. Chrysler ended the year with a return on assets below 10 per cent for the first time in six years, and Iacocca sounded the call for reducing costs. He wanted to take $1 bn. out of the annual costs of the organization, which had become bloated through its acquisitions. He sold off Gulfstream Aerospace and Electrospace Systems, and he decreased Chrysler's share of the Diamond-Star joint venture. As the decade of the 1990s began, Chrysler found itself in a position similar, though not nearly as serious, to the one it occupied at the beginning of the 1980s: financially constrained, caught in the middle of a recession, and in urgent need of a successful new platform launch.

Hence the second half of the 1980s saw Chrysler beginning to put into place the pieces of what appeared to be a new industrial model or operating system which would extend the complete length of the value chain from product development to suppliers to manufacturing to distribution. The influences of Honda and AMC on Chrysler's revised product development function can be seen in the introduction of cross-functional teams for the development of the L/H cars which were to compete with the Honda Accord, Toyota Camry, and the Ford Taurus. A second key component of the new industrial model could be observed in the area of manufacturer/supplier relations. Having closed down numerous Chrysler-owned supplier plants (though not its Acustar subsidiary), Chrysler had to begin developing relationships with more external suppliers, and began to use what might be called a 'development model' in which the company would work with suppliers with the idea of creating a long-term relationship based on mutual reliance on each other. This relationship-oriented model contrasted with the traditional model by which suppliers were selected, in which the manufacturer scanned the market annually looking for the best price for each part. Thirdly and finally, Chrysler laid the groundwork for improving manufacturing quality through the redesign of the workplace that was the foundation of the 'modern operating agreement' with the UAW. Chrysler increased its productivity as measured by the number of vehicles per employee in the USA from twelve in the 1970s to fifteen in the 1980s, an improvement which may be attributed to both the reduction of employee numbers due to lay-offs and the redesign of the workplace which improved the assembly process.

## 10.4. CONSOLIDATION OF A NEW INDUSTRIAL MODEL IN THE 1990S

### *10.4.1. Finance*

As at the start of each of the preceding three decades, at the beginning of the 1990s Chrysler found itself in financial difficulty. Yet although the company was financially constrained with $3 bn. in pension liabilities and excess capacity owing to its purchase of AMC, it had strong light truck sales due to continued and growing consumer interest in minivans, sport-utility vehicles, and trucks. Iacocca believed that one way for Chrysler to escape from its dilemma was to form a partnership with another manufacturer. Despite its financial difficulties, Chrysler was on the verge of becoming a much stronger and more capable manufacturer than it had been in 1980, and could therefore make a stronger case for a partnership. Accordingly, Iacocca spent the first half of 1990 in negotiations with Fiat and then Ford before both rejected the partnership because of Chrysler's precarious financial position and high pension liability.

During 1991, Chrysler needed to raise cash to support the company until the L/H cars were launched. An emergency line of credit was secured from a combination of 152 US banks. Thirty-three million shares, originally re-purchased in the late 1980s at a price of over $20, were sold at $10–11, generating $350 m. This, combined with the emergency credit, allowed the company to survive until the launch of the L/H cars in summer of 1992.

The issue of Iacocca's retirement now became a serious distraction. Having rebuilt the company once, he again wanted to play a significant role in a second revival. The Chrysler board, however, wanted a younger leader to rebuild the company and prepare it for the next century. In 1990 Iacocca, now 66 years old, spent months considering a possible successor, refusing to name either of the two main internal candidates, Robert Lutz or Robert Miller, the vice-chairman. Only when the Chrysler board firmly announced that Iacocca would retire at the end of 1992 did he finally select Robert Eaton, GM's European head, as a possible candidate. Fortunately for Chrysler, when the choice of Eaton was announced, Lutz decided to stay.

*10.4.2. Product Development*

What distinguished the early 1990s from the early 1980s for Chrysler was the company's ability to avoid the need for government intervention for financial aid, and their team-based product development. Chrysler's survival in the 1980s was based on its products, while the company's survival in the 1990s was primarily to be based on process. This was the focus of Chrysler's reinvention during the 1990s. Although since the launch of the L/H cars in 1992, the product range was meeting or exceeding customer expectations, it was the new product development process, based on the platform teams, that enabled the company to produce such successful vehicles in a short period of time.

Iacocca gave his blessing to the reorganization of product development into platform teams in 1989, and the L/H platform team (the Intrepid/Concorde/Vision) was created, as were teams for the Z/J platform (the Jeep Grand Cherokee), a new pick-up truck platform (the Ram truck), a new subcompact platform (the Neon), and a new compact platform (the Cirrus/Stratus). In 1990 Iacocca even approved the Viper, a high visibility sports car that Lutz saw as an image builder for the company. The L/H platform, if successful, would also serve the next generation minivan. Glenn Gardner, who had led the original minivan project in 1981 and was the senior Chrysler executive for the Diamond-Star joint venture, was named project leader for the L/H platform team. His knowledge of Japanese techniques gleaned from his tenure at Diamond-Star proved invaluable in the new development process, which not only reduced the length of the development process from fifty-four to forty-two months, but also delivered a new powertrain and structural design. This new 'cab forward' design permitted 75 per cent instead of 65 per cent of the vehicle's length to be dedicated to passenger and cargo.

Having the manufacturing function as part of the development team was new for US vehicle makers. Typically, workers were not given the specifications of how to build the vehicle until some twenty-two weeks before the start of production. In Chrysler's case, workers were allowed to observe pilot assembly techniques two years in advance. The focus was on design for manufacture, in which parts were designed for ease of assembly. This display of genuine interest in workers' views of vehicle assembly also helped show management's vision of teamwork to the workers.

Another innovation borrowed from the Japanese was closer relations to suppliers, and reforms in this area were now rapid. During the 1980s Chrysler sold off many of its internal suppliers and was forced to rely more on outside suppliers. During the 1990s Chrysler took the relationship a step further by cooperating closely with suppliers during the design stage of development. Instead of receiving the detailed specifications of the part they were to provide directly from Chrysler, suppliers were provided with the basic requirements for the part and told to use their own expertise to design it. Suppliers were also given long-term contracts to supply parts and components, which gave them an incentive to provide innovative designs based on a more secure relationship.

All of these improvements in the product development process came to fruition in 1992 when the L/H cars were launched. The L/H models succeeded in proving to the industry that Chrysler could build cars that could compete in the mid-sized segment with the Ford Taurus. They also restored Chrysler's reputation for engineering and design as well as setting the standard, at least for the Big Three, of how manufacturer–supplier relations should be conducted. But what followed the L/H models was perhaps even more impressive. All of the other new platform teams were able to design and produce vehicles that competed successfully. The Jeep Grand Cherokee sold well within the four-door sport-utility segment; the distinctive styling of the Ram truck made it a success; the Neon proved that Chrysler could compete in the subcompact segment; and the Cirrus/Stratus proved a worthy competitor for the Honda Accord. With the introduction of the new minivan in 1995, Chrysler had almost completely refreshed its vehicle line in a matter of three years. By 1994 Chrysler was widely believed to have become the lowest cost producer of vehicles in the USA, with a profit per vehicle of $1,259, while at the same time the number of vehicles per US employee had increased to a company high of twenty-one.

## 10.5. CHRYSLER'S FUTURE: CONCLUSIONS AND CAUTIONS

Because of the relatively short time period over which Chrysler transformed its organization, starting from near bankruptcy, with the key transformations occurring only in the 1990s, it is not necessarily safe to conclude that Chrysler consciously developed a new industrial model. First of all, many of the initiatives were adaptations of Japanese organizational and manufacturing techniques to an American management and labour environment. Secondly, many of Chrysler's changes during the 1980s and 1990s were undertaken in a 'need-to-survive' mentality. There was no overarching plan to integrate new techniques and practices over a long period of time in order to develop a new model. Finally, a review of the company's recent history reveals that many decisions that were to affect the organization over a long period of time were made by individuals who tended not to look past the next five years, much less try to mould a corporation into a model for the future. And yet if Chrysler continued to succeed as it did in 1993 and 1994, and if the company actually developed, refined, and codified their innovations at each point of the value chain, in a further ten years we might be discussing an industrial model that allowed a full line vehicle manufacturer controlling about 20–30 per cent of the vehicle market to survive and prosper.

But if Chrysler were to continue to prosper, it was now imperative not to repeat the strategic mistakes of the past, such as alignment with weak overseas manufacturers, diversification into businesses outside core competencies, and failure to maintain sufficient cash reserves for full product development to continue even during industry downturns. The strength of any vehicle manufacturer lies in its product, and Chrysler now had to continue to provide products that stimulated buyers. In this context, Chrysler faced significant challenges because of the relative novelty of its new product development process. The company would need to prove itself again with its next generation of vehicles. For instance, would Chrysler now follow Honda's example of refining a vehicle instead of replacing it? Refinement instead of replacement would allow Chrysler to focus on improving the drivetrain and structure instead of focusing as much on the design of the chassis. To that end, by the mid-1990s Chrysler was switching over a plant to make a series of new V–6 and V–8 engines, which would replace 30-year-old engine technology.

Focusing on model refinement instead of replacement might also allow Chrysler to improve its quality ratings. Although Chrysler vehicles now scored extremely high on styling, their scores were average or below average on customer initial quality ratings. To be sure, the quality gap between US and Japanese vehicles had narrowed, and differences between Chrysler and others on the J. D. Powers initial quality ratings were small. The new car buyer of the 1990s, however, would no longer accept an inferior product. There were now well-documented, quality alternatives in the marketplace. Some consumers were suffering the psychological effects of past experiences with a poor quality product. Moreover, the average number of weeks' earnings needed to pay for a new car in the USA had increased from nineteen in 1980 to twenty-seven in 1995. The result was that new car buyers increasingly sought vehicles that offered both value and quality.

In principle, Chrysler's platform teams, which include design and manufacturing functions together, represented a good way to address quality issues by creating an interface between the two functions. What might still have been missing, however, was a process of learning and sharing best practices between teams and between plants. Given its smaller size, improving lines of communication might prove easier for Chrysler, and might offer the company a distinct advantage, if it could harness and distribute information and use it to improve the products.

As stated earlier, one of Chrysler's challenges from a labour–management perspective was to convince workers of management's goal of total quality through continuous improvement and by reducing the time it takes to change from one model to another within the same factory. A further future challenge for both management and labour was to make the manufacturing system flexible enough to be able to shift production quickly from models that were not selling well to those that were, and to be able to shift production to any of the company's plants. As production was set up in the early 1990s, it would take months to re-tool and redesign a plant to build a totally different vehicle from the one currently produced. But if Chrysler were fully to take advantage of market opportunities such as the increased demand for sport-utility vehicles in the mid-1990s, the company needed flexible factories, manufacturing processes, and workers.

One noteworthy aspect of Chrysler's size as a medium-sized automotive company remained its potential to be taken over or controlled by investors able to buy large amounts of their stock, such as Kirk Kerkorian. In 1990, the 73-year-old investor, through his company, Tracinda Corporation, purchased 9.8 per cent of Chrysler's stock, worth about $277 m. Though Iacocca was assured by Kerkorian, in 1991, that he had purchased the stock because of his confidence in Iacocca, in 1995 Kerkorian bid to buy a controlling interest in Chrysler for $55 a share or close to $23 bn. He expected $5.5 bn. of Chrysler's own cash to be used to secure borrowings, an idea pioneered by the leveraged buyout specialists of the 1980s (Levin 1995: 325).

Because of the number of hostile takeovers that had occurred in the US economy during the 1980s, Chrysler adopted a 'poison pill' defence against takeovers, which allowed shareholders to buy new shares of Chrysler stock at steep discounts when an unwelcome investor's stake in the company rose to 10 per cent. The intent was to dilute a buyout specialist's share of the company, and make a takeover too expensive to pursue (Lapham 1995: 1). The 'poison pill' defence was not used in this instance, but Kerkorian (who had recruited Iacocca and former Chrysler vice-president of finance Gerald York to support him) continued to try to push the company in directions Chrysler executives believed unwise. In particular, Tracinda argued that Chrysler did not need the $7.6 bn. the company had saved and set aside for product development, capital improvements, and funding to support the company during the next industry downturn.

This challenge to Eaton and his management team might well have distracted them from their goal of excellence as an automotive manufacturer. From their perspective, their strategic alternatives for future growth were to focus on the USA or to attempt a renewed expansion into the rest of the world. Because very limited or no growth was predicted for the US market for the remainder of the 1990s, the potential for growth within the USA was limited to taking over other companies' market share, a severe challenge because of the extremely competitive market in the USA. Chrysler's poor historical experience with following the GM and Ford models of globalization now made it cautious about expanding its manufacturing outside North America. And yet Chrysler might also have felt obliged to take part in the race for market share in emerging markets such as Central Europe, where their assembly of minivans in Austria supported entry into Western Europe and could begin to support Eastern Europe, or China where the company was building Jeeps. Chrysler had also signed agreements to produce vehicles in Vietnam and Brazil. Past experience, however, suggested that Chrysler was unable to extend its production to overseas environments successfully, and the benefits of these 1990s ventures remained to be seen.

Chrysler's high reliance on its suppliers also brought new risks. Some major suppliers might not want even long-term commitments for Chrysler's lower volumes if they could obtain higher volumes from the other companies such as GM or Toyota. Moreover, by retaining long-term contracts with suppliers, there was concern that Chrysler would lose out on new technology being developed by suppliers Chrysler did not contract with. Chrysler therefore had to carefully pick suppliers who offered innovation as well as reliability and quality, or they had to blend innovative suppliers into their organization by allowing each one to supply one platform instead of the whole product line.

Finally, as revealed by the battle with corporate raider Kerkorian, Chrysler remained potentially vulnerable to takeover. This vulnerability might yet be translated into a need to merge with another company to create a stronger, more viable unit. At one time or another, Chrysler had been mentioned as a good match for almost every car company in the world (indeed, during the late 1970s either GM or Ford could have taken over Chrysler if their own companies had not been in such disarray). No automobile company seems to want Chrysler when the latter is weak, and Chrysler is not interested in merging with other companies when it is strong. Will a merger be the next reinvention of Chrysler? Automobile industry analysts, together with analysts of emerging industrial models, will have to wait for the next chapter of Chrysler's history to be written to find out if a merger occurs, if the product development process will keep its edge, if quality will improve, if globalization will succeed, if relationships with suppliers will prove fruitful for both parties, and, finally, if Chrysler will continue to succeed.

**STATISTICAL APPENDIX 10: Chrysler**

| Year | Canadian Production[a] | US Production[b] | Employment[c] | Return on assets (%)[d] | US export ratio (%)[e] | Turnover[f] (In $ M.) |
|---|---|---|---|---|---|---|
| 1960 | 56,345 | 1,089,600 | 105,410 | 8 | 3.7 | 3,007 |
| 1961 | 53,222 | 713,556 | 74,337 | 4 | 3.6 | 2,127 |
| 1962 | 56,950 | 812,911 | 77,194 | 11 | 2.6 | 2,378 |
| 1963 | 96,496 | 1,158,709 | 90,752 | 18 | 2.8 | 3,505 |
| 1964 | 117,796 | 1,377,792 | 104,845 | 19 | 3.2 | 4,287 |
| 1965 | 153,186 | 1,611,005 | 126,000 | 19 | 1.0 | 5,300 |
| 1966 | 189,651 | 1,598,755 | 133,114 | 15 | 1.2 | 5,650 |
| 1967 | 202,812 | 1,505,561 | 127,894 | 13 | 1.0 | 6,213 |
| 1968 | 235,724 | 1,759,360 | 140,204 | 17 | 0.8 | 7,445 |
| 1969 | 216,620 | 1,557,587 | 140,454 | 7 | 0.8 | 7,052 |
| 1970 | 249,080 | 1,452,043 | 129,127 | 3 | 1.0 | 7,000 |
| 1971 | 249,759 | 1,518,072 | 129,531 | 7 | 0.7 | 7,999 |
| 1972 | 289,402 | 1,692,073 | 136,620 | 13 | 0.6 | 9,759 |
| 1973 | 283,292 | 1,933,932 | 152,560 | 13 | 0.8 | 11,774 |
| 1974 | 250,847 | 1,538,670 | 135,782 | 4 | 1.4 | 10,971 |
| 1975 | 285,980 | 1,222,308 | 103,342 | 3 | 2.0 | 11,598 |
| 1976 | 329,604 | 1,775,252 | 129,186 | 13 | 1.6 | 15,538 |
| 1977 | 332,068 | 1,710,860 | 133,572 | 8 | 1.5 | 16,708 |
| 1978 | 255,500 | 1,615,328 | 131,758 | 1 | 2.3 | 13,618 |
| 1979 | 194,958 | 1,231,612 | 109,306 | −9 | 2.6 | 12,002 |
| 1980 | 125,228 | 758,206 | 76,711 | −16 | 2.7 | 9,225 |
| 1981 | 149,154 | 847,888 | 68,696 | 0 | 2.1 | 9,972 |
| 1982 | 235,751 | 722,418 | 58,607 | 6 | 1.5 | 10,057 |
| 1983 | 348,979 | 1,051,222 | 65,832 | 17 | 0.7 | 13,264 |
| 1984 | 364,377 | 1,460,941 | 80,233 | 28 | 0.4 | 19,573 |
| 1985 | 390,120 | 1,479,987 | 84,804 | 17 | 0.4 | 21,256 |
| 1986 | 404,789 | 1,449,965 | 87,414 | 10 | 0.4 | 21,937 |
| 1987 | 422,797 | 1,332,300 | 94,352 | 8 | 0.9 | 25,381 |
| 1988 | 493,813 | 1,718,652 | 100,732 | 9 | 1.8 | 30,790 |
| 1989 | 590,965 | 1,516,359 | 92,029 | 5 | 2.2 | 31,039 |
| 1990 | 516,664 | 1,168,744 | 79,123 | 1 | 3.3 | 26,965 |
| 1991 | 406,866 | 1,002,032 | 69,276 | −1 | 3.7 | 26,707 |
| 1992 | 464,501 | 962,549 | 77,878 | 6 | 4.5 | 33,409 |
| 1993 | 643,050 | 1,388,662 | 79,494 | 15 | 4.0 | 41,247 |
| 1994 | 695,630 | 1,659,921 | 79,953 | 18 | 4.4 | 49,534 |
| 1995 | 538,627 | 1,701,582 | 98,000 | 10.3 | 3.5 | 53,195 |
| 1996 | 705,467 | 1,702,256 | 107,000 | 14.7 | n/a | 61,397 |

*Notes and Sources*:
[a] Chrysler Canadian light vehicle production (passenger cars and trucks under 10,000 lbs. GVW) (Ward's Automotive, AAMA).
[b] Chrysler US light vehicle production (passenger car trucks under 10,000 lbs. GVW) (Ward's Automotive, AAMA).
[c] Chrysler's total US employment, including subsidiaries (Chrysler Annual Reports, 1960–94).
[d] Return on assets = operating profit/total assets (Standard and Poors, Schroder Wertheim, Inc.).
[e] Export ratio = US light vehicle exports/production. Figures are for US manufacturers combined and do not include exports to Canada (MVMA, Ward's Automotive).
[f] Turnover = sales of manufactured products (Standard and Poors, Schroder and Wertheim, Inc.).

# BIBLIOGRAPHY

'America at the Wheel: 100 Years of the Automobile in America', *Automotive News Special Issue*, 19 (1993), 211–13.
American Automobile Manufacturers Association (AAMA), *Economic Indicators: The Motor Vehicle's Role in the US Economy*, 3 (1995), 24–5.
Flynn, M. S., and Andrea, D. J., 'Corporate Learning from Japan: The Automotive Industry', The University of Michigan Transportation Research Institute, Report Number: 94-14 (Ann Arbor, 1994).
Halberstam, D., *The Reckoning* (New York, 1986).
Harbour, J. E., *The Harbour Report, 1995: Manufacturing Productivity Company by Company—Plant by Plant* (1995), 98.
Lapham, E., 'Iacocca Once Led Takeover Defense', *Automotive News*, 17 (1995).
Levin, D. P., *Behind the Wheel at Chrysler: The Iacocca Legacy* (New York, 1995).
Thomas, C. M., 'Time of Reckoning', *Automotive News Special Issue*, 19 (1993), 162–5.
Toga, M., 'Mitsubishi: Apron Strings to Boot Straps', *Nikkei Weekly*, 4 (1995), 9.
US Department of Transportation, 'Our Nation's Highways: Selected Facts and Figures', Federal Highway Administration, Office of Highway Information Management, Publication no. FHWA-PL-95-028, HPM-40/5-95(50M), 14.
Ward's Communications AutoInfoBank, 1995, computer database (Detroit, 1995).
Yates, C., 'Public Policy and Canadian and American Autoworkers: Divergent Fortunes', in M. Molot, *Driving Continentally: National Policies and the North American Auto Industry* (Ottawa, 1993).

# PART III

# EUROPE'S DILEMMA: TRANSFER, ADAPT, OR INNOVATE?

# 11

# The Development of Volkswagen's Industrial Model, 1967–1995

ULRICH JÜRGENS

## 11.1. THE VW INDUSTRIAL MODEL AND ITS CONTEXT

Analysis of the trajectory of Volkswagen[1] (VW) between the late 1960s and the mid-1990s reveals two prominent events: the crisis of 1973/4 and the crisis of 1993/4. Both years mark the end-points of periods of internal crisis, indecision, trial-and-error approaches, and accompanying internal controversies and uncertainty, which were of extended length: from around 1968 until 1974 in the first case, and from around 1988 until 1994 in the second. Both of these periods were basically related to issues of product strategy, that is, to 'market uncertainty'. These, in turn, affected the firm's governance structure.

The configuration that made up Volkswagen's industrial model after the war had three basic elements: the concept of a standardized, affordable 'people's car', the governance structure, which ensured strong state and union influence, and a very special labour relations situation, which allowed a high degree of jointness and co-determination even in areas not stipulated by German law.

*The people's car concept*, embodied in the Beetle, was a clear continuation of the socio-economic concept behind Ford's Model-T car: through product standardization, the car was to become affordable for the mass consumer. The technical concept of the Volkswagen was developed by the Austrian engineer Porsche, who had already approached the German government in 1932 in this regard, and later received full support from Hitler, who regarded the Volkswagen as a corner stone of his programme for a 'national socialism'. The new company and new production facilities were established, but when they opened in 1940 it was to produce military vehicles and aeroplane components (for the pre-war and war period, cf. Mommsen and Grieger 1996; Doleschal 1982; Wiersch 1974, Klauke 1960). Starting up as a repair shop for the military vehicles of the British Armed Forces in 1945, VW was given a second chance to become a car manufacturer. The first 2,000 vehicles based on Porsche's prototype were manufactured in 1945, and production volumes grew slowly to 50,000 cars by 1949, when the company was handed over to German ownership.

*Volkswagen's governance structure* has often been discussed as an example of German neo-corporatism. Here it is important to analyse briefly the founding of the company and its relationship to its shareholders. The Volkswagen works were run as a state company after the British allied force withdrew in 1949. In 1960 Volkswagen Works Limited turned into a public limited company (AG) and was partially privatized. State institutions now held 40 per cent of the shares, with the State of Lower Saxony and the German Federal Government each holding 20 per cent. The remaining shares were thinly spread across banks and insurance companies and private shareholders, among them many Volkswagen employees. As a consequence state representatives remained a dominating influence on the supervisory board, the body which determined long-term strategies and personnel selection for the executive board. Traditionally, these representatives were the ministers of finance and of economic affairs, and sometimes the minister of social and labour affairs. There were times when both Lower Saxony and the Federal Republic were governed by Social Democrat majorities, and thus 'labour' controlled the majority of Volkswagen's supervisory board. The State's influence on VW remained strong, even after 1988 when the federal state sold its Volkswagen shares, since the state of Lower Saxony remained the single most important shareholder (17.6 per cent in 1988). Hence the composition of the supervisory board in 1993 was as follows: on the side of labour, three representatives of IG Metall (among them the leader of IG Metall) and six representatives of Volkswagen's works council; on the side of capital, three representatives from other companies, four persons representing banks and shareholder associations, two of them high level politicians, and two representatives of the state of Lower Saxony, the Prime Minister and the Minister of Social Affairs who had previously been leader of Volkswagen's general works council.

In addition to the presence of politicians on the board there is the presence of union representatives. Traditionally the chairman of the IG Metall union has been a member of Volkswagen's supervisory board. The chairmen of the works councils of most of VW's German plants are also members of the board. Thus, labour's perspective has always been strong on the board and has greatly influenced the selection of the chief officers who run the company. This influence was only strengthened when in 1976 a law was passed which extended the principle of co-determination from the steel and coal companies to all companies with a certain minimum number of employees. While the law (of 1976) specified formal parity on the supervisory board between representatives of labour and capital, the chairman has a casting vote in a second voting which in most cases is used in favour of the shareholders' side. But it was possible for a pro-labour majority to form, and this was the case at Volkswagen for long periods of its history.

*Labour relations* are characterized by a high degree of 'jointness' between management and works councils in company policy, which exceeds the formal co-determination framework of industrial relations in Germany. The foundations of this partnership were laid in the early post-war period when the works council system was given strong support by the British allied command. The partnership developed over the decades, not least owing to the long and continuous reign of

## The Development of Volkswagen's Industrial Model

both the first German chief executive, Nordhoff, and the leader of the works council, Bork. Nordhoff, a manager at General Motors German Operations before the war, held his position from 1949 to 1968, and Bork, who was the first IG Metall candidate to win the top position in the works council system, held this position from 1951 until 1971. As a social democrat he was also mayor of the city of Wolfsburg during the 1960s and early 1970s. Close cooperation between management and works council and the vision of a happy 'VW family' fostered by Nordhoff did not reduce the influence of IG Metall. On the contrary, IG Metall obviously profited from this configuration. Its influence grew slowly from a rather low level, especially among white-collar employees in the early 1950s. It did not achieve a clear majority in works council elections before 1955, and it was not until 1967 that individual IG Metall membership reached the 50 per cent mark (Koch 1987: 5). The process of a gradual strengthening of IG Metall's influence among white-collar employees continued until the 1980s when it stabilized at a high level. In 1995 about 96 per cent of the employees at VW's main location, the Wolfsburg plant, including almost 90 per cent of white-collar employees, were members of IG Metall. It was hardly surprising therefore that IG Metall candidates received the lion's share of votes in the works council elections—in 1994 IG Metall candidates won 85 per cent of the works council positions of VW AG. Thus, strong union influence and a strong position for the works councils turned out to be complementary and self-reinforcing characteristics of VW's system. Its stability was further strengthened by long spells of continuous leadership. The position of the chief works council representative was held by four persons during the years from 1951 to 1995. During this time the company had six chief executives on the management side. The long periods of continuity and at times close personal relationship between the CEO and the head of the general works council helped to solve problems in many cases before they could turn into conflicts or even threaten industrial peace. The influence of the works council at VW has often been cited as an example of 'co-management'. The flip side of the strong role of the works councils in the system was that it worked against initiatives that might have been taken among rank and file employees and did not support any evolution towards direct participation and grass-roots activities on the shop-floor (Koch 1987).

These three elements remained central to the industrial model as it evolved right up until the mid-1990s. A prerequisite of its stability was the fact that no major changes in the political regulatory environment took place during this period. We refer here to the constitutive elements of the West German neo-corporatist system, particularly to the system of co-determination. The system is sufficiently independent of party platforms and alliances that it cannot be attributed to social democracy alone. Conservative governments also shaped and supported this structure. Of particular relevance in this regard was the system of co-determination. The influence of works councils exercised through elected officials representing blue- and white-collar interests at different levels within the company was the foundation of this system. The Works Constitution Act of 1952 (substantially revised in 1972) gave the works council varying levels of influence, depending on the subject-matter, extending from the right to information concerning economic data and business performance, to the right to consultation, up to the right of co-determination—that is, the right to veto certain measures proposed by management. Plant agreements between management and the works councils further extended the influence of the elected counsellors defining work organization, the introduction of new technology, the delivery of training, and other matters. Through their legally mandated representation on company boards, the delegates from the union and the works councils can, under certain conditions, exert strong influence on the nomination of the executive board and on strategic company decisions.

Collective bargaining takes place between national unions and employer associations with collective agreements reached on a regional level encompassing the whole range of metal-related industries, and not just the automotive industry. Volkswagen is an unusual case in this regard, however. As it has not become a member of the employer association Gesamtmetall, collective agreements have to be bargained directly with the union, the IG Metall. As a consequence, Volkswagen became a unique case of a company-wide collective bargaining system. At VW the 'dual system' of industrial relations in Germany—in which the union concludes general agreements over wages and salaries, and the works councils at the plant level deal with the details of wage differentiation, working conditions, and grievances—was therefore 'short circuited' to a certain degree.

A further characteristic of the neo-corporatist system in Germany is the system of vocational training. In principle this gives every secondary school graduate the opportunity to receive a three to three and a half year period of training in a broad spectrum of crafts and occupations. During this time the trainees are temporary employees of the company which enters into a training contract with them; their employment ends formally when they complete their training. The training takes place in part in public vocational schools, in part at the company itself, and it thus offers companies the opportunity to train workers for their specific needs. At the same time, the universal training certification allows for labour market mobility. The training regulations are developed in a tripartite manner between the corporate associations, unions, and governmental institutions and adapted to changing requirements by means of revisions. After training, many of the young skilled workers from the workshops attempt to find employment in industrial firms with attractive earning opportunities and accept the risk of being deployed in jobs classified as unskilled or semi-skilled. As a consequence, a significant proportion of the skilled workers who are deployed at levels below their qualifications have completed the same vocational training as those who are formally classified as 'skilled workers'. This situation had already begun to develop as early as the 1950s, for labour market reasons (Koch 1987: 77 f.). In later years the disequilibrium between the supply and demand for skilled workers was further exacerbated when companies—in order to meet the threat of youth unemployment—offered more training slots than their specialist departments considered necessary. A surplus of skilled workers could be found above all in plants located in rural areas, and thus at the majority of the VW sites.

# The Development of Volkswagen's Industrial Model

Indeed the rural environment of most of Volkswagen's production plants should be mentioned as a specific factor of influence in the case of VW. All of the production sites, particularly the main location in Wolfsburg, are situated in rural areas, and, with the exception of Emden, they were for most of the period under consideration located close to the Iron Curtain. These areas had little industrial tradition, and during the 1950s and 1960s for many displaced persons and migrants from the East, VW offered the first chance to earn a living in the West. Work attitudes and labour relations during the early decades were strongly influenced by these factors.

This chapter follows the trajectory of Volkswagen starting with the first crisis of the post-war industrial model and the period from 1967 to 1974 (Section 11.2). It then analyses the reconfirmation of the model and the attempts to achieve coherence between technical, organizational, and managerial structure, in the period 1975–88 (Section 11.3). By the late 1980s VW had entered another period of uncertainty, again closely linked to its product strategy but exacerbated by tensions in its governance structure and by the economic recession (Section 11.4). The final section deals with the reshaping of VW's industrial model during the 1990s.[2]

## 11.2. THE POST-WAR MODEL IN TROUBLE, 1967–1974

The year 1967 marked the first recession after the early 1950s, and its impact was felt by Volkswagen. It was not the recession, however, that caused the most trouble for Volkswagen's post-war model. There was only a short interruption of the fast growth VW had experienced during the *Wirtschaftswunder* years. Between 1950 and 1966 sales volume had grown from DM 405 m. to almost DM 10 bn., while employment levels had increased from 15,000 to almost 125,000. During these years Volkswagen had become the dominant mass manufacturer of passenger cars with a German market share of 30 per cent in 1967. Almost every second car produced in Germany was produced by the Volkswagen company (see Wellhöner (1996) on VW's development during the post-war years until around 1960).

In 1967 the Volkswagenwerk AG still made up the dominating core of the Volkswagen Group. Almost 85 per cent of all the vehicles produced worldwide by the Group came from its German plants. Auto Union, which had been bought from Daimler-Benz in 1965 produced barely 3 per cent, and the two foreign affiliates with major manufacturing operations, Volkswagen do Brazil and Volkswagen Australia produced slightly more than 10 per cent (with Brazil producing nine out of ten of these cars). The rapid growth of Volkswagen was export-led from the start, with two-thirds of the sales of the German Volkswagenwerk AG coming from exports, and almost half of these sold in the USA, by far VW's most important foreign market.

The explanation for Volkswagen's great success during the 1950s and 1960s is clearly to be found in its Fordist strategy. It was type 1, the Beetle, with its basic design dating back to the mid-1930s, which lay behind the success. All the 5 million passenger cars produced cumulatively during the 1950s had been Beetles. When the 10 million cars mark was reached in 1965 and the 15 million mark in 1968, almost all of them were still Beetles. By 1972 15 million Beetles had been produced cumulatively, more than Ford's Model T, and of the 40 million Volkswagens produced by 1981, half of them were still Beetles, although a quarter were now Golfs, revealing that the Golf had successfully taken over the role of the Beetle. Table 11.1 shows the development of VW's model range between 1962 and 1972.

While Volkswagen emulated Ford's Model-T strategy of mass production it did not share its fundamental philosophy. As early as 1949, the independent body maker Karmann had started to produce a cabriolet version of the Beetle, and in 1950 type 2 of VW's model range, the VW transporter, was introduced. It took VW until 1961 to come out with a further line of cars, type 3, a limousine with close technical resemblance to the Beetle. Type 3 never came close to the market success of the Beetle, however: while the latter reached volumes of 1 million and more per year during the 1960s, the former barely exceeded the 200,000 mark. In 1968 type 4 was launched, the 411/412 model series, but it was even less successful, with sales of below 100,000 cars per year. The flop of type 4 and the limited success of type 3 indicated severe restrictions on Volkswagen's development capacity. All models relied on the same technical concept. With the increase in engine horsepower the concept of the air-cooled engine had reached its limits. The purchase of two southern German car makers, Auto Union and NSU in 1965 and 1969, helped to overcome this limitation. Thus, the first Volkswagen model with a water-cooled engine, the K70, had been developed by NSU before it was purchased by VW.

Despite the weaknesses in its development potential revealed at the end of the 1960s, the dominance of the Beetle and Volkswagen's monoculture was primarily a matter of strategy. Top management wanted VW to be identified with the Beetle. At the same time, 'just-in-case' development projects for new models were pursued during the 1950s and 1960s. Some of these prototypes—the development projects 158 and 266—even reached the final stage with prototypes and die sets ready for mass production. They never saw the light of the market because of the success of the Beetle, which absorbed all production capacities at Volkswagen during this period.

The purchase of Auto Union, the company which had thrived on its two-stroke engine cars, the DKW models, in the post-war years, was not a deviation from this strategy. The basic design of the DKW stemmed from the pre-war years, too. Daimler, which had bought the company in 1958, had therefore insisted on the development of a new car line equipped with a four-stroke Otto engine. Auto Union's top management, however, continued with the two-stroke principle. Frustrated by the stubbornness of its affiliate, despite shrinking market shares and rising losses, Daimler-Benz sold Auto Union to Volkswagen, which was primarily interested in the additional production capacity of the newly built plant

TABLE 11.1. *The development of VW's model range 1962–1972: VW and Audi in Germany[1] and 'overseas'*

| VW and Audi Models (no. of units and %) | Lower Range Type 1 Beetle (1200/1300/1500) | Middle Range Type 3 (1500/1600/1700) Type 4 (411/412) K70, Audi 60/80 | Upper Range Audi 90/100 | Vans and Trucks Transporter Military Vehicle etc. | Total[2] |
|---|---|---|---|---|---|
| 1962 W. Germany | 819,326<br>73.7 | 127,324<br>11.5 | — | 164,596<br>14.8 | 1,111,974<br>100 |
| Overseas | 57,688<br>79.8 | 14,563<br>20.2 | — | — | 72,251<br>100 |
| 1967 W. Germany | 818,416<br>70.2 | 250,615<br>21.5 | 23,813<br>2.0 | 72,257<br>6.2 | 1,165,657<br>100 |
| Overseas | 106,898<br>83.5 | — | — | 21,172<br>16.5 | 128,070<br>100 |
| 1972 W. Germany | 906,944<br>52.7 | 353,961<br>20.6 | 159,216<br>9.2 | 269,159<br>15.6 | 1,722,352<br>100 |
| Overseas | 294,000<br>70.3 | 87,000<br>20.8 | — | 35,458<br>8.5 | 418,034<br>100 |

*Notes*:
[1] Includes VW Brussels.
[2] Includes other special vehicles and obsolete models from the previous NSU product range (1972).

*Source*: VDA (Verband der Automobilindustrie, Tatsachen und Zahlen aus der Kraftverkehrswirtschaft, VDA: Frankfurt am Main, various editions, Table 11.3).

in Ingolstadt and the manpower of 12,500: indeed the manpower shortage had been the main bottleneck in VW's expansion for many years. Nordhoff and most of his management did not expect Auto Union to become another brand. From 1965 to 1968 the Ingolstadt plant produced mostly Beetles. The existence of a separate sales organization and a separate product development staff seemed justified only as long as the old product range was still produced (in ever smaller numbers). The Auto Union development staff were ordered to refrain from initiating new car projects. The fact that Auto Union's technical director and chief executive did not comply with this policy would later turn out to be very fortunate for Volkswagen, however.

Fordist principles of production organization dominated throughout the Volkswagen system. In 1967 the huge production plant in Wolfsburg still produced eight out of ten of the passenger cars Volkswagen made in Germany. The Wolfsburg plant, like Fiat's Mirafiori Plant, was modelled after Ford's River Rouge plant in Detroit. Until the mid-1950s the Wolfsburg site integrated all Volkswagen's operations. However, the principle of a standard dedicated production line and the increasing volume of the Beetle made it soon necessary to farm out production to other sites. The production of the van, type 2, and of light trucks, was shifted to the newly built Hanover plant in 1956, which, from 1958, also hosted engine production, transferred from the Wolfsburg plant. In the same year a new plant for transmission and components production was opened in Kassel. In order to be able to further standardize production on the Beetle lines a new plant dedicated for US exports was set up in Emden in 1964. Finally, when the K70 was brought to market in 1970, it was made at the newly built assembly plant in Salzgitter (this plant was later refurbished for engine production). With the exception of Kassel, which is located in nearby Hessen, all of these plants, including the components plant at Brunswick, are located in Lower Saxony, and the sites were chosen to foster industrial development in areas of high unemployment.

Throughout this process, Wolfsburg became increasingly dedicated to Beetle production. The type 3 model, introduced in 1961, was built in a separate area with a different production layout. With the export model shifted to the smaller Emden plant, Wolfsburg could fully reap the benefits of standardized high volume production. After machining operations had been automated to a high degree at the Hanover plant, press shop and body shop operations for the Beetle became highly automated in Wolfsburg in 1962. In the body shop the degree of automation was increased to about 85 per cent, with most of the spot welding done automatically on 'welding presses' fully dedicated to the standard Beetle. The new equipment was fully developed and produced in-house by Volkswagen's process engineers and toolmakers. Process technology was regarded as a major asset in terms of the 'dynamic capabilities' of the company, and employees in these areas regarded themselves as an elite.

One reason for mechanizing body shop operations was to eliminate an area of poor working conditions which had experienced troubled labour relations for quite some time. Other reasons were clearly economic. The yearly output of Beetles at the Wolfsburg site by 1963 had reached almost a million cars, and the capacity of the body-welding transfer lines was 4,000 units per day (in two shifts). Mechanisation led to enormous savings in space and personnel, the two major bottleneck factors in the early 1960s.

Assembly operations were reorganized, too, and the degree of mechanization was increased to about 10 per cent. While the plant produced one Beetle every eight seconds, the production cycle in the assembly areas was 1.5 minutes. This was achieved by splitting up the assembly process into seven parallel lines, each with a capacity of 300 cars per shift. The cycle time for the individual worker was three minutes owing to the fact that in general work was performed over two workstations. Thus Fordist production organisation at Volkswagen did not mean shortening cycle times to the one-minute ideal pursued by other car makers.

The effect of mechanization and high volume production was enormous cost savings. In addition came the cost savings owing to the product policy of model continuity. These savings were reflected in the price of the Beetle and this proved that VW was truly following the Fordist paradigm. The price of the 1967 base model of the Beetle was 7 per cent below the price the consumer paid for the standard model of 1950 (4,800 DM).

The benefits for the consumer were highlighted by an advertisement of VW USA in 1969, which contrasted VW's strategy with the model change policy of the Big Three:

We've been making the same basic VW for so long now you'd be bored with the whole thing. But the fact is, we're still learning. For no matter how perfect we think one year's model is, there's always an engineer, who wants to make it more perfect. You see, at the Volkswagen factory we spend 100 per cent of our time making our car work better and 0 per cent making it look better. Any change is an improvement. And when we do make new parts we try to make them fit older models. So there's nothing to stop a Volkswagen from running forever (which may explain why Volkswagens are worth so much at trade-in time). Starting from scratch each year can get in the way of all that. Just when they've ironed out the kinks in the current model, they have to face the kinks in the next. We'll never understand all the hoopla over the 'big changes' for the next year's models. Weren't they proud of this year's? (Text of an exhibit posted next to a Beetle at the Ford Museum, Dearborn)

Labour relations were not a great concern in the years around 1967. That year VW AG employed 92,000 employees in Germany and 26,000 overseas, while Auto Union had 11,000 employees. Among Volkswagen's German workforce 40 per cent were deployed as skilled workers (*Facharbeiter*), 49 per cent as semi-skilled, and 7.5 per cent as unskilled workers; 11 per cent were salaried employees. The high proportion of jobs requiring skilled workers can be explained by VW's efforts to increase the level of mechanization, and by the fact that the new process technology was developed and manufactured in-house.

Five and a half percent of VW's workforce in 1967 were foreigners. The company began to recruit foreign workers (*Gastarbeiter*) in larger numbers starting in 1962. In 1973 the peak was reached with around 20,000 workers, 16 per cent of VW's workforce. Different German companies at this time specialized in certain nationalities, with Volkswagen concentrating on workers from southern Italy. At Wolfsburg the share of Italians among the foreign workers exceeded 80 per cent by the beginning of the 1970s. There were no particular labour conflicts related to this foreign workforce comparable to, for instance, the strikes organized by Turkish workers at Ford. This was the result not least of the efforts of the works council to integrate this part of the workforce by, for instance, nominating persons to look after the foreign workers on the shopfloor.

If there were conflicts in the field of labour relations they were mainly about money. But this did not cause particular headaches to the company. In the 1950s and 1960s a one-day strike would cost the company more money than the wage increase necessary to avoid such a strike. Thus, trouble makers could easily be 'bought off'. The memorable 'October revolution' of 1970 can serve as an example. It was called by the toolmakers, who regarded themselves as the skilled workers' elite. The toolmakers questioned why they could not advance into the same wage category as the foremen. They stopped working and demanded discussions with management, and their demands were met. But then groups of direct workers followed by demanding additional pay, which they received. The 'revolution' of 1970 was no more than an incident, but in a company which by 1995 had not experienced any strikes (there was one strike ballot, which did not result in a strike, in 1978), this was a memorable event.

As a consequence of the policy of acquiescing to labour demands by means of monetary compensation, management had only a loose grip on wage costs. Wage drift was high. The works council led the committee which decided on issues of pay grades, until, in 1970, a new arrangement was introduced, which granted management parity in this regard. The new committee was nominated by the company and IG Metall and consisted of three representatives each from the company and the employees. The employees' side was *de facto* represented by leading works council members. The wage system for the direct workforce was based on piece-rate with a built-in incentive to increase the degree of effort beyond the 100 per cent 'normal effort' level. By 1970 the average degree of effort which formed the basis for the piece-rate calculation had reached 134 per cent. Time standards had lost their function as controls for efficiency. Later in the 1970s these were issues which led to a major reform of the wage system, which is discussed below. However, until the early 1970s the system helped to keep labour acquiescent.

The year 1967 marks the beginning of a crisis of the post-war model at VW. The recession demonstrated the vulnerability of VW's product strategy. The decrease in demand in Germany affected VW more than the other car manufacturers. Costs arose as an issue for the first time. It became evident that product differentiation had its price when even a slight deviation from the standard product required a separate process and work organization. Thus labour costs also became an issue for the first time after the boom years. After sales had picked up again in 1968 the dependency on the Beetle remained. None of the new car models (the 1500/1600 and 411/412 series) became a success in the marketplace. In America Ralph Nader's and the consumer movement's critique of the Beetle as an unsafe car and the strong revaluation of the Deutschmark against the dollar after the end of the Bretton Woods system made prospects ever more bleak. The various interest groups represented in VW's governance structure were unable to reach a consensus to build a plant in the USA, a plan which had been on the agenda since the mid-1950s.

Other events around the year 1967 supported the widespread impression that times were changing at Volkswagen. In 1968 Nordhoff, the chief executive since 1948, died. Throughout his leadership he followed the Fordist profit strategy, relying on one model produced in a standard manner in large quantities. But now the company seemed trapped by the same industrial model which had fuelled its growth for such a long time. The company lacked a coherent strategy. One of the first measures taken by the new chief executive, Lotz, was to set up a product planning department to develop such a strategy. The acquisition of Auto Union in 1965 and of NSU in 1969 played a central role now. The strategy worked out by the product planning department aimed at reducing Volkswagen's dependency on one product champion. As it could hardly be expected that the car model succeeding the Beetle would emulate its success, three new car projects were initiated: Volkswagen should develop the Beetle successor, which became the Golf, and the new mid-sized model, the Passat; Audi was given the task of developing the small car model, the later Audi 50/Polo model range.

Audi was established as a subsidiary in 1969 by merging Auto Union and NSU. NSU, the car maker based in Neckarsulm, was renowned as the pioneer of the Wankel engine technology. However, VW was not so much attracted by the technological capabilities of the company in this regard as by the prototype of a front-wheel-drive mid-sized car with water-cooled engine developed by NSU. This car was immediately incorporated into VW's model range as the K70 and produced at the VW Salzgitter plant, whereas production of the Wankel engine car Ro 80 was continued at NSU. However, Wankel technology played no role in the further development of Audi, the cornerstone of which became the Audi 100 and—even more important—its water-cooled engine introduced in 1968. With the Audi 100 Audi broke the mould of the Auto Union trajectory based on two-stroke-engine low-end cars and established itself in the high-end upper market as a potential competitor even for Daimler-Benz. This was not without irony as Daimler-Benz itself had laid the ground for this achievement. When it had purchased Auto Union, Daimler-Benz management had seen no future in Auto Union's two-stroke-engine car range and had urged the company to develop a new car model with an up-to-date four-stroke water-cooled engine. In the early 1960s a group of engineers was transferred from Daimler-Benz to Auto Union in order to help develop this engine. This group became the core which was to develop the Audi 100, which was said to have a certain Mercedes-Benz flair. The engine developed by this group of engineers and first installed into the Audi 100 was taken over and adapted for the Golf, Passat, and Polo ranges in later years, and thus played a vital role in Volkswagen's recovery. Another irony in this was that Volkswagen's management did not expect Auto Union to develop a new Audi car line. It had given orders that technical development should be restricted to the adaptation of existing models. Thus, the Audi 100 was a result of non-compliance and of undercover activities by Audi's technical staff: approved, however, by Audi's chief executive, Leiding, who was later chief executive at Volkswagen.

The oil crisis of 1973 hit Volkswagen when the Golf, the cornerstone of this new product strategy, was not yet ready, and sales of the Beetle plummeted. Volkswagen again suffered more from the recession than most of its competitors. Tensions were exacerbated by difficulties in reaching agreement in the 1974 wage bargaining round.

The leader of IG Metall, Loderer, is said to have taken the initiative at this stage, complaining in a letter to the head of the Federal government, Prime Minister Schmidt, about the 'autocratic style' of Volkswagen's chief executive: 'in the current crisis situation at the Volkswagen works there is nothing more urgently needed than to stand together and cooperate. There is nothing of the sort happening, however' (quoted by Endres 1990: 82). Urged on by the leader of IG Metall and by the VW works council, the head of the German government obtained the resignation of Volkswagen's chief executive. As his successor, Schmücker, a previous labour director at a steel company, was recommended by the chairman of IG Metall because of his good relationship with the union (ibid.). Schmücker was eager to re-establish a close relationship with his opposite number on the works council side, who had succeeded Bork in 1971. The new pair of leaders developed an even closer relationship. Whenever a difficult situation came about, the chief executive and the chairman of the works council sought an informal meeting to solve the problem by consensus. A further long period of close cooperation and mutual respect between the leading management and works council representatives began.

As we have seen, VW's governance structure played a decisive role in this critical period. Close cooperation between management and works council was an essential ingredient of Volkswagen's industrial model. The neo-corporatist governance structure resulting from the specific history and ownership of Volkswagen helped to reconfirm this relationship. This was the basis for a new phase of jointness in its future-oriented strategies.

It should be noted at this point that the year 1971 marks the peak employment level of the core company Volkswagen AG. As Statistical Appendix 11 shows, it was only in 1986, and again in 1991 and 1992, that the employment level slightly exceeded this peak. In all other years it remained below the level reached in the early 1970s. It is clear that from this time on, employment issues became 'zero sum games' within a stagnating organization which at the same time continued to fuel the growth of the overall group. Indeed from then on the share of Volkswagen AG in employment in the Volkswagen Group started to decline. In 1971 the share was 64 per cent, and in 1993, 48 per cent. An even greater drop would have taken place but for the reduction of working time, a point to which we will return later.

## 11.3. RENEWING THE MODEL FOR A NEW GROWTH CYCLE, 1975–90

In the year following the events described above, the principles of VW's strategy for the post-Beetle era were formulated. The basis of this strategy was a compromise between Fordist and Sloanist principles: while the company hoped to emulate the high volume production of one dominant model, it did not put all its eggs into this one basket. The establishment of Audi in 1969 was already seen by some observers as a step towards a 'GM-ization' of Volkswagen. The product range established for the 1970s assigned one distinct car model to each of the two brands, the Golf to Volkswagen and the Audi 100 to Audi. In addition each brand received two car lines with twin models, the Polo/Audi 50 and the Passat/Audi 80. While the Golf inherited the huge capacity of the Wolfsburg plant, which was set up again with dedicated process equipment, the production facilities for the two twin models were planned to be flexible between VW and Audi. (Owing to a unilateral decision taken by Audi's management to change the engine design of its Audi 80, interchangeability was undermined, however—a measure taken by Audi's management to strengthen its independence from VW.)

It did not take long before the Golf took over the Beetle's role at Volkswagen. Fifteen million Beetles had been built by VW up to 1972, of which almost 11 million were built by the Wolfsburg plant until its changeover to the Golf in 1974. In 1994, twenty years after Golf production began, the Volkswagen Group produced its 15 millionth Golf. The Golf and its derivative Jetta, later the Vento, achieved yearly output volumes of around 1 million units throughout most of the 1980s with daily production rates of more than 3,500 at Wolfsburg, which remained its main production site.

However, dependency on a single product was reduced considerably. Table 11.2 shows that the VW/Audi product range became more balanced in this period compared to the Beetle era. The divergence becomes even more obvious when the number of variants and options are taken into account. The number of models and engine displacement sizes of VW's German passenger car fleet had increased from eleven variants in 1982 to forty-eight (in addition to twenty-one at Audi) and in 1992 the number had risen to fifty-nine at VW and forty-four at Audi. Thus, the strategy of what one might call a 'diversified Fordism' led to a broadening of the product range and, particularly, an increase in variants and options.

In view of the policy of product differentiation the concept of a 'people's car', which lay behind the Beetle, now meant something very different. Management now wanted the Golf to be regarded as a 'classless' car, 'the only automobile

TABLE 11.2. *The development of VW's model range 1977–1992: VW and Audi in Germany[1] and 'overseas'*

| VW and Audi Models (no. of units and %) | Lower Range | | Middle Range | | Upper Range | Vans and Trucks | Total[2] |
|---|---|---|---|---|---|---|---|
| | Beetle | Polo, Derby, Audi 50 | Golf, Scir., Jetta, Vento | 1600, Passat, Santana, Audi 80 | Audi Cabrio, Coupé, Avant 100/200 | Transporter, Military Vehicle etc. | |
| 1977 W. Germany | 32,762 | 240,340 | 646,531 | 380,783 | 195,021 | 190,214 | 1,687,539 |
| | 1.9 | 14.2 | 38.3 | 22.6 | 11.6 | 11.3 | 100 |
| Overseas | 178,187 | — | — | 296,870 | 7,152 | 45,611 | 530,916 |
| | 33.6 | | | 55.9 | 1.3 | 8.6 | 100 |
| 1982 W. Germany | — | 175,244 | 667,160 | 396,227 | 123,479 | 138,691 | 1,501,880 |
| | | 11.7 | 44.4 | 26.4 | 8.2 | 9.2 | 100 |
| Overseas | 84,087 | — | 233,690 | 113,096 | 4,512 | 86,738 | 522,123 |
| | 16.1 | | 44.8 | 21.7 | 0.9 | 16.6 | 100 |
| 1987 W. Germany | — | 116,138 | 1,028,227 | 472,401 | 126,936 | 146,535 | 1,890,964 |
| | | 6.1 | 54.3 | 25.0 | 6.7 | 7.7 | 100 |
| Overseas | 8,273 | 116,020 | 182,834 | 144,012 | 3,384 | 164,107 | 619,525 |
| | 1.3 | 18.7 | 29.5 | 23.2 | 0.5 | 26.5 | 100 |
| 1992 W. Germany | — | 84,268 | 1,023,205 | 662,231 | 173,061 | 206,752 | 2,150,036 |
| | | 3.9 | 47.6 | 30.8 | 8.0 | 9.6 | 100 |
| Overseas | — | 137,954 | 287,322 | 150,347 | — | 148,607 | 724,230 |
| | | 19.0 | 36.9 | 20.8 | | 20.5 | 100 |
| Europe | | 370,201[3] | 105,514[4] | | | | |

*Notes:*
[1] Includes VW Brussels.
[2] Includes other special vehicles and obsolete models from the previous NSU product range (1972).
[3] SEAT Marbella, Ibiza; Škoda, Favorit and Forman.
[4] SEAT Toledo, Malaga.

*Source:* VDA (Frankfurt am Main, various editions, Table 11.3).

which both the manager and his doorman could drive' (Hahn, quoted in *Industriemagazin Report*, 4, September 1991, 14). This statement was made when the Golf variant equipped with a six-cylinder engine was introduced as part of a general strategy to move the product upmarket into the higher price segments (see below).

The fact that the Polo and Passat remained marginal in terms of their sales and output volume at Volkswagen was partially determined by process layout and capacity planning. But this explanation does not suffice. The Polo, in particular, which had been developed at Audi, did not get the attention in terms of product development and marketing at VW necessary to fully exploit the market potential in its segment.

The interpretation of a 'diversified Fordism' also holds true for the new process layout, which corresponded to the new product range. In the case of the Golf a completely new process was set up. Flexibility of process equipment in the body shop was increased in order to cope with model variation, but still within the limits of the Golf/Jetta range. Compared to the manufacturing process used for the Beetle the degree of automation in the body shop was slightly reduced now, to 75 per cent. The degree of automation in assembly operations was also reduced (from 10 to 5 per cent); the principle of parallel assembly lines was maintained. The major changes in assembly operations were caused by the different product concepts; whereas the Beetle had different lines for chassis and body assembly, the Golf was the first product at VW with a unitary body construction.

The Wolfsburg Polo line was given much higher process flexibility, however. Spot welding operations, for instance, took place with the body parts positioned on automated guided vehicles, which moved through production islands where robots performed the spot welding operations. These lines had flexibility to produce, in principle, the whole range of VW's products. This allowed special versions of the Golf to be made here which could not be manufactured on the dedicated Golf lines. The process layout of the Golf became the focus of a major restructuring effort in the early 1980s (see below). With regard to the other product lines, again the principle of one process per model was pursued. In Wolfsburg the separate Polo process was established, and after the Beetle was moved from the Emden plant, Passat production was located there. The plant in Brussels which VW had acquired in 1971 did not receive its 'own' model and produced the Passat or Golf or a mix of both as required.

Implementing the new product strategy and production layout was not sufficient, however. Other critical issues had to be addressed, the most urgent one being in the field of labour relations. A range of measures were adopted to stabilize VW's new strategy. Many of them turned out to be ineffective or even costly failures. These negative results should be regarded as part of the dynamics of the industrial model in process. They demonstrate the need to continuously correct, adapt, and optimize the model, although they do not question its basic principles. The first urgent issue to be addressed was employment security.

## 11.3.1. Personnel Reduction and Employment Security

From the beginning of the 1970s employment security became a central issue at Volkswagen. The poor sales of type 4 and the decreasing demand for the Beetle required a reduction of the workforce even before the oil crisis. As a consequence, the Wolfsburg and Emden plants began to produce Audi models in addition to their own products. By 1974, however, it was evident that a major reduction of personnel had become inevitable. In fact, on the basis of the reconfirmed cooperation between management and works council, VW's workforce was reduced by more than one-quarter within a few months. It is particularly noteworthy that this reduction was achieved without a single worker having to be dismissed, other than for personal reasons (Dombois 1982*a*). This achievement was an important milestone in the reconfirmation of the VW industrial model: 'The fact that the restructuring of production as well as the adjustment of personnel levels could be coped with so smoothly on a co-operative basis and, from the perspective of subsequent developments, so successfully, led to a significant and long-term reinforcement of the pattern of industrial relations at Volkswagen' (Brumlop and Jürgens 1986: 87).

Mass dismissals were avoided by the following means. First, contracts with foreign workers were terminated. Of the total of 33,000 employees who left Volkswagen between early 1974 and the end of 1975, over 13,000 were foreigners; the proportion of foreign workers was reduced to a little over 7 per cent and did not rise much higher later. Secondly, there was severance pay. Voluntary redundancies were facilitated through financial incentives (usually three to four months pay) independent of age. Many of those workers found work in other car companies which were not hit so badly, while others returned when Volkswagen began to recruit personnel, just a few months later. Thirdly, there was early retirement. According to an agreement reached in 1975, employees aged 59 could retire on a voluntary basis. With the early retirement age of 60 under the state-regulated social security system, these employees could receive unemployment pay for one year. In principle, the State labour exchanges ought to have offered them a new job but there was a tacit understanding that these persons received unemployment benefits until they reached the age of 60. To ensure that they did not lose money compared to their previous employment the company made their income up to their previous income level. Fourthly, there was a hiring freeze, allowing 'natural attrition' by regular retirement, dismissal for personal reasons, and voluntary departures to take its course.

The principle of carrying out personnel adjustments in a 'painless' and 'acceptable' manner remained a cornerstone of personnel policy in later years. In particular, the early retirement option was kept open even in years when the company hired new staff in larger numbers. By tapping the State labour market programme for 'structural unemployment', from the mid-1980s the threshold could even be lowered to include the 57-year-old cohort, and from 1993 the majority of the 55-year-old cohort left the company. Early retirement thus became a major tool for adjusting personnel. The works council particularly stressed the point that in this way young skilled workers had a chance of finding a job at VW after having finished their vocational training. Hence the 51,000 employees who left the company for early retirement between 1975 and 1996 made room for 26,000 young skilled workers. The cost of this policy was largely met by social security funds. However, changes to the regulatory structure in 1996 made early retirement very costly for employers in future.

By 1975, with the success of the new product programme, especially the Golf, the immediate pressure to reduce personnel was removed. After the experience of the economic downturn, however, strategic measures were taken to prevent a similar impact on personnel levels in future. A role model for the policy of employment stabilization was Daimler-Benz. Traditionally, Daimler-Benz customers had to wait at least a few months for their car to roll off the line. If demand increased, the waiting time simply lengthened. As a consequence, the sphere of production was protected from the uncertainties of the market. Volkswagen wanted to copy this model to a certain extent. A policy guideline stated that a middle line should be drawn between demand forecasts based on pessimistic assumptions and those based on optimistic assumptions and that capacity and manning levels should be set accordingly. It was made explicit that the company would accept losing the peaks of demand in order to stabilize the workforce. But the 'Personalpolitik der mittleren Linie' was promulgated only as long as the company was recovering from the downturn. After sales prospects had brightened up again, the good principles were forgotten.

The guarantee of employment security remained high on the agenda at Volkswagen, however. VW did not use a buffer of overtime hours or recruit temporary employees as other companies did in order to safeguard its core employees against economic downturn, either in the crisis of 1973–4 or at any time up to 1995. When the downturn at the beginning of the 1980s necessitated (slight) personnel reductions at Volkswagen, the strategic orientation on the labour side was to reduce regular working hours in accordance with IG Metall's demand for a 35-hour week.

*11.3.2. Labour Flexibility and Wages*

Another labour relations issue which had to be resolved was related to the wage system (Hildebrandt 1982; Brumlop 1986). Both management and labour were interested in reforming the wage system, albeit for different reasons. Management wanted to obtain control over wage drift, eliminate the cumbersome administration of the existing system of wage differentiation, and meet the need for greater flexibility in labour deployment. Labour was interested in protecting the existing wage level and preventing the downgrading of wages, which was an imminent threat for workers who had shifted from Beetle to Golf production. It took several years for a set of agreements to be reached, and for a new wage system to be created in 1980, one which would considerably influence the future direction of work organization at Volkswagen. Volkswagen was only able to undertake the wage reform, which was a major social innovation in the field of German industrial relations, because of its unique collective bargaining situation. If VW had been a member of the employer's association it would probably have taken a different direction. Indeed the association condemned the agreement immediately.

The old system of wage differentiation, introduced in 1951, was based on an analytical scheme of job evaluation which attached a monetary value to each microtask of each individual job. Evaluation criteria included skill requirements of the task, difficulties of performing the task, strain and stress factors, environmental conditions, and so on. Even slight changes of process and, of course, any new job assignment, affected pay under these conditions. Besides the difficulties of administering this system it was difficult to control wage drift as pay rates were permanently changing for a multitude of reasons.

Another problem was that the wage incentive related to individual work effort had become obsolete. For years the rate of extra pay had been frozen at an 'effort level' of 134 per cent, which had in fact become part of the standard wage in a measured-day work system.

The new wage formula abolished the system of evaluating individual jobs on the basis of seemingly objective criteria. Rather, it grouped similar tasks together into job systems (*Arbeitssysteme*), which were graded by the joint labour—management committee. These wage committees were established at the plant level with an equal number of representatives nominated by management and by the works councils. The members of these committees, however, were fully authorized to take decisions, and in this role did not have to follow instructions from their superiors in the company hierarchy. In this way a new subsystem of co-determination had been created, one which dealt with the function of wage differentiation rather autonomously. The new formula grouped all hourly jobs within the company into about 3,000 systems. The systems for the jobs of specialists might extend across plants but most of the systems comprised jobs within the same area, on average about twenty-five jobs. The jobs were grouped according to the criteria of similarity of skill requirement and spatial proximity. In assembly areas, for instance, some of the job systems cut across the parallel assembly lines, while others covered a longer stretch on the same assembly line. The new wage system fully met the requirements it was designed for:

- Wages were stabilized, and the risk of downgrading minimized, as wages no longer reacted so sensitively to changes in the process and environment.
- Flexibility was achieved as transfers within the system could be expected at any time. In addition transfers to other job systems were regulated: in this case an allowance was paid, which varied according to whether the transfer was to another job system at the same pay grade or to a higher graded job

system. In any case transfer could be expected at short notice and for a period of up to three months, in both cases including to other areas or even other company plants.
- Wage drift was reduced drastically, and the new system was easy to administer. It was also easy to adapt to the needs of high technology production areas where it was no longer possible to specify job requirements in the way done in the old system.

The new wage system became a cornerstone of VW's renewed industrial model and contributed to a rather low level of conflict at shopfloor level from this time on.

One might argue however, that the wage reform of 1980 hampered the progress of group work at Volkswagen. Indeed, some of the goals of group work were achieved by the system, thereby reducing the pressure for reform: the smallest unit for regulating work was no longer the individual job but 'groups' of jobs between which flexibility was obtained. The new system, however, organized job relations only, not social relations. Moreover, it did not support co-operative relations among workers of different pay grades and specialization. And it did not foster small group activities of the *kaizen* type within the 'job systems'.

*11.3.3. Assembly Automation*

Another important element of the industrial model which evolved in the mid-1970s was the technology strategy concerning, specifically, assembly operations. From an engineering standpoint, process innovation could be regarded as the next logical step after the product innovations introduced in the early 1970s. Now management envisaged a revolutionary step in process technology based on the potentials of microelectronics for flexible production systems. This technology orientation was in line with the then dominant explanation of the success of Japanese companies. These companies were said to be highly automated, deploying a multitude of robots.

It was soon evident that the introduction of the new process technology would coincide with the first major model changes of the Golf and later the Passat. In the case of the new Golf model, which was launched in 1983, the automation project focused on final assembly operations. Whereas the degree of mechanization in the press and body shops was raised by 5 per cent only, the assembly operations leapt from 5 to 25 per cent, and it was technically possible for it to be raised still further to 30 per cent. The automated area was set up in a new building at the Wolfsburg production site: building no. 54. This building became a place of pilgrimage for technology planners worldwide. Many of them, however, commented negatively on the costs and complexity of the new facility.

The new technology was installed in an area called the 'mechanized section'. Because of the large volume and the high speed of throughput the planners wanted to achieve, the flexibility of the equipment was restricted, which meant that it focused on the Golf and the majority of its variants. The mechanized section was spread over two areas: the first consisted of the pre-assembly of the drivetrain module; after pre-assembly this module was transported to the second mechanized section which was part of the main assembly line. Similar to the previous layout, the assembly line before and after the mechanized section was split up into five parallel lines. While one of the lines remained as a 'manual' bypass line, the other four merged into two parallel lines, each of which had parallel automation complexes which carried out about ten operations in strict sequence and rigidly coupled within a cycle time of 20 seconds. Most of these operations were final assembly jobs which had required overhead work and heavy lifting before the change (Jürgens et al. 1993: 353–62).

In contrast to the automated production lines for the Beetle, the equipment and the process used to build the new Golf had a certain degree of flexibility. Yet flexibility was limited to one model and its variants. The manual 'bypass' line was an essential complement to compensate for problems with certain model variants and, of course, machine breakdowns. Owing to the policy of model differentiation, 'teething problems' never ended. Over the years, however, they were mastered better, and the mechanized section of building 54 became an efficient high volume mass production machine. Up to 3,000 cars per day ran through the automated lines on peak days, much more than its originally planned capacity, with another 700 on the bypass line. In Fordist terms this automation of assembly automation was therefore a success. But the Wolfsburg plant became further dependent on the Golf owing to the investment that had been made, and the level of demand became more critical. There is no doubt that building 54 contributed to raising the break-even point at Volkswagen, which was to become a major problem for the company in the early 1990s.

Obviously the introduction of new technology affected the organization of work and labour relations. In this regard a set of New Technology agreements were reached, which provided the works council with the right to receive information early, and co-determination in all aspects related to work organization and training issues. With further development of the principles of the new wage system, a method for determining performance standards in mechanized work cells was introduced and the job profile of the *Anlagenführer* (system regulator) was created. This job profile received much attention as a result of the work of the industrial sociologists Kern and Schumann, who quoted it as an example of a general trend towards the reskilling of production work (Kern and Schumann 1984, Schumann et al. 1994). Process control, computer programming and trouble shooting were the most important functions of the system regulators, but simple tasks such as quality control, cleaning, or machine feeding were also typical. In contrast to the traditional production operator or maintenance worker, the system regulator was responsible for the equipment as well as for production. Although not a formal requirement at Volkswagen, the job of the system controller was normally filled by certified skilled workers (*Facharbeiter*). The human resources of highly trained skilled hourly workers available for manufacturing operations were not the least important factors explaining why, despite all of the problems, the mechanized section operated relatively successfully.

However, cost and manpower savings targets were not achieved by assembly automation. The flexibility of the high technology lines was overestimated. Each minor change in product design and logistics caused major problems of re-adjusting the technology. Large buffer areas were necessary to cope with breakdowns. Shopfloor management in the automated areas tended to adopt a 'fire-fighting' approach. In short, by the middle of the 1980s the high technology strategy was losing support in the company.

*11.3.4. Production in the USA*

Another element of the 'pact' of 1974 was the decision to produce cars in the USA. Construction of the Westmoreland, Pennsylvania assembly plant began in July 1976 and production started only eighteen months later, in April 1978, with a workforce of about 6,000 and a rather steep launch curve. As already mentioned, the idea of locating a production site in the USA had been controversial since the mid-1950s. The revaluation of the Deutschmark against the US dollar and the general fluctuation in the Deutschmark–dollar relationship after the termination of the Bretton Woods system had made the export strategy increasingly risky. In addition there was also a growing protectionist trend in American politics.

When Volkswagen finally decided to build a production site in North America, it was ahead of the Japanese automobile manufacturers which were at the time hesitating to take this step. By the time the Westmoreland plant was closed down in 1988, however, only ten years after opening, seven Japanese transplants had started production in North America.

With the exception of Dombois's research report (Dombois 1982*b*), there is little empirical evidence on why the Westmoreland project failed. According to Dombois, Volkswagen did not transfer its German concepts of apprenticeship training, skilled workers (*Facharbeiter*), a works council system for conflict regulation, and so on to its North American plant. Volkswagen thus did not follow the kind of 'transplant' strategy adopted by the Japanese in the 1980s (Jürgens 1992: 75 f.).

Several explanations have been given for the decision to close the plant, which remained a subject of controversy among Volkswagen's management for many years. One was that the company had relied too much on management it hired from the American Big Three. These managers were said to have 'Americanized' the plant, which above all affected product quality. Another reason was that VW started its transplant in a rush. In contrast, no Japanese company started its transplant operations with such a huge initial workforce in such a short time before full-scale manufacturing. A third reason was that the product was not designed for the American market. Yet it is hard to find an explanation for why so little was done to resolve the problems at the Westmoreland plant. There were no major efforts, for instance, to introduce Japanese-oriented concepts, as many Big Three plants were attempting to do during this period. The actors involved appear never really to have fought for the plant.

At the end of the 1980s, machinery and equipment from the Westmoreland plant were shipped to China for Volkswagen's plant at Changchun. Within Volkswagen the closure of the Westmoreland plant and the withdrawal from the North American production site was seen with regret and stupefaction by many. It became one ingredient in an atmosphere of uncertainty and crisis about future direction which was building up in the late 1980s. The strategy to supply the North American market from Mexico and Brazil was not supported by a decisive upgrading of the capability of these production sites. Thus they were unable to overcome the negative image the cars made in America had acquired. Immediately after the Westmoreland plant had closed, another round of Deutschmark revaluation made the decision even more questionable. Exports to North America had to be subsidized heavily, and by the mid-1990s plans to stage a comeback with a North American production site were on the agenda again.

### 11.3.5. Diversification of Business Activities

In order to reduce its dependence on the automotive industry VW decided to strategically broaden the scope of its business. In 1978 Volkswagen purchased the Triumph-Adler Group and consolidated it as an affiliate in business machine and information technology. The board of management declared this to be part of the long-term company policy:

The further long-term aim of company policy is the assurance of continued profitability through involvement in growth areas whose business cycle is counter-cyclical to that of the automobile industry. The concept of balancing out economic fluctuations through diversification is to be viewed from the perspective of long-term strategy. (The Board of Management, *Annual Report* 1978: 55)

The new Triumph Werke AG employed around 14,000 persons in 1979, and had sales of DM 1.2 bn. The general perception was that the acquisition was more than just a financial manoeuvre. It was seen rather in the context of Volkswagen's technology orientation, and it was anticipated that the whole group would benefit from the technological potential of its computer affiliate. But these hopes were soon shattered. The Triumph-Adler Group lost money and was in need of a long-term strategy itself. Volkswagen's management sought a way out and, in 1986, sold these businesses to Olivetti.

The failure of Volkswagen's diversification policy cannot be analysed in detail here. The general weakness of the European computer industry is part of the explanation. Also notable again was Volkswagen's own short-term orientation to its supposedly 'long-term strategic' commitment. Volkswagen's top management decided to get out of the computer field less than five years after entering it. The strategy of diversification was laid to rest, replaced by renewed focus on the core business.

*11.3.6. Volkswagen's Industrial Model as a Paradigm in the 'German Model' Debate*

The development of Volkswagen's industrial model from the mid-1970s to the mid-1980s might well be interpreted as a sequence of failures. Paradoxically, the company fared very well in most of its major markets (except North America) during much of this time period. The second oil crisis, which was strongly felt by many companies in North America and Europe, had little impact on Volkswagen. The company now profited from strong sales of its product range, especially in Europe. Net financial losses in 1982 and 1983 resulted from losses from its North and South America operations and those of the Triumph Group. The core of VW's operation was doing well. This was the period in which Volkswagen gained its position as Europe's largest automobile company, when it bought SEAT and started its activities in China. Sales increased almost threefold within a decade and an expansionist mood took root.

A closer look at Statistical Appendix 11 reveals that the increase in sales by value was in fact much higher than the increase in output volume. Volkswagen in particular benefited from a market trend towards more expensive model variants during the 1980s as customers tended to move up the product range. A continuous increase in value/price per unit lasted into the 1990s. Growth seems to have taken another path, not by volume but by value. This brings us to the crux of what came to be regarded as the basis of the 'German model', or 'the strategy of diversified quality production'. According to Streeck, countries with the highest wage costs and with strong unions and workers' interest representation like Germany and Sweden adopted this strategy and fared better than countries like the USA and the United Kingdom which instead sought competitiveness through low costs. 'Diversified quality production' was said to respond flexibly to consumer desires and take advantage of the market trend towards higher quality products and customization. Flexible production technology, a high skill level, and an 'intelligent' form of work organization were prerequisites for such a strategy (Streeck 1986, Sorge and Streeck 1987). Streeck regarded the adoption of diversified quality production as a validation of the effectiveness of the German system of co-determination, humanization of work, and centralized collective bargaining in securing growth and employment (Streeck 1986: 120). This 'German model' approach seemed to open up an alternative to the low-cost strategy of catching up with the Japanese. It stood in perfect harmony with West Germany's 'ingrained social and productive principles' (Boyer and Freyssenet 1995).

## 11.4. THE AMBIGUITY OF SUCCESS: THE INDUSTRIAL MODEL IN A NEW PHASE OF CRISIS, 1988–1993

Why then does our discussion return to crisis? The crux of the German model debate was that its supporters did not recognize that special political and economic conditions were necessary to support the strategy of product upgrading, specifically a favourable currency relationship between the Deutschmark and the Dollar. A second problem was that it overlooked the fact that flexibility and quality production were very expensive under the German system of work. Moreover, the search for 'intelligent' forms of work organization could not be described as very intense, outside the framework of high technology areas. German companies were very reluctant, and thus started quite late, to introduce group work or involve suppliers in integrated systems of product development.

The problems of the product upgrading strategy first became apparent in the American market where German cars became increasingly regarded as too expensive. Volkswagen ought now to have feared that, with the opening up of the European markets, the high costs of the German model would no longer be accepted by customers in Europe either. The MIT study (Womack *et al.* 1990, German edition 1991) came as welcome support for all those who had for some time already been advocating a change of trajectory. This view was also widely shared within Volkswagen, along with a feeling of increased uncertainty, indecision, and ambiguity about future strategy. This is the sense in which a crisis developed. This crisis was not evident in Volkswagen's economic performance. On the contrary, the Volkswagen Group, specifically Volkswagen AG, thrived on the (bubble) boom of the early years after German reunification when output and employment reached peak post-war levels. But the irony was that at the height of this success, the profit rate became negative, the break-even point increased beyond 100 per cent of capacity, and the company suddenly lost its economic viability.

It is not easy to untangle the various elements of the second crisis of VW's industrial model. The strategies adopted to further develop it and make it viable in changing macroeconomic and social circumstances in many cases had not succeeded. And yet the company had grown rapidly on the basis of this model. By the end of the 1980s the Volkswagen Group included SEAT, and after the fall of the Iron Curtain, the Czech company Škoda too. Volkswagen had also taken over the major production plant of the East German car industry, and had expanded into China. In terms of production and employment as well as in terms of brand structure Volkswagen had evolved into a *highly* multinational and multidivisional company. As Table 11.1 shows, the production volume abroad overtook the German production volume for the first time in 1992. The ingredients of a growing feeling of uncertainty and latent crisis which was emerging at VW by the late 1980s will be discussed in the following.

## 11.4.1. A Loss of Control over Costs

The high cost base was cutting into profit margins, particularly as competition stiffened and prices had to be lowered to protect market shares. In 1991 the VW AG, in its consolidated annual accounts, published the highest net earnings figures in its history, DM 1,114 m. Yet compared to the year 1979, which was one of the most profitable years of the 1970s, it becomes evident that profitability had deteriorated. The DM 667 m. net earnings of 1979 were earned by a volume of sales of about DM 31 m., but the net earnings of 1991 were earned by a sales volume of DM 76 m. Even the chairman of IG Metall, as a member of Volkswagen's supervisory board, demanded higher profit rates at this point. Rationalization was necessary, but there was uncertainty as to how it could be achieved without violating the principle of jointness with labour.

## 11.4.2. Internal Competition within the Group

With the integration of SEAT and Škoda into the Volkswagen Group, the modernization of its production systems, and new production sites from Martorel (Spain) to Mosel (former East Germany) to Changchun (China), new production capacities were built up which had to share the bread-winning products of the Group. Competition between plants within the Group became increasingly bitter. The struggle of the Volkswagen works council in Germany to retain some Polo production as a buffer against the risk of dependency on the Golf made this evident. After the acquisition of SEAT in 1986, part of Wolfsburg's Polo production was shifted to SEAT's Pamplona plant in order to free space for Golf production. Over the following years VW's management insisted that the entire Polo production should be shifted to Pamplona. The works council resisted, arguing that experience had shown the risk of dependency on just one model line. In 1992 a decision to end Polo production at the Wolfsburg plant was taken, but in 1994 in the context of a good sales performance by the new Polo, this decision was revised and Polo production re-established at Wolfsburg. When in 1995 Golf sales declined, Polo production had come just in time to secure employment levels at the plant; by 1995 every fourth car in Wolfsburg's output of 591,000 cars was a Polo.

Rivalry between plants and competition over the allocation of order volumes also sprang up between Wolfsburg and the new East German production site at Mosel, as well as between the Mosel and Škoda sites. The plans for future production sites for the Golf model—the Golf III was introduced in 1991—indicated that this was only the beginning. As of 1991, it was planned that the new generation of Golfs would be built at the German Wolfsburg and Mosel plants and by VW's foreign plants in Brussels, Belgium, in Sarajevo, Yugoslavia, in Puebla, Mexico, and in Changchun, China (the Golf Cabrio by Karmann). The rise in internal competition and rivalries—which was partially intended by corporate management, of course—had put the question of corporate governance on the agenda.

### 11.4.3. Product Strategy Increasingly Controversial

The official heir to the throne, Goeudevert, was at the centre of this controversy. Goeudevert stood for a policy of redirecting Volkswagen's product line towards green products and towards 'traffic systems'. He became chief executive of Volkswagen, one of the four divisions within the newly established group structure (Volkswagen, Audi, SEAT, Škoda). He brought to the board a former minister of environmental affairs in the state government of Hessen to promote further activities in this direction. As a consequence Volkswagen moved to the forefront of car manufacturers in recycling, and a range of green car alternatives was developed, including a hybrid car and low fuel consumption cars based on diesel technology. Moreover, Volkswagen had been working on new models for integrated traffic systems which included banning private car utilization in inner city traffic. According to Goeudevert:

> If we want to save the car from its biggest enemy, congestion, it is necessary both to develop new concepts for traffic in general and also to accept that certain solutions require new types of automobile, namely new propulsion systems. The automobile industry carries the responsibility for both aspects. Volkswagen is committed to do pioneering work. (*Industriemagazin*, Report 4, September 1991, 120)

Obviously, the commitment to redefine the business of car manufacturing and the shift towards new product lines, for example, green cars, implied a paradigm shift in product development. Such a shift did not take place. Fundamentally, the traditional technical user concepts for passenger cars prevailed and conflicts were avoided. Thus the organization remained in a state of uncertainty as to its future direction.

For some years the feeling of uncertainty about future direction and the sense of crisis coincided with excellent market performance. Post-war records in sales were achieved in 1992. Yet in the second half of 1992, sales and order volumes began to decline and the economic recession reached Volkswagen. The company entered another stage of acute crisis. The Group's net profits reached a record loss of almost DM 2 bn. in 1993. VW's fundamental weakness had become apparent the year before. While sales had increased by about 12 per cent for the Group (likewise for the VW AG), the economic result was negative and profits plummeted by 87 per cent (70 per cent for the VW AG). All of a sudden, the weakness of the colossus became manifest. The organization was shaken to its bones and its underlying industrial model again entered a trial by fire.

In a similar way to the situation twenty years before, the actual crisis resulted from the coincidence of problems of company governance and market problems. And like 1974 the solution was sought through the nomination of a new chief executive manager by the supervisory board. While Goeudevert was regarded as Crown prince and would have been the logical choice for a successor, the supervisory board decided differently. The choice fell on Piëch, the grandson of the father of the Beetle, Porsche. Piëch had been chief executive at Audi. In this role he was a controversial figure, autocratic in his management style and technocratic in his product conception. His nomination was widely regarded as a decision against the new product and business strategy that Goeudevert stood for, in favour of a policy of re-establishing profitability and focusing on the core business of car production in the classical sense.

Once again the importance of the governance structure and its embeddedness in the German political economy was demonstrated. With regard to what the name Piëch stood for, it was by no means self-evident that the representatives of labour, the head of IG Metall, and the leading works council representatives and the prime minister of the state of Lower Saxony, a social democrat, would opt for Piëch. Obviously they saw the need for a tough policy of rationalization and a return to profitability.

Just a few months after Piëch took charge, Goeudevert left the company and, with this, the new concepts and structures he had set up to reorient the Volkswagen brand were weakened decisively. The hiring of López from General Motors together with the forced exit of most of the executive board's old guard demonstrated Piëch's determination to shake off 'path dependencies' rooted in the fiefdoms of the past.

The nomination of López as corporate executive for production optimization and procurement indeed indicated a fundamental change in the Group's relationship to its suppliers. López had been the purchasing 'Czar' of General Motors Europe before he transferred to General Motors' US headquarters. He was famed for his success in forcing down the price levels of suppliers. Supplier prices had shown a 'López effect' in that suppliers differentiated their prices and offered GM the same products for lower prices, a practice that upset Volkswagen and other companies when it was made public. By hiring López himself, Volkswagen signalled that it was now going to be tough on suppliers.

## 11.5. RESHAPING THE INDUSTRIAL MODEL IN THE MID-1990S

When we compare the VW of the mid-1990s with the company in 1967 the fundamental changes but also the continuities become obvious. The VW Group of 1994 had a sales turnover of DM 80 bn. versus DM 6.4 bn. in 1967, but the return on sales was much higher in 1967: 5 per cent versus 0.18 per cent in 1994. The market share of the Group in Germany was almost unchanged: 29 per cent. Foreign markets still played a vital role for the company, but compared to 1967 they had lost some ground: 59 per cent of income from sales resulted from foreign markets in 1994 as opposed to 73 per cent in 1967. The role of the foreign affiliates and production locations, however, had fundamentally changed. In 1994,

53 per cent of output was produced outside Germany. This shift of balance in favour of overseas production first materialised in 1992. However, it was not reflected in employment and investment patterns: in 1994, 59 per cent of the employees were still domestic, as were 69 per cent of investments.

The balance between production and sales in the world regions had also changed completely. While in 1967 the American market accounted for 45 per cent of all vehicles sold outside Germany, the customer base of 1994 was to a much higher degree German and European. Germany still dominated the VW Group in terms of production: 66.3 per cent of sales value was produced there, with 41.1 per cent also sold there. The other European countries were more important as markets than as locations of production. These countries produced 16.6 per cent of sales value but realized 33.7 per cent of sales value. In North America, 2.6 per cent of sales value was produced and 6.7 per cent realized, in Latin America 13.1 per cent was produced and 12.3 per cent realized. Other countries produced 1.4 per cent and realized 6.8 per cent of the sales value (*Annual Account 1994*: 72).

The changes in corporate structure, the increased complexity, together with the problems of profitability, made it necessary to take decisions that would mobilize synergies within the Group. These decisions were taken in 1994–5. We focus on the three critical elements of VW's industrial model: governance, product strategy, and labour relations.

### 11.5.1. Corporate Governance

By 1994 the formal structure of the VW Group had been established, and it was differentiated between four brands and three regions (Table 11.3). The four brands were Volkswagen, Audi, SEAT, and Škoda, each of which remained independent companies represented on VW's corporate board. In 1995 a fifth brand was added, which combined light truck production in Hanover with heavy truck production at VW of Brazil. The regions became profit centres represented by other board members: the region, South America/Africa, for instance, was represented by Lopez, who was also the executive director of production optimization and logistics.

### 11.5.2. Product Strategy

In the context of the new corporate structure the new product strategy represented the core of Volkswagen's revitalized industrial model for the 1990s. It also represented a new profit strategy compared to the Fordist strategy of the first period and the diversified Fordism of the second period described above. The new strategy was Sloanist in its market orientation, and this would require major adjustments of the other components of the industrial model, specifically production organization, in the future.

TABLE 11.3. *Car manufacturing divisions in the VW Group, 1994*

| | Share of VW AG (%) | Production No. of Vehicles | Employees[a] |
|---|---|---|---|
| *Volkswagen* | | 1,392,558[b] | 125,237 |
| VW AG | | 1,246,392 | 108,963 |
| VW Sachsen GmbH | 100 | 90,100 | 3,198 |
| VW Bruxelles S.A. | 100 | 169,930 | 5,820 |
| VW Navarra S.A. | 100 | 145,784 | n/a |
| VW Europa Automoveis Lda. | 50 | under construction | |
| VW Bratislava | 100 | 5,873 | 808 |
| *Audi* | | 354,610 | 31,588 |
| Audi AG | 98.99 | 354,610 | 31,588 |
| *SEAT* | | 313,690 | 15,383 |
| SEAT S.A. | 100 | 313,690 | 15,383 |
| *Škoda* | | 173,586 | 15,985 |
| Škoda, automobilovà s.a. | 60.3 | 173,586 | 15,985 |
| *Region Northamerica* | | 256,317 | 15,652 |
| Volkswagen de Mexico, S.A. de C.V. | 100 | 256,317 | 14,057 |
| *Region Southamerica/Africa* | | 552,482 | 35,052 |
| Autolatina Brasil S.A., and | 42.58 | | |
| Autolatina Argentina S.A. | 51 | 504,310[c] | 14,057[c] |
| Volkswagen of South Africa (Pty). = Ltd. | 100 | 48,172 | 7,265 |
| *Region Asia-Pacific* | | 115,326 | 7,275 |
| Shanghai-Volkswagen Automotive Company Ltd | 50 | 115,326 | 7,275 |
| FAW-Volkswagen Automotive Company Ltd | 40 | (8,219) | n/a |
| VW Concern total | | 3,042,383 | 243,638[d] |

*Notes*:
[a] Yearly average, excluding apprentices in vocational training.
[b] The consolidated figure is lower than the aggregate output figures of the individual companies because CKD-based vehicles are double counted.
[c] Theoretical share of VW.
[d] Includes 6,137 apprentices in vocational training.
*Source*: Compiled from Volkswagen AG, *Annual Report 1994*.

A major element of the new strategy was the reorganization of product development and rationalization of the product range. Until 1995 each of the lead companies of the former brands (and to a certain extent Brazil) had their own product development organization and, despite corporate coordination, there was a high degree of overlap and redundancy. In 1995 the Group produced sixteen different model lines of passenger cars. The new strategy was based on a platform approach by which the number of platforms in the Group was to be reduced to four. In future these platforms would be developed by VW and Audi only. The VW brand would be in charge of the small car platform (for instance, Polo) and the compact car platform (for instance, Golf). Audi would be responsible for the mid-range car (for instance Passat) and the upper-range cars (for instance the A6 and the A8). On the basis of these platforms all brands would develop their own 'hats'. A new product development organization and a new master schedule for the development of new products was to decrease the 'time to market' to less than four years after concept approval. At the same time the number of models was to be increased: between 1995 and 1998, twenty-five new models were planned (Hillebrand 1995). By commonizing parts and reducing development costs the profit rate would thus be increased drastically.

The expected cost savings were enormous and included savings in purchasing and manufacturing. Compared to the profits of DM 150 to 400 earned per Golf in 1995, the target for a Golf 4 slated for 1997 was DM 1,000 to 3,000 (Hillebrand 1995: 66). There was the danger, of course, that Volkswagen would encounter the same problem that General Motors had experienced with its Sloanist strategy: the problem of differentiation between the car lines and of avoiding the 'look-alike' syndrome. But it seemed likely that Volkswagen would fare better than General Motors in this regard, since the four brand companies had not gone through a process of levelling, and were still very different in terms of their image and cultural background.

In order to underline the distinctiveness of the brands, the sales organization at Volkswagen was to be reorganized. Traditionally, VW operated a franchise system, with independent dealers linked to the company via contracts and supplied by wholesale and import companies. There were distribution companies for certain markets: the USA, Canada, France, Sweden, and later Italy and the UK. In the late 1960s Audi was able to defend the independence of its dealer network. Later, however, most of these dealers added VW cars to their sales offer and VW dealers added Audi cars. In 1978, the VAG Sales and Service Organization was created. The Group sales division became responsible for 152 sales centres and importers in charge of regional supply and servicing for the 10,600 retail dealers and service workshops with a total of 211,000 employees, (slightly) more than the 206,000 employees of the VW Group (in that year). From the viewpoint of Audi, this sales organization provided little opportunity to differentiate the Audi products and meet the specific needs of its customer base. The reorganization of 1995 specifically addressed this concern. In the future, VW and Audi products were to be sold by different types of dealership: one-brand sales centres in charge of only VW or Audi products; brand-exclusive dealers with separate showrooms and separate contracts for the VW and Audi products, general dealers selling VW and Audi products. Whereas the general dealers had handled most of the sales volume in the past, the aim now was to reduce their share to around 50 per cent, thus giving the separate brands a more visible and exclusive representation. SEAT and Škoda retained their independent dealer organizations, making it easier for them to maintain their brand identity.

### 11.5.3. Group Work and Improvement Activities

Benchmarking studies in the late 1980s clearly demonstrated that VW's plants were lagging in terms of productivity and quality performance compared to the 'best practice plants' of the industry. Among the new concepts considered for improving performance, group work played a prominent role. Group work had been under discussion at VW since around 1987. The concept was particularly promoted by the works councils. The praise heaped on the 'dynamic team as the heart of the lean factory' by the book *The Machine that Changed the World* helped to promote the issue. In 1991 works council and management agreed on the introduction of group work—but on a pilot basis for selected areas only. Overall, group working retained a low profile at VW compared to other companies in Germany like Opel or Mercedes-Benz. This was in stark contrast to the avant-garde role which Volkswagen played in other fields, particularly with regard to work organization and training linked to high technology. Scepticism towards teamwork seemed to be particularly deep seated at Volkswagen. The sheer size of the Wolfsburg production site, with its, 60,000 employees as of 1993 and its very complex horizontal and vertical division of labour and deep specialization structures, may explain part of this resistance towards teamwork. As of autumn 1993 there were only ten projects involving new forms of work at Wolfsburg covering around 1,000 employees, twenty-six group work projects in Hanover with around 1,500 employees, 100 employees at the Brunswick component plant working in pilot projects, around 600 in the Kassel transmission plant, and around 3,000 in the Emden assembly plant. The only plant with full-scale group work organization was the Salzgitter engine plant with about 8,000 employees. Between the plants there was considerable variety of approach towards group work and in some cases insiders even denied that it was group work at all. Thus group work cannot be said to have received special attention as a strategic focus in the restructuring of the production organization. Restructuring measures within Volkswagen's core organization and existing plants instead focused on: the introduction of decentralized management structures through the establishment of profit centres with local management empowered to act as 'entrepreneurs'; the reorganization of the product development process to ensure that quality and efficiency targets were designed into the product; production optimization measures called $CIP^2$ (continuous improvement process squared).

The $CIP^2$ process clearly stole the march on the teamwork process. It was Lopez's key project. $CIP^2$ was not linked to group work. It was a one-off intensive activity drawing together experts from different areas including rank and file workers, and was a structured process with trained moderators and set procedures. The quality circle movement, which never gained strength in Volkswagen, was further reduced in importance by the $CIP^2$ programme, as were individual suggestions, also never a mass phenomenon at Volkswagen. Sceptical comments on the $CIP^2$ process were soon being voiced, specifically in regard to the total savings resulting from the workshops. Nevertheless, no single measure involved the shopfloor and initiated visible changes to the extent of the CIP² process. In 1993, 2,000 workshops were held at VW alone with 16,000 employees involved, in 1994 it was 6,300 with around 60,000 employees. After 1994 the CIP² process was applied in almost all functional areas and affiliated companies within the Group.

While it was true that these measures were influenced by the 'Japan model' debate, Volkswagen did not adopt an explicit after-Japan approach. There was no 'learning from Japan' movement, and even the opportunities for organizational learning offered by its joint ventures with Japanese companies were hardly recognized. This was particularly true for the joint venture with Nissan starting in 1981 to build the Passat Variant at the Zama Plant in Japan. When VW started to build up Toyota pick-ups at its Hanover plant in 1987, the Japanese production concepts received more attention, but learning still took place in a haphazard way.

The next potential step in the process of learning from Japan was the new SEAT assembly plant in Martorel. Martorel was planned in the late 1980s. Originally the process planners at corporate headquarters in Wolfsburg followed traditional principles for plant layout, but at a certain stage the decision was made to introduce 'after-Japan' concepts and, specifically, to learn from the first European transplant of a Japanese firm, Nissan in the United Kingdom. Accordingly, the production manager was wooed away from Nissan. A few years later when the new Mosel plant in East Germany was built, Martorel was in turn regarded as its model. An expectation of feedback from the foreign affiliate was a fundamental element of Volkswagen's new internationalization strategy. In this sense, Volkswagen did not pursue a 'transplant' strategy so much as a modernization and restructuring strategy, the key vector of which ran to the centre from the periphery (Jürgens 1992).

The vision of future working structures shared by the process planners in the mid-1990s was oriented towards the concept of a fractal factory (Warnecke 1995). This concept was linked with the name López and his 'dream factory' in the Basque region. The concept was of a factory for the assembly of cars, which assigned most of the direct activities to systems suppliers. These suppliers were to assemble their modules in assigned areas of the factory and also be responsible for the installation of 'their' part into the car on the main lines.

*11.5.4. Labour Relations: A New Pact for Employment Security*

Employment security had been a central issue at Volkswagen since the mid-1970s, and the policy of avoiding dismissals was a centrepiece of VW's industrial model from that point onwards. A former labour director entitled a book he edited on the employment risks of new technology *Working without Fear* (Briam 1986). His successor in 1993 took up this motto in his book on the work-sharing agreement in 1993: *Each Work Place has a Human Face* (Hartz 1994).

## The Development of Volkswagen's Industrial Model

When, in autumn 1993, plans, forecasts, and commitments were tallied up for the next two years it became evident that in total 30,000 employees would have to become redundant, a good quarter of VW AG's 108,000 employees. This situation was due to the economic recession, but also to VW's investment in new capacity since the late 1980s, a reduction in the degree of vertical integration, and the various measures to improve productivity and performance. In regard to the programmes like CIP[2] and group work specifically, all actors were aware that the company would no longer be able to count on shopfloor participation, if dismissals on such a scale took place. Moreover, industrial peace at VW itself would be threatened if management insisted on mass dismissals. It was quite characteristic of Volkswagen's governance structure that quick joint action was possible in this situation. An agreement between IG Metall and Volkswagen was struck to reduce working hours in order to share the work among the existing employees. The reduction of weekly working hours by 20 per cent (from 36 to 28.2 hours) secured 20,000 jobs; the remaining 10,000 were to be secured by additional measures. The reduction of the work time was paid for by a reduction of income on an annual basis (between 10 and 11 per cent) and by reducing the number of special holidays. In return for these concessions the company guaranteed an employment level of 100,000 employees for Volkswagen AG. At Audi a similar agreement was reached, though with only a 10 per cent reduction of weekly working hours. A major element of the work-sharing agreement was that everybody in the company should share the burden of income reduction. Thus managers up to the executive board level also made the same monetary 'concessions'. The principle of equality was also upheld with regard to work time reduction for blue- and white-collar employees throughout the company. In this regard, however, an exception was made: the group of employees whose work contract was not regulated by collective bargaining, senior white-collar employees and managers, were expected to work the hours they deemed necessary.

The work-sharing agreement was designed as a measure to cope with the difficult employment situation, but it was not perceived as temporary. Its duration was limited to two years with the understanding that it would be continued. In September 1995 a new agreement was reached, however, with some modifications. Thus, for instance a 'performance contribution' of the indirect areas was introduced raising the weekly work time for the indirect hourly and salaried employees by 1.2 hours up to 30 hours per week. Another modification was the flexibilization of the working time to a maximum of 38 hours per week as long as the 28.8 (resp. 30 for the indirect and salaried employees) hours average was reached over a one-year period. In a follow-up agreement of June 1996 this one-year limitation was given up and the concept of an individual 'employment check' was created whereby overtime work (exceeding the 28.8 resp. 30 hours) could be saved on an individual-time bank account for longer leave or absence or for early retirement. The flexibilization of the weekly work time was regarded by management as an important step to give factories space to 'breathe' in response to fluctuations in order volumes (Hartz 1996). The new work-sharing agreement will be in force until the end of 1997, but it is generally expected that a follow-up agreement will be reached after that.

The agreement on work sharing was a bold step. Weaknesses of the agreement became apparent, however, when implementation began. Different local conditions had to be taken into account, and the uneven development of demand required transfers between plants and led to variations in actual work time (from 29.2 to 32.6 hours per week) between plants. Two problems specifically stood out, with potential negative long-term effects: work-time reduction, especially in the field of product and process development activities, fostered a tendency to contract out work to engineering service companies, which might lead to a hollowing out of the company in regard to its technical competencies; the 'dualization' of work-time patterns, with the general workforce working shorter hours while white-collar specialists and managers worked ever longer hours carried the danger that a new kind of division between those who structure and plan the work and those who execute it might develop.

The strike of 1995 could turn out to be an early sign of a weakening of the very foundations of this model. The work-sharing agreement in any case did not fully solve the employment problem, since there was still overcapacity, and the pressure to improve remained. Despite work sharing, the Volkswagen workforce, especially the Wolfsburg plant, continued to shrink. This might eventually affect the power base of Volkswagen's works council system, which has been, as we have seen, the essential ingredient to VW's industrial model. Thus, the work-sharing agreements provided relief, but uncertainties as to the future development of VW's industrial model remained.

## 11.6. CONCLUSION

This analysis of Volkswagen's trajectory over twenty years has revealed an outstanding level of stability in the initial post-war model. The product strategy and the governance structure were the two major pillars of VW's industrial model. Whenever one of these central elements was in danger, VW underwent a period of uncertainty and instability. When both of them were threatened, VW fell into crisis. The two overt crisis situations at Volkswagen, in the years 1974 and 1994, occurred at a time of economic recession. The recessions were not the primary causes of crisis, however. Rather, they acted as catalysts to encourage decisions to be taken which were necessary to end periods of hesitation, internal struggles, and uncertainties within the organization.

The findings of the Volkswagen case reveal the importance of the internal dynamics of an industrial model. The model must be considered as a process rather than as a rigid structure. Measures taken to flesh out the first set of strategic decisions and make them viable and feasible in different economic and social circumstances proved to be more difficult than was assumed at the beginning. There were a number of failures in the implementation, fleshing out, and substantiation of the original concepts at Volkswagen. In the end, such failures, together with the complacency resulting from economic success, led to situations of open crisis. In these situations of crisis a new consensus emerged and new arrangements were laid out which became the basis of a new cycle of growth.

# NOTES

1. The core organization, with headquarters in Wolfsburg, Germany, became a private limited company in 1949 and a public limited company in 1960, the Volkswagenwerk AG (*Aktiengesellschaft*: public limited company). The name Volkswagenwerk AG was changed to Volkswagen AG in 1985. The Volkswagen Group, established 1969, encompassed, as of September 1996, the Volkswagen AG, Audi AG, SEAT S.A., Škoda Automobilovà a.s., and VW Commercial Vehicles.
2. Information is based, in addition to the literature, on several research projects carried out by the author since 1982. For the purpose of this chapter further interviews were made with H. Amtenbrink on process layout and organization of the Beetle and Golf production (Amtenbrink was an area manager of paint and assembly operations of the Wolfsburg plant in the 1960s and later became the plant manager) and with W. H. Pallasch, Department on Collective Agreements, and W. Widuckel Mathias of the General and Corporate Works Council on labour relations issues.

## Statistical Appendix 11: Volkswagen

| | Production Total[a] | Production Domestic[b] | Production Abroad[c] | Export[b] | Turnover Total[d] | Turnover Domestic[d] | Turnover Abroad[d] | Workforce Total[e] | Workforce Domestic[e] | Workforce Abroad[f] | Net Earnings[g] |
|---|---|---|---|---|---|---|---|---|---|---|---|
| 1970 | 2,215 | 1,889 | 326 | | 15,113 | 4,911 | 10,202 | 190 | 155 | 35 | 407 |
| 1971 | 2,354 | 1,867 | 487 | | 16,473 | 5,135 | 11,338 | 202 | 160 | 42 | 147 |
| 1972 | 2,193 | 1,673 | 520 | | 15,996 | 5,035 | 10,961 | 192 | 149 | 43 | 206 |
| 1973 | 2,335 | 1,720 | 615 | | 16,982 | 5,364 | 11,618 | 215 | 161 | 54 | 330 |
| 1974 | 2,068 | 1,359 | 709 | 976 | 16,966 | 5,161 | 11,805 | 204 | 142 | 62 | −807 |
| 1975 | 1,949 | 1,229 | 720 | 736 | 18,857 | 6,552 | 12,305 | 177 | 118 | 59 | −157 |
| 1976 | 2,166 | 1,436 | 730 | 838 | 21,423 | 8,068 | 13,355 | 183 | 124 | 59 | 1,004 |
| 1977 | 2,219 | 1,561 | 658 | 877 | 24,152 | 9,714 | 14,438 | 192 | 133 | 59 | 419 |
| 1978 | 2,385 | 1,569 | 816 | 847 | 26,724 | 11,229 | 15,495 | 207 | 139 | 68 | 574 |
| 1979 | 2,542 | 1,558 | 984 | 824 | 30,707 | 12,499 | 18,208 | 240 | 157 | 83 | 667 |
| 1980 | 2,574 | 1,499 | 1,075 | 845 | 33,288 | 11,850 | 21,438 | 258 | 159 | 99 | 321 |
| 1981 | 2,246 | 1,410 | 836 | 827 | 37,878 | 12,064 | 25,814 | 247 | 160 | 87 | 136 |
| 1982 | 2,130 | 1,381 | 749 | 856 | 37,434 | 12,027 | 25,407 | 239 | 158 | 81 | −300 |
| 1983 | 2,116 | 1,413 | 703 | 787 | 40,089 | 14,453 | 25,636 | 232 | 156 | 76 | −215 |
| 1984 | 2,148 | 1,474 | 674 | 928 | 45,671 | 14,638 | 31,033 | 238 | 160 | 78 | 228 |
| 1985 | 2,398 | 1,635 | 763 | 1,092 | 52,502 | 16,171 | 36,331 | 259 | 170 | 89 | 596 |
| 1986 | 2,777 | 1,654 | 1,123 | 1,023 | 52,794 | 18,839 | 33,955 | 276 | 169 | 107 | 580 |
| 1987 | 2,771 | 1,666 | 1,105 | 988 | 54,635 | 22,555 | 32,080 | 260 | 170 | 90 | 598 |
| 1988 | 2,848 | 1,694 | 1,154 | 1,033 | 59,221 | 22,653 | 36,568 | 252 | 165 | 87 | 780 |
| 1989 | 2,948 | 1,783 | 1,165 | 1,142 | 654,352 | 23,682 | 41,670 | 251 | 161 | 90 | 1,038 |
| 1990 | 3,058 | 1,816 | 1,242 | 1,137 | 68,061 | 26,929 | 41,132 | 261 | 166 | 95 | 1,086 |
| 1991 | 3,238 | 1,814 | 1,424 | 927 | 76,315 | 36,360 | 39,955 | 277 | 167 | 110 | 1,114 |
| 1992 | 3,500 | 1,929 | 1,571 | 1,105 | 85,403 | 39,508 | 45,895 | 273 | 164 | 109 | 147 |
| 1993 | 3,019 | 1,411 | 1,608 | 799 | 76,586 | 34,326 | 42,260 | 253 | 150 | 103 | −1,940 |
| 1994 | 3,042 | 1,425 | 1,617 | 844 | 80,041 | 32,907 | 47,134 | 238 | 141 | 97 | 150 |
| 1995 | 3,595 | 1,526 | 2,069 | 973 | 88,119 | 34,504 | 53,615 | 257 | 143 | 114 | 336 |
| 1996 | 3,977 | 1,591 | 2,386 | 1,013 | 100,123 | 36,419 | 63,704 | 261 | 139 | 122 | 678 |

*Notes*:
[a] VW group: VW, Audi, SEAT since 1986, Škoda since 1992, in thousand units.
[b] From domestic VW and Audi operations, in thousand units.
[c] In thousand units.
[d] Adjusted for internal transactions between the individual consolidated Group companies; SEAT was consolidated into the group in 1986, Škoda in 1992 (million DM).
[e] At year's end; since 1986, year's average. Excluding apprentice trainees (1974: 6,137), in thousand employees.
[f] In thousand employees.
[g] The difference between the income and expenses of the year before transfers to or from reserves and after taxes (million DM).

# BIBLIOGRAPHY

Boyer, R., and Freyssenet, M., 'The Emergence of New Industrial Models: Hypotheses and Analytical Procedure', *Actes du GERPISA*, 5 (1995).
Briam, K.-H., *Arbeiten ohne Angst: Arbeitsmanagement im technischen Wandel* [Working without Fear: Labour Management in Times of Technological Change] (Düsseldorf/Wien, 1986).
Brumlop, E., *Arbeitsbewertung bei flexiblem Personaleinsatz: Das Beispiel Volkswagen AG* [Job Evaluation and Personnel Flexibility: The Example of the Volkswagen AG] (New York, 1986).
—— and Jürgens, U., 'Rationalization and Industrial Relations: A Case Study of Volkswagen', in O. Jacobi, B. Jessop, H. Kastendiek and M. Regini (eds.), *Technological Change, Rationalization and Industrial Relations* (London, 1986).
Doleschal, R., 'Zur geschichtlichen Entwicklung des Volkswagen Konzerns' [On the Historical Development of the Volkswagen Group], in R. Doleschal and R. Dombois (eds.), *Wohin läuft VW? Die Automobilproduktion in der Wirtschaftskrise* [Which Direction is VW Heading to? Car Production in the Economic Crisis], (Reinbek bei Hamburg, 1982).
Dombois, R., 'Arbeitsplatz Volkswagen' [Workplace Volkswagen], in Doleschal and Dombois, *Wohin läuft VW* (Reinbek bei Hamburg, 1982*a*).
—— 'Volkswagen in USA', in Doleschal and Dombois, *Wohin läuft VW* (Reinbek bei Hamburg, 1982*b*).
Endres, E., *Macht und Solidarität: Beschäftigungsabbau in der Automobilindustrie: Das Beispiel Audi/NSU* [Power and Solidarity: Employment Cutbacks in the Car Industry: The Example of Audi/NSU (Hamburg, 1990).
Etzold, H.-R., Rother, E., and Erdmann, T., *Im Zeichen der Vier Ringe 1945–68* [Under the Sign of the Four Rings], ii (Verlag Audi AG, 1995).
Hartz, P., *Jeder Arbeitsplatz hat ein Gesicht: Die Volkswagen-Lösung* [Each Work Place has a Human Face: The Volkswagen Solution] (New York, 1994).
—— *Das atmende Unternehmen: Jeder Arbeitsplatz hat einen Kunden* [The Breathing Company: Each Work Place has a Customer] (New York, 1996).
Hildebrandt, E., 'Der VW-Tarifvertrag zur Lohndifferenzierung' [The VW Agreement on Wage Differentiation], in Doleschal and Dombois, *Wohin läuft VW* (Reinbek bei Hamburg, 1982).
Hillebrand, W., *Mit verdeckten Karten* [With Cards Played Close to the Chest], in *Capital*, 11 (1995), 52–67.
Jürgens, U., 'Internationalization Strategies of Japanese and German Automobile Companies', in S. Tokunaga, N. Altmann, and H. Demes (eds.), *New Impacts on Industrial Relations*, Monographs of the German Institute for Studies on Japan of the Philipp-Franz-von-Siebold Foundation, iii (Munich, 1992), 63–96.
—— Malsch, T., and Dohse, K., *Breaking from Taylorism: Changing Forms of Work in the Automobile Industry* (Cambridge, New York, and Oakleigh, 1993).
Kern, H., and Schumann, M., *Das Ende der Arbeitsteilung? Rationalisierung in der industriellen Produktion* [The End of the Division of Labour? Rationalization of Industrial Production] (Munich, 1984).

Klauke, P., 'Hitler und das Volkswagen Projekt' [Hitler and the Volkswagen Project], in *Vierteljahreshefte für Zeitgeschichte*, no. 4 (1960).

Koch, G., *Arbeitnehmer steuern mit—Belegschaftsvertretung bei VW ab 1945* [The Employees Steer, too—Workers' Interests Representation at VW since 1945] (Cologne, 1987).

Mommsen, H., and Grieger, M., *Das Volkswagenwerk und seine Arbeiter im Dritten Reich* [Volkswagen Works and its Workers in the Third Reich] (Düsseldorf, 1996).

Schumann, M., Baethge-Kinsky, V., Kuhlmann, M., Kurz, C., and Neumann, U., *Trendreport Rationalisierung: Automobilindustrie, Werkzeugmaschinenbau, chemische Industrie* [Trend Report Rationalisation: The Automotive Industry, Machine Tool Industry, Chemical Industry] (Berlin, 1994).

Sorge, A., and Streeck, W., 'Industrial Relations and Technical Change: The Case for an Extended Perspective', Wissenschaftszentrum Berlin für Sozialforschung, dp IIM/LMP 1987–1 (Berlin, 1987).

Streeck, W., 'Neue Formen der Arbeitsorganisation im internationalen Vergleich' [An International Comparison of New Forms of Work Organization], in Industriegewerkschaft Metall, Verwaltungsstelle Wolfsburg (ed.), *Zukunft der Automobilindustrie: Symposium der IG Metall Wolfsburg in Zusammenarbeit mit dem Betriebsrat der Volkswagen AG, Werk Wolfsburg* [The Future of the Automobile Industry: Symposion of the IG Metall Wolfsburg in cooperation with the works council of the Volkswagen AG, plant Wolfsburg] (1986).

Warnecke, H. J. (ed.), *Aufbruch zum fraktalen Unternehmen: Praxisbeispiele für neues Denken und Handeln* [Towards the Fractal Company: Real-life Examples of the New Mind Set and Behaviour] (Berlin, 1995).

Wellhöner, V., *'Wirtschaftswunder'—Weltmarkt—westdeutscher Fordismus: Der Fall Volkswagen* ['Economic miracle'—World Market—West German Fordism: The Case of Volkswagen] (Münster, 1996).

Wiersch, B., *Die Vorbereitung des Volkswagens* (The Preparation of the Volkswagen), Dissertation (Hannover, 1974).

Womack, J. P., Jones, Daniel T., and Roos, D., *The Machine that Changed the World* (New York, 1990).

# 12

## Making Manufacturing Lean in the Italian Automobile Industry: The Trajectory of Fiat

ARNALDO CAMUFFO AND GIUSEPPE VOLPATO

The objective of this chapter is analysing the strategic and organizational evolution of Fiat Auto. During the 1980s Fiat's strategy consisted in a refinement of the traditional manufacturing model based on a few extraordinarily successful products, the production efficiency provided by process automation technology, the work flexibility granted by an industrial relations strategy based on managerial unilateralism and concession bargaining, and the prevalent concentration in the domestic market (Dealessandri and Magnabosco 1987; Locke and Negrelli 1989; Kochan et al. 1990).

However, the tremendous success, obtained by Fiat between 1983 and 1989, in some respects disguised or at least delayed the perception of the competitive challenge which would erupt in the subsequent years, calling for a new organizational strategy and manufacturing model in which human resources and skills were to play a central role. At the turn of the decade market difficulties and international competition revealed Fiat's strategic imbalance, organizational weaknesses, and the short-sightedness of its industrial relations policy. Fiat reacted by designing a comprehensive strategic and organizational change consisting of new relationships with suppliers and dealers, massive investment in new product development, a new organizational model for manufacturing plants, and innovative industrial relations and human resource management policies.

From a theoretical standpoint, studying and interpreting such a massive change is interesting in terms of assessing the characteristics of the emergent organizational paradigm. The main thrust of this paper is therefore to highlight the fact that external pressures, both institutional and competitive, explain by themselves only a part of the changes taking place. In fact these changes are filtered and interpreted by the firm on the basis of given and historically determined resources and codes, such as technologies, organizational knowledge, professional competences and routines accumulated in the past, managerial cognitive 'schemes'. As a consequence, internal factors (especially in terms of specialization and the scope of competence endowment built up by the firm throughout its history by means of costly and risky investment in technologies, organizational routines, labour skills, and so on) greatly influence the firm's evolutionary process, shaping the emerging organizational paradigm, determining the speed of the transformation process, and characterizing the difficulties and obstacles hindering change.

Given these three points, this chapter tries to discuss some intertwined sets of questions related to the history of the firm. Before the ongoing transformation, what was Fiat's organizational model? Was it a modified version of a mass production or Fordist approach? If so, when did Fordism emerge at Fiat, and what were its specificities?

Since 1990 Fiat seems to be moving towards a different organizational paradigm. What changes are taking place in terms of its manufacturing model and relationships with suppliers and dealers? Does the new model converge towards 'lean manufacturing'? Is Fiat 'japanizing' or are there significant differences?

## 12.1. FORDISM AT FIAT

The first issue that must be analysed in order to understand recent changes in FIAT manufacturing strategies and its possible shift towards a new organizational paradigm is if and when Fiat began mass production and adopted a manufacturing model based on the principles of scientific management. At Fiat, the introduction of the Fordist paradigm was gradual and did not reach a meaningful threshold till the 1950s. Although the US automobile industry experience inevitably represented a kind of benchmark to which all European car makers had to refer to when developing their competitive positions at the beginning of the twentieth century, the prospect of 'doing like Ford' was largely hindered by the specific characteristics of the European setting, especially in terms of the size of the market, general socio-economic conditions (quantity and quality of workforce available), and union hostility towards the implementation of Fordist schemes. For at least three decades, instead of focusing on expanding the market through massive investments aimed at achieving economies of scale, European producers firmly maintained high product differentiation, believing that a broad product line would be the most effective strategy in the Old World highly fragmented market.

Fordism was instead fully applied at Fiat after World War II, with the 600 model, whose production began in 1955 and the new 500 model (1957). Production growth was striking: it soared to 100,000 units in 1950, 500,000 in 1960, and 1 million in 1966. These volumes allowed the introduction of technical innovations and organizational solutions consistent with mass production. Considering a rough measure of labour productivity as the number of cars per employee, no relevant variation in the ratio can be seen from 1901 to 1950 (it varies between 1 and 2), meaning that any efficiency gains had been compensated by the increased technical complexity of cars. But since 1950, with the comprehensive restructuring of Mirafiori, the mass production scheme entailed a constant productivity growth, up to 7.5 cars per employee in 1960. Throughout the 1960s Fiat basically completed its 'organizational trajectory', becoming an efficient mass producer of small cars, mainly targeted at the domestic market. But this 'forced specialization'[1] although enhancing Fiat's short-term performances, hindered the development of technological innovations, since these are generally experimented with and introduced first on more expensive cars. In other words, Fiat's 'forced' focus on the domestic market and lower car segments entailed a specialized heritage of competencies, narrowing the scope of future strategic options.

In the post-war period, with Vittorio Valletta replacing Giovanni Agnelli at the top of the company, Fiat developed the mass production schemes, fully drawing its implications in terms of management style, human resource management, and industrial relations. Fiat's strategy, based on high volumes, cost efficiency, and Fordist work organization, required a strict control of workers' behaviour. Valletta therefore implemented a set of human resource management and communication policies explicitly or implicitly aimed at reducing uncertainty and lowering the unions' power. The labour movement emerged in Italy as highly politicized and centralized. Throughout the 1950s and 1960s the politics and strategies of the three confederal unions (CGIL, CISL, and UIL) were shaped by their political affiliations and rivalries (Locke 1992). Until the mid-1960s, unions were weak and divided because of ideological, political, and economic factors. Valletta exploited these weaknesses and divisions with an opportune mixing of *realpolitik* and rigidity of rules. His strategy, supported by a paternalistic management style and ad hoc policies for middle managers and shop stewards, alternated timely openings to union groups willing to cooperate, and systematic confinement and ostracism of militant union members and representatives. Though effective in reducing union intervention, these 'ghetto-like' practices inevitably amplified the tensions which would erupt in the late 1960s.

With the complete shift to mass production Valletta realized that the Fiat he was building was as powerful on the production side, as it was fragile in its mechanisms. The company became extremely complex in its internal links and therefore rather delicate; its competitive strength had grown enormously, but it was highly conditioned by environmental stability. During those years, the turbulence owing to the shift to mass production was such that Valletta intended to neutralize any other form of turbulence. He sought a 'normalization' of the working environment because he felt that the ongoing stage was of key importance for the consolidation of Fiat, and he feared that his complex scheme could falter if the turbulence arising from product and process evolution was intensified by constraints derived from increased workers' power. This normalization first aimed at defeating the antagonist union. In other words, during these years Fiat management neglected the issue of consensus-seeking not just with regard to the workers themselves, but also in terms of institutionalized dialogue with the union.

Along with the implementation of the mass production scheme, Fiat experienced huge growth, expanding the marked vertical integration processes which had begun in the 1920s. Fiat tended towards a structure similar to that of General Motors, on a smaller scale. This vertical development was allowed by large financial resources, but was an entirely necessary step, because until the 1960s the Italian industrial setting was undeveloped and few component suppliers had the managerial, technological, and financial resources required to produce according to the standards demanded by Fiat. The organizational structure was to remain centralized and highly vertically integrated until the end of the 1970s. The 1950s and 1960s had represented a period of strong development for vehicle motorization worldwide, and Italy could promptly take advantage to become, in 1968, the world's fifth automobile manufacturer behind the US Big Three and Volkswagen. Such a major expansion was partly owing to the growth of domestic demand for vehicles, which Fiat in 1968 was able to meet by a 78 per cent share. But Fiat had undoubtedly managed to enter foreign markets, where it directed a 38 per cent share of its total production, totalling 547,000 vehicles (1968). In this way Fiat managed to take up a relevant share of the whole automobile industry production (6.6 per cent in the world, 15.7 per cent in Europe, 21.2 per cent in the EEC). Since it was established, Fiat had paid considerable attention towards the just-developed international market, both by a relevant flow of exports and by the establishment of assembly plants abroad. During post-war reconstruction this interest grew stronger. A first effort was participation in the Spanish company SEAT, whose manufacturing agreement became operational in 1953. Another production effort abroad had begun in 1954, through an agreement with the Yugoslav company Zavodi Crvena Zastava (ZCZ), and a third one in Argentina (Fiat Concord, operational in 1960). Then there was the even more important plan of a financial and manufacturing integration with Citroën. It began in 1968 with the acquisition of 15 per cent of Citroën shares and increased to 26.9 per cent in 1970. But the plan was not carried out owing to hostility by the French government.

## 12.2. The Crisis of Mass Production

The 1970s represented a decade of profound crisis for Fiat, as social, market, and internal factors contributed to the worsening of competitive and financial performances. The union-restraint strategy produced a lot of trouble during a period of full employment and social pressures, so industrial relations played a major role in Fiat crisis. The changing of the composition of the workforce within Fiat was an important element in this process. Between 1968 and 1969 another element affecting the equilibrium of the factory took place, in the city of Turin itself: a massive new wave of recruitment (20,000 workers), for the opening of the new Rivalta plant. New personnel mostly came from the south, and this generated additional demand for housing which Turin was not able to satisfy.

At the end of the 1960s the union movement regained strength and unity. The unsettled nodes of the Valletta era, together with the new political and social context, burst out in the 'hot autumn' (*autunno caldo*) conflicts. The growing discontent among workers, enhanced by a wave of political resentment, brought the unions together. During the 'hot autumn' of 1969 strikes amounted to an equivalent of 273,000 cars lost, and in 1972 a wave of acts of terrorism aimed at union representatives and managers began.[2] Fiat management blamed the union for not collaborating to discipline all abuses, and the renewed bargaining power of the union, together with a unilateral approach to labour relations by Fiat management, was unable to mature into advanced forms of industrial relations. Union pressures on union rights,[3] reductions of job categories,[4] information, and work organizations systematically increased in the early 1970s. For example, in August 1971, after strikes and long negotiations, Fiat management and unions signed an agreement introducing major constraints on work organization, eliminating discretionary job assignments, posing limits to worker saturation levels and cycle times, and so on. There is no doubt that the contents of the agreement seriously infringed efficiency requirements, at least in manufacturing.[5]

Another massive negative influence upon the economic situation of Fiat was the 1973 oil shock. The oil shortage triggered inflation. In Italy the impact of price increases was more severe and prolonged than in the other European countries. As a consequence, there was a dramatic slump in the car market. To appreciate the difficulties which challenged Fiat in the 1970s one must examine step by step the main building blocks of work organization and productivity. A first element to consider is the ratio between contractual hours and hours actually worked. Owing to new union agreements, contractual hours were reduced from 2,022 (1969) to 1,870 (1977). During that same period, the hours of actual work, net of absenteeism, were reduced by 15.7 per cent (more than twice that ratio), from 1,800 to 1,518. Clearly one must also take into account the reduced production level owing to labour unrest, whose climax was reached in 1969, when production lost for strikes was estimated at 273,000 vehicles. During the following years, too, the strong conflictual climate negatively affected production levels.

Another relevant issue was the rise of labour costs and salaries. The 1970s featured high inflation. However, salary increases, thanks to wage indexation mechanisms (*scala mobile*), allowed reasonable increases to Fiat workers. With the exception of 1974, the average rate of salary increase had always been above the average rate of cost of living. Between 1970 and 1977 the cumulative variation of salaries reached +141.3 per cent, compared to +102.7 per cent of the cost of living. Of course, the main part of these salary increases was pulled by automatic adjustment mechanisms (responsible for 53.8 per cent of the adjustment), while the remainder was due to the new collective labour agreements and inter-confederal or company agreements (for 38.1 per cent) and to the overall average evolution of job classifications (for 8.1 per cent).

In 1973 the company clearly entered a situation of financial tensions. In that fiscal year it formally broke even, by giving up accelerated depreciations of 30 bn. lire, and writing off to the following balance sheet 36 bn. lire in investment in equipment. In 1974 the financial situation became even more serious. On the cost side the extremely low capacity utilization (at about 65 per cent of the total) yielded a strong impact on manufacturing cost per unit. On the demand side the reduction in registrations translated into an increase in unsold stocks with extremely high financial costs. During the first part of the year the excess production was outweighed by 1973 order backlogs (when loss of production due to labour unrest amounted to about 150,000 vehicles), but at the end of August vehicle stocks amounted to 258,000 units, and at the end of September they had reached the 300,000 level, totalling over 450 bn. lire in stocks, with huge management costs for interest, maintenance, and reconditioning. If the production level had continued unchanged as in the previous months, at the end of the year stocks would have totalled 450,000 units, an incredible amount considering the storage space required, apart from the economic and financial effects. Hence the use of the *Cassa Integrazione Guadagni* (State redundancy fund) to reduce unsold stocks by about 200,000 units, pushing them back closer to the 180,000 level (believed to be the 'physiological level').

Apart from the unsold stocks, the implementation of the internationalization policy was causing several more difficulties. The plan to establish a vehicle manufacturing plant in Brazil had entered the implementation stage, and it required huge investments which were impossible to put off without incurring considerable losses. Hence a reduction of available resources which, coupled with a phase of disorientation on the future of the automobile industry, slowed down product line renovation. To this extent, by comparing the timing of new product launches across the main European car manufacturers, one can note a pause between 1974 (launch of 131 model) to 1978 (launch of the Ritmo).

The reduced international competitiveness of Fiat was also due to the ageing managerial structure. At the beginning of the 1960s Fiat looked like a massive agglomerate of industrial activities revolving around vehicle manufacturing. The degree of vertical integration was very high, not only from a manufacturing standpoint, but also from a legal and social perspective, since the bulk of activities lay within Fiat SpA. This tendency was also reinforced, apart from the expansion policy on vehicle component activities, by fiscal norms which by then affected purchase and sale and not added value (Volpato 1983).

When the presidency was handed over by Vittorio Valletta to Gianni Agnelli, nephew of Senator Agnelli (in 1966), the reorganization process began, further fuelled by the appointment of Umberto Agnelli (Gianni's younger brother) to chief executive officer (CEO) in 1970. Umberto Agnelli was highly receptive to managerial culture ideas which had evolved in the USA. He was responsible for the establishment of the *Istituto di Sviluppo Organizzativo* (ISVOR), based in Marentino (a few kilometres away from Turin), which aimed to carry out training activities for Fiat Group managers. Umberto Agnelli realized that the concern for expansion shown by Vittorio Valletta and the absence of competition in the national market had reflected an organizational structure which by then appeared inadequate. He then decided to undertake a profound reorganization process for the company structure. It was a difficult task, especially because a specific decisional structure inevitably ends by selecting people who have features consistent with the structure itself.

A few documents reporting the first hypotheses of company reorganization had begun circulating since 1967, but the structuring of a new organizational chart began only in April 1970. The goals were: to achieve a more manageable structure by a gradual divisionalization; to achieve a more flexible company structure by breaking up Fiat SpA into many companies linked but legally independent, while also envisioning some sales and acquisitions, and the loosening of financial constraints on the development of group activities.

The first step towards a new Fiat structure did not represent a true divisionalization (which would be achieved by 1979), but it represented an intermediate stage with the establishment of some operation groups. The operation groups formed during this first restructuring still encompassed very heterogeneous activities. There were just three: automobiles, industrial vehicles, and diversified activities. The potential overlapping of competencies and the duplication of functions and structures within this scheme were clear.

An accelerated turnover among the Fiat managers meant that a feeling of discouragement and general uncertainty was pervasive even at the top. In order to avoid such compelling pressure the new top management on duty in the second half of the 1970s[6] implemented a new strategy based on financial restructuring of the Group and technological improvement of products and processes through a strong application of automation whose first applications were put into effect in the first half of the decade.

## 12.3. THE AUTOMATION STRATEGY

The automation strategy is most significant for the understanding of the new Fiat trajectory. It was oriented towards achieving higher levels of quality and productivity, reducing conflicts through the abolition of the most dangerous and tiring manual operations (ergonomy was a major cause of workers' conflicts), bypassing union control on work organization (the disruption of traditional lines implied deskilling workers), reducing union influence which was extremely high in the traditional Fordist assembly line, where social rules determined working standards and rhythms.

Fiat began investing heavily in more advanced and labour-saving manufacturing technologies. Through automation, Fiat aimed at making production to a certain extent independent of workers' consensus and participation. This technology strategy was based on the progressive automation of the manufacturing process and can be interpreted as a two-phase process: a first phase (1972–8) of rigid automation, and a second phase (1978–88) of more flexible, partial, or full automation.

The first robots were introduced at Mirafiori in 1972 (for the 132 model line), but a conspicuous leap in manufacturing automation was made with the Digitron system two years later (for the 131 model line). Digitron, implemented at the Mirafiori plant in 1974, was a computer-controlled system of docking. The chassis was loaded on a robocarrier, and conveyed to automated screwing stations where the body was assembled. Tiring and wearisome work operations (like working hands-up) were eliminated. As already mentioned, the major purpose of these innovations was clearly to solve ergonomic problems by eliminating the kind of jobs more associated with episodes of industrial conflict (Camuffo and Volpato 1997). Robogate, a fully automated body welding system, was implemented at the Rivalta plant in 1978, and served this purpose. Theoretically, it allowed a body framing '360° flexibility' in terms of: market response, product lines renewal, and process affordability. Put in practice, the system was designed for five different bodies (but Fiat applied it only for two), 80 per cent of the investment was reusable and adaptable for new product lines, and the non-rigid sequence of manufacturing operations (computer-regulated asynchronic movements) allowed systematic prevention of complete breakdowns. Thanks to Robogate, welding was highly automated (more than 90 per cent of operations), so that it required only inspection and service by maintenance workers: indirect workers quadrupled, as direct workers dropped to a quarter. Even though these sophisticated technological features were never completely exploited, as the system entailed an excess of costly manufacturing flexibility, Robogate represented an advanced solution worldwide. It was subsequently installed by Comau, a subsidiary of Fiat, in other European and US plants.

Another example of this technology strategy, based on progressive automation within an organizational context characterized by industrial relations constraints, is the implementation of LAM at the Mirafiori plant in 1979. The LAM (Lavorazione Asincrona Motori) system consisted in the partial automation (manual operations still existed) of the automobile engine assembly process: computer-guided minitrailers loaded the engine pallet and transported it along the different engine assembly areas. The restructuring of the engine manufacturing process (in terms of both machining and assembly) implied a reorganization of work together with job redesign, safer and better working conditions, job enlargement with extended working sequences and larger phase time, a certain degree of job enrichment, the separation of manual operations and working sequences, increased flexibility (engines of the same 'family' were manufactured for four models).

LAM did not represent a fully automated solution for engine assembly. Rather, consistently with the context, it introduced elements of automation allowing job redesign and ergonomy. Its impact on productivity was good but not striking, partly because of intrinsic features, partly because it was applied to the manufacturing of pre-existing engines produced since the end of the 1960s and not specifically designed to be assembled with LAM, but it got a deservedly hearty reception from workers. Although it cannot be considered a complete success, the experience in developing this human-friendly system of engine assembly would be a relevant background in designing, at the beginning of the 1990s, the Pratola Serra plant.

## 12.4. TURNAROUND AND COMEBACK IN THE 1980S

In the 1980s Fiat managed to face the vast crisis which affected its performances and its industrial relations setting. This section describes the emerging industrial relations strategy based on managerial unilateralism. Such a strategy seems consistent with the strong automation-based technology strategy pursued in the 1980s. On 11 September 1980 the management announced the imminent lay-off of 14,469 workers. After a decade of exacerbating union–management frictions this event inevitably led to an extensive blockade of the firm. But on 14 October 1980 the 'silent majority' of workers (the 'march of the 40,000') stood against union militants and the protest strike that had shut down the firm for five weeks, proving themselves willing to end a period of violence and anarchy. This was the moment for Fiat to adopt a tough line with unions and state in negotiations the labour implications of the restructuring. In 1980 about 15 per cent of the workforce (20,500 workers) were on the state-supported redundancy fund (*Cassa Integrazione Guadagni*—CIG), and tough concession-bargaining policies were negotiated with the weakening unions. After decades of personnel policies focused on employees as a whole and on collective (firm–union) transactions, new attention was paid to specific segments of the workforce and to differentiating and customizing personnel policies. Ad hoc human resource management (HRM) policies were designed for shop stewards, professionals, and middle managers; merit-based promotions and career progressions, pay-for-performance, increased autonomy, and responsibility were applied to cadres in order to speed the generation of organizational information and to create a new and homogeneous strategic vision through middle-management commitment (Camuffo and Costa 1993*a*, 1993*b*). Moreover, new emphasis was generally put on individual transactions, that is, on the relationship between the firm and each employee at every level of the organizational structure. Fiat management tried to take a step forward, conceiving internal relations as customer–supplier relationships, and managing them also with internal marketing tools.

In the mid-1980s the recovery owing to the growing demand for cars (especially the Uno model) and the parallel return of workers who were previously in the redundancy fund, had a strengthening effect on the bargaining power of unions. Along with that, the rigidities embodied in automation and information-based technology (despite the higher 'theoretical flexibility') required a different attitude in the workforce, different HRM policies, as well as a different approach to manufacturing quality (the 1988 Cassino plant is an example of that). As a consequence, the period following 1985 saw the introduction of a new industrial relations policy based on union concessions to the strong managerial pressures for internal flexibility in order to face growing demand. On the whole, the concession-bargaining model still appeared grounded on collective relationships, and the union was involved in the flexibility strategy with an instrumental role, that is, when its agreement was necessary (i.e. third shift introduction, collective overtime, mobility). By the last years of the decade the redundancy fund was cleared, absenteeism and strike levels reached their minimum, while the company enjoyed a flourishing competitive and financial situation.

The above-mentioned motivations of Fiat's technology strategy based on progressive automation of production resulted in a series of manufacturing systems or non-integrated computer automated manufacturing 'islands'. Only with the Termoli 3 facility (1985) was the idea of a fully automated factory conceived and implemented. With Termoli 3 and the FIRE (fully integrated robotized engine), for the first time Fiat designed and developed jointly the product, the plant, and the manufacturing system. The FIRE manufacturing system at the Termoli plant produced engines from 1,000 to 1,300 cc, and represented the first example of HAF (highly automated factory)[7] to be recognized worldwide. The FIRE system was less flexible in terms of product mix and plant convertibility. But its relatively dedicated lines allowed higher productivity and efficiency. This is the reason why it represented an important shift from an over-abundant concept of flexibility to a more moderate one, determined by market requirements in terms of product variety and variability. As a consequence, if compared with Robogate and LAM, FIRE marked a return to a linear operational sequence. The only manual operations maintained in the cycle consisted mainly of some minor assembling activities (filters) and fluid insertion, with an automation ratio as high as 85 per cent.

On the whole, the Termoli plant undoubtedly conferred on the Fiat Group an outstanding technological and manufacturing status, still recognized by most of its competitors. The new FIRE engine enabled Fiat to reduce by 10 per cent the list price of its best-selling Uno model, previously equipped with the traditional 903 cc engine. FIRE was much better also in terms of both quality and performance. These benefits came from such features of the FIRE manufacturing system as reduction of the number of components by 30 per cent, employment cutback of nearly 40 per cent, a halved manufacturing lead-time[8] and a production schedule of 1,000 engines per shift on three daily shifts. As is easily understandable, the FIRE engine and the Termoli 3 plant were a great success for Fiat. The competencies in terms of automation technology, organizational solution, and human resource management policies, as developed at Termoli, led Fiat management to ask the question whether the technological and organizational know-how so successfully developed and implemented at the Termoli plant in engine manufacturing could be applied (by analogy) to a final assembly plant. Fiat management was thus induced to redesign and restructure the Cassino assembly plant. In other words, the set of core competencies associated with the HAF philosophy were adapted to the Cassino plant, where the Tipo model bodies were originally manufactured and assemblied. The Tipo had been produced there ever since its appearance in September 1987, but the managerial decision to implement the HAF in Cassino dates from January 1988. The 2,000 bn. lire investment was intended to go far beyond a single model life-cycle, as Fiat Auto CEO Vittorio Ghidella said. But Ghidella's idea that automation was the crucial weapon to keep up with the Japanese, together with the not completely successful results of such models as the Fiat Tipo and Tempra, and the Lancia Dedra, did not prove to be correct. The adaptation of the HAF concept to the Cassino assembly plant entailed huge implementation problems. It was soon clear that logistics was going to play a crucial role in the whole system, and that the degree of complexity to be governed was too high. Considering that there were many more manual operations than in Termoli[9] and notwithstanding the 450 robots installed (out of a total of 1,933 robots in Fiat Auto in 1988), it is understandable why Cassino productivity has been systematically lower than expectations, even when market conditions were favourable.

The organizational problems encountered, combined with the sales of the Tipo, which did not meet Fiat's high expectations, prevented Cassino and the Tipo from reaching the status which Fiat expected. Overall, it seemed that the systemic and holistic approach adopted in designing and implementing the HAF at the Cassino plant, on the one hand, underestimated the complexity of the processes to be governed, and on the other hand, hindered the possibility of fine-tuning the manufacturing process and triggering organizational learning. Labour flexibility is critical (in assembly plants even more than in engine or other part manufacturing) in order to diagnose and solve quality-related problems. In fact, line workers not only truly embody the tacit knowledge necessary to react immediately and simultaneously to emerging troubles and problems, but can also codify these competencies, transforming them in working practices and shared knowledge (Adler 1993).

But this attention to human resources and labour relations was lacking until 1989. This awareness that human resources are, especially in assembly plants, not merely instrumental to plant performance, but that they can be, within a new organizational framework, trouble shooters rather than trouble makers came only with the 1990s (Magnabosco 1994).

## 12.5. THE CHALLENGE OF THE 1990S

Fiat's weaknesses at the turn of the decade have already been pointed out in the first section: somewhat too high a degree of vertical integration, an outdated product line (and a certain slowness in model renovation), a too heavy reliance on its domestic car market, a marked dependence on low segments. Apart from this dangerous concentration of sales, an even more serious issue concerned Fiat profit margins, which were strongly dependent on the domestic market. As sales abroad required higher logistic costs and aggressive marketing policies (lower prices, more advertising, and so on), domestic sales had to finance expansion abroad. Since on many occasions aggressive export policies relied on internal market profits and Fiat market share in Italy (higher than 60 per cent) seemed impregnable, few expected that it would be eroded so rapidly and so far in advance of the opening up of the Italian car market to greater Japanese competition. Industry analysts agree that it was Ford's decision in 1989 to pursue greater volume sales in Italy at the expense of margins that really started the rot. Prior to that, most competitors in the Italian car market had appeared willing to shelter under the high margin umbrella provided by Fiat, traditionally a price-maker, at least for the first four market segments.

All these circumstances moved Fiat to launch an immense reorganization process which involved all the group's activities, the relationships with suppliers, and the dealers' network. This wide effort began in autumn 1989, but some steps had been taken some time before. Quality circles, for example, originated first in 1982, and in 1987–9 as much as 1,000 bn. lire were invested in quality improvement projects. But it was with the Fiat Manager Convention in October 1989, that the whole restructuring project was launched officially. It was a five-year programme (Total Quality Programme), articulated in a number of operational projects which were grouped at three levels. At the first level Fiat put four projects regarding the whole company: time-to-market reduction, new product development decision-making, product-process carryover, product quality information systems aimed at cross-plant quality measurement. At the second level, Fiat developed eighty projects regarding organizational processes in different areas. The most important among them was the integrated factory project. At the third level, over 200 projects were developed on microprocesses and specific activities.

In 1991, in connection with the appointment of Paolo Cantarella as CEO of Fiat Auto, the Total Quality Programme was reorganized into a series of twenty plans called Competitive Improvement Projects with specific assignment of responsibility for times and results. Before carrying out a detailed analysis of the integrated factory programmes, it is necessary to describe briefly the other components of this comprehensive restructuring project, as instances of its size and scope. As regards new product development, Fiat decided to invest 25,000 bn. lire in the 1992–6 period and 40,000 bn. lire until the year 2001. This investment should lead to twenty new models before the end of 1996. It is at the midway stage that the most important part of this plan must come into being with the complete renewal of the product ranges of the three makes, Fiat, Lancia, and Alfa Romeo and the return to important market niches as the production of spider and coupé models by Fiat and Alfa Romeo and a minivan produced in conjunction with Peugeot. This plan, in the past, would have appeared overambitious, considering that the process of designing and testing a new model took on average four to five years. But recently Fiat has restructured its product design organization according to Japanese principles (Clark and Fujimoto 1991; Cusumano and Nabeoka 1992). New emphasis was put on 'simultaneous engineering', 'co-design', and 'carry-over'. The internal organizational structure was revised according to matrix principles. Project/product managers got wider responsibilities, tasks, and autonomy.

On top of these fundamental aims the Range/Product Plan (RPP) looks to revamp Italian cars' style, whose form has suffered at the hands of a somewhat conservative approach to design and a policy of assimilating models to others in the same range. An important step in this direction is the new cooperative relationship between Fiat's in-house designers and outsiders. One example of this is the new Fiat Coupé that has recently been launched commercially with the cooperation of Fiat and Pininfarina.

TABLE 12.1. *Fiat Auto vertical integration*

|                 | 1987  | 1992  | 1997  |
|-----------------|-------|-------|-------|
| Make            | 38.0  | 35.0  | 30.0  |
| Buy captive*    | 14.0  | 18.0  | 17.0  |
| Buy non-captive | 48.0  | 47.0  | 53.0  |
| Total           | 100.0 | 100.0 | 100.0 |

*Note:* *Bought from captive companies.
*Source:* Fiat Auto.

## 12.6. LEAN ORGANIZATION WITH SUPPLIERS AND DEALERS

In its attempt to recover its competitive edge, Fiat is targeting the sphere of component manufacture. The fact that Fiat, until the second post-war period, was the only domestic example of a large automobile manufacturer, combined with the low technological level of the Italian mechanical industry, has made it necessary for the evolution of the company to be based on markedly vertical integration. This type of development has already undergone profound changes, above all through making the group's (captive) component suppliers more autonomous, not only with the aim of making themselves more competitive, but also to increase the volume of direct sales to automobile producers worldwide. This aim can be said to have already been accomplished for the leading firms that produce components, such as Magneti Marelli and Teksid; a noticeably large portion of their turnover comes from sales outside the Fiat group.

What remain outstanding are decisions regarding the ratio of components bought from outside suppliers to those produced in-house and, above all, the establishment of some form of cooperation with these suppliers. These moves are working towards a relationship based on 'partnership' with the supplier. The outcome of this partnership-building, that began in 1990 and is now in an advanced phase, can be seen in the sharp rise in the number of components purchased (as opposed to produced), but above all in the amount of component design and innovation assigned to the outsiders. Fiat Auto already buys the majority (around 65 per cent) of components from outsiders, of which 50 per cent come from businesses in no way financially linked to the group. These figures make Fiat appear a less vertically integrated company, much like the Japanese car producers (Table 12.1). This picture is partially altered by the fact that a large proportion of these components, despite being produced outside, are actually designed by Fiat itself, who then commissions the pieces to be produced exclusively for Fiat, based on its own designs. In fact, if one looks at the value of the parts supplied in terms of their different sources, the degree of verticalization will fall to 30 per cent at the end of 1997 (Table 12.2).

TABLE 12.2. *Fiat Auto components design*

|                 | 1991 | 1992 | 1993 | 1994 | 1995 | 1997 |
|-----------------|------|------|------|------|------|------|
| Internal design | 78.0 | 70.0 | 60.0 | 50.0 | 40.0 | 30.0 |
| External design | 22.0 | 30.0 | 40.0 | 50.0 | 60.0 | 70.0 |

*Source:* Fiat Auto.

These new partnerships which Fiat intends to form with suppliers[10] operate through making reciprocal concessions, which requests not only innovative ability but also risks (investment) and sacrifices (in terms of progressive cost reduction). Moreover, Fiat offers the best suppliers the opportunity for steady, sustained growth, be it in terms of acquiring know-how through co-design or in terms of economies of scale. In the case of the Uno model, launched in 1983, the proportion of components externally designed was 30 per cent. This figure has risen to 45 per cent for Punto and is roughly 60 per cent for the K model of the Lancia released at the end of 1994. Fiat's process of deverticalization can be characterized as a concentration. In the future, the Turin-based company will not only purchase a higher fraction of the value of the automobiles it manufactures and sells, but will also concentrate its purchases on a more restricted circle of partners, whose task will be that of coordinating second-tier suppliers (subcontractors).

By far the most important part of this process is called guided growth. The guided growth scheme aims at inducing quality improvement on several fronts (from design to adaptability in response to orders) and control of costs. Within the guided growth scheme the supplier is appointed and can assume the title of 'Fiat partner', and long-term contracts follow. The conclusion of this guided growth project can be seen in the move towards providing self-regulated supplies (*autocertificazione*). This means that the components will be guaranteed by the supplier itself. This makes just-in-time deliveries possible, since the *autocertificazione* eliminates the process of quality control upon arrival at Fiat, and the pieces can be immediately employed in assembly. For new models this practice is applied, as in the case of the new Punto and the new family of engines manufactured at Pratola Serra. As a consequence, the overall fraction of components supplied according this *autocertificazione* approach has reached 90 per cent by 1995.

At the beginning of the project, given Fiat's long tradition of high vertical integration, the shift to JIT practices encounters many technical, social, and cultural resistances. These issues are particularly critical during recessions, when market difficulties may incite opportunistic or at least elusive behaviour by both car makers and suppliers. For example, the first-tier suppliers (partners) sometimes dispute with Fiat about who should take care of JIT malfunction-related costs. But all these kinds of difficulties, easily understandable in such a huge transformation process, seem to be on the way towards resolution. In the new Melfi plant the suppliers are servicing the assembly line of the new model Punto with a 'synchronous *kanban*' system.[11] Synchronous *kanban* means that the delivery of parts must arrive in the same sequence as the bodies processed by the assembly line. The supplier has only three to four hours in which he can make the delivery: the time a specific body takes to go from the paint shop to the point when the component must be utilized in the assembly. The value of the fraction of components supplied through synchronous *kanban* will be approximately 42 per cent of the total, an amount similar to, if not higher than, that realized at the new Toyota Miyata plant in Kyushu. Many partnership programmes are targeting the dealers' network. The starting-point is the realization of customer satisfaction, regarding both Fiat cars and dealers' service. While in the past the incentives for dealers were based exclusively on volumes of sales, today Fiat aims at relating incentives to the Customer Satisfaction Index. The main objectives are reduction in stock, order processing time, dealers' turnover, and an improved concern for quality among dealers by defining ad hoc improvement programmes. Many of the considerations proposed in the previous section for the relationship between Fiat and its suppliers can be applied to the relationship between Fiat and its dealers.

During the 1980s, Fiat implemented a commercial strategy based on a systematic expansion of the number of dealers in the domestic market, and in other European countries. Underlying this policy was Fiat's assumption that its products were competitive. As a consequence, the efficiency of the dealer network was to result from strong competitive confrontation. But competitive relationships existed not only within the networks of competitors, but also within Fiat network. As an implication, dealers had to focus their efforts on an exclusive target: market share gains. In fact, Fiat's incentive structure was centred on bonuses based on volumes. This strategy allowed a spectacular growth of market share for Fiat in the second half of the 1980s, when models were still quite new and appealing. In 1988 and 1989 Fiat became the market leader in Western Europe. But this leadership, together with the difficulties of governing a much wider dealers' network, also caused some pitfalls, especially in terms of quality and customer satisfaction.

Since 1990, Fiat distribution strategy has been oriented towards two objectives. The first was creating, developing, and maintaining customers' satisfaction in order to win their loyalty, and the second was studying, testing, and diffusing new procedures for effective partnerships with dealers. These objectives have been pursued by means of a reduction in the number of dealers in the various national markets. For example in Italy, from 1989 to 1994, the number of Fiat-franchised dealers decreased from 852 to 617 and the same trend affected Lancia and Alfa Romeo networks. There was an opportunity for small Fiat Auto dealers, operating abroad, to become multi-franchise dealers, adding to the franchise of Fiat that of Lancia or Alfa Romeo. Moreover the Fiat Marketing Institute had the task of systematically searching for excellence in commercial activities and then transferring this to the network through integrated educational efforts. But the most relevant lever of the partnership programme is represented by the redefinition of the incentive mechanisms. Instead of centring them on market shares, Fiat targeted customer satisfaction as a key variable. As a consequence, the dealers' incentive scheme is linked to the actual level of service provided to the customers.

## 12.7. THE INTEGRATED FACTORY

In order to understand how Fiat's comprehensive strategic change is affecting industrial relations strategies, human resources practices, and work organization concepts, it is necessary to focus attention on the most important organizational project: the Integrated Factory (IF). The IF originated as a necessary aftermath of (and as an attempt to overcome) some of the mismatches which emerged, especially at the Cassino assembly plant in the highly automated factory (HAF). The basic concept of this new plant organizational model is 'integration', that is, the emphasis is put on organizational flexibility, coordination, and improvement, as automation ceases to be the only driver of productivity (Camuffo and Volpato 1995). With IF, organizational, industrial relations, and human resource management choices are rediscovered as key constituents of the firm's competitive performance.[12]

The IF model was initially designed in 1990. Fiat chose to implement it almost simultaneously in its plants rather than test it in a pilot plant. During 1991 only partial applications took place at the Termoli and Cassino plants, but Fiat extended the model to all its plants during 1992 (beginning with the Rivalta assembly plant). With the integrated factory, substantial delayering and decentralization is supposed to take place in the plant organizational structure: 'macro-sets' of activities are allocated to operational units (*unità operative*), by aggregating previous machine shops (*officine*). Functions which were formerly centralized, such as maintenance, material planning, and so on, are decentralized at the level of operational units. In fact, each operational unit is articulated and specialized in two areas: operations and production engineering. Production engineering is therefore a rather decentralized pool of technical competencies, working as a support to operations. Through delayering, two out of seven hierarchical layers were eliminated, including the so-called *capireparto*: the most traditional shop steward figure. Plants should achieve a flatter organizational structure. This should enable both cost reductions and, when associated with integrators, more horizontal (less hierarchical) coordination.

The key element of the integrated factory is the Elementary Technological Unit (ETU),[13] defined as a unit which governs a segment of the process (a technological subsystem), in which such activities as prevention, variance absorption, self-control, and continuous improvement are carried on, in order to achieve the firm's goals in terms of quality, productivity, costs, and service. The main thrusts of the ETU are appointing the solution of problems at the lowest possible level (resources and skills are placed so that problems may be solved as and where they occur) and facilitating product and process quality improvement by systematically incorporating organizational learning developed in the workplace. At the same time, the ETU should allow leaner and smoother manufacturing processes, reductions in cycle times, flexibility enhancement, and lower costs. ETUs are upstream/downstream related to one another with a supplier/customer-like relationship. Within the IF model, teams are therefore conceived as integrators, that is, as temporary organizational units aimed at trouble shooting (e.g. line breakdowns) or organizational devices enhancing horizontal coordination, information sharing, and organizational learning. In order to understand what role teams play in this organizational model, it must be pointed out that, within the production engineering unit (staff of the operational unit) other new jobs and skill profiles emerge with the integrated factory: the line technologist, whose task is that of supporting the ETU chief in training and respecting time and cost targets, the technology specialist, whose task is that of technical anomaly diagnosis and solution (referred to a given technology area, e.g. electronics, and so on), the product/process technologist, whose task is that of continuous improvement, process reliability, and product quality, as well as participation in new or modified product engineering.

At the shopfloor level, a team is therefore made up of the three technologists, the ETU chief, a procurement manager, and a maintenance and quality manager. The team's task is problem identification and solving. Once the problem has been discussed, the line technologist will then cooperate with technology specialists. These teams are not autonomous work groups (like those in the socio-technical tradition or similar to those adopted at Alfa Romeo during the 1970s, or at Volvo during the 1980s) but organizational mechanisms of interfunctional integration, aimed at governing complex interdependences, at activating lateral relations on purpose, that is, when specific problems arise on the process segment considered. Another interesting aspect of Fiat plants' restructuring is the relocation of white-collar workers and offices. Especially in the new, greenfield, plants, most of the white-collar operatives and services have been moved to the shopfloor level, close to operations. On the new organizational approach this should imply better integration and improved internal services.

The most recent research, seems to suggest that the organizational innovations underlying the IF model cannot be successful if the unions do not accept them (Bonazzi 1993; Pessa and Sartirano 1993; Rieser 1993, Camuffo and Volpato 1995). Innovative industrial relations are necessary not only to facilitate and speed up the implementation of the IF, but also to avoid the negative implications of the recession of the early 1990s hindering the massive restructuring effort. At the moment, there seem to be hints and clues of new bilateral relationships between management and unions. This new attitude is the result of an evolutionary process rooted in the 1980s, when managerial unilateralism and concession bargaining dominated Fiat's industrial relations. A chronological threshold can be found in the 18 July 1988 firm-level union agreement, linking wage increases to firm performance. The accord on PPG (*Premio Performance Gruppo*) largely reflected the strategic, organizational, and social context characterizing Fiat Auto and the whole group at the end of the 1980s: a flourishing financial situation (in 1990 FIAT Auto return on investment was still 17.1 per cent), a good employment situation (the state redundancy fund had been cleared and plants were working at almost full capacity), and union pressure for wage increases.

The PPG accord represented a turning-point since, as analysed above, the industry scenario and Fiat competitive situation rapidly worsened after 1990. Industrial relations became more complex for a number of reasons. The need to lower break-even points and to enhance competitiveness led to cost reductions, externalization of activities, and personnel downsizing not only in manufacturing units but also, and above all, among white-collar workers, in central staff and services (Auteri 1994). The need to manage a situation in which the industrial strategy simultaneously called for restructuring and downsizing in some plants and investment in new plants required the design and implementation of different and segmented IR policies. Italian unemployment made Fiat decisions concerning redundant workers critical, especially in some areas (Turin, Milan, Naples)[17] where their social impact could be disruptive.

Fiat's industrial relations in the early 1990s were characterized by a certain degree of innovation. The landmarks were the 27 March 1993 accord at Mirafiori for the introduction of a third (night) shift on the production line of Punto; the 11 June 1993 accord on industrial relations at the new greenfield plants at Melfi and Pratola Serra; the 20 February 1994 accord on the implementation of a Fiat five-year industrial plan (workforce rightsizing). At this time of labour redundancy and lay-offs, staffing does not represent a problem, except for the new plants. At Melfi and Pratola Serra, Fiat workers' recruiting policies are innovative. The selection criteria, used as 'port of entry' into the internal labour market, are extremely rigorous. First of all, the recruits are young. No new employee can be older than 32. Both for line workers and plant level professionals, the emphasis is not only on skills and learning potential, but also on the psychological and social traits of the candidates. Loyalty, a cooperative attitude, the capability to interact and absorb stress are only a few of the new abilities Fiat is seeking. These recruiting policies are aimed at avoiding mismatches between the new organizational philosophy (which emphasizes workers' flexibility, teamwork, cooperation, and so on).

This focus on recruiting is associated with a new effort in training. In fact, as focused selection of candidates allows turnover reduction and higher workers' motivation, investment in training tends to be, according to Fiat, more effective and less risky. Training expenses have been boosted by the introduction of new work organization practices. For example, training expenses as a percentage of salaries were 0.75 per cent in 1992 (0.55 per cent in 1990), while, on average, in 1992, 47.9 per cent of the employees were involved in training programmes (compared with 43.3 per cent in 1990). Moreover, the new recruiting policies are instrumental to this cultural shift, given their prevalent focus on a relatively young and well-educated workforce. Nevertheless, this creates a generational gap which can jeopardize the climate and social equilibrium within the firm.

At the Rivalta assembly plant, for example, there has been a certain degree of turnover among Integrated Process Conductors (IPC), Integrated Process Operators (IPO), and middle managers. Some young and educated people, recently recruited, have been appointed ETU chiefs. Moreover, the most important selection criterion for IPCs and IPOs, apart from age, is relational ability (dialogue, communication, collaboration, and so on) with other workers. These jobs require a great deal of flexibility and much overtime. Strong pressure is also put on ETU chiefs, who are often subject to a sort of selection by survival.

Throughout the 1980s, Fiat labour costs have consistently grown, in part as a result of wage indexation, in part because of collective bargaining at the national and firm level, and in part because of Fiat compensation policies. These labour cost increases were covered by the productivity gains realized with the implementation of the highly automated factory (at least in engine manufacturing). But in the late 1980s, as the first signs of weakening emerged, Fiat management began thinking about how to make compensation and wages more flexible. Owing to the rigidities inherited from the 1970s, Fiat had, throughout the 1980s, put much effort into widening wage differentials among the classes of the national level job classification scheme. Ad hoc compensation policies were designed for shop stewards, professionals, and middle managers with the explicit intention of 'reconstructing' a hierarchy among jobs.[18] But now, the IF organizational model seems to require very different policies. Delayering implies less hierarchy and thus less 'vertical' wage differentiation. The new professional figures, and the continuous changes in the organization of work, together with the application of the team concept, make it impossible to rigidly attach compensation to jobs. Knowledge, competencies, and performance are becoming the new drivers of the reward system. Moreover, the need for highly skilled workers implies a concentration of the workforce in the central and higher classes of the job classification scheme.

Bonuses, merit pay, and pay for performance are used at Fiat plants. However, they still follow the guidelines that emerged during the 1980s (Cerruti and Rieser 1991), and do not seem to represent a relevant fraction of compensation. For blue-collar workers this element of the compensation packages depends on shop stewards' performance appraisals, which are based on such criteria as seniority, loyalty, flexibility, willingness to do overtime, to change shifts, to intensify work. In some cases wage increases take the form of *una tantum* (spot bonuses), in others that of permanent increases (*superminimi*). Other reward policies are linked to quality circle activities. The number of workers participating in the suggestion system (as a percentage of the total population) is low if compared with Japanese car makers. However, at least in some plants such as Cassino and Rivalta, results are promising if the recent introduction of the system is taken into account.

Union representatives also point out that suggestion-related incentives are inadequate in effectively stimulating participation. However, confederal unions did sponsor the suggestion system. For example, at the plant level, unions autonomously instituted prizes (mountain bikes, television sets, and so on) for those

330                    *Europe's Dilemma*

suggestions covering particular issues: especially quality of work life and safety issues. This 'parallel' incentive structure was meant as complementary, and not in contrast with that of Fiat.

12.8. WHAT IS LEAN AND WHAT IS NOT

As previously pointed out, the IF can be seen as a crucial component of the manufacturing paradigm-shift taking place at Fiat. From a theoretical standpoint, an interesting question is how much of this change resembles a move towards the lean production model. Another interesting problem is whether this change is the mere result of an attempt to adapt to institutional or economic/competitive pressures, or whether it depends and is shaped by internal factors, rooted in the firm's organizational heritage and endowment of competencies. From this standpoint, it is important to look at those areas in which the adoption of the new organizational model generates tensions and contradictions, and which factors are causing these difficulties. It goes without saying that, given the stage at which Fiat is, evaluations be suspended awaiting developments. This will allow a deeper understanding of the dynamics taking place. Considering the organizational structure implied by the IF model, delayering requires a great deal of cultural shift. The traditional shopfloor authority-based setting (vertical communication, hierarchical coordination, and so on), which generally inhibited cooperation and horizontal information flows, has been replaced by integrating mechanisms such as teamwork within ETUs and Technological Teams.

Another key aspect of the IF is decentralization. Decision-making is supposed to take place at the lowest possible organizational levels. Direct workers should take on responsibilities and be autonomous as regards some aspects of production. Despite this approach and intentions, a number of contradictions arise. For instance, the IF model entails the Japanese practice according to which workers could and should stop the assembly line in order to anticipate or solve problems as they occur. But this does not happen systematically, given the resistance of both workers and plant managers, who tend to stick to the traditional hierarchical practice of activating upper levels. In practice, lines are seldom stopped by workers. Rather, plant managers continue to decide about line-stops, and their traditional concern about quantity goals tends to dwarf attention to quality. For example, at the Rivalta assembly plant little decentralized decision-making is taking place. The ETU chief is the lowest hierarchical layer at which line stoppages can be decided. However, in the assembly line set up for the production of the Lancia Dedra (which was moved from the now closed Chivasso plant in 1992), workers frequently stop the line because of quality problems. These breakdowns, although implying production inefficiencies, eliminate some typical low-quality related activities and costs: for example, those jobs taking care of defects and poor quality-related problems at the end of each workstation. Moreover, another factor hindering decentralization is that the average worker does not want to take responsibility for stopping the line. The ETU chief himself often tends to avoid the responsibility and calls the operational unit manager or the plant manager.

The implementation of the IF model is not facilitated either by the layout of older plants. Contrary to those who see the IF as a model able to maximize organizational transparency and to eliminate information asymmetries, the complex system of information displayed in every ETU is often ignored by blue-collar workers, probably because of time pressure and difficulties in understanding and updating (Bonnazi 1993). The workload is perceived as being too heavy to allow systematic use of all these instruments which, in some cases, are seen as 'useless paperwork'. Moreover, especially when the workforce is old and unskilled, the interpretation of all these fancy management techniques causes frustration and misunderstandings.

Cerruti and Rieser (1992) pointed out that quality data are sometimes disguised or not detected, and that workers' or ETU's quality self-regulation and certification is only theoretical. These frustrations sometimes emerge too in IPCs and IPOs, first of all because they trigger and activate teams in order to solve the problems they detect on the line, but are not actively involved in problem diagnosis and solving, which is carried out by technologists and ETU chiefs. Secondly, sometimes IPCs and IPOs tend to misinterpret their role, acting like quasi-hierarchical figures. Overall, structural changes are conspicuous from a quantitative standpoint. From a qualitative standpoint, hierarchical coordination has been only dampened by integrators and other horizontal coordinating mechanisms (teams, management by sight, and so on). The new model impacts, in terms of job design and work organization, mainly on higher technical-hierarchical levels (middle managers and ETU chiefs), while only a small percentage of workers (IPCs and IPOs) see a significant transformation of their roles. None the less, in the last fifteen years, significant changes in skill profiles and jobs have taken place as a result of the cumulative effects of the HAF and the IF implementation.

The turnaround programme announced by Fiat Auto at the end of 1989 featured an exceptional investment effort. The thrust of this investment programme translated into a substantial renovation in the product line offered by different makes within the Group. Renovation of the three main makes (Fiat, Lancia, Alfa Romeo) featured the launch of twenty new models, and was completed in 1996. The most important product has clearly been the Punto. It strengthened the company's resolution to carry on along this path, which was not without risks given the financial resources at stake, thanks to the immediate success achieved both in the domestic and in the international markets (Volpato 1996).

The recovery, arising from the rationalization across the components industry, design, manufacturing, and distribution is visible in the lower break-even point, that is, in the reduction in the number of vehicles necessary in order to balance costs and revenues. Before the cure implemented in 1990–3, the break-even point was set at 1.9 million vehicle sales. Now it is set at 1.4 million. This outcome has been achieved both by outsourcing some components (i.e. joints and engine shafts) and by considerably reducing fixed costs. This has also yielded a sizeable reduction in employment. This took place partly among blue-collar workers (both through more efficient plants and by profound changes in work organization), but the main reduction took place among white-collar workers, where retired personnel have not been replaced. The overall downsizing involved 26,000 workers, with 15,000 retired, of which 3,000 were agreed retirements and 8,000 early retirements, plus 6,000 exits after some operations were sold. With the reduced overheads in 1994, it was already possible to achieve a little more than 100 bn. lire in net profit. Shortly, looking at partial results for 1996, a net profit for 1,000 bn. lire can be estimated.

However, these data still underestimate the Group's potential for expansion since models launched after the Punto had not yet yielded their complete results. There are great expectations for the vehicles derived from project 182, meant to replace the Tipo. These are two models sharing the same chassis and design principles, but featuring different bodies with three doors (Bravo) and five doors (Brava). These two models required a 1,500 bn. lire investment, with sales objectives of 400,000 units per year (60 per cent exported). At the end of 1996 they are going very well, so Fiat should regain the European market share it conquered in the past, which was then eroded in 1992 and 1993.

One of the most significant dimensions of the Fiat turnaround is linked to the process of internationalization of the Group. In the past, the company's fate was necessarily linked to the domestic market. It was the Italian demand for vehicles which led sales volume and profits. Since 1994 this picture appears to have changed. In 1991 Italian sales were over 50 per cent, while sales outside Europe were lower than 18 per cent. In 1994 Fiat domestic sales were 35 per cent, while 30 per cent of production was sold in the rest of Europe, and 35 per cent outside Europe (mainly Argentina, Brazil, and Turkey). Outside Europe, the main strength of Fiat's strategy is linked to investments for the 178 project. The specialized press has talked about 178 as a utility passenger car, of similar size to the Punto, but engineered specifically for these developing markets. As a matter of fact, the 178 project is more than a vehicle. It is actually a family of models, sharing the same chassis, specifically reinforced to endure a 'raw' use in difficult road conditions in such countries, and which will cover a range of different body and engine combinations, including a van for light transport. The first model of the family, named Palio has been initially manufactured in the Brazilian plant of Belo Horizonte, but it will also be produced in Argentina, where a new plant is being established. Anyway, intentions aim at making it a *world car* to be marketed (and possibly manufactured) in a variety of countries, with an estimated production level of 800,000 units.

## 12.9. CONCLUSION

Leaving aside the ambiguities related to the definition of lean manufacturing, the organizational model based on the IF concept can be considered as a hybrid model, with strong differences as far as its implementation is concerned, across Fiat plants. These divergences from the Japanese manufacturing model derive from differences in the institutional and cultural context (which can nevertheless be reduced as proved by successful Japanese transplants in the USA and Europe), in the difficulties of adapting 'Japanese' techniques generally owing to misinterpretations and biases based on management and union cognitive schemes and problem framing, the non-compatibility between Fiat's past trajectory (and the related background of competencies) and a new evolutionary avenue purely and merely based on Toyotaism.

This last point seems particularly intriguing. Firms' trajectories are not only the result of adaptation to exogenous pressure. Rather, they are based on internal factors and are history dependent. Enterprises differ from one another in terms of organizational knowledge, that is, of resource endowment and repertoires of capabilities (Nonaka 1991). Organizational knowledge is incorporated in tangible and intangible assets (sunk costs and routines), and results from learning, that is, from the development by means of risky choices and actions of competencies *vis à vis* the institutional and competitive pressures posed by the environment. The learning processes shaping organizational knowledge are firm-specific and, at least in part, cumulative and non-reversible. As a consequence, each firm develops a given set of capabilities, different in terms of scope and depth. But this peculiar and unique set of competencies largely determines also the scope of the strategic and organizational alternatives available. In other words, firms' evolutionary patterns are path dependent.

The emerging organizational paradigm largely depends on the configuration of Fiat's organizational knowledge as derived from its history. Moreover, the Turin-based company shaped its competitive strategy in the 1980s according to the 'smart machine' approach (Zuboff 1988) or, as recently argued by Bonazzi (1993), by 'information technology based' neo-Fordism. Many scholars criticized this technology strategy and pointed out its excess emphasis on automation (Cerruti and Rieser 1992), and this chapter has itself shown some of the excesses and inconsistencies underlying Fiat's manufacturing automation. Nevertheless, the HAF model (built on the experiments of Digitron, Robogate, LAM, and so on), allowed the development of state-of-the-art knowledge in process technology, that is, the acquisition of top know-how and capabilities in engine manufacturing and body assembly. And these capabilities can now be applied to the new plants in South Italy. In fact assembly operations at the new Pratola Serra engine plant can be seen as an intelligent evolution of the LAM system.

Fiat's industrial relations strategy based on managerial unilateralism and concession bargaining was a key success factor during the 1980s. However, this policy left a heritage of mistrust and a legacy of tensions and conflict in union–mangement relationships that is only now beginning, slowly and painfully, to fade. Middle managers (especially at the plant level) were a privileged target of HRM during the 1980s. Ad hoc policies in terms of compensation, incentives, careers, and so on were designed and implemented in order to regain middle management's loyalty and commitment, and rebuild a solid and reactive hierarchy.[14] But now these policies represent, paradoxically, one of the largest obstacles to implementing organizational change. Middle managers (but also workers, especially those with long tenure within the firm) still behave as they have always done, and resist change. Problems continue to be framed in terms of the old cognitive schemes which were consolidated and reinforced by incentive and success during the 1980s. However, the design of the IF organizational model proves that Fiat has learned the lesson that advanced technology had to be carefully matched with innovations in organizational and human resources practices. As a result, the key issue for the Turin-based company is having 'smart people around the machine' and, even more importantly, participating people. The IF model also shows that Fiat put much emphasis on organizational design and engineering, focusing on maximizing information diffusion throughout the organization. In other words, the Integrated Factory is a comprehensive attempt to make the organization transparent. Its guiding principle is the systematic reduction of information asymmetries: among workers, between management and workers, and between management and the unions.

To summarize, Fiat has envisioned the restructuring process as an 'organizational revolution' or 'radical organizational change'. This may entail all the positive outcomes of a major leap, but all the troubles associated with a holistic and comprehensive plan. In order to catch up, Fiat is pursuing a leapfrogging strategy, and is facing a phase of upheaval which partially contradicts the Japanese formula based on incremental processes and gradual changes.

# NOTES

The authors wish to thank Leonardo Buzzavo, research assistant, for valuable help, and members of the GERPISA international network and MIT-IMVP for stimulating discussions. They also gratefully acknowledge comments by Harry Katz, Tom Kochan, and Richard Locke. Financial support for this research was provided by the GERPISA programme and by Consiglio Nazionale delle Ricerche.

1. The high fiscal pressure on cars and gasoline drove Fiat to specialize in lower car segments. See Volpato (1993).
2. The 1970s saw a strong wave of terrorism throughout Italy—i.e. the Red Brigades.
3. In 1970 a labour law called *Statuto dei lavoratori*, inspired by the US Wagner Act, legitimated new union rights.
4. With the *Inquadramento Unico* there was a reduction of job categories from thirteen to seven with a unique grading for white- and blue-collar workers. And each category was made wider by the metalworkers national agreement of 1973.
5. The rise in labour costs, generated by the introduction of wage indexation (*scala mobile*) and national agreements with unions (1969, 1973, and 1976), affected Fiat's productivity

to a greater extent than other European car manufacturers. A major role in this decreased efficiency was played by complementary compensation (social security). According to Fiat calculations it accounted for about 45 per cent of labour costs (1977) in comparison to 32 per cent for France, 32 per cent for West Germany, and 16 per cent for the UK (Fiat 1978).
6. In the late 1970s, Cesare Romiti became CEO of Fiat Group, while Vittorio Ghidella was CEO of Fiat Auto.
7. FAA in the Fiat jargon stands for 'Fabbrica Altamente Automatizzata'.
8. From 231.5 minutes for the previous 903 engine to 107.5 minutes for FIRE.
9. According to the Fiat figure, Cassino automation in assembly activities was 22 per cent (1,000 automated operations out of 4,500).
10. Fiat policy regarding choices between make or buy and relations with suppliers has known various phases. See Enrietti (1987); Enrietti et al. (1991).
11. In the development of the new assembly plant at Melfi (South Italy) approximately 700,000 sq. m. are dedicated to suppliers' facilities.
12. This is, according to Fiat HRM director Maurizio Magnabosco, a radical change and a discontinuity in Fiat's management philosophy. Since work organization becomes a key competitive factor, the role of industrial relations and human resource management policies tends to be strategic.
13. In Fiat: Unità Tecnologica Elementare (UTE). These manufacturing cells have various sizes according to the operational unit in which they are included (25–30 workers in mechanics, 40–70 in assembly). They are related to one another in a supplier–customer fashion, are relatively autonomous and self-regulating, and are not something peculiar to Fiat. Rather, they resemble the *kumi* used at Toyota and the *ka* at Mitsubishi (although the size of these organizational units is smaller than that of UTE). Toyota's *kumi* are, in fact, relatively autonomous and self-regulating organizational units, performing a consolidated task.
14. The ratio between the average salary of the workers included in the highest level of the job classification scheme (7th super) and the average wage of the workers included in the lowest (3rd) was approximately 1:3 in 1977, and 1:72 in 1993.

STATISTICAL APPENDIX 12: **Fiat**

| Year | Production automobiles (000 units) | Exports (from Italy) (000 units) | Employees (units) | Net sales (billion lire) | Net income (billion lire) |
|---|---|---|---|---|---|
| 1974 | 1,417 | 686 | | | |
| 1975 | 1,182 | 661 | | | |
| 1976 | 1,336 | 696 | | | |
| 1977 | 1,338 | 644 | | | |
| 1978 | 1,412 | 640 | | | |
| 1979[a] | 1,437 | 647 | 169,457 | 5,810 | −59 |
| 1980 | 1,442 | 511 | 155,418 | 8,017 | −566 |
| 1981 | 1,248 | 405 | 141,076 | 9,579 | −364 |
| 1982 | 1,297 | 437 | 125,970 | 10,433 | −95 |
| 1983 | 1,371 | 491 | 116,397 | 11,888 | 110 |
| 1984 | 1,410 | 480 | 107,681 | 12,878 | 458 |
| 1985 | 1,420 | 449 | 99,764 | 14,392 | 766 |
| 1986 | 1,698 | 603 | 98,976 | 16,384 | 1,533 |
| 1987[b] | 1,967 | 641 | 129,818 | 22,142 | 1,535 |
| 1988 | 2,141 | 686 | 130,899 | 25,454 | 1,764 |
| 1989 | 2,246 | 694 | 134,270 | 28,424 | 1,909 |
| 1990[b] | 2,163 | 742 | 133,431 | 27,675 | 751 |
| 1991 | 1,963 | 638 | 128,925 | 27,506 | 386 |
| 1992 | 2,201 | 550 | 125,378 | 27,446 | 3 |
| 1993 | 2,101 | 453 | 120,338 | 25,049 | −1,756 |
| 1994 | 2,312 | 520 | 123,699 | 32,992 | 103 |
| 1995 | 2,700 | 608 | 114,386 | 39,093 | 583 |
| 1996 | 2,725 | 675 | 116,144 | 42,502 | −193 |

*Notes*: Fiat Auto consolidated data.

[a] Fiat Auto SpA was incorporated in 1979; financial data and employees not available for previous years.

[b] Alfa Romeo was acquired in 1987, Innocenti and Maserati in 1990.

*Source*: Fiat Auto.

# BIBLIOGRAPHY

Adler, P., 'Time and Motion Regained', *Harvard Business Review*, 1 (1993).
Auteri, E., 'Nuove basi organizzative per l'impresa eccellente', *L'impresa*, 1 (1994).
Bonazzi, G., *Il tubo di cristallo* (Bologna, 1993).
Camuffo, A., and Costa, F. G., 'Human Resource Management and Strategic Change: The Italian Case', in A. Pettigrew, L. Zan, and S. Zambon, *Perspectives in Strategic Change* (Dordrecht, 1993a).
—— —— 'Strategic Human Resource Management: The Italian Style', in *Sloan Management Review*, winter (1993b).

—— and Volpato, G., 'Building Capabilities in Assembly Automation: Fiat's Experiences from Robogate to the Melfi Plant', in K. Shimokawa, U. Jürgens, and T. Fujimoto (eds.), *Transforming Automobile Assembly* (Berlin, 1997).

—— —— 'The Labour Relations Heritage and Lean Manufacturing at Fiat', *The International Journal of Human Resource Management*, December (1995).

Cerruti, G., and Rieser, V., *Fiat: qualità totale e fabbrica integrata* (Roma, 1991).

—— —— 'Fiat: aggiornamenti sulla Fabbrica Integrata', *Quaderni di ricerca*, 1 (1992).

Clark, K. B., and Fujimoto, T., *Product Development Performance: Strategy, Organization and Management in the World Auto Industry* (Boston, 1991).

Cusumano, M. A., and Nabeoka, K., 'Strategy, Structure and Performance in Product Development—Observations from the Auto Industry', *Research Policy*, 21 (1992).

Dealessandri, T., and Magnabosco, M., 'Contrattare alla Fiat', in C. Degiacomi, *Quindici anni di relazioni sindacali* (Roma, 1987).

Enrietti, A., 'La dinamica dell'integrazione alla Fiat Auto SpA (1979–86)', in *Economia e politica industriale*, 55 (1987).

—— Ferrero, V., and Lanzetti, R., *Da indotto a sistema: la produzione di componenti nell'industria automobilistica* (Turin, 1991).

Fiat, *Fiat Raffronti 1969–77* (Turin, 1978).

Kochan, T. A., Locke, R. M., and Heye, C. H., 'Industrial Restructuring and Industrial Relations in the US and Italian Automobile Industries', in D. Lessard and C. Antonelli, *Managing the Globalization of Business* (Naples, 1990).

Locke, R. M., 'The Demise of National Union in Italy: Lessons for Comparative Industrial Relations Theory', *Industrial and Labor Relations Review*, 2 (1992).

—— and Negrelli, S., 'Il caso Fiat Auto', in M. Regini and C. Sabel, *Strategie di riaggiustamento industriale*, Il Mulino (Bologna, 1989).

Magnabosco, M., 'La Fabbrica integrata: nascita e caratteristiche', *Personale e Lavoro*, 372 (1994).

Nonaka, I., 'The Knowledge Creating Company', *Harvard Business Review*, November–December (1991).

Pessa, P., and Sartirano, L., *Fiat Auto: ricerca sull'innovazione dei modelli organizzativi*, Fiom CGIL, working paper (Turin, 1993).

Rieser, V., 'La Fiat e la nuova fase della razionalizzazione', *Quaderni di sociologia*, 1 (1993).

Volpato, G., *L'industria automobilistica internazionale* (Padua, 1983).

—— 'Le Système automobile italien', in E. de Banville and J.-J. Chanaron (eds.), *Vers un système automobile européen*, CPE Economica (Paris, 1991).

—— 'L'internazionalizzazione dell'industria automobilistica italiana', in various authors, *L'industria italiana nel mercato mondiale dalla fine dell'800 alla metà del '900* (Turin, 1993).

—— *Il caso Fiat—Una strategia di riorganizzazione e di rilancio* (Turin, 1996).

Zuboff, S., *In the Age of the Smart Machine* (New York, 1988).

# 13

## Peugeot Meets Ford, Sloan, and Toyota

JEAN-LOUIS LOUBET

Over the past thirty years, the Peugeot group has been marked by the most radical changes in its history. The company had long been a specialist automobile producer, and had only entered the mass market towards the end of the post-war boom. In 1974 a strategy of external growth was launched, with the purchase of Citroën and then the European subsidiaries of Chrysler. This was a vertiginous ascent which boosted *Peugeot Société Anonyme* (PSA: the Peugeot Limited Company) into temporarily becoming the largest automobile manufacturer in Europe just as the second oil crisis arrived. However, the extraordinary difficulties of reorganizing and harmonizing three distinct automobile companies were soon revealed, to the point that management became embroiled in a completely different struggle: that for survival. Growth, reorganization, restructuring, and strategic change came in quick succession, as the company encountered the ideas of Ford, Sloan, and Toyota respectively. Perhaps the most unusual aspect of the Peugeot group is precisely its capacity to change. Indeed it may be precisely this capacity for self-questioning that has enabled Peugeot to adapt to an ever changing economic environment for more than one hundred years.

### 13.1. THE PERIOD OF MASS PRODUCTION

Was Peugeot initially a Fordist company? With only one model produced over a period of several years in a single factory coupled with a constantly expanding market, one might think so. Yet if this were the case, Peugeot did not become Fordism's prisoner. With an internal logic based on profitability of investment, a focus on rigorous management, and the maintenance of its own strong culture, Peugeot was capable of very quickly changing direction.

#### 13.1.1. Peugeot and Mass Production

Peugeot was relatively late to adopt mass production. Until 1964 the company had remained a specialist manufacturer of mid-sized models, positioned at the margins of the automobile industry. Since the end of the World War II, Peugeot had clearly opted for profits rather than production volume. For twenty years, the company had offered models which gained a reputation for quality and low maintenance costs as well as profitability, with its 203 (1948), 403 (1955), 404

(1961), and soon to come the 504 (1968). The products were increasingly profitable, since with each model replacement the company shifted to slightly larger, more expensive, and therefore more profitable automobiles. From 1950 to 1965, Peugeot was a model of the successful company; its financial results were the best in the French automobile industry. All investments were self-financed, and Peugeot had no debts. It paid its suppliers immediately and so obtained prices 2 per cent below those of its French competitors. On the social front too the company was exemplary. In 1955, simultaneously with Renault, Peugeot proposed the first 'company accord' with its employees, thus instituting a new contractual strategy of cooperation with the trade unions.

Despite this indisputable success, the group's chief executives, Jean-Pierre Peugeot and Maurice Jordan, remained vigilant. Aware of constraints on market growth, as well as of the tendency towards industrial concentration within the automobile sector, they had analysed the limitations of their strategy, namely the problem of relying, over the long term, on products that were shifting further and further towards the upper end of the market. Peugeot failed to evolve into a French Mercedes-Benz for several reasons: inadequate purchasing power in France, heavy taxation of automobiles, a still non-existent motorway network, and poor exports. The company's health and its continued existence were instead to depend upon a new strategy in which productivity increases had to be achieved through volume production. In 1965, Peugeot launched the 204, a small car model which repositioned the company at the heart of the national and European markets. This was a cultural revolution for Peugeot, which thus became a generalist company and eventually a mass producer. The launch of the 304 (1969), 104 (1972), and 605 (1975) revealed that Peugeot now planned to base its strategy on a complete range of different models.

Peugeot's production system evolved commensurately. In 1971, a second assembly site at Mulhouse (Alsace) was opened, in the same region as the original site at Sochaux-Montbéliard (Doubs), in Eastern France. Production capacity rose from 280 to 350 automobiles per day, the minimum thought necessary if the company was not to be swept aside by continually intensifying competition. Yet it proved difficult to realize substantial economies of scale without some form of association with another manufacturer. Having failed between 1963 and 1965 to convince Citroën that the two companies should join forces as equal partners to form a French General Motors, in 1966 Peugeot opted instead to link up with Renault. If marriage between a private company and a nationalized company might at first have appeared unnatural, early results were quite promising: a joint engine factory at Douvrin (FM, or Française de Mécanique), an automatic gearbox factory at Bruay-en-Artois (STA, *Société de Transmissions Automatiques*), and an assembly factory at Maubeuge, not counting production the two groups undertook for each other in their respective factories. To further deepen their cooperation, in 1972 Peugeot and Renault merged research and development activities to create a new four-cylinder aluminium engine and shared drivetrain for their latest models, the Peugeot 104 and the Renault 14. For Peugeot this experience was positive, perhaps even decisive. The transition to mass production and a full product range would never have been as rapid without cooperation with Renault. None the less, difficulties remained. Production capacity was still inadequate (to the point that it was planned to construct a factory in Brazil to supply European markets), quality was often difficult to maintain as output grew, the arrival of a foreign workforce changed social relationships and practices, and financial requirements became such that Peugeot was for the first time obliged to fund its investments from external sources. The volume strategy had revolutionized the company's practices.

### 13.1.2. The First Oil Crisis and the Purchase of Citroën

With widespread inflation, disorder in international monetary markets, and in particular the energy crisis, economic growth appeared to come to a halt in late 1973. Was the volume strategy now inappropriate? The question was posed acutely for Citroën, which was now on the verge of collapse. A prisoner of heavy investment programmes to retain the technological aura of the marque (notably investment to develop the rotary engine), obliged to leave its factory sites in Paris and relocate to the suburbs (in Aulnay-sous-Bois), for too long unable to renew a product range so incoherent that it had lost the company more than 10 points in the French market (falling from 29.2 per cent in 1965 to 18.7 per cent in 1973), and proud to the point of refusing to surrender control to Fiat between 1968 and 1972, Citroën found itself in a disastrous situation, incapable of funding a modernization programme which was now vital. In this context, the downturn of the market in 1974 was catastrophic. Between the first half of 1973 and the first half of 1974, the production of the GS (6–7CV) and DS (11–12CV) models fell by 22 per cent and 47 per cent respectively. By August 1974 Citroën's stock of unsold cars had reached 91,000. Further losses accrued and indebtedness mounted. In April that year, Michelin, which had owned Citroën since 1935, announced a financial shortfall of F 500 m. to the state authorities. One billion francs would be needed by September. The fear of state intervention—that is, of Citroën being taken over by Renault, thus reinforcing the growth of the public sector—led Michelin to resume contacts with Peugeot.

Peugeot and Michelin had always cooperated to defend a shared vision of free enterprise. In 1974, immediately following the oil crisis, both companies remained convinced the automobile industry would now more than ever be based on economies of scale. If production was to be increased in a shrinking market, partnerships between manufacturers were indispensable in order to lengthen production runs, design and manufacture common components, and augment purchasing volumes. Hence the goal was to transfer to Citroën what Peugeot has been creating with Renault since 1966, but on a larger scale, with control in the hands of Peugeot and Michelin owning a minority share. For both companies, this reorganization—more defensive than offensive—was more an adaptation to economic change than a questioning of fundamental principles. The rescue of Citroën helped avert serious problems for suppliers, subcontractors, and dealers throughout the French automobile industry. It was particularly important for Peugeot that it succeed. The failure of Citroën might have attracted foreign buyers (American or Japanese), and, if not, certainly the state. Managers at Peugeot were convinced that 'the direct or indirect takeover of Citroën by the state, with the consequent introduction of imbalances into our country's automobile industry, would inevitably have brought a loss of independence for our company'. However, opinions on the merger were far from unanimous. While the sales department was delighted, financial managers were concerned, and technical staff remained silent. George Taylor, managing director of Peugeot and one of those most opposed to the merger, was despatched to Citroën precisely because he understood the difficulties involved. The company had to be entirely rebuilt without effacing its outward character, for there was no question of merging the two brand names, nor the distribution networks. Citroën and Peugeot were to preserve their distinct identities, clients, and product ranges.

In fact Citroën was restructured far more rapidly than had been expected. During 1974, prior to the official arrival of Peugeot, the company had already reduced its workforce by more than 7,000 workers, from 60,000 to 52,600. From early 1975, all unprofitable projects were eliminated. A link with Maserati and SM (a luxury coupé launched in 1970), the GS-Wankel (1973: which sold fewer than 850 models) and the rotary engine itself were cancelled. The industrial reorganization that had been planned by Michelin quickly delivered encouraging results. With the construction of the Aulnay factory and the abandonment of the factory at the Quai de Javel in Paris, Citroën possessed excellent production facilities, just as two new models, the CX (1975) and LN (1976) were launched. If the former was a pure product of Citroën's technological culture, the latter was a curious 'Meccano set' project and an unusual departure for the company. Because it used the body and suspension systems of the Peugeot 104 coupé, a long-serving two-cylinder Citroën engine, and the gearbox of the GS, the LN permitted major economies of scale at group level at the same time that it gave Citroën a small, attractive model to sell alongside the venerable 2CV (1949) and Dyane (1967). The design of the Visa, destined to replace the Ami 8 (1969), confirmed that a new technological strategy had been adopted. The Y prototype built by Citroën was quickly set aside by Peugeot and replaced by a new design which borrowed the platform of the 104 and combined it with various mechanical components, some from Citroën but mainly from Peugeot. The recovery of national and international markets from 1976 marked the end of Citroën's convalescence. While the company succeeded in maintaining its market share in France at around 16 per cent, exports grew significantly, from 45 per cent of production in 1976 to 52 per cent in 1979. Turnover doubled between 1976 and 1979, rising from 11 to 22 bn. f. In 1978, the accounts revealed 1.4 bn. f. in profits and 6.5 bn. f. in investments. As early as 1977, when 'normal financial avenues were re-opened for Citroën', the Peugeot group was able to repay the 1 bn. f.

loan it had received from the state when Citroën was purchased, twelve years before it was due.

There was now widespread optimism. In 1975 PSA became the largest of the French automobile manufacturers. In 1976 management spoke of 'an excellent year . . . contrary to all predictions, an exceptional recovery'. The crisis seemed long behind, especially since Peugeot itself had resisted it well, thanks in particular to its highly profitable diesel-powered automobiles, sales of which had risen dramatically. Indeed with the launch of the 305 (1977), there was insufficient production capacity at Sochaux to satisfy demand. Growth appeared to have returned. Even lengthening the working week to 42.5 hours proved inadequate. Hiring new workers became an imperative, particularly given the high level of employee turnover; of a workforce of 67,600, 7,300, or 10.8 per cent, left Peugeot in 1978. That year Peugeot hired 11,900 new workers, of whom 8,400 were under 25 years of age. The situation had been the same at Citroën since 1976, although there was one distinguishing factor in terms of the composition of personnel; Citroën had hired many more immigrant workers: 30 per cent of its workforce compared to 18.3 per cent at Peugeot. With the move to Aulnay in 1974, Citroën had replaced many of its older workers with immigrant workers, since the 'old workers' did not want to leave central Paris. Thus the automobile industry was still creating jobs: more than 9,000 new jobs at Peugeot between 1975 and 1978, and about 5,000 at Citroën between 1975 and 1977. It was boldly announced in 1977 that 'the French automobile industry has confirmed its role as a driving force in the economy'.

Yet all was not well. Analysis of the financial accounts reveals that while output was growing, profitability was in decline. Financial managers at Peugeot confirmed as early as 1977 that there was 'a significant shrinking of margins', with growth slower than expected. Citroën's financial managers recognized in 1978 that 'growth in profits was due only to exceptional factors'. They complained that interest payments were too high, reducing the capacity for self-finance and increasing indebtedness, which in turn was threatening to curtail investments.

Would the further harmonization of the two marques, Peugeot and Citroën, be impaired by a shortfall of financial resources? Despite an accelerated pace of investment, by 1979 PSA was far from having completed the reorganization of its two automobile divisions. Beyond the merger of the purchasing departments, little had been achieved, particularly in production and in design and development. Two examples are illustrative. Whereas the Citroën LN and Visa shared components with the Peugeot 104, thus risking a dilution of Citroën's identity, Peugeot's 305 and 505 (1979) were direct descendants of its 204–304 and 504 respectively, without making the least concession to any group policy. If the fact that much time had already been spent on designing and developing the 305 explained that model's situation, the same argument could not be used for the 505, which was launched four years after Citroën's CX, a model similar in size. It appeared that harmonization worked in one direction only, a conclusion which can also be drawn from the initial transfers of manufacturing activity. In 1976, the Citroën factory at Metz in Lorraine began to produce gearboxes for Peugeot. The following year the Aulnay plant started assembling the Peugeot 104 coupé while the Citroën-Espagne factory at Vigo, Spain, produced the Peugeot 504 model. This situation created tensions which PSA quickly tried to resolve. Harmonization of the model range was to be based around modern mechanical components which only Française de Mécanique (FM), the Peugeot-Renault subsidiary, was capable of producing in sufficient volumes at low cost. It could therefore only be organized around the four-cylinder engine of the Peugeot 104, Renault's 2 litre engine, and (marginally) the FM-produced V6 PRV engine (jointly developed by Peugeot, Renault, and Volvo), which proved impossible to fit under the CX's bonnet. But this range of engines remained inadequate for a generalist producer, and the company was therefore obliged to continue with old engines. What was needed was to create a new factory capable of restructuring FM around products specifically designed to meet the needs of the group. By early 1977 PSA was therefore envisaging a long-term and particularly costly investment programme. While the market recovery between 1975 and 1978 legitimated this strategic choice, it also tended to shift priorities and focus all efforts on expanding volumes. The risk was that other underlying difficulties would not be faced up to, if they could be masked by strong growth.

### 13.1.3. The Purchase of Chrysler-Europe

The 1978 announcement that PSA was purchasing Chrysler's European subsidiaries (Simca, Rootes, Sunbeam, and Barreiros) for 1,8 million new PSA shares (equivalent to 15 per cent of PSA's value) did not really come as a surprise. While clearly, as in the case of Citroën, the event had not been planned in advance, the new purchase was a logical evolution of the desire to prepare for fierce international competition.

Faced with a highly concentrated American automotive industry, a Japanese industry in which companies have the added dimension of being endowed with particular socio-cultural advantages with a protected domestic market and an under-valued yen, the European industry appears dispersed and vulnerable, with seven mass producers, manufacturing which is hardly standardised, and a productive apparatus which is insufficiently concentrated. (Loubet 1995*b*)

For PSA and its new president Jean-Paul Parayre, Europe was still the key market, given the considerable size of that market as well as Europe's technological capabilities.

The automotive industry can only fully thrive in a well developed industrial and cultural environment. The fact that the lowest production costs are obtained in the developed countries with high wages reveals that the industry will continue to play a role, even if it is in the interest of some developing countries to create, under the shelter of strict protectionism, an automobile industry to supply their own markets. (Loubet 1995b)

With this highly optimistic reasoning, in which growth and volume still dictated strategy, and in which the rich countries remained the masters of the global industrial game and competitiveness, Jean-Paul Parayre led the PSA group further into a logic based on size, 'the size needed to confront the biggest competitors'.

During the 1976–8 period, PSA's strategic choices were complex, because they contained elements of offensive and defensive strategies simultaneously. The offensive strategy consisted of the more rapid introduction of new models thanks to three distinct brands which increasingly utilized shared components permitting the realization of improved economies of scale. It also included a determination to focus on the European markets. Strategy was defensive, however, in the decision to prevent foreign competitors from establishing themselves in France in place of Simca, and in preventing the potential further growth of the public sector (that is Renault). This dual strategy was unanimously welcomed internally, and through growth outside France Peugeot temporarily became Europe's largest European automobile producer, third in world terms behind General Motors and Ford (measured in terms of unit output). This was an exceptional turnaround in the company's evolution, which Jean-Paul Parayre summarized as follows;

it was in 1974, at the time of the first oil crisis, that Peugeot decided not to delay in adapting itself to a profoundly changed environment. The aim was to guarantee the continued life of the company and to preserve its independence. This was to be achieved by expanding in size to acquire the resources for competitiveness. . . . Initially a medium sized producer, Peugeot had, in the span of a few years, achieved the rank of the third largest automobile producer in the world, multiplying its turnover fivefold without compromising its traditional financial equilibrium.

Was this, in 1978, a model of success *à la française*?

### 13.2. PSA: A FRENCH GENERAL MOTORS?

Sloanism now clearly took the place of Fordism. With three automobile divisions, had PSA become a French General Motors? From 1979, PSA management had to organize a company comprising more than thirty factories that produced twenty-six different models (eight Peugeot, nine Citroën, seven Simca and Sunbeam, and two Matra) with a workforce of nearly 220,000, maintain three sales networks, and above all ensure that each of the three marques, Peugeot, Citroën, and Chrysler-Europe (now renamed at great expense Talbot, after a producer purchased by Simca in 1958) would have distinct and separate characters: serious and traditional for Peugeot, innovative for Citroën, and luxury and sports-oriented for Talbot. The final remaining optimists understood how profound the problem was going to be when the second oil crisis arrived.

## 13.2.1. Structural Reforms

The first difficulties related to the reorganization of PSA lay in the human realm. Peugeot lacked the managerial staff necessary to lead its new automobile divisions, a direct consequence of an absence of planning to train and prepare new managers. It was not possible to wait twenty or thirty years for candidates to reach high positions. PSA was forced to draw from deep within the organization, but also—a new factor—to look outside the company (high-level government bureaucrats, for instance) to find many of its upper-level managers. This was clearly 'the end of an era', as François Gautier, the last of the 'old presidents', explained, 'the era of one man rule is over . . . young people must be got ready to facilitate the transition when the absolute monarchy comes to an end'. The new order had already been symbolized by the arrival of Jean-Paul Parayre in 1974, brought in to the company from the state to become president of PSA in 1976 when he was only 40 years old!

Harmonization of the three automobile divisions started with purchasing and product development. By combining the purchasing activities of the different marques, PSA expected to obtain better prices based on a volume effect. The task was entrusted to Talbot, which already had wide experience of the issues and had learned methods from Chrysler in order to organize the purchasing of its various European factories. Jean Peronnin, formerly managing director of Simca, aided by Roger Lansard of Peugeot, prepared the ground for the 1980 creation of SOGEDAC, a special purchasing company for the PSA group. SOGEDAC worked in tandem with product development staff to encourage parts standardization, which pleased suppliers who could be sure of longer production runs and would therefore offer better prices. The reorganization of the research and development departments was more complicated as a result of their deeply rooted traditions and cultures. In order to help it harmonize investments and encourage standardization, the Technology Department at PSA saw its functions broadened. It no longer sufficed to restrict activities to the design and development of shared mechanical components. From now on it would design the vehicles produced by each of the three marques, up to the first prototype. The plan was that the design centres at Peugeot, Citroën, and Talbot, with 5,700 engineers and technicians, would then take over the product development work, plan the product launch, and deal with quality issues. However, the weakness of Talbot's design centre forced it to be merged with the other design centres. Lastly, to build up the overall research capacity, and based on the experience of the Advanced Design Department set up with Renault when the two companies began to cooperate, PSA created a Department for Science and Research at La Garenne, responsible for working on new automobile technologies independently of the Technology Department.

With a merger strategy in which internationalization was an important challenge, PSA also reorganized its activities abroad. Schematically, each automobile division was linked to different regions. Thus Peugeot was to focus on large-scale exports to its traditional markets in Africa, North America, South America. For Citroën the aim was to establish relations with Eastern Europe in addition to its traditional ties with Spain and Portugal. As for Talbot, this subsidiary was to permit PSA to better establish itself in Northern Europe, especially in the United Kingdom, the second largest market in Europe, where neither Peugeot nor Citroën had had much success to date. Largely through Citroën, between 1976 and 1980 PSA also increased its foreign tie-ups significantly. In 1976, an agreement was signed with the Romanian company OLTCIT which envisaged the construction of a local factory capable of manufacturing 130,000 Axel vehicles per year, half of which were to be reimported into Western Europe via the Citroën distribution network. In 1978 Citroën created a joint venture in East Germany for a factory to produce more than 675,000 constant velocity joints per year, a contract worth 1.6 bn. f. annually. Finally, there was a serious attempt to resume relations with Fiat, Citroën's former partner, but this time in close association with PSA. Several projects were planned. There were to be shared factories, to produce a small commercial vehicle at Pomigliano d'Arco in Italy (1980), and later a minivan at Valenciennes in France (1994). These models used Fiat or PSA mechanical components depending upon the marque under which they were sold. The two companies researched a small engine together, the FIRE or TU, which could be produced in very large volumes at very low cost. The Peugeot and Fiat subsidiaries in Argentina were merged in an attempt to maintain the viability of their activities in a particularly difficult market where Peugeot had a solid reputation.

Indeed, Peugeot was very reliant on large-scale exportation. The factory in Nigeria, with components supplied entirely by air, produced 56,000 vehicles in 1977. If its growth could be continued, it would permit the expansion of sales in other African countries. In the same year a contract was signed with Iran National Manufacturing to produce 100,000 units of the 305 model annually; Talbot-UK was already active in Iran, and its Stoke factory assembled more than 70,000 CKD automobiles in 1980. In North America Peugeot now hoped to gain a firmer foothold, particularly since its sales—which had grown from 7,700 to 12,000 504 and 604 models during 1976 thanks to the diesel engine—could not continue to expand without the support of a local distribution network. By 1977 contacts were being developed with AMC (American Motors Corporation), which had been looking for a partner able to revive its own distribution network by supplying new models. By 1978 PSA and AMC were on the point of signing an agreement under which the Americans would distribute Peugeot and Citroën models in the USA in exchange for distribution of Jeeps in Europe. Only one problem remained: how was the operation to be financed? While PSA wanted it restricted to a purely commercial agreement, AMC had hoped that the French company would inject new investment. As a result, PSA held back, and Renault moved in to accept AMC's conditions. However, only a few months later Peugeot announced its arrival in the USA. An agreement signed with Chrysler in 1978 envisaged the distribution of French models in the USA by Chrysler's distribution network, two contracts to deliver engines (1.6 l. and the diesel), and lastly a letter of intent that a small car would be produced in North America from around 1985, targeted at the USA and Canada. Internationalization was clearly expected to play a prominent role in the new PSA.

*13.2.2. Reforming Manufacturing*

A second set of reforms dealt with manufacturing and the means of production. Jean-Paul Parayre had pursued advanced ideas since purchasing Citroën. The replacement of models had to be accelerated and a stock of mechanical components created. Time was a major constraint, since five years were required to design a new automobile, and new engines and gearboxes were still at an embryonic planning stage. The situation was all the more difficult since customers had expectations about the successive mergers, and particularly that there would be a concrete outcome to the birth of the Talbot marque. PSA was confronted by the dual problem of technological constraints linked to the harmonization of the product ranges produced by the three marques, and, more immediately, by a lack of projects at Simca. Ten years had elapsed since Chrysler's design centre had designed mechanical components. The engineers had little creative capability except in stylistic design. Indeed the technical contribution of the former American company was limited to electronic systems and air conditioning. From this viewpoint, the purchase of Chrysler-Europe had been a bad deal. Only two products were being designed, the Solara (7–8CV), which was a 1308 model equipped with a boot, and the Tagora, an upmarket product which PSA did not need. Worse yet, there were no plans to replace the old Simca 1000 (5CV, 1960), and therefore to compete in one of the most important market segments. The first Talbot models did not appear until 1980–1, the time lag necessary to prepare PSA components for them. The Solara's (1980) gearbox was made by Citroën. The Tagora's (1981) axles and some engines were made by Peugeot. A small car, the C1 project, or Samba (1984) was imposed on Talbot by Jean-Paul Parayre. The recipe was unchanged; on the basis of a Peugeot platform and engine, Talbot offered a fourth version of the small PSA automobile to add to the 104, the LN, and the Visa. Citroën and Peugeot were simultaneously producing new variants of existing models: a GS equipped with a fifth, hatchback door (GSA), variants of the CX with petrol and diesel engines, the French-Romanian Axel, estate versions of the 305 and 505, diesel and turbo diesel engines, and so on. Old models were reinvented in new guises to create 'events' and maintain market share before genuinely new models could be launched in 1982–3.

Planning for the Citroën BX and Peugeot 205 started in 1977. The group's growth was to focus on the middle of the product range, where the market remained strongest. These were the first two models of a new generation of products, using new mechanical components around which PSA was to construct new automobile ranges. Manufacturing was organized at two Citroën factories, at Metz-Borny in Lorraine, which specialized in gearboxes, and a new foundry at Charleville in Ardennes. These factories, constructed during the final years under Michelin control, were to replace the old Paris factories. In 1980 PSA opened two new factories at Trémery, near Metz (engines) and at Valenciennes (gearboxes). These factories were built and started operations in less than eighteen months, at a cost of 6 bn. f. Each of the four factories utilized advanced technologies; all were automated and robotized. At Trémery output reached 1,000 engines per day during the first year of operation and 2,000 by the end of 1983. At Valenciennes, automated equipment reduced production times by 40 per cent. The Charleville foundry, with only about a hundred workers, was completely automated and controlled by computer. This new factory permitted the Peugeot group to experiment with a new approach to manufacturing in which robotization and automation transformed working practices. With the reduction of stocks and fixed costs, and increases in productivity, by 1980–2 PSA was beginning to envisage a new way to manage its mechanical components factories.

The last focus for manufacturing reform was the factories which undertook subcontract work for the PSA group. These were old linkages, closely tied to the Peugeot marque. Two of them, Aciers et Outillages Peugeot (AOP) and Cycles Peugeot, had started to specialize in supplying the automobile industry. PSA wanted to strengthen this cooperation, and permit these companies to take responsibility for the group's diversification strategy. In 1977, AOP and Creusot-Loire merged to form Peugeot-Loire, specializing in sheet steel and specialty steels. At the same time, the new company further developed its activities in plastics, the use of which was spreading throughout the automotive industry, purchasing Quillery, a major French plastics specialist. AOP expanded its interests to small mechanical parts, electric engines, safety belts, and the recycling of diesel engines when it took over SEMAS (Société d'Exploitation Mécanique), its main competitor in France, in 1977. Cycles Peugeot pursued the same strategy. In parallel to two-wheeler production, the company sought to diversify into the automotive industry, focusing on all products involving tubes, such as exhaust systems and seat frames (delivered to AOP and Quillery). Hence, alongside the development of PSA's automobile production activity in the narrow sense, there was a refocusing of the group's subsidiaries on the automotive industry more broadly.

### 13.2.3. Towards a New Social Policy

The takeover of Citroën and Simca caused numerous difficulties on the social policy front. It seemed that Peugeot and Renault had been correct to deliberately exclude such issues from their 1966 agreement. Renault's desire, in 1969 when the shared factories were opened, 'to align social policies within the cooperative agreement' had been a point of fundamental disagreement. From 1975, a very different situation emerged within the Peugeot group. The strategy was for each company to retain its own social policy, which had generally evolved from practices, traditions, or victories closely linked to the history of each particular company. But this strategy was made more difficult by the magnitude of the differences among them, in terms of classification systems, wages, pensions and even social benefits. For example, wages at Simca were markedly higher than at Citroën or Peugeot; in 1978, the average wage of a worker at Poissy (Simca) was 28 per cent higher than his counterpart at Sochaux. The founder of Simca, Henri-Théodore Pigozzi, had always been careful to mould its wages policy on that of Renault. In exchange for these advantages, and with the help of a strong internal security force, Simca had been able to muzzle the most radical trade unions and establish an almost complete social peace. Yet while there may have been few discussions and no disputes, internal resentments accumulated in a company in which employee turnover was very low (less than 8 per cent of Simca employees were aged under 25) and the number of foreign workers very high (32 per cent of Talbot employees in 1979). Accordingly, the social climate at Simca was explosive. High wages had not resolved the inherent problems of blue-collar work which had been afflicting many automobile industry companies for twenty years. In the United Kingdom the situation was even worse, to the extent that in 1974 the Americans had already threatened to close the factories of Chrysler-UK, a path they would have followed in the absence of subsidies from the British government. In 1979, workers at the Ryton, Stoke, and Linwood plants in the United Kingdom struck over demands for 20 per cent wage increases. The claim was impossible to meet in factories where productivity was 30 per cent lower than in the French factories, and was suicidal for the Linwood factory, which for several months had only been operating for two to three days per week and by 1980 had been closed down. The social inheritance of PSA's newest automobile division was particularly onerous.

For Jean-Paul Parayre, who took a firm stance in the British conflict, a new social policy for the group as a whole was indispensable. 'We must implement', he declared, 'an imaginative social policy, which goes beyond merely concerns over wages'. However, it would be difficult to change the practices of a lifetime overnight. Despite the new intentions, wages remained the focus of both union claims and management policies. In 1977, unions and management at Peugeot agreed to progressively establish a bonus equivalent to one month's pay for all personnel. Until 1980–1 negotiations concentrated on the creation of bonus schemes (from bonuses for heads of families to holiday bonuses to new model launch bonuses and to end of year bonuses) and increased purchasing power. At Citroën the issues had already changed. Long before the PSA purchase, the personnel had won a partial reduction of working time. In 1976–7, 41 hours were worked per week at Aulnay, in contrast to 42–42.5 at Peugeot. From 1976, Citroën had resolved to adjust its work times, attempting to 'organise around anticipated leave and other days off' in order to give extra holidays. In 1977 Peugeot adopted a similar policy. However, results came slowly. Advances on other fronts were needed if real progress towards a new social policy was to be made. The task was made all the more difficult by the fact that it was assembly-line work that was the underlying focus of conflict. From 1978 onwards, new efforts were being made in terms of communication, information, and training. In 1979, PSA agreed with the unions to create fifty 'operational groups' in which workers would be responsible for helping to improve working conditions in their factories. The 1979 introduction at the Lille factory of automated guided vehicles on engine assembly lines, as well as the development of automation in the production of mechanical components and the robotization of body assembly from 1980 onwards revealed the importance of a well-trained workforce.

There was an acute absence of training and learning throughout the French automobile industry. Mechanization in combination with a high level of task decomposition had led the company to hire an unskilled workforce which often not only lacked qualifications but was also in some cases illiterate. Although initial experiments led to the recognition that it was necessary to grant operators greater autonomy, hence greater responsibility, in their work, the diffusion of this principle remained impracticable in the short term. The changes necessary were so significant that it remained uncertain whether the social partners (management and unions) were truly prepared for them. Within the factories, the managerial hierarchy as a whole was not ready for reform, notably for reform which threatened the most deeply rooted habits and local power bases. Since 1977, the company had undertaken surveys and questionnaires in the factories. By surveying worker attitudes, management was well placed to gauge the pervasive high levels of discontent. Morale was low even among managers. Working teams, which had been organized to facilitate meetings and dialogue, were not proving successful. The company was unable to win the confidence of a workforce troubled by an eroding economic context. The 'imaginative social policy' called for by Jean-Paul Parayre had come to a halt.

### 13.2.4. PSA and the Crisis of the French Automobile Industry

The newly reorganized PSA found itself facing the consequences of the second oil crisis, the impacts of which would last much longer than expected. 'After the model of earlier crises, especially in 1974–5', acknowledged J.-P. Parayre, 'at the end of 1980 we had expected the market to recover around Autumn 1981.' This did not occur, however, and during 1981 a double change became manifest: a reduction in sales and a shift in the balance of sales towards smaller cars. Alarm bells sounded. The group's share of the market in France fell from 42.6 per cent in 1979 to 30.3 per cent in 1982. Similarly, in other European countries, the share of the three PSA marques slipped from 17.2 per cent to 12 per cent. Production plummeted by close to a third, from 2.3 million units to 1.6 million, with severe consequences for productivity, which between 1979 and 1983 fell from 9 to 8.3 automobiles per employee per year, compared with a rise from 9.9 to 11.8 at Fiat and from 10.4 to 13.2 at Ford-Europe. Whereas Fiat employed

6,000 workers to build 1,200 cars per day, Citroën required 8,000. The financial repercussions were dire. With accumulating financial losses, 8.5 bn. f. between 1980 and 1984, PSA was forced to borrow in order to survive. Long-term indebtedness grew considerably, reaching 1.7 per cent of the company's value (compared to 0.95 per cent at Fiat and 0.3 per cent at Volkswagen). In 1982, finance costs represented close to 4 per cent of turnover, compared to 2 per cent for other European producers. Under such conditions it was difficult to invest and modernize. Investment expenditures were reduced, from 5.8 bn. f. (8.2 per cent of turnover) in 1979 to 4.6 bn. f. (6.3 per cent of turnover) in 1981. Industry analysts expressed doubts about the group's ability to respond; PSA shares dropped in value from 560 f. to 140 f. on the Paris stock exchange between 1978 and 1980.

The disintegration of the group is best explained in terms of a series of interlinked factors, both external and internal. The rise in the cost of energy, inflation, increased interest rates, monetary fluctuations, stagnation or decreases in buying power, all joined to reduce activity levels in the automobile market. This in turn led to a relentless competitive struggle, price competition, and a squeezing of margins which accentuated losses. Moreover, some of the measures undertaken by the new government during the 1981–2 period aggravated the situation by worsening the group's finances at the worst possible moment. The nationwide institution of 39 hours work for 40 hours pay, together with a fifth week of paid holiday, cost PSA 1 bn. f. The adoption of austerity measures in 1982 (including price freezes) forced PSA to sell its 1983 models at 1982 prices in spite of improvements to the products. Furthermore, guarantees that consumer prices would not rise for three months prolonged the consequences of the price freezes until the end of the year. As a direct result, Peugeot lost 300 m. f. during 1982.

Of course the company also had many problems of its own. The spectacular drop in sales was primarily explained by the absence of model replacements at Citroën, Peugeot, and Talbot. It was Talbot that collapsed fastest, with sales plummeting from 640,700 in 1979 to 351,300 in 1982. Its range was incomplete and too close to Peugeot's, its new identity was poorly defined, and its distribution network was not competitive and was too big in relation to sales volume. The merger of Peugeot and Talbot was inevitable as the death-knell for the third marque sounded. It was no longer a question of reorganization but of restructuring. The financiers returned in force to recommend drastic new measures. This in turn stirred up the social unrest which had been brewing for so long. There was violent unrest at Aulnay and Poissy, revealing the profound malaise of the French automobile industry.

## 13.3. PSA AND THE JAPANESE MODEL

'Of all the automobile manufacturers, Peugeot and Toyota are the closest' (author interview). If Henri Combe, president of Toyota-France and once a senior manager at Peugeot, is to believed, the two companies shared a number of characteristics in common. They were both provincial companies and still strongly family based, even as they opened up to managers from outside. Both viewed financial rigour as of the utmost importance. They shared similar values in terms of their products, with robustness and profitability linked tightly together. Both companies were prudent, yet determined to evolve continuously in order to survive. Thus both were capable of questioning themselves. As the Japanese model became increasingly significant, did PSA now need to adopt Toyotaism?

*13.3.1. Reducing the Break-Even Point and Returning to Fundamental Equilibrium*

In 1982, the urgency of the situation, along with the weight of PSA's financial tradition, forced Peugeot to call upon the services of Jacques Calvet, a financier and former president of the Banque Nationale de Paris. His arrival, like that of Georges Besse at Renault (1985), accelerated the disappearance of managers consumed by a passion for the motor car and their replacement by managers whose specialty was management. In order to re-establish the basic equilibrium the company needed as quickly as possible, Jacques Calvet proposed a novel strategic approach to the automobile industry. Previously, the practice had been to set ever higher production targets each year so as to make profits through volume. Now the objective would be to balance the accounts around production levels adapted to a smaller market. At the start of the 1980s, 2.2 million vehicles had to be produced in order to cover expenses and begin to make profits. While in 1979 the group's output reached 2.3 million, it stabilized at around 1.6 million between 1982 and 1985. A reduction of the break-even point was imperative to stem the financial losses. The goal was to bring it down to 1.5 million by 1984. In fact the break-even point would be as low as 1.2 million in 1988, just as the market began to recover. To attain these results Jacques Calvet had instigated a restructuring programme around three principal axes: adapting the size of the workforce to output, increasing productivity, and reducing finance costs.

The workforce reduction was the most spectacular measure. In ten years, PSA's workforce was reduced by 27 per cent. The company most affected was Peugeot-Talbot, which lost a third of its workforce, principally at Sochaux and especially at Poissy. Management employed a number of measures to adapt the workforce to the size of the automobile market. Having ceased to hire workers in 1980 (Peugeot) and 1981 (Citroën), voluntary redundancies were proposed and factories were temporarily closed. In 1981, agreements were signed with the *Fonds National de l'Emploi* (FNE) and the *Office National d'Immigration* (ONI). The first permitted volunteers aged over 56 years and 2 months to take early retirement with benefits similar to those of workers retiring at 60. The second offered a financial package to foreign workers who agreed to return to their country of origin, to help them reintegrate themselves. However, these measures soon proved inadequate. In 1983 PSA calculated the surplus of workers to be 9,000 workers at Peugeot-Talbot and 6,000 at Citroën. This was problematic for the Mauroy government, which was making the fight against unemployment a priority. The struggle between the state and PSA was a hard one, with social conflicts at Talbot (Poissy) and Citroën (Aulnay), and the public authorities apparently engaged in a rearguard action. In 1991, Minister for Labour Martine Aubry was to recognize that 'workforce reductions were necessary if the automobile industry was to reach productivity levels closer to those of their competitors' (*Rapport de la Commission*, Official News, 13 June 1992). During the 1982–6 period, the automobile producers were given state aid to reduce their workforces. In 1984, out of 5,182 departures, including 'voluntary redundancies' and 'early retirements' from Citroën, 3,689, or 71 per cent, were redundancies 'for economic reasons'. The proportions were similar at Renault (73.5 per cent) and Peugeot (70 per cent) in 1985.

The steady shrinkage of the labour force was an essential factor in the re-establishment of financial equilibrium. It was a precondition for reducing the break-even point. The smaller workforce permitted spectacular increases in productivity. The vehicle per worker ratio rose from 7.7 in 1982 to 13.9 in 1990. Between 1985 and 1991 turnover per worker soared from 567,279 f. to 1,029,498 f., an increase of 80 per cent (50 per cent in real terms), with annual growth averaging nearly 7 per cent. These statistics provide a rough approximation of the improvements made, and PSA itself confirmed that Peugeot's productivity rose by 50 per cent between 1985 and 1989. A further financial advantage of a smaller workforce was the decline, then relative stabilization, of expenditure on personnel, from 31.4 per cent of turnover in 1983 to no more than 18.8 per cent in 1990, despite an absolute rise of 4.7 bn. f. due to wage increases and the new skills of the workforce. Within this period, from 1984 to 1987 these costs remained stable, at around 25 bn. f. whereas production and turnover rose by 19 per cent and 30 per cent respectively. And while expenditures on employees rose later, they simply kept pace with the increase in turnover.

Yet the reduction of the workforce was only one factor in the overall strategy to reduce expenditure. Other measures were adopted, one of the most important being limitations on purchases and the management of new vehicle stocks. From 1984 to 1990, the PSA group's purchases rose from 61 bn. f. to 111 bn. f. Their share of turnover was stable, at around 66 per cent. While purchases per vehicle rose from 38,159 f. to 42,957 f. between 1984 and 1988, this was not due to increasing costs but to improvements in the models. Another measure devised to control or reduce expenditures was control over new vehicle stocks. From 1983 for Peugeot, and 1985 for Citroën, the distributors became the owners of new cars upon delivery. This measure removed financial responsibility for stocks from the producers, while burdening the funds of the dealers. Peugeot had already reduced its stocks of new vehicles by more than 50,000 units in 1981, and this measure made it possible to save a further 200 m. f.

Expenditure cuts permitted operating margins to be increased, from 118 m. f. in 1982 to 3.3 bn. f. in 1984, 7.5 bn. f. in 1986 and 13.5 bn. f. in 1989. With healthier finances, PSA was able to accelerate its investments. Maintained only with difficulty at a 4 bn. f. level during 1983–4, investments climbed steadily to reach 7 per cent of turnover (7.3 bn. f.) in 1987 and 9.4 per cent in 1990 (15.1 bn. f.). This provided the means to modernize the production apparatus, and above all to accelerate model replacement.

### 13.3.2. New Strategic Focus

As part of its new approach to the automobile industry, PSA also envisaged a new strategic focus. The growth of the group required profound adaptations to the new economic context. The crisis made the diversification strategy more uncertain, for it was difficult to finance and modernize activities as diverse as these. Thus PSA began to divest itself of its parallel activities. First, several of its subsidiaries were linked to world-leading companies. In 1980–1 Cycles Peugeot signed technical and commercial agreements with the Italian company Piaggio to produce motorcycles and with Honda to supply an engine to Honda's Belgian motorcycle factory. In 1981 a sales agreement was signed between Outillages Peugeot and AEG-Telefunken Second; Peugeot sold some of its shares to new partners, notably the German company of Fichtel & Sachs and the American group Stanley Works (1986). In 1991, PSA sold its subsidiary Engrenage et Réducteurs, after having sold off its popular food chain, RAVI. PSA was withdrawing from other sectors in order to concentrate on the automobile industry.

The automotive-based activities of Cycles Peugeot and AOP were merged in 1987 to create the group ECIA (Equipments et Composants pour l'Industrie Automobile). ECIA was a group of international proportions focused on four product lines: exhaust systems, seating, driver controls, and bumpers and structural body components. This reorganization permitted research and development to take place on the scale needed to ensure absolute independence in product design. In 1988 ECIA shifted onto the offensive, signing several cooperative agreements (Zastava, Leistritz, Luchaire, Tréca), buying up some of its competitors (PCG-Silenciadores, Stollberg, Tubauto), and building three new factories in proximity to the assembly plants: a factory at Cernay for just-in-time deliveries to Peugeot's Mulhouse (Alsace) factory, one at Crevin (Brittany) near Citroën-Rennes, and one at Marines near Poissy. PSA's strategic reorganization was not, therefore, limited to creating ECIA, but also involved new forms of production organization.

Another indication of new strategic focus was the closure of old factories. The restructuring of manufacturing facilities included the closure of five outdated Citroën factories between 1983 and 1992 and nine Chrysler-Europe factories between 1981 and 1987. The outcome was reorganization around four components factories (Charleville, Metz-Borny, Trémery, and Valenciennes) and five final assembly factories (Aulnay, Mulhouse, Poissy, Rennes, and Sochaux), supported by the overseas sites in Spain (Vigo, Villaverde), the United Kingdom (Ryton), Portugal (Mangualde), and Italy (Orense). PSA clearly chose to focus on production, and therefore markets, in Europe, the world's largest market. Several foreign factories were closed between 1980 and 1984, in Belgium (Citroën), South Africa (Citroën-Peugeot), Argentina (Citroën), Morocco (Citroën), Yugoslavia (Citroën), and Romania (Citroën). Besides several small plants, only three substantial factories remained active; in Argentina, Iran, and Nigeria. All were Peugeot factories, all playing a secondary role in the group's strategy while awaiting improved local economic and political conditions.

### 13.3.3. New Product Ranges

The restructuring of PSA was dependent on the introduction of new model replacements. This was made more difficult when, after 1980, it was deemed necessary to make deep cuts in Peugeot and Talbot ranges which competed with each other, and to harmonize the plans of Peugeot and Citroën. Following the launch of the Citroën BX (1982) and Peugeot 205 (1983), PSA began to replace the Peugeot-Talbot range with a single Peugeot range and create a complete range for Citroën (Fig. 13.1). Jacques Calvet gave each marque a range of four models able to cover the main market segments: a small car (segment B), a small-medium vehicle (M1), a medium-large vehicle (M2), and a luxury range model (H). By 1990, with the launch of Citroën ZX, the goal had been attained; not only did each manufacturer possess its own complete range, but for the first time in its history Citroën had become a generalist producer. A side-effect was that in focusing all its technical, human, and financial potential on the creation of generalist ranges, PSA was ignoring niche markets which might be lucrative and important for the company's image. The 205 cabriolet, launched in 1986, was the only niche model, and in 1982 Peugeot rejected the French company Matra's proposal

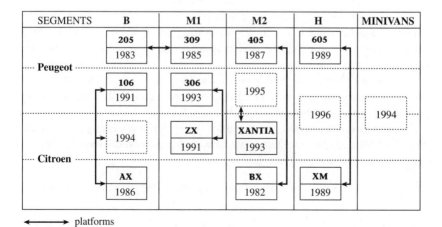

FIG. 13.1. *Peugeot-Citroën product range, 1982–1996*

to jointly build a minivan, a proposal Renault quickly took up, launching the vehicle in 1984 as the highly successful Renault Espace. 'We were misunderstood', insisted PSA's management. 'We make coupes and cabriolets if we have the resources . . . but only as complements to the basic models. Our future is as a generalist producer'.

The Peugeot and Citroën ranges were based on technical complementarities. The vehicles had to share as many components as possible. As a general rule, the models in each segment used the same platform, such as the 306 and ZX or the 605 and XM. These pairs employed the same drivetrains and steering systems, and utilized similar braking systems and similar mechanical parts. With the 205 (B) and 309 (M1), Peugeot went even further, and these two models shared components such as doors and seats. For minivans, where the market remained limited, in 1994 PSA launched a single model shared by the two marques, the Peugeot 806 and the Citroën Evasion, a model also sold under Fiat and Lancia marques. More audaciously still, in 1996 PSA replaced the Citroën AX (B) with the Saxo, based on the Peugeot 106 (1991). With this twin range strategy, design centres had to plan model replacements very carefully. Peugeot and Citroën had to launch new models every two years, and sometimes every year including estate and commercial variants. The rate of model replacement became a vital factor in competitiveness. To reduce lead-times in design and development, PSA reorganized its testing and product launch phases and tightened the link between product design and production engineering services (which had been working closely together since 1984); an objective accelerated by the Product Development Plan established in 1991. Between the Peugeot 104 (1972) and the 106 (1991), the period from design to manufacture of bodies had been reduced from twenty-seven to eighteen months. While the D40 project (Peugeot 405) lasted five years (October 1982 to June 1987), the D80 project (Peugeot 406) took 208 weeks, from June 1991 to October 1995. The planned reduction towards three years for the T1 project (the Peugeot 206) seemed likely to succeed. These improvements resulted from ten years of achievements with computer-aided design and computer-aided manufacturing, as well as new organizational forms, notably *Projet-Plateau* teams. From 1988, the design and development departments no longer intervened sequentially in a project, at the risk of undermining the preceding stage thereby wasting time and increasing costs. They now worked together to plan product, process, and manufacturing resources simultaneously. This marked the end of separated functions in an exclusively hierarchical organization, the elements of which acted sequentially, and was a revolutionary change that the engineers had learned through their contacts with Japanese and American firms.

### 13.3.4. New Forms of Manufacturing: Flexible and Polyvalent Factories

The launch of new models offered opportunities to modernize the production apparatus, including the components and assembly factories. Originally designed to improve productivity, the factories producing the group's mechanical components established by Jean-Paul Parayre in 1977 were continually modernized to increase capacity and broaden their product ranges. In 1988, the factory at Metz-Borny raised its production capacity from 600 to 2,600 gearboxes per day, and then to 5,000 in 1991. Similar increases were achieved in the production of engines: Trémery, which began production in 1982 at a rate of 1,000 engines per day, expanded to 4,600 units in 1989. Production planning at Trémery was made more difficult since it produced both petrol and diesel engines, demand for which fluctuated according to government policy. To adapt themselves, the factories focused on the polyvalency of their assembly lines and the flexibility of their automated lines. The same focus on automation and polyvalency was prevalent in final vehicle assembly. While there were 114 robots (programmable automation units) at PSA assembly factories in 1981, by 1986 there were 798, and by 1989 1,492, 60 per cent of which were used in welding. The 205 was the first model to be welded fully automatically. On the basis of experiments conducted at Citroën's Vigo factory in Spain, which was capable of assembling two distinct families of models on the same line, from 1984 onwards emphasis was placed on polyvalent production equipment. The first site in France to benefit from this was Poissy, where 1,225 m. f. was invested between 1984 and 1985 in preparation for the launch of the 309. Most progress was made on the stamping lines, with 280 m. f. invested in automated presses, including tri-axis presses capable of stamping 960 parts per hour, compared with 400 parts previously. A polyvalent welding line comprised of 134 robots, 106 of them six-axis robots, manufactured the bodies of the 309; the number of welds having been reduced by 50 per cent compared to the Simca Horizon. The line was able to manufacture one 309 body per minute. Poissy had not been restructured on this scale since its reconstruction in 1957. Yet five years after this, further change was under way. Driven by a desire to increase production rates and productivity, yet limited in terms of physical expansion and investment funds, management at Peugeot decided to experiment with a new way of organizing working time. From the technical standpoint, the factory could operate for twenty consecutive hours, with four hours allotted for maintenance work. Hence it was proposed to move to a ten-hour working day, with two shifts, combined with a four-day week. Although it was criticized by the unions because of the increased fatigue of operators, the plan permitted the factory to operate for 96.25 hours per week as opposed to the previous 77 hours, thus increasing equipment operating time by 25 per cent. Production capacity rose from 1,200 to 1,500 cars per day, and 1,800 new jobs were created.

Following the Poissy example, other final assembly sites (Sochaux, Mulhouse, Rennes, and Aulnay) were given polyvalent equipment as new models were launched. Between 1985 and 1987 Sochaux and Aulnay were equipped for the launch of the 405 and AX. Aulnay was given four polyvalent assembly lines able to assemble models as different as the AX and CX. Polyvalency was the key to the reorganizing PSA's factories. Each factory was now capable of producing a supplementary model on top of its basic product to enable it to operate at full capacity. Thus from 1984 Poissy assembled not only the basic Peugeot 309 but also the 205, and by 1991 the Citroën ZX and the first Peugeot 306 models, all without halting production of the 309. At Aulnay, the AX and 106 were produced alongside the 205 and ZX model. This factory, situated in the suburbs of Paris, became one of the group's most automated plants. Several assembly operations for the AX were entirely automated, including waterproofing and soundproofing, windshield assembly, installation of headliners in the roof of the vehicle, and even dashboard installation. With the launch of the Peugeot 405 in 1987, numerous subassemblies were produced away from the main line, which improved assembly quality: especially the front bumper/lights module, the dashboard module, the doors, and various mechanical components. The quality of these subassemblies was controlled before installation, which in most cases was undertaken by robots.

This drive towards quality and automation increased in intensity with the launch of the group's two upmarket models, the Peugeot 605 and Citroën XM, in 1989. Their respective final assembly factories at Sochaux and Rennes were further modernized. They were equipped with new anti-corrosion treatment lines and automated paint installations. With these two models, PSA also increased the supply of components just-in-time, starting with large components like dashboards, front and rear bumper assemblies, carpets, fuel tanks, seats, interior door panels, and wheels. Up to this point the manufacture of these various components had been integrated into the group's own factories, but it was now handed over to first-tier suppliers (the total number of which contracted by SOGEDAC, PSA's purchasing arm, was cut from 1,430 in 1986 to 690 in 1992) which were located in proximity to the factories they served; for Citroën-Rennes or Peugeot-Mulhouse this meant within a radius of 15 minutes to 1.5 hours.

Were Peugeot and Citroën now employing Japanese production techniques? Based on its own experiences and those of Japanese companies, since 1981 PSA had been re-examining its own industrial organization. While the group did not choose to move away from the model of assembly-line work, preferring instead to rely on a form of 'flexible Taylorism', it did introduce a number of fundamental changes. Within the PSA group Citroën progressed furthest, particularly through its *Plan Mercure*, a programme designed entirely by company employees. Accepted by top management and implemented from 1985, this programme combined a new way of organizing manufacturing in the factories with innovative social policies which increased the active involvement of workers in their work. Concepts introduced included flexibility, just-in-time, the *kanban* system, the Japanese 'no defects, no breakdowns, no stocks' approach, *poka-yoke* and *kaizen*. The merits of taking responsibility, teamwork, and the enrichment and polyvalency of work were praised.

Change also involved social reform. The *Plan Mercure* attempted to involve first-level supervisors in production management, as well as to entrust workers with greater responsibilities. The managerial hierarchy was reduced from seven layers to three. The positions of lower-level factory manager, foreman, and work group leader all disappeared. This type of reform required new skills, such as better technical knowledge, better people management, and better aptitudes for working in groups. Moreover, reorganization included training schemes to give managers the new competencies needed to perform their new responsibilities as well as to enable workers to undertake a broader array of more complex tasks. This approach underlay the agreement on classifications signed by Citroën in 1983 and Peugeot in 1985. The drive for training was unprecedented, for the first time extending to embrace even production workers. The share of overall employee costs due to worker training rose from 1.8 per cent to 4 per cent at Peugeot between 1982 and 1990. At Citroën, with its larger immigrant labour force, the increase was even greater: from 1.9 per cent in 1982 to 5 per cent in 1990. Hours devoted to training rose from 710,000 to 1.4 m. hours over the same period. By 1988, 64 per cent of Citroën's personnel were participating in practical and theoretical training courses, more than one-third related to the introduction of computerization. The same year, Sochaux signed an agreement with the Ministry for National Education to further develop continuing education. The group's intent was clearly to reskill the workforce, but this was not without its difficulties, since some workers were incapable even of reading bus or metro plans.

The enormous difficulty of getting these workers to acquire better skills forced the company to start hiring new personnel. This time, however, the emphasis was placed on previous training. All workers hired between 1987 and 1990 held at the very least a *Certificat d'Aptitude Professionelle* (CAP), and 80 per cent of technicians and staff had at least two years of post-school study. The most intense training effort remained linked to the launch of new models. For the ZX, Citroën spent 8 m. f. to train the whole of its production personnel, delivering 167,000 training hours. A mini-assembly line, costing 4 m. f., was installed at Aulnay to faithfully reproduce the chronology of assembly, involving the same types of work as on a real assembly line. There were several reasons for doing this. The personnel had optimal conditions in which to learn their tasks, and product development technicians were able to put the finishing touches to the product under conditions close to mass manufacture. This system permitted the much faster attainment of full output levels when production began for real. It took five months to reach 1,000 ZX automobiles per day, compared with eight months for the AX in 1987 and twelve months for the BX in 1983. The system was hardly new, however, since Citroën had practised it as early as 1962 when it opened its factory at Rennes, where peasants were brought in off the land to work in the factory.

*13.3.5. New Ways to Sell: Distribution Network, Servicing, and Image*

The last step in PSA's restructuring involved distribution and sales. After 1980 the merger of the Peugeot and Talbot distribution networks had been the key element in the merger of the two companies. The progressive rapprochement of the distribution networks—over 6,000 sales outlets, of which 913 were dealerships —was made necessary by declining sales. The networks of the three marques were no longer profitable when Talbot sales began to plummet. Talbot's difficulties afforded an opportunity to rejig an organization which had so far managed to avoid reform. There were long and difficult negotiations since 250 dealerships ceased trading and 200 garages disappeared in the competition. This was the end of a fifty-year era in which it was believed that the density of a distribution network alone could resolve all difficulties. The distribution networks were now a focus for profitability, an essential factor in a company's ability to respond more effectively to the new aspirations of the clientele.

Changes in long-standing sales practices began in 1982, and more clearly in 1988. From PSA's viewpoint this had started with a major offensive on the part of the American producers (GM and Ford), who were attempting to compensate in Europe for flagging sales in the USA owing to the success of Japanese producers in American markets. There were attempts to outbid other marques with price cuts, rebates, discounts, and preferential terms for financing. This placed considerable financial burdens on dealerships and on producers too, which were obliged to intervene financially themselves. The Peugeot and Citroën networks were fragile. The 1980–4 period may have covered the merger of Peugeot-Talbot, but there was also a decline in overall sales, reductions in margins, and above all price controls on automobile repairs in France. Citroën had even been compelled to repurchase some of its dealerships and turn them into subsidiaries in order to safeguard the coherence of its network. A decline of demand after 1990 made matters worse. To extricate PSA from these difficulties, the sales managers of Peugeot and Citroën decided to introduce reforms aimed at improving the services offered to their clients, since it was believed that by the 1980s and 1990s clients were more sensitive to the service provided by their garage than to the technical aspects of the automobile. Given that the distribution network was incapable of restructuring itself, the producers themselves took the initiative in the plans.

Training continued to play an important role. In 1990 Citroën opened a facility for sales training at Villepinte in the suburbs of Paris, and was followed by Peugeot at Cergy-Pontoise in 1994. Specialized training and refresher courses were given to strengthen the sales potential of dealerships, especially those which had so far hardly been concerned with the sales profession. In 1988 there was a reorganization aimed at equipping sales agents with real sales skills. However it was in after-sales service that the greatest efforts were made. With the technical development of models, particularly in electronics, it was necessary to increase the number of courses if garages were to cope with this profound transformation which meant that mechanical engineering was no longer the core skill. Garage owners too wanted to retain their clients, who were tempted by offers from franchise holders. Producers had an interest since they hoped for another way to cultivate loyal customers. Thus producers and garage owners went on the offensive, by simplifying maintenance operations, service without prior appointment, maintenance contracts, breakdown services, and payment facilities; not to mention the fact that the producers required their distribution networks to dispense with their traditional practices, increasing the number of days open for trading, ensuring a permanent presence, and even night-time opening at some Paris repair garages. In exchange, producers negotiated the resources: bank loans at preferential rates, purchase of tools and machines at discounted prices, special rates for emergency vehicles, and always broad support for training. Lastly, since 1986 each dealership had been obliged to computerize his business along lines established by the producer, and this too brought progress: diagnoses and estimates, precise knowledge of stock levels for parts and new vehicles, faster offers of credit, faster provision of vehicle immatriculation (registration) cards, and better control over orders. All this was designed to attract clients while creating a new image for the automobile repairers and producers. Image had become vital, in fact: a crucial factor in competitiveness. Peugeot and Citroën both wished to be perceived as young and enterprising companies. The above changes, combined with the use of advertising and sponsorship of automobile racing (Le Mans, Formula One), helped to counter any negative images related to restructuring.

13.4. CONCLUSION

Over a thirty-year period, the Peugeot group has experienced the most profound change in its history. It is not so much the change in size, accomplished largely during the 1970s, that is remarkable, but rather the leap made by an industrial group of disparate companies on the verge of crumbling, which, after a difficult period of restructuring, was capable of competing with the most powerful automobile producers. This capacity for change has been most significant, and appears to be related to PSA's ability to first draw on its own know-how before copying from experience elsewhere, whether the adoption of Fordist production techniques over a long period without really relying on volume, or the adoption of Toyotaist techniques without adopting Toyotaism. Indeed the most painful period in the group's history was precisely the Sloanist period when Peugeot distanced itself from its traditional values (notably financial) to create a 'General Motors à la française', seeming to lose its bearings and sense of direction.

It is true that during the recent period in which Volkswagen added SEAT and Škoda to Audi, or Fiat-Lancia assumed control of Alfa-Romeo, or Jaguar and Saab were taken over by Ford and GM, PSA seemed very isolated, rather like Renault, which failed in its attempt to merge with Volvo. Yet does this constitute failure? 'Given that these are the major European producers . . . [which are forging alliances], any movement towards merger will entail a loss of market share for the new partnership', warned Jacques Calvet, 'this was our experience during the merger with Talbot'.

PSA now preferred to rely on separate cooperative agreements with other producers, namely with Fiat, Heuliez, Renault, and even Japanese producers for assembly or distribution in Asia. Indeed in contrast to previous years, by the mid-1990s PSA was no longer exclusively focused on Europe. Although the company had partially withdrawn from its traditional African markets, it had intensified its presence in Latin America and, above all, had been aggressive in East Asia since the end of the 1980s, notably in China where both its marques were now present. In 1990 Citroën agreed to construct two factories, which by 1996 would be capable of producing 37,500 ZX models per year; addition of the Saxo was to boost output to 150,000 cars per year in 1999. It was these ZX and Saxo models, supported by the Xantia, which were hoped to provide Citroën with the opportunity to recapture the place it had held as a major producer for many years following World War II. The four model range, upon which Citroën had been able to base itself since 1991, gave Citroën the opportunity to achieve the same growth as Peugeot.

Finally, there is the social climate. Strikes at Sochaux in 1989, like those at Renault in 1991 and 1995, revealed the extent of worker unrest in France as far as industrial restructuring was concerned, as well as the determination of company managers to stay the course. The pressure exerted by Japanese producers, increased by the 1991 accord between the European Community and Japan to gradually open European markets, in tandem with a stagnating economic environment, brought the risk that the financial managers would force pursuit of a strategy based on inflexible rigour, a rigour which had for many years formed a key pole in Peugeot's strategy and model prior to the 1970s.

# NOTE

Translated by Sybil Hyacinth Mair.

STATISTICAL APPENDIX 13: Peugeot-Citroën

| Year | Production[a] | Exportation[b] | Employees[c] | Turnover[d] | Profits[d] |
|---|---|---|---|---|---|
| 1973 | 765,930 | 365,922 | 60,645 | 9,024 | 215 |
| 1974 | 730,770 | 401,264 | 58,735 | 9,920 | 52 |
| 1975 | 659,777 | — | 58,565 | 11,820 | 110 |
| 1976 | 1,437,800 | 670,000 | 176,500 | 35,066 | 1,428 |
| 1977 | 1,568,400 | 810,700 | 184,500 | 41,885 | 1,251 |
| 1978 | 1,665,300 | 843,100 | 190,170 | 47,810 | 1,382 |
| 1979 | 2,310,400 | 1,356,800 | 263,000 | 71,034 | 1,800 |
| 1980 | 1,961,000 | 1,160,300 | 245,000 | 71,103 | −1,504 |
| 1981 | 1,716,000 | 1,003,000 | 218,000 | 72,389 | −1,993 |
| 1982 | 1,602,600 | 895,700 | 206,000 | 75,263 | −2,148 |
| 1983 | 1,680,600 | 933,100 | 203,000 | 85,207 | −590 |
| 1984 | 1,600,000 | 877,700 | 187,500 | 91,111 | −341 |
| 1985 | 1,630,800 | 902,000 | 176,800 | 100,295 | 543 |
| 1986 | 1,707,100 | 951,800 | 165,000 | 104,946 | 3,590 |
| 1987 | 1,901,100 | 1,037,100 | 160,600 | 118,167 | 6,709 |
| 1988 | 2,080,700 | 1,201,500 | 158,100 | 138,452 | 8,848 |
| 1989 | 2,232,500 | 1,251,200 | 159,100 | 152,955 | 10,301 |
| 1990 | 2,208,200 | 1,276,800 | 159,100 | 159,976 | 9,258 |
| 1991 | 2,062,900 | 1,283,700 | 156,800 | 160,171 | 5,526 |
| 1992 | 2,049,800 | 1,306,700 | 150,800 | 155,431 | 3,372 |
| 1993 | 1,751,600 | 1,186,700 | 143,900 | 145,431 | −1,413 |
| 1994 | 1,989,800 | 1,255,200 | 139,800 | 166,195 | 3,102 |

*Note*: Growth through acquisition means that it is not possible to make precise comparisons for the period before 1979. From 1973 to 1975, the statistics refer to Peugeot, from 1976 to 1978 to PSA Peugeot-Citroën, including the Chrysler subsidiaries.
[a] Passenger cars, commercial cars, and light trucks. World production.
[b] Exportation, included foreign production.
[c] World employees.
[d] Turnover is pre-tax, and profits are net: in millions of francs.
*Source*: PSA, Annual Reports.

# BIBLIOGRAPHY

Automobiles Citroën, Automobiles Peugeot, Talbot Cie, PSA Peugeot-Citroën, *Rapports annuels d'activité*.
Caracalla, J.-P., *L'Aventure Peugeot* [The Peugeot Adventure] (Paris, 1990).
Célérier, S., 'Le Plan Mercure de la Société des Automobiles Citroën' [Citroen's Mercure project], in *Gestion industrielle et mesure économique, Approches et applications nouvelles* (Paris, 1990).
Célérier, S., 'La Machine humaine, un flux de fabrication et ses gestionnaires' [The Human Machine, the Production Flow, and their Manager], unpublished, CEREQ (1993).
Ciavaldini, B., and Loubet J.-L., 'La Diversité dans l'industrie automobile française: Hésitations et enjeux: Regards croisés de l'Historien et de l'Ingénieur' [Diversity in

the French Automobile Industry: Uncertainties and Challenges: Views from a Historian and an Engineer], *Annales des Mines, Gérer et Comprendre* (1995).

Corouges, C., and Pialoux, M., 'Chronique Peugeot' [Chronicle of Peugeot], *Actes de la recherche en sciences sociales*, 52–3 (1984).

Dalle, F., *Rapport sur l'industrie automobile française* [Report on the French Automobile Industry] (1984).

Dollet, C., and Dusart, A., *Les Sorciers du lion: Un siècle dans le secret du Bureau d'Etudes de Peugeot* [The Lion's Sorcerers: A Century of Secrecy in the Peugeot Design Department] (Paris, 1990).

Gandillot, T., *La Dernière Bataille de l'automobile européenne* [The European Automobile Industry: The Final Battle] (Paris, 1992).

'L'Industrie automobile' [The Automobile Industry], *Annales des Mines, Réalistés Industrielle* (Paris, 1991).

Hatzfeld, N., 'De Peugeot 1945 à PSA 1987: Croissance, crise et remise en cause d'une grande entreprise' [From Peugeot in 1945 to PSA in 1987: Growth, Crisis and Question for a Major Company], unpublished document, GERPISA (1989).

Henri, D., 'Peugeot: une histoire de famille' [Peugeot: A Family History], *L'Histoire* (1987).

Jemain, A., *Les Peugeot: Vertiges et secrets d'une dynastie* [The Peugeots: Secrets and Successes of a Dynasty] (Paris, 1987).

Loubet, J.-L., *Automobiles Peugeot, une réussite industrielle, 1945–1974* (Paris, 1990).

—— 'Comment Peugeot a épousé les classes moyennes?' *L'Expansion*, 461 (1993).

—— 'Peugeot, de l'entreprise familiale à la multinationale', in J. Marseille (ed.), *Les Performances des entreprises françaises au XX° siècle* (Paris, 1995a).

—— *Citroën, Peugeot, Renault et les autres: Soixante ans de stratégie* (Paris, 1995b).

—— 'L'Industrie automobile française d'une crise à l'autre', in J. Marseille (ed.), *Les Crises économiques du XX° siècle, Vingtième siècle* (1996).

Masdeu-Arus, J., 'Rapport de la Commission d'enquête chargée d'étudier la situation actuelle et les perspectives de l'industrie automobile française', *Journal officiel* (1992).

Rutman, G., 'Rapport du groupe de stratégie industrielle "automobile": l'automobile, les défis et les hommes', Commissariat Général au Plan (1992).

# 14

## Renault: From Diversified Mass Production to Innovative Flexible Production

MICHEL FREYSSENET

A cursory review of the history of Renault, combined with a glance at a few statistics and quotations, might lead one to believe that towards the end of the 1980s the French car maker made the transition from a Fordist model to a 'lean production' model, which is proposed by the authors of *The Machine that Changed the World* as the model for the twenty-first century. Yet Renault tried to be a Fordist company, but never succeeded. Still less did it now conform to the idea of 'lean production'. In fact during the 1960s the company had adopted the Sloanist industrial model, named after Alfred Sloan, who developed it and later theorized it while chief executive officer (CEO) and then president of General Motors. The Sloanist model can best be summarized as follows: the mass production of a diverse automobile product range which shares many components in common, within a framework of a compromise (called in this text company compromise) negotiated with trade unions which guarantees regular increases in the purchasing power of the workers, in return for which the latter accept the particular form of work organization associated with it. Renault adopted this model in an environment characterized by strong growth in demand, shortage of labour, and a crisis of work. Yet the model was never really mastered, and the attempted transition towards it brought a significant rise in production costs. The difficulties Renault encountered at this point might potentially have led to innovative solutions. But the abandonment of fixed exchange rates and the first oil crisis led to a breakdown in economic growth in several industrialized countries and prematurely transformed the automobile market in France into a replacement market.

During the following years, Renault sought to re-establish profitability without questioning its industrial model, that is, by following a strategy based on volume and a broader product range. After a brief attempt to seek renewed growth in the countries producing primary materials, the company attempted to take market share from other producers in the industrialized countries, particularly in the USA, at the same time diversifying its activities beyond the automobile sector. Fleetingly, Renault became the largest car maker in Europe. However, this was attained at the cost of growing indebtedness, and with no solution having been found to the crisis of work and the new problems of product design and manufacture which stemmed directly from the broader product range. Measures adopted to transform the company compromise and reduce costs came too late to prevent a situation of near bankruptcy for Renault.

After Renault had drastically reduced its break-even point in 1985 and 1986, management got the French State to play its role as the sole shareholder. The company then adopted a profit strategy in which the company's products were to be positioned at the upper end of each market segment, based on their quality and price, and of designing commercially innovative products, each with its own 'personality'. To create the conditions necessary for this strategy, a new social compromise was created. On the one hand, working time became flexible and workers became more fully involved in the improvement of performance. On the other hand, the content of work was enriched, careers were guaranteed, and employees gained a financial interest in the company's results. Renault thus began a transition towards an industrial model characterized by innovation and flexibility (Boyer and Freyssenet forthcoming). The group was profitable for nine consecutive years, significantly improved the quality of its products, designed a new product range, eliminated its debt, and transformed itself into a private enterprise. And yet the 1993 recession and the failure of a planned merger with Volvo were to pose severe tests for the strategy and forced decisions to be taken about the product range which might well prove contradictory with it.

## 14.1. RENAULT SELECTS THE SLOANIST MODEL FROM THE MID-1950S, AND SUCCEEDS IN ADOPTING IT DURING THE 1960S

Renault was nationalized in 1945 by the government of General de Gaulle. The company was now given the name *Régie Nationale des Usines Renault* (RNUR). Pierre Lefaucheux, the first managing director, clearly followed a Fordist strategy. However, he did not manage to make Renault's production system Fordist, if Fordism is understood as the mass production over several years of a basic, perhaps single model, according to continuous, stable, and standardized processes, by unskilled workers paid by the day at fixed rates that are high compared to the local average. He immediately opted for the mass production of a small 'popular' car, the 4CV, several prototypes of which had been made secretly during the war (Fridenson 1979). Specialized machine tools, and particularly transfer machines for machining, were purchased or made by the *Régie* itself. He also tried to obtain, for Renault's steel-making subsidiary, a wide rolling mill to make sheet steel. On the other hand, he was unable to install a continuous assembly line in the labyrinthine workshops of the Billancourt factory. The reorganization of work involved the enormous task of measuring the times needed to complete 25,000 tasks, which was undertaken during the two years prior to the launch of the 4CV in August 1947. Piecework was abandoned and replaced by an hourly production rate and not a daily wage. Half of this hourly wage was based on the rate of production, calculated for each worker or group of workers depending on the nature of their work, and comparing the actual level of production with the theoretical level established by the production engineering department (Labbé 1990). This major upheaval in the organization of work and the wage system, in a context of high inflation which reduced the purchasing power of workers, lay behind a series of strikes against the production-based salary and in favour of equal wage rises for all. On this occasion it proved impossible to maintain the cooperation between management and the principal union, the CGT (Confédération Générale du Travail), which had started when the *Régie Nationale* was established as a means of assuring its economic success. Cooperation would not be re-established for a further ten years, this time on the basis of a clear division of roles.

The 4CV was a commercial success. Yet Renault did not restrict itself to a single standard product. Before launching the 4CV, the company had restarted production of a pre-war model, the Juvaquatre, which had independent front wheels and a monocoque body, in saloon and van versions and later an estate version, for small businessmen and artisans. This model was produced until 1959. Renault also launched two further models. The Frégate was a large saloon for its better-off clientele, the four versions of which were produced between 1951 and 1959. The Colorale was a 14CV hybrid vehicle which was half passenger car and half commercial vehicle, the three versions of which were made between 1950 and 1956, and one of which, targeted at rural markets, had a rear door and flat floor, thus prefiguring one of Renault's principal designs of the 1960s and 1970s. While neither of these two models was as successful as expected, they did bear witness to the external pressures and internal compromises that led the *Régie* not to restrict itself to the mass production of a 'popular' standard model. Hence Renault's industrial model was not Fordist in product strategy, wage system, or production organization. At Renault, the production system and the employment relationships[1] remained Taylorist. Output was diverse, and as a result volumes varied widely. Wages depended on production and the continuous and mechanized production line had not yet been adopted widely in machining and assembly .

Many of Renault's engineers were fascinated by the success of Volkswagen, and wanted the *Régie* to focus its efforts on a single model. However, Pierre Dreyfus, who became managing director in 1955 after the accidental death of Pierre Lefaucheux, pushed for the idea of a progressive shift upwards of the model range combined with frequent replacement of models while at the same time reusing as many parts and components from the previous model as possible (Freyssenet 1979). The poor sales of the Frégate persuaded him that the demand for larger cars was still too limited among their potential clientele, which was composed largely of employees. He decided to maintain a close watch over the incomes of this part of the population, which was rapidly becoming the majority. Renault launched the Dauphine in 1956, when the 4CV was at the height of its production. This somewhat larger model won over a new clientele, permitting the 4CV to survive a further four years. The Dauphine was produced until 1970, with two derivative models, the sporty Dauphine Gordini and the better equipped Ondine. In 1959 two further models were introduced, the Estafette, a front-wheel-drive

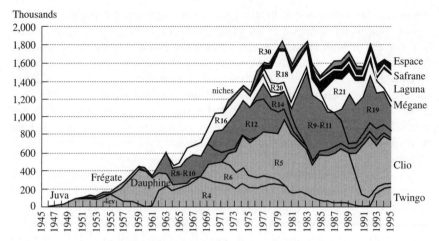

FIG. 14.1. *Total production of Renault passenger cars by model*
*Source*: Renault annual reports.

commercial van, and a 'niche' model, the Floride, which had coupé, cabriolet, and convertible versions as well as a version for the American market, the Caravelle.

To this classic Sloanist product strategy, Pierre Dreyfus added a strategy of differentiating products from those offered by the competition. He sought to design functional models that could be used both to transport people and to transport small loads. The first of these models was the replacement for the 4CV, the R4, a front-wheel-drive model with a flat floor and five doors, one of which was a large rear door. Produced until 1995, total output of the R4 was to exceed 8.5 million units (Fig. 14.1). When the Dauphine reached the middle of its life, in 1962, Renault launched the R8, just above it in the small-medium car segment. In 1965, the company introduced the R16, which overlapped the small-medium and medium-large segments. This model was part saloon, part estate, and had a sloping rear door, a body type that was later copied by many car makers. Believing that the gap between the R8 and the R16 was still too big for owners of the R8, Renault introduced the R10, a derivative of the R8, between them; this was the company's last rear-wheel-drive car (Loubet 1995). By 1965 there were six models in the range, with three platforms, and an average volume of 160,000 units per platform (Lung and Jetin, forthcoming).

Pierre Dreyfus also made exportation a priority of the *Régie Nationale*. Sales outside France rose from 25.9 per cent of overall sales in 1954 to 54.9 per cent in 1960. Indeed sales abroad had reached 61.1 per cent in 1959 with a rapid increase of exports to the USA. However, since the Dauphine was not adapted to the US market, the distribution network was underdeveloped, the American car makers reacted to its introduction, and sales declined equally rapidly. Exports gradually rose again to 44.8 per cent in 1965, largely owing to the growth in foreign factories and agreements to assemble vehicles abroad, of which there were twenty in 1965 (there had been five in 1955). By 1965, the main production sites for Renault vehicles abroad were in Spain (47,300), Belgium (46,100), Argentina (23,400), Brazil (13,464), Mexico (4,150), Venezuela (2,460), South Africa (2,350), and Canada (1,250). In 1964 Renault created a holding company in Switzerland, Renault Holding, in order to finance its investments abroad by borrowing abroad, and hence to avoid weighing down its balance sheet in France and being penalized by the depreciation of the French franc (Fridenson 1993).

Renault's Sloanist orientation in the development of its product range was complemented by its relations with its workforce. In 1955, management and unions signed the agreement that would form the basis of the employment relationships at Renault until the end of the 1970s. Significantly, this agreement guaranteed employee purchasing power, which was overseen on a monthly basis by a round-table committee that examined the cost of living. The agreement also introduced a third week of paid holiday, and the parties agreed not to resort to the lockout or the strike until all other possibilities for finding solutions had been exhausted. Management had entered into a tacit alliance with the CGT, the main union, which would last more than twenty years. The alliance was centred around the objectives of Renault, which had simultaneously to be competitive, to offer innovative products in mass quantities, position itself at the cutting edge of social advances, and respond as far as possible to the needs of France as a country, in terms of exports, industrialization, regional planning, and the 'French role' in world affairs.

The workforce at Renault doubled between 1947 and 1965, reaching nearly 67,000 employees, despite the spinning off of certain activities and the growth of subcontracting. In 1947 employment was concentrated in the Paris agglomeration (83 per cent), largely at the Billancourt site. By 1965 only 48 per cent of employees worked in the Paris region. The *Régie* had rapidly spread its activities down the valley of the River Seine in search of appropriate conditions for mass production—space to build continuous production lines, road, rail, and river links to support components supply and the distribution of finished cars, and a sufficient quantity of unskilled labour—at the same time responding to the wishes of the public authorities in France to reduce congestion in the Paris agglomeration (Freyssenet 1979). Two principal factories were built: a body and assembly factory at Flins in 1952 and a mechanical components factory at Cléon in 1958. But growth was so rapid that Renault had to retain a significant workforce at Billancourt, where employment in 1965 was at the same level as in 1947, with approximately 30,000 employees. Billancourt had to bring in increasing numbers of immigrant workers, who by 1967 accounted for 20.1 per cent of the workforce there (compared with 13.2 per cent at the *Régie* as a whole). The widespread adoption of the continuous assembly line, combined with the complete replacement of the stock of machinery by specialized machine tools and transfer machines, modified the socio-professional composition of the workforce. Mechanization combined with reductions in cycle times, moreover, meant that

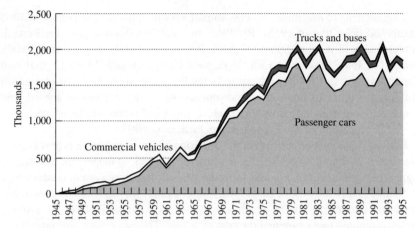

FIG. 14.2. *Total production at Renault by category of vehicle: passenger cars, light commercial vehicles, trucks and buses*
Source: Renault annual reports.

the worker had less and less control over his production activities. In the end, the timing of tasks, which led to conflicts each time production was changed, gave way to numerous local arrangements which substantially modified the 'production rates' that served as the basis upon which wages were calculated, thus creating significant disparities which led to further conflicts and complications in the management of wage slips. Between 1954 and 1960, the timing of tasks was gradually replaced by time and motion studies, while the production-based rate was replaced by an hourly rate which varied depending upon the classification of work post occupied. For the first time, there was no direct link between the amount of the wage and output. However, the output level that had been set had to be maintained, or the worker would be moved to another post, generally one less well remunerated (Labbé 1990). The management of Renault, concerned to maintain the social compromise that had pacified industrial relations, granted higher wages than those paid by other producers, as well as a fourth week of paid holiday in 1962.

The Sloanist orientation of the Renault management was also reflected in the development of the company's organization. The *Régie* inherited numerous activities connected to automobile production from the company of Louis Renault. In order to identify and decentralize responsibilities, between 1955 and 1965 Pierre Dreyfus had created a number of divisions: agricultural machinery, railway products, general machinery, and machine tools. In 1955 Renault merged its heavy goods vehicles activities (over 5 tonnes) into SAVIEM, a company created with two other producers, Latil and Somua, of which Renault would take control in 1965 (Fig. 14.2). There were also two steel-making subsidiaries created before the war, SAFE and SAT, and a subsidiary, SNR, created after the war at the request of the government to mass produce ball bearings, which French industry was lacking. Renault also eliminated some of its activities by subcontracting them. By 1960 there were about 5,000 suppliers. Subcontractors were also used to manage production capacity for some bodywork and for the assembly of some special versions of car models. The result was that the level of vertical integration (estimated by comparing value added to turnover) declined gently, from 48.9 per cent in 1956 to 41.4 per cent in 1965.

Hence by the mid-1960s Renault possessed some of the basic characteristics of the Sloanist industrial model. The company's worldwide production of passenger cars and small commercial vehicles (under 5 tonnes) reached 590,431 units. The average break-even point during this period was an enviable 38.2 per cent (Boyer et al. forthcoming). The quality of the vehicles was average, according to Swedish tests and the criteria of the time. Renault was ahead of its competitors in France, was the third largest producer in Europe, and the sixth largest in the world.

## 14.2. THE CRISIS OF WORK AND PROBLEMS IN PRODUCTION, IN A CONTEXT OF STRONG GROWTH, INTERNATIONALIZATION AND THE DIVERSIFICATION OF OUTPUT, 1966–1973

*14.2.1 The Doubling of Volumes and the Broadening of Activities is Accompanied by a Rise in the Break-Even Point*

The market for passenger cars and commercial vehicles in France grew from 1.21 million vehicles in 1965 to 2.02 million in 1973. During this period Renault also grew rapidly, with only two slower years, 1968 and 1971, both affected by long and hard strikes. Worldwide production expanded 2.4-fold, reaching 1.41 million units in 1973. Renault's share of the French passenger car market grew from 26.5 per cent to 30.1 per cent, even though customs barriers between Common Market countries were eliminated in 1968. Two-thirds of the increase in production was due to exports and to vehicles assembled or manufactured abroad, the volumes of which expanded 3.1-fold to 809,255 units (Fig.14.3).

While these results were facilitated by the devaluation of the franc in 1968 and by the various difficulties being encountered by some of Renault's foreign competitors, especially Fiat, British Leyland, and Volkswagen, they confirmed the validity of the strategy of moving up market and bore witness to the company's competitiveness. During this period, Renault began to replace its product range for the first time, at the same time broadening it. At the lower end of the range two five-door models were launched, the R6 in 1969 and a small urban car, the R5, in 1972. In the small-to-medium segment, a saloon, the R12, replaced the R10 in 1970. Renault pursued its niche strategy in 1971 with two coupés, the R15 and the R17. By 1973 the *Régie* was offering eight models on four platforms, maintaining the 2:1 ratio. The average volumes per platform for

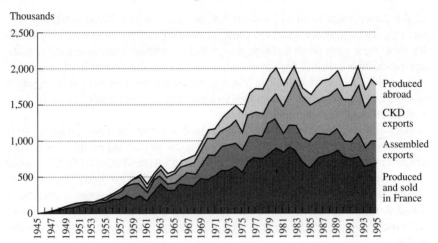

FIG. 14.3. *Renault's worldwide production of passenger cars and light commercial vehicles by place of production and sale*
*Note*: CKD: author's calculations from 1945 to 1955.
*Source*: Renault annual reports.

production in France had risen from 160,000 in 1965 to 275,000. In other words, Renault had strictly applied Sloanist recipes. Moreover, in 1966, the company joined forces with Peugeot, with which it had already cooperated in export markets, in order to organize joint research projects, to undertake production for each other so as to better utilize their respective production equipment, to construct shared factories that neither company could afford alone, to jointly design and produce mechanical components that both companies could use, and to help each other further in export markets. The Renault and Peugeot design departments began to cooperate. The two companies manufactured components for each other. Each invested 25 per cent in the company Chausson-Carrosserie in order to permit construction of a new factory at Maubeuge which would produce their low-volume models. Most significantly, they created two subsidiaries, Française de Mécanique, to make engine parts, and Société des Transmissions Automatiques to make gearbox parts. This partnership was joined by Volvo in 1971. The three car makers jointly designed a V6 engine, first produced in 1975 (Freyssenet 1979).

Renault was able to increase its production volumes by assembling CKD (completely knocked down) kits abroad, and to a lesser extent by sending 'small sets' of components to countries where a high level of local content was required. By 1973 there were a total of twenty-five factories abroad or local assembly agreements. The operations in Spain and Belgium were the key poles, and in 1973 they produced 173,000 and 114,000 Renault vehicles respectively, nearly two-thirds of the 450,000 vehicles assembled or produced abroad that year. Renault abandoned Brazil in 1968, but the same year signed an agreement with Romania to produce the R12 model in exchange for mechanical components, and started operations in Yugoslavia in 1969, Colombia in 1970, and Turkey in 1971. The company was obliged to substantially increase its participation in the capital of the companies that produced its vehicles abroad, sometimes having to assume control of them when other partners could not afford new investments. Renault's share of FASA in Spain reached 50 per cent in 1967 when the body and engine factories at Valladolid were opened (Charron 1985).

Renault diversified not only its range of vehicles but also its other activities. When invited by the government, Renault purchased companies, often those in difficulty. This was only done, however, when these companies supported Renault's own activities and capabilities. On the other hand, these acquisitions weighed heavily on the company's short-term investments. Between 1965 and 1973 the number of industrial subsidiaries in France in which Renault controlled the majority of the capital rose from four to sixteen, and employment at such subsidiaries rose from 14,931 to 32,532. Some new subsidiaries were related to automobile production, including Renault-Gordini, Alpine-Renault, and STA (gearboxes), while others were involved in industrial products, including CPIO (rubber and plastic products), SBFM (cast-iron parts), and SMI (screw cutting). Renault was now involved in transportation equipment with SMV and SNAV, engines with Renault Marine Couach and Bernard Motors, and machine tools with SMC and Acma-Cribier.

However, the increase in average volumes per platform and mechanical components, a strategy pursued in order to compensate for the higher costs entailed by a wider product range, were not accompanied by a reduction in the break-even point. On the contrary, this rose from 48.8 per cent in 1967 to 18.0 per cent below added value in 1973. The rise was owing to a 116.6 per cent increase in the overall wage bill in constant francs and a 26.7 per cent increase in the average wage (including benefits) as well as a 3.4-fold increase in depreciation. These increases in turn resulted from both a crisis in the company compromise and difficulties in coordinating and retaining flexibility that were linked to the diversification of the product range and of Renault's activities as a whole.

## 14.2.2. The Crisis of the Company Compromise

The workforce in France rose from 62,902 employees in 1965 to 101,415 in 1973, a 61.2 per cent increase. Renault had to expand its factories. It was decided to create a fourth body and assembly factory at Douai in France, in two stages. The body factory was opened in 1971, and the assembly plant in 1974. The company was none the less surprised by a sudden increase in demand in 1969, and therefore had to set up two-shift work in the older factories, especially at Billancourt and Flins, at short notice. For this, Renault recruited large numbers of immigrant workers for the Billancourt and Flins factories and large numbers of rural and young people for the factories in the regions. Immigrant workers made up 21.5 per cent of the total Renault workforce in 1972, compared to 13.2 per cent in

1965. They accounted for 40 per cent of recruitment between 1968 and 1970, a proportion that rose to 60 per cent at Billancourt and Flins. Young employees, under 25 years old, made up 16.9 per cent of the workforce in 1973.

These new workers, whose prospects for career advancement towards skilled jobs were less favourable than those of the previous generation, threatened the company compromise that had been introduced during the 1950s. More frequent changes of assignment to posts that were remunerated differently introduced monthly variations of salary that employees found increasingly hard to accept. This refusal to accept the prevalent form of work organization in a context of full employment was reflected in the difficulty in recruiting French workers and by a rise in the number of workers leaving the company through resignation or dismissal: a number which increased from 5,500 in 1965 to 8,000 in 1970. On the other hand, absenteeism (absences of less than one month) among production workers remained relatively stable, at between 5 and 6 per cent. Above all, however, the questioning of work organization and of the company compromise manifested itself in an explosion of social conflicts. From 1968 to the first oil crisis, there was not one year without a major conflict in one of the factories, conflict that was generally initiated by the unskilled workers.

These conflicts usually ended in an agreement that focused on a wage settlement, without any real solution being found to the problems related to work organization and the content of work. At the same time, certain changes were made to the wage system based upon classification of work posts, which paved the way for future developments. The number of classes of post was reduced from twelve to five between 1968 and 1973, and the number of pay rates within each class was reduced from five to three, including a starting rate, defined according to precise criteria, that was the same across the company (Labbé 1990). From 1971, the class of production posts with the highest remuneration was called 'skilled production worker', which corresponded to posts that required somewhat more experience. The workers, under the principle of 'equal work, equal pay' which was inscribed in the labour code, used new conflicts to ensure that most workers now came under this category. From 1972 to 1975, 15,000 'unskilled workers', or 25 per cent of the total production workforce at Renault, obtained 'skilled production worker' posts (Fig. 14.4). This was the first major slippage of jobs in the classification schema that did not correspond to a change in work content. Formally, remuneration remained linked to work post. Yet in introducing the idea of complexity into the definition of production work posts, Renault opened the door to a system of classifying jobs according to experience, a system for which unions and workers continued to press. Moreover, monthly pay was spread step by step to the whole workforce between 1968 and 1974, and was a root cause of the sudden increase of absenteeism among production workers, which rose from 6.3 per cent to 8.1 per cent from 1972 to 1973.

The second response to the crisis of work was to reduce working time and develop skills training. Between 1968 and 1977 the length of the working week was gradually reduced from 47 hours and 30 minutes to 39 hours and 10 minutes

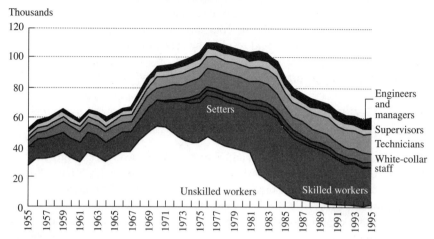

FIG. 14.4. *Renault employees by category*
*Source*: Renault annual statistics, social reports.

for employees doing shift work, and from 48 hours to 40 hours for all others. The government's 1971 law on skills training proved controversial as management sought to utilize it to prepare for the introduction of automation whereas the unions sought programmes based on general culture, literacy for immigrant workers, and skills training for unskilled workers through acquisition of the official diploma awarded by the National Education Ministry, the Certificate of Professional Aptitude.

It was some of the company's managers, rather than the unions, who became preoccupied with changing the content and organization of work. Experiments with lengthening the cycle of work and enriching tasks took place from 1971 onwards. In 1972, a central working group, including representatives of the Personnel Department and the Production Engineering Department, was given the objective of 'not only humanizing technology, but also investigating new ways of organizing production areas and designing equipment and buildings, from the perspective of the degree of freedom and initiative permitted to the personnel, the utilization of its competencies, and opportunities to work in teams, [and] measuring the economic value of each operation' (Renault 1973).

Several new production operations resulted: notably the assembly of an entire engine by a single worker at a fixed post, the adoption of four short assembly lines in place of a single long line at Douai, and the widespread adoption at the Le Mans factory of modular assembly of front and rear axle assemblies by self-organizing groups of three to five workers. In this last case, the groups not only undertook the entire assembly of the unit but also took on all the controls and maintenance of the equipment used for control, accepting responsibility for quality, undertaking any rework, and accepting responsibility for the cleanliness of their machines and work areas. Yet these new operations proved controversial at the highest levels within the company. In 1974 Pierre Dreyfus publicly declared that 'none of this would resolve the problems of production work', the sole solution being the progressive reduction of working time: it would be 'utopian to recreate a sort of craft work in the factory' (Dreyfus 1977).

### 14.2.3. The Difficulties of Flexibilization and Coordination Caused by the Variety of Production and Diversification of Activities

The diversity of a firm's products and activities poses two types of problem: the first concerns its apparatus for design, production, and its labour force, while the second concerns its overall organization and investments. The demand for different products evolves at different rates. The risk is that some factories or lines will run under capacity, with others over capacity, and that design departments will become fragmented. Moreover, the diversification of activities runs the risk of dispersing and partially duplicating investments.

During this period, Renault took no particular measures to make its production apparatus more flexible, with the exception of certain modular work experiments, the principal aim of which was in fact to modify work. However, the company did possess a relatively polyvalent labour force, since the wage system based upon classification of posts permitted operators to be moved about as necessary. In terms of product planning, an office for economic planning and research was set up in the company's headquarters, under the management of Bernard Hanon. In 1970, this office became a full department, the Department of Data Processing and Planning (DIP, Département de l'Informatique et de la Planification), which would eventually supervise the whole set of activities involved in the automobile industry from product definition to after-sales (Freyssenet 1979). Until the mid-1960s, the design department moved successively from one project to the next. The managing director decided to start the design of a new model, making decisions more or less intuitively. At the lower rungs of the process, the designers and technicians from different departments and services coordinated their work informally. Their senior managements only intervened when they were unable to agree upon a technical solution, or at the major stages in the process when the dossiers were passed on. The profusion of models now required teams to work in parallel on different projects. The number of employees in the design and research department rose from 752 in 1956 to 1,488 in 1967 and then to 2,876 in 1973 (Poitou 1988). The system then began to fragment. While the managerial hierarchy became more and more absorbed with administrative tasks, it grew increasingly incapable of making the necessary decisions. In 1972, a project leader was chosen for each new vehicle, but this leadership from above was still not accompanied by coordination at the lower echelons (Midler 1993).

Renault pursued the decentralization of its organization while at the same time creating points of coordination. A new autonomous division was created in 1968, Renault Machine Tools, to supervise the manufacture of machines. In 1967 a secretariat for the Renault Group had been created with the goal of coordinating the profit and investment budgets and improving the use made of each Group company's potential for design, production, and sales. Yet the fourfold increase in the number of subsidiaries was too fast for the secretariat to be able to assess the merits of all the investments made. In 1971, an International Affairs Department was established to manage and coordinate both export sales and the overseas factories. Renault added to its international financial apparatus by establishing Renault Finance at Lausanne, Switzerland, in 1973, to manage the Group's international financial affairs and to minimize exchange rate risks, which had increased since President Nixon had decided to end fixed exchange rates, and by setting up *Renault Crédit International* in Paris in 1974, which oversaw consumer credit subsidiaries that were established abroad. The financial subsidiaries were expected to supply additional income, especially during the difficult phases the company was soon to experience.

On the eve of the first oil crisis, then, neither the employment crisis nor the difficulties of coordination and flexibilization had been resolved. However, there were no doubts that the pursuit of growth would permit solutions to be found within the context of the industrial model that the company had adopted, which seemed at the time to be the only one possible. Nobody foresaw that the reduction in incomes for the raw-materials exporting countries brought about by the depreciation of the dollar would cause severe reactions in the oil producing countries to political events in the form of an agreement among them to raise the price of crude oil. The first oil crisis stopped growth in its tracks.

## 14.3. THREE ATTEMPTS TO RE-ESTABLISH PROFITABILITY, STILL BY INCREASING VOLUMES AND DIVERSIFYING ACTIVITIES, 1974–1984

The domestic French market declined by 12.8 per cent with the first oil crisis, and reached a plateau of 2 million for the next ten years. Yet, in contrast to other European producers, Renault pursued growth through volume, at the global as well as national level, except for two years: 1975, when two strikes led to the loss of more than 90,000 vehicles, and 1981, when the European market declined. Renault's worldwide production rose from 1.41 million in 1973 to 2 million in 1980, and maintained this level in 1982 and 1983. The company's share of the domestic market for passenger cars increased from 30.1 per cent to 40.5 per cent. While, overall, exports stagnated during this period, the number of vehicles assembled or manufactured abroad rose by 79.8 per cent between 1973 and 1980. The Spanish subsidiary FASA-Renault saw its production grow by more than 50 per cent between 1977 and 1980 with the opening of a third assembly plant in Palencia.

Renault owed this growth to the fact that it possessed mass market vehicles to which its clientele turned during the latter half of the 1970s. The R4 model was attractive again, and the recently introduced R5 became the 'vehicle of the crisis'. On the other hand, the models launched between 1974 and 1984 all experienced difficulties. At the upper end of the market, the R30, launched in 1975, failed, and was not replaced by the R25 until 1984. In the upper-middle range, sales of the R20 were eaten into by the R18, launched two years later, which then went into decline before its replacement, the R21, was ready. In the lower-middle segment, the R14 was added in 1976 to the R12, then in mid-life, but had to be replaced prematurely by the R9/R11 in 1981 and 1983. Lastly, the premature announcement of the launch of the 'Supercinq' provoked a severe drop in sales of the R5. Elsewhere, Renault pursued its policy of 'niche' vehicles with the Alpine, a sports car, and the Fuego, a coupé. The Renault Espace was first marketed in 1984, after having been offered to Renault by Matra, and was a new type of vehicle which became Europe's first minivan. In commercial vehicles, the Estafette was replaced by two more powerful models, the Trafic and the Master, which were made by a new subsidiary and new factory at Batilly, in Lorraine, in the context of the 'Steel Plan' for the restructuring of this declining steel region.

The ratio of models per platform remained about 2:1. The average volume per platform continued to rise, oscillating between 250,000 and 400,000. Never before had Renault achieved such economies of scale. Despite this performance and the recency and diversity of its model range, Renault's gross profits were structurally negative for more than ten years (see Statistical Appendix 14). The company was only able to break even thanks to other income, largely financial. Renault made three successive attempts to make its car manufacturing activities more profitable while pursuing its strategy of increasing both the volume and the breadth of its model range and overall activities, each attempt made in a different way.

### 14.3.1. The Search for Markets in Countries Producing Raw Materials, the Pursuit of Diversification, and the Deterioration of Relations with the CGT, 1974–1977

Managers at Renault believed that, in the future, demand for automobiles in the industrialized countries would increase less than before. With competition becoming more intense, it would be necessary to identify client expectations and make the production system flexible. Conversely, countries producing raw materials were likely to become areas of strong market growth for automobiles, and diversification of activities might allow compensation for short-term losses in the automobile markets. Renault therefore attempted to invest or expand in the oil producing countries. It was unsuccessful in the cases of Algeria and Venezuela. In 1975 the company negotiated a contract with Iran for a factory producing 100,000 vehicles per year, but the Islamic revolution thwarted its completion. It soon became clear that the new markets were very uncertain, and had already been entered by the Japanese. Industrial diversification meant an increase in the number of subsidiaries (from sixteen to twenty-two between 1974 and 1978) and their employees (from 32,532 to 57,242). However, this did not always occur in the sectors that might have been anticipated. In 1974, Renault was driven to assume control of the heavy commercial vehicle maker Berliet, which was merged with its own subsidiary, SAVIEM, in 1977, under the name of Renault Vehicules Industriels (RVI). The oil crisis had dealt fatal blows to Citroën and Berliet. At a time that was hardly opportune, Renault and Peugeot found themselves competing to purchase these companies, in the knowledge that one of them would then succeed in becoming France's principal car producer. In the end, Citroën was taken over by Peugeot and Berliet by Renault, following discussions with the State. The Peugeot-Renault partnership could not withstand the battle between the companies, and became limited to the production of shared mechanical parts. The expansion of Renault's activities led Bernard Vernier-Palliez (successor to Pierre Dreyfus in 1975) to create the Renault Group, which was comprised of three sectors: automobiles, industrial vehicles (heavy trucks, buses, and coaches), and industrial firms. This reorganization, however, was accompanied by duplication at group level of the core functions that were also shared by the sectors, and so failed to resolve problems of coordination, which were in fact amplified (Fridenson 1993).

Neither one of the two strategies produced the desired results, and nor had the crisis of work and the rigidities of the design and production apparatus been surmounted. Indeed these caused a fall in the quality of Renault vehicles compared to those of competitors. In the short term it was not believed necessary to reduce the labour force, and by 1977 there were 110,485 employed at the *Régie* itself, the most in the company's history, and 243,456 employed at Group level (see Statistical Appendix 14). Policies on wages, however, became more restrictive and social conflicts ensued. The strikers demanded uniform salary increases and for all production workers to accede to the 'skilled production worker' category. While management conceded a new reduction in the number of classes of work posts and wage rates (to four and two, of which one was a rate for the newly hired), it refused to question the principle underlying the classification system, namely payment according to the class of work post held. The sequestration of Bernard Vernier-Palliez by strikers from Department 38 in Billancourt in November 1977, and the subsequent firing of the CGT secretary at the factory, revealed the deterioration of relationships between the management and the principal union. Short-term absenteeism among workers rose to 9.9 per cent in 1976 and remained above 9.5 per cent until 1978. Yet the rapid rise of unemployment had already reversed power relationships between employees and employers in France, as revealed in the case of Renault by a significant fall in the voluntary turnover rate in 1974 and 1975, and its steady decline in the years following. Some production managers, who found it increasingly difficult to manage the diversity and the increasing variation in production, saw in autonomous group work the possibility of modifying the volume and variety of the production programme more easily and at lower cost. However, the production engineering department openly opposed this (Midler 1980). Their solution to the malaise of the unskilled worker was rapid automation, which was supposed to eliminate arduous and repetitive tasks and to create more attractive and skilled operation-supervision-maintenance functions, while simultaneously making the production apparatus more flexible and raising product quality. This is the path that would be pursued.

The poor results obtained by Renault in its automobile activities, despite greater volumes, were ultimately the result of the price strategy and the failure of three of the company's models. The increase in production costs due to the increase in raw materials prices could only partially be carried over to the price of the product, given market conditions and the consumer price controls that had been established by the state. The R30 was launched despite having twelve major defects, in order to avoid delays. Renault had to pay high warranty costs as well as see the image of its cars deteriorate. In 1977 a new quality department was created, which attempted to focus on quality at each stage of product development (Fridenson 1995). However, the company's primary preoccupation remained production volumes. The R20 suffered from being too closely related to the R18. The R14's failure was even greater since its objective was to repeat, in the middle of the range, the success achieved by the R5 in mass market vehicles, and thus to increase Renault's European market share from 11 to 13 per cent. The model had been designed by a team from DIP to be very 'liveable' while remaining very compact, thus opening Renault's path towards the concept of designing 'living cars', during the latter half of the 1980s. The same DIP team suggested thinking about the replacement for the R4 as a whole, in order to achieve significant savings and reach the difficult target of making a mass market car profitable. This was different from the conventional approach in which each department involved in design would make marginal improvements to the parts for which it was responsible. Both the resulting conflict, and the failure of the R14, led to a return to a form of industrial classicism and realism, in terms of product strategy and an abandonment of project-based design (Midler 1993).

### 14.3.2. *A New Direction: The End of the 1950s Company Compromise and Attempts to Invest in the USA, 1978–1981*

The period of stagnation in 1978, combined with the increased uncertainty over exchange rates, oil prices, international trade, and host country policies towards foreign investment, convinced Renault's management that the crisis did not prefigure a displacement of the world's poles of economic growth, but instead a crisis in the global economic system. Only one thing was certain, wrote Bernard Vernier-Palliez in early 1979, and that was increasingly intense competition, 'since each company only gains at the expense of others'. For the first time Renault was concerned about the growing strength and competitive pressures exerted by Japanese producers. In Europe the *Régie* profited from the difficulties experienced by its direct competitors, notably PSA and Fiat. However, Renault watched, not without some disquiet, as Ford, then GM, took advantage of changes in European regulations to open factories in Spain, a market where Renault's subsidiary FASA had become the main producer. Moreover, Renault had no presence in the world's largest automobile market: the USA. It was decided to reinforce the Iberian investments by constructing assembly and engine plants in Portugal. The moment was seized to acquire a 15 per cent share in Volvo Car. Above all, in 1979 Renault signed an agreement with the American Motors Company (AMC), which found itself unable to renew its models, in order to manufacture Americanized versions of the R9, R11, and the R21 (Alliance, Encore, and Medallion) in AMC's Kenosha factory. In addition, an engine plant was built in northern Mexico. In the same year, Renault's heavy commercial vehicle subsidiary, RVI, acquired shares in the American company Mack trucks, later acquiring a majority holding. Unfortunately, the deterioration of the American market between 1980 and 1982 weakened AMC and obliged Renault to increase its capital share from 22 per cent to 46 per cent. 'Even before the launch of the Alliance, AMC's operating costs had exceeded previously forecast levels by more than 50 per cent' (Charron 1985).

This attempt to acquire market share required better control over costs and improved productivity and quality. The workforce at Renault fell steadily by 1,500–2,000 workers per year between 1978 and 1983, with only partial replacement of leavers. Lower demand led immediately to lay-offs. The government's advice to companies to abandon index-linking of salaries to prices was followed by Renault. The very foundations of the earlier company compromise were now under assault. A new system linking remuneration and promotion to attendance at work was instituted in 1978. Complemented by various controls, this encouraged lower rates of short-term absenteeism, which fell from 9.5 per cent in 1978 to 8.2 per cent in 1979 and then to 7.1 per cent in 1983. Management no longer hesitated to crush local strikes, requesting police intervention and firing dozens of workers. But it was difficult for the company to maintain the fiction of a system of classifying personnel according to the post occupied. The fall in turnover and the reduction of the workforce restricted opportunities for promotion, while the ageing of the workforce only intensified demands for the creation of career paths for workers. The election of a Socialist government in 1981 and the conflicts that took place later that year led management to sign an agreement with the unions which for the first time instituted a system of classification and promotion linked to an individual's skill and not the classification of the work post itself. This system anticipated progression to the higher class after five years, the period needed for the worker to acquire the capability to hold various posts (at the same level) and meet relevant quantity, quality, and safety standards. It reconciled the worker's demands for careers and the management's objective of making polyvalency official. More immediately, many workers already fulfilled the conditions established for advancement into the higher category. Hence in 1982 57,763 and in 1983 12,747 of the workers changed categories. With the highest level, that of skilled production worker, attainable after fifteen years' seniority, the system led to the quasi-disappearance of the unskilled production worker ten years later (Fig. 14.4). Automation was management's second agenda item. Major programmes were launched to robotize welding operations and automate machining, with the goal of, it was hoped, improving productivity and quality simultaneously. Those who still supported group work, tiring of the opposition of production engineers to reorganizing manual work, saw in the automation of the factories opportunities for new ways of working (Freyssenet 1984).

Both increased capacity and the modernization of the factories entailed a substantial rise in physical investments, the annual level of which doubled between 1976 and 1980 (constant prices). Renault's excellent financial management, its international reputation, and the *de facto* State guarantees made it easier for the company to obtain credit when its ability to finance itself declined (Fridenson 1993). Despite the success, 1981 confirmed Renault's growing fragility. In contrast to the first oil crisis, the second shock had a particular impact on Renault, and the company's worldwide output shrank by 235,000 vehicles, or 11.8 per cent. Renault's results at the end of the year were negative. And now, soaring interest rates led the company heavily into debt.

*14.3.3. Further Changes of Direction: Plans to Increase Capacity and Reduce the Break-Even Point, but they Come too Late, 1982–1984*

Bernard Vernier-Palliez was replaced by Bernard Hanon at the end of 1981. The new managing director set ambitious targets in terms of volume, range, and internationalization, summarized thus: 'one model, one factory, one new country per year', with the intention of attaining an output level of 2.5 million vehicles in 1985. At the same time he demanded a reduction in the break-even level to 20–30 per cent below added value. The problem, he believed, was not so much capacity increases and productivity investments, but lack of control over costs.

By 1982 Renault had regained its record outputs of 1980, benefiting from the stimulus to consumption given by the first leftist government and the difficulties being experienced by PSA (Peugeot) following its purchase of Chrysler-Europe (Loubet 1995). And yet by the end of the year the company had announced a deficit of 2.56 bn. f. Bernard Hanon attempted to mobilize the whole company through a wide-ranging debate on 'industrial change and economic and social dynamics' (the MIDES debate), in which all unions accepted to participate. In light of the new, more favourable political context, the CGT sought to reopen dialogue with company management. Discussions revealed a range of wealth of experience (notably in terms of group work) in parts of the company that was often unknown by other factories, or indeed other departments. Changes were in fact started throughout the company. The difficulties and opposition encountered in trying to implement project-based organization of design from the top down did not lead to a rejection in principle of this mode of organization. Between 1980 and 1985, several study tours were organized to visit foreign producers, notably the Japanese, with the aim of comparing experiences and seeking possible solutions. During the early 1980s, relationships with suppliers also began to evolve. Renault became more demanding in terms of the quality of the components it purchased and increased the number of visits to suppliers, pressing them to renew their stocks of machinery and improve their methods of quality control (Gorgeu and Mathieu 1996). The *kanban* system of organizing components supply was adopted in 1983. Automation had permitted a number of organizational innovations and a modification of relations between the factories and the central production engineering department, given the problems personnel management and machine reliability automation had caused within the factories. A wide set of ways to organize work emerged, ranging from the old system, to the formation of groups of production workers who through job rotation looked after not only parts supply, running of machines, quality control, and tool changes, but also primary maintenance and repair, to mixed groups composed of production workers and maintenance workers who together were responsible for all work, maintenance, and improvements to reliability, as at the new robotized welding area for the Supercinq at Flins (Freyssenet 1984).

Despite encouraging results in the USA, where production of Renault vehicles reached 200,000 units in 1983 and 1984, the measures to reduce costs came too late, or were inappropriate to reverse the slide. The reduction of the workforce by 4,375 employees in 1984 had no effect, for the market in France contracted by 12.5 per cent that year. By the end of 1984, the company's accumulated losses had reached 17.6 bn. f. The financial crisis was accelerated by soaring interest rates and upheavals in currency markets. Renault's indebtedness became substantial, reaching 46.1 per cent of turnover in 1984. Bernard Hanon was forced to announce to the board that 15,000 jobs would be eliminated in six months, even at a time when the government had just reminded Renault that it was expected to play a leading role in the French economy. Bernard Hanon was therefore obliged to resign in the first days of 1985. He had nevertheless been able to implant his concept of the automobile in the company, popularized through the slogan 'living cars'. He had established the plans for the R19, which in 1988 would become the model that relaunched Renault. He had also been given the opportunity, and had had the intuition to bet on the Espace, the minivan that Matra had designed and proposed that Renault should market.

## 14.4. FINANCIAL RECOVERY AND THE RAPID TRANSITION TO AN INDUSTRIAL MODEL FOCUSING ON QUALITY, INNOVATION, AND FLEXIBILITY INSTEAD OF VOLUME AND DIVERSITY, 1985–1994

### *14.4.1. The Ten Years that Changed Renault*

The year 1985 was again very difficult. Although the decline of the domestic market came to a halt, registrations of Renault cars in France fell by 6.8 per cent, and market share dropped to 28.7 per cent, a fall of 11.8 points in only five years.

Accumulated debt was equal to 55.6 per cent of turnover by 1985. Georges Besse was appointed to head the company. His decisions radically changed Renault's trajectory: the abandonment of a volume-oriented strategy in favour of a substantial reduction of the break-even point to 1.2 million vehicles per year, the decision not to negotiate the reduction in the workforce with the unions and to confront the CGT in order to reduce its influence, a refocusing of Renault on the automobile industry and on Europe, re-establishing unity of command by eliminating the overlap of functions between the Group management and that of the sectors, and lastly calling for the State to fulfil its role as shareholder by contributing to reducing the company's debt. Just as the initial results of his actions were emerging, Besse was assassinated by members of a small extreme left-wing group with no employee or union connection to Renault. At the start of 1987, he was replaced by Raymond Lévy. On the basis of the recovery that had been initiated, Lévy's task became to make Renault a company just like any other, to endow it with a coherent, viable, and durable strategy, and to prepare it for the future.

From 1986, gross profits became positive once more. Net profits the following year reached 2.3 bn. f., and approximately 7 bn. f. in the following two years. The net debt of the Group declined from 55.6 per cent of turnover in 1985 to 10.1 per cent in 1989, with the State contributing a quarter of the debt reduction. By 1989 Renault regained its 1983 sales level of 2 million private and small commercial vehicles. However this time the company made a profit of 6.9 bn. f., whereas six years previously it had lost 1.8 bn. f. The difference had to be ascribed to factors other than market recovery. Under pressure from the late 1980s financial bubble, the domestic market expanded from 2.1 million private and light commercial vehicles in 1985 to 2.7 million in 1990. Moreover, both the government's deregulation of prices in 1986 and the lowering of value added tax enabled the automobile producers to improve their margins. Yet Renault owed its recovery predominantly to a complete change in its profit strategy. The Renault Group experienced nine consecutive years in profit between 1987 and 1995, with worldwide production (passenger cars and commercial vehicles) varying between 1.7 and 2 million annually (Fig. 14.1). Renault's recovery, shifts in the political context, pressures exerted by the European Community, Raymond Lévy's own wishes, along with the prospect of a partnership with Volvo, led the French National Assembly to pass a law that would transform the *Régie Nationale des Usines Renault* into a limited company—Renault SA—thus opening up ownership of its capital. By 1995, the State's share had been reduced to 52.97 per cent.

In 1986, the R21 replaced the R18 in the upper-middle segment, the last of the old generation of Renault cars. That same year a small commercial vehicle, the Express, which shared the Supercinq platform, was launched in place of the R4 commercial van. The R19, which in 1988 replaced the R9–R11 in the lower-middle segment, symbolized the new range in its design and in its high quality at the time of launch. In the mass market range, in 1990 the Clio substituted for the Supercinq, which had not lasted as long as the R5. At the upper end of the market the Safrane replaced the R25 in 1992. The success of the Espace led Renault to conceive of a range of minivans with the Twingo in 1993 at the lower end of the range and the Mégane-Scenic in 1996 for the middle segment. In 1994, the R21 was replaced by the Laguna. The following year, a fourth replacement of the entire range began, with the introduction of the Mégane family of models to replace the R19.

### 14.4.2. Recovery through Debt Reduction and Lowering the Break-Even Point, 1985–1987

Renault not only refocused its efforts on the automobile industry *per se*, but also on what it considered to be the essential competencies of the car maker: the design of models, the production of the principal mechanical parts and car bodies, final assembly, and distribution. The company was obliged to sell off its subsidiaries or to withdraw from financial participation even when this was profitable and strategic, since it no longer had the means to invest in them, and withdrawal could raise significant sums of capital. Hence the number of subsidiaries in France was reduced to eleven by 1995. As early as 1985, Renault had resold to Volvo the 15 per cent of shares it had held in that company since 1980. The most significant withdrawal was, of course, from AMC. Although the American subsidiary made its first profits in 1984, the switch of the American market towards more powerful cars, combined with the slowness of the design departments to adapt the R21, led sales to plummet in 1985. When it proved impossible to find a partner with whom management and financial costs could be shared, Renault sold AMC to Chrysler in 1987 and refocused its efforts on Europe.

The rapid reduction of the break-even point was obtained primarily through drastic job cuts, which the Socialist government no longer prevented. The move away from the 'job for life' was not a matter for negotiation, for Georges Besse, who believed it impossible to expect unions to bargain over redundancies. When the announcements of job cuts led to occupations of factories by workers, the occupiers were immediately suppressed by police intervention. Ten CGT representatives, accused of participation in attacks on managers and of damaging property, were fired. For the new management, Renault's recovery required the restoration of management authority, which in turn meant breaking ties with the CGT and restricting this union's power in the company. The workforce at the *Régie* fell from 98,153 in 1984 to 71,898 in 1988, a reduction of 26.7 per cent. Most of the departures took place within the framework of agreements with the State (Fonds National de l'Emploi), which allowed workers over 55 years of age to take early retirement.

The reduction of stocks in the sales network was obtained by obliging dealers to pay immediately for vehicles they ordered, and by linking dealers to a single computer centre that centralized orders. The network itself was reduced in size and reorganized. The different versions of each car model were regrouped by 'family' in order to simplify the programming of orders and create less disruption in production. Between 1984 and 1987, stock rotation in the factories was halved, from 20.3 to 10.4 days, improving the company's finances by 2.6 bn. f. Moreover, draconian conditions were imposed on all suppliers, including advertising agencies, requiring them to lower their prices. The ratio of purchases to total turnover fell significantly, from 83.4 per cent in 1984 to 75.2 per cent in 1988. Investments were restricted to the absolute minimum—what was needed to replace models—thus falling from 3.94 bn. f. in 1983 to 1.79 bn. f. in 1986, although this figure would climb back to 4 bn. f. by 1988.

### 14.4.3. A New Profit Strategy, the Search for an Industrial Model to Support it, and Preparing for the Future through Partnership with Volvo, 1988–1992

Since the early 1980s the various departments at Renault had been undertaking activities which improved performance in their fields of expertise, but which implied visions of the company and modes of recovery which were not necessarily mutually coherent. Now Raymond Lévy pursued the elimination of non-productive costs. But he also arrived with two key concepts, quality and project-based organization, which were soon integrated into a 'total quality' programme which sought to improve quality while reducing costs and lead-times.

The purpose of improving the quality of Renault vehicles was to position each model at the top of its market segment. Instead of seeking profits through volume, the aim was to make higher profits on each car by using quality to attract clients who would pay somewhat higher prices. Accordingly, the marketing strategy was reoriented towards northern European markets, more sensitive to quality and better able to pay for it. This strategy became married to the 'living car' concept which favoured liveability and conviviality over aggression and speed (Pointet 1997). This concept, which was supported by the success of the Espace, led to overall performance, equipment, comfort levels, internal and external design, colours, materials, and even sounds and odours being designed together, so as to respond coherently to client expectations. Renault therefore set out to make not only high quality vehicles but different vehicles. Innovation came through the range of minivans and in introducing a range of six versions of the Mégane with different bodywork (five-door saloon, coupé, cabriolet, four-door saloon, minivan, and estate), responding to the desires of clients previously limited to saloon cars and estate cars. On the other hand, giving each model its own personality led each to be given its own platform, which reduced substantially the commonization of components. Although the number of models per platform remained between 1.8 and 2 after 1962, even reaching a high of 2.3 in 1983, it rapidly tumbled to 1.3 in 1986, then declining gently to 1.1 in 1992. The average volume per platform fell from 408,518 to 209,000 units.

Although not without risk, this profit strategy would be viable under certain conditions. The company had to be capable of regularly generating its own innovative teams so as not to lose the commercially significant creativity of the original team. It also had to have sales personnel sensitive to client demands and able to transmit these to the designers. Flexibility would be essential from design to production, permitting a rapid response to new expectations in the market and to variations in demand which were potentially greater for the targeted segments. The company would also need a low break-even point, a low level of vertical integration, and the financial independence to be capable of withstanding the inevitable setbacks the strategy would entail. Lastly, the company would need to create employment relationships supportive of the required employee creativity and flexibility. Did Renault succeed in creating the internal preconditions for its new profit strategy?

As far as innovation in product design was concerned, an industrial design department was created in 1988, and more preparatory studies were undertaken. Yet Renault did not appear to have established a system capable of ensuring the continuation of current levels of creativity. In contrast, quality immediately became a leitmotif at company headquarters. In 1987, Raymond Lévy appointed a quality director, who came from the sales side rather than manufacturing, to coordinate and control activities aimed at improving quality at each stage of production. He was given the authority to refuse to sanction the launch of a model of inadequate quality. The same year he invoked this power to delay the launch of the R19, creating a seismic shock in the company. Also in 1987, suppliers began to be selected and evaluated according to a process called 'Supplier Quality Assurance', used by both PSA and Renault. In the factories, quality initially implied mastering the consequences of product diversity, which lay at the root of numerous mistakes and disruptions. To this end, continuous assembly lines were gradually replaced, as the models themselves were replaced, by a shorter central assembly line supplied by feeder subassembly lines (known in France as a 'fish bone' process). This system permitted diversity to be shifted upstream, and it simplified, even standardized, final assembly. Self-supervision of quality by operators, which consisted of drawing attention to defects they had been unable to avoid, had diffused throughout the company by the end of the 1980s. Statistical Process Control (SPC) to improve process quality, and Total Productive Maintenance (TPM) to increase the reliability of machines, were both introduced in 1991. Finally, quality was viewed in terms of service provided. A quick repair service was soon established and payment facilities were both extended and personalized (Monnet forthcoming). In demanding markets such as those of Germany and Sweden, the quality of Renault cars, in relation to that of competitors which themselves continued to improve, rose perceptibly.

The flexibilization of the design and production processes improved step by step. In principle, operators became polyvalent in 1981, with the new wage and classification system based on an operator's capacity to undertake all work posts at his level. Both the factories, and gradually the support services, were organized into Basic Work Units (BWU), with the aim of increasing the skills, involvement, and responsibility of the personnel. The BWUs in production areas were defined as teams averaging twenty employees, delimited by the product they made, which controlled and analysed the realization of the objectives they were given, maintained client–supplier relationships with other teams upon which they depended both upstream and downstream of the production process, and with a front-line hierarchical employee as their moderator. The hierarchical line was reduced from eight to five levels, namely the production operator, the BWU chief, the shopfloor manager, the departmental manager, and the factory manager (Freyssenet 1995; Allam and Decoster forthcoming). The reform of the suggestion systems led to an increase in the average number of suggestions per person per year from 0.4 in 1991 to 2.4 in 1995, and to savings of 480 m. f. that year. Development of personnel competencies was the goal of a programme aimed at providing less skilled employees with the ability to participate in teamwork and enable some of them to develop towards 'skilled' positions as a function of the company's needs. Agreements to make working time flexible were signed in several factories. When demand rose, Renault systematically called on 2,000–5,000 contract workers (depending on the year). Permanent workers were immediately laid off (and compensated) if orders declined. From 1987, the European factories located in Belgium, France, Spain, Portugal, and Slovenia were integrated into a single industrial system, with each model assembled at at least two different sites, with the exception of the larger cars and 'niche' vehicles (made by subsidiaries Alpine and Chausson). A system of dividing up production and sharing markets permitted subtle modifications to production and distribution plans depending on the loading of each factory and the evolution of demand. Deliveries on a daily or several-times-per-day basis replaced the weekly deliveries which in 1986 had accounted for 90 per cent of deliveries made to a factory such as Sandouville. The number of direct suppliers had been almost halved between 1987 and 1995: from 960 to 527. The project-based organization of design based on a matrix, which had been sketched out during the first half of the 1980s and generalized under Raymond Lévy, had its geographical and architectural reflection in the Technocentre, at Guyancourt in the western suburbs of Paris, which gathered on one site the product and process designers as well as factory and supplier representatives. The processes involved in future vehicles would now be tested properly, so that they could be rapidly perfected in the factories. Project leaders were directly responsible to Renault's managing director in terms of meeting cost, lead-time, and quality objectives. Yet difficulties in keeping to these objectives persisted, and investments remained high. While productivity increases enabled the workforce at Renault SA to fall to 60,608 in 1992, a reduction of 40.9 per cent compared to 1983, and the total wage bill to fall from 17.1 bn. f. to 15.7 bn. f. (in 1994 francs) between 1987 and 1992, conversely the annual charges for paying off investments rose from 1.2 bn. f. to 4.2 bn. f. over the same period. Hence the break-even point rose noticeably again, from 49.6 per cent of value added in 1987 to 22.9 per cent in 1992, even with domestic output higher by 160,000 vehicles.

This change in Renault's industrial model was made possible by the 1989 signature of a 'living agreement' with all the unions except for the CGT. This agreement created the basis for a new company compromise, in which flexibilization of work and involvement in achieving the company's objectives for improving performance were exchanged for an enriched work content, career prospects, and a sharing of the financial results. While Renault's workforce accepted the

## Renault: Towards an Innovation Strategy? 389

demands made of them, in 1991 they vigorously reacted against what they then considered to be an inadequate sharing of the financial benefits and against the continued reduction in employment. A strike and occupation of the Cléon mechanical parts factory served as a reminder that the new social compromise was not yet stable. Following this conflict, the wage system was changed. A bonus, rising by level of skill, was now to be awarded by the supervisor as a function of criteria related to sound execution of work, application of acquired competencies, and know-how developed.

Even as it sought an organizational form and employment relationship coherent with its strategy, Renault had endeavoured to prepare for the future in 1991 by forming a partnership with Volvo, with the intention of a later merger. Negotiations were conducted by Louis Schweitzer, who had succeeded Raymond Lévy in 1992. Renault purchased 25 per cent of Volvo Car and 45 per cent of Volvo Truck; in return Volvo purchased 20 per cent of Renault and 45 per cent of Renault Véhicules Industriels (RVI). For Renault this partnership had several facets. The first related to heavy commercial vehicles. RVI's principal market was in France, and was subject to wide fluctuations which regularly plunged the subsidiary into the red. The partnership comprising RIV and Volvo Truck ranked second in the world after Daimler-Benz. In terms of passenger cars, the partnership permitted the companies to plan to combine a generalist product range with a specialist range, much as Volkswagen and Audi had done (see Chapter 11). It gave the two producers, each small and with a single marque, greater size and greater diversity in the market. It also enabled Renault, which was too focused on markets that were temporarily protected, to penetrate northern European markets better, and to prepare for a return to the North American market. The two companies began to create a stock of shared components, to design a platform together, and to share their suppliers. Lastly, cross-share ownership was a means for Renault to move gradually towards privatization (Lévy 1994). Having consolidated its position in Western Europe, Renault made an offer to purchase Škoda, but then refused to engage in a bidding war with Volkswagen.

### 14.4.4. Future Prospects: Renault's Profit Strategy is Tested by Recession and by the Failure of the Volvo Merger

The French market for passenger cars and commercial vehicles fell by 450,000 registrations in 1993 compared with 1992 and by 730,000 compared with 1990. The European market (seventeen countries) had ridden the crest of a 15 million vehicles per year wave until 1992, before falling abruptly in 1993 by 2.5 million. The recovery in 1994 and 1995 was not enough to avert a price war among the car producers. Renault's gross profits dipped into the red once more. When the merger with Volvo failed, moreover, Renault was deprived of its plans for the future. On the other hand, the difficulties that were now being encountered differed in nature from those which had driven the company to brink of bankruptcy in 1984. The net financial debt of the Group was only 1.8 per cent of turnover, with its capacity to finance its own investments rising to 6.3 per cent in 1995.

The net financial results remained in the black. In contrast to Renault's venture with AMC, the relationship with Volvo did not incur financial losses. In the belief that the company had attained a stable position in Europe (10–11 per cent of the domestic European market), after having acquired majority control of its Turkish subsidiary in 1992, and having disengaged from its subsidiaries in Columbia, Venezuela, and Chile, Renault invested in Brazil, constructing a body and assembly plant with a capacity of 120,000 vehicles per year, in partnership with the State of Parana.

It was the structural change in the European market which accompanied the fall that was perhaps the more troubling for Renault. The pattern of demand split in two after 1991, with on the one hand the maintenance of the upper end of the market, and on the other hand growth in the lower-middle and mass market segments, with the upper-middle segments declining by 6.7 points between 1990 and 1995. Was Renault's profit strategy, based on offering innovative models targeted in terms of quality and price at the top of each segment, still viable? Renault embarked on a cost reduction programme of the order of 3,000 f. per vehicle before the end of 1997 'without impacting on the quality the client perceives and notices', for both existing models and new models. While guaranteeing suppliers their profits, in 1996 Renault asked them to make overall savings of 500 m. f. Industrial complexity was now being reduced by decreasing the number of parts and adopting identical modes of assembly for different versions and models. This programme none the less seemed to be bringing about a significant change in orientation, since it was now decided to design only two platforms for the whole future product range, one for models in the small and small to medium market segments, and the other for the upper and upper-middle market segments. Was this decision consistent with the constantly expressed desire to offer different vehicles with their own personalities? Lastly, if the privatization of Renault was no longer a political problem, the company compromise established at the end of the 1980s still had to demonstrate its capacity to withstand the new efforts being demanded, in the context of stagnating employee purchasing power and the continued pursuit of labour force reductions.

# NOTES

Translated by Sybil Hyacinth Mair.

This chapter has benefited from the Renault working group set up under the GERPISA international programme. Robert Boyer, Frédéric Decoster, Patrick Fridenson, Christophe Midler, Jean-Claude Monnet, and Giuseppe Volpato have also given their comments and suggestions. The author thanks them all.

1. see Chapter 1, note 2.

## STATISTICAL APPENDIX 14: Renault

| Year | Production (passenger cars and light commercial vehicles) | | | | Workforce[a] | | Turnover[b] | | | Net income[c] | | Gross income[c] |
|---|---|---|---|---|---|---|---|---|---|---|---|---|
| | worldwide | domestic | of which exports[d] | production abroad | Group, worldwide | automobile sector | Régie Renault SA | Group | automobile sector | Régie Renault SA | Group | Régie Renault SA | Régie Renault SA |
| 1945 | 12,036 | 12,036 | | | | | 23,250 | | | 36 | | | |
| 1946 | 28,842 | 28,842 | 12,614 | | | | 29,050 | | | 77 | | | |
| 1947 | 44,484 | 44,484 | 26,059 | | | | 36,471 | | | 123 | | | |
| 1948 | 65,317 | 65,317 | 26,375 | | | | 39,770 | | | 305 | | | |
| 1949 | 106,079 | 106,079 | 37,658 | | | | 44,233 | | | 474 | | | |
| 1950 | 131,903 | 131,903 | 46,590 | | | | 48,519 | | | 570 | | | |
| 1951 | 163,944 | 163,944 | 48,316 | | | | 52,470 | | | 959 | | | |
| 1952 | 169,543 | 169,543 | 36,437 | | | | 52,138 | | | 1,178 | | | |
| 1953 | 160,102 | 160,102 | 39,830 | | | | 50,337 | | | 1,151 | | | |
| 1954 | 198,932 | 198,932 | 52,078 | | | | 50,400 | | | 1,338 | | | |
| 1955 | 219,622 | 219,622 | 64,887 | | | | 52,235 | | | 1,424 | | | 227 |
| 1956 | 264,044 | 259,825 | 68,868 | 4,219 | | | 57,467 | | | 1,696 | | | 269 |
| 1957 | 317,443 | 313,425 | 112,744 | 4,018 | | | 58,981 | | | 2,162 | | | 486 |
| 1958 | 409,185 | 405,436 | 165,947 | 3,749 | | | 62,010 | | | 2,532 | | | 376 |
| 1959 | 494,160 | 487,044 | 297,287 | 7,116 | | | 65,657 | | | 3,131 | | | 480 |
| 1960 | 542,927 | 521,969 | 276,563 | 20,958 | | | 61,432 | | | 3,227 | | | 112 |
| 1961 | 393,163 | 353,218 | 183,970 | 39,945 | | | 58,313 | | | 2,962 | | | 379 |
| 1962 | 565,555 | 536,955 | 229,949 | 28,600 | | | 65,036 | | | 3,703 | | | 317 |
| 1963 | 668,867 | 639,797 | 219,514 | 29,070 | | | 63,575 | | | 4,438 | | | 419 |
| 1964 | 551,755 | 497,555 | 164,058 | 54,200 | | | 58,899 | | | 4,268 | | | 598 |
| 1965 | 590,431 | 551,904 | 226,305 | 38,527 | | | 62,902 | | | 4,536 | | | 529 |
| 1966 | 737,979 | 648,354 | 243,566 | 89,625 | | | 66,171 | | | 5,534 | | | 596 |
| 1967 | 777,468 | 695,148 | 295,586 | 82,050 | | | 66,882 | | | 5,886 | | | 821 |
| 1968 | 807,407 | 714,314 | 339,635 | 93,093 | | | 76,060 | | | 6,468 | | 22 | 593 |
| 1969 | 1,009,372 | 898,486 | 415,211 | 110,886 | | | 86,348 | | | 8,539 | | 20 | 737 |
| 1970 | 1,159,745 | 1,040,112 | 561,006 | 119,633 | | | 97,261 | | | 10,674 | | 151 | 350 |
| 1971 | 1,174,314 | 1,040,321 | 527,181 | 133,993 | | | 98,091 | | | 10,078 | | 5 | −474 |
| 1972 | 1,318,327 | 1,155,507 | 549,777 | 162,820 | | | 100,001 | | | 12,087 | | −197 | 720 |
| 1973 | 1,414,563 | 1,209,342 | 604,034 | 205,221 | 175,000 | | 101,415 | 20,659 | | 13,777 | | 74 | 870 |
| 1974 | 1,487,528 | 1,291,196 | 649,044 | 196,332 | 185,436 | | 100,478 | 25,674 | | 16,173 | | 57 | −1,069 |

## STATISTICAL APPENDIX 14: (cont'd)

| Year | Production (passenger cars and light commercial vehicles) | | | | Workforce[a] | | | Turnover[b] | | | Net income[c] | | Gross income[c] |
|---|---|---|---|---|---|---|---|---|---|---|---|---|---|
| | worldwide | domestic | of which exports[d] | production abroad | Group, worldwide | automobile sector | Régie Renault SA | Group | automobile sector | Régie Renault SA | Group | Régie Renault SA | Régie Renault SA |
| 1975 | 1,391,948 | 1,128,972 | 562,707 | 262,976 | 222,436 | 156,846 | 103,614 | 33,539 | 31,286 | 18,264 | | -551 | -956 |
| 1976 | 1,659,973 | 1,365,442 | 640,905 | 294,531 | 241,259 | 163,663 | 110,406 | 44,351 | 34,321 | 25,778 | 579 | 610 | -382 |
| 1977 | 1,737,707 | 1,398,550 | 624,106 | 339,157 | 243,456 | 170,632 | 110,485 | 48,589 | 39,770 | 28,696 | -111 | 12 | -1,196 |
| 1978 | 1,718,398 | 1,372,084 | 613,927 | 346,314 | 239,447 | 167,229 | 108,586 | 56,215 | 49,850 | 34,011 | -102 | 158 | 27 |
| 1979 | 1,899,470 | 1,544,995 | 730,771 | 354,475 | 233,408 | 169,794 | 106,740 | 68,535 | 58,006 | 42,185 | 1,016 | 470 | 699 |
| 1980 | 1,999,591 | 1,659,099 | 760,879 | 340,492 | 223,450 | 164,461 | 105,319 | 80,118 | 63,669 | 49,864 | 638 | 303 | -1,102 |
| 1981 | 1,764,702 | 1,479,691 | 640,156 | 285,011 | 215,844 | 157,402 | 103,613 | 87,971 | 76,272 | 53,620 | -690 | -875 | -1,773 |
| 1982 | 1,921,307 | 1,674,416 | 757,954 | 246,891 | 217,269 | 152,202 | 103,759 | 104,145 | 82,271 | 65,752 | -1,281 | -2,563 | -1,577 |
| 1983 | 2,035,133 | 1,842,801 | 979,425 | 192,332 | 219,805 | 161,643 | 102,528 | 110,274 | 85,379 | 73,560 | -1,576 | -1,875 | -2,579 |
| 1984 | 1,740,737 | 1,607,441 | 887,177 | 138,264 | 213,725 | 157,696 | 98,153 | 117,584 | 89,634 | 72,105 | -12,555 | -11,324 | -7,246 |
| 1985 | 1,637,634 | 1,499,979 | 881,149 | 137,636 | 196,414 | 144,961 | 86,122 | 122,138 | 101,824 | 72,644 | -10,897 | -11,241 | -4,242 |
| 1986 | 1,754,332 | 1,537,123 | 779,867 | 217,209 | 196,731 | 139,313 | 79,191 | 131,060 | 114,375 | 82,992 | -5,847 | -7,355 | 1,635 |
| 1987 | 1,831,390 | 1,612,146 | 809,589 | 219,244 | 188,936 | 136,046 | 75,911 | 147,510 | 123,495 | 93,333 | 3,254 | 2,314 | 7,580 |
| 1988 | 1,850,667 | 1,630,786 | 807,739 | 219,876 | 178,665 | 135,010 | 71,898 | 161,438 | 135,717 | 99,802 | 8,834 | 7,316 | 9,222 |
| 1989 | 1,966,724 | 1,717,279 | 837,608 | 249,445 | 174,573 | 129,699 | 70,720 | 174,477 | 129,230 | 113,731 | 9,289 | 6,932 | 7,268 |
| 1990 | 1,776,717 | 1,571,264 | 784,112 | 205,184 | 157,378 | 114,516 | 68,713 | 163,620 | 129,230 | 110,694 | 1,210 | 1,223 | 3,282 |
| 1991 | 1,790,709 | 1,587,787 | 829,298 | 202,922 | 147,185 | 106,232 | 63,644 | 171,502 | 133,206 | 112,297 | 3,078 | 2,467 | 2,530 |
| 1992 | 2,041,829 | 1,777,401 | 987,932 | 264,428 | 146,604 | 106,912 | 61,075 | 184,252 | 143,387 | 129,972 | 5,680 | 3,251 | 4,361 |
| 1993 | 1,713,633 | 1,459,188 | 817,788 | 254,445 | 139,932 | 103,148 | 60,608 | 169,789 | 130,179 | 116,776 | 1,071 | -5,225 | -438 |
| 1994 | 1,850,267 | 1,618,831 | 923,485 | 231,436 | 138,279 | 102,358 | 59,346 | 178,537 | 135,506 | 130,875 | 3,636 | 1,463 | -1,452 |
| 1995 | 1,761,643 | 1,610,216 | 900,077 | 151,427 | 139,950 | 102,213 | 59,264 | 184,065 | 136,444 | 132,050 | 2,139 | 944 | -3,657 |
| 1996 | 1,741,161 | 1,602,632 | 961,940 | 138,529 | 140,905 | 111,523 | 58,528 | 184,078 | 145,962 | 135,658 | -5,266 | -190 | -3,794 |

*Notes:*
[a] The workforce is as at 31 December, comprising permanent and short-contract employees.
[b] Turnover is in million francs. It is pre-tax from 1971.
[c] Net and gross incomes are in million francs.
[d] Exports include CBU (completely built up) and CKD (completely knocked down).

*Sources:* Renault Annual Reports.

# BIBLIOGRAPHY

Allam, D., and Decoster, F., 'Organisation du travail', in M. Freyssenet and P. Fridenson (eds.), *Les Données économiques et sociales de Renault* [Economic and Social Data on Renault] (Paris, forthcoming).

Boyer, R., and Freyssenet, M., *The World That Changed The Machine* (forthcoming).

—— —— and Jetin, B., 'The Profit Strategies of Carmakers' (forthcoming).

Charron, E., 'La Stratégie internationale de Renault' [Renault's International Strategy], *Annales de la Recherche Urbaine*, 29 (1985).

Dreyfus, P., *La Liberté de réussir* [Freedom to succeed] (Paris, 1977).

Freyssenet, M., *Division du travail et mobilisation quotidienne de la main d'oeuvre: Les cas Renault et Fiat* [Division of Labour and the Daily Mobilization of the Workforce: the Case of Renault and Fiat] (Paris, 1979).

—— 'La Requalification des opérateurs et la forme sociale actuelle d'automatisation' [Re-skilling of Workers and Current Social Forms of Automation] *Sociologie du travail*, 4 (1984) 422–33.

—— 'The Origins of Team Work at Renault', in Å. Sandberg (ed.), *Enriching Production* (Aldershot, 1995).

Fridenson, P., 'La Bataille de la 4CV' [The battle for the 4CV], *L'Histoire*, 9 (1979).

—— 'Renault face au problème du franc et du risque devises (1957–1981)' in M. Aglietta, C. de Boissieu, M. Levy-Leboyer and A. Plessis (eds.), *Du franc Poincaré à l'écu*, Comité pour l'histoire économique et financière de la France (Paris, 1993).

—— 'Fordism and Quality: The French Case, 1919–93', in H. Shiomi and K. Wada, *Fordism Transformed* (Oxford, 1995).

Gorgeu, A., and Mathieu, R., *L'Assurance qualité fournisseurs des constructeurs automobiles français, PSA et Renault* [Supplier Quality Assurance for the French Automobile Producers], Centre d'Etudes de l'Emploi (Paris, 1993).

Kipping, M., 'Les Tôles avant les casseroles. La compétitivité de l'industrie française et les origines de la construction européene', *Entreprises et Histoire*, 5 (1994).

Labbé, D., *Travail formel et travail réel: Renault-Billancourt, 1945–80* [Formal Work and Real Work: Renault-Billancourt, 1945–80], DEA Histoire des civilisations, EHESS, (Paris, 1990).

—— 'Renault: les trois âges de la négociation' [Renault: Three Eras of Negotiation], *In Travail*, 26 (1992).

Lévy, R., *Le Cas Renault* [The case of Renault], Notes de la Fondation Saint Simon (Paris 1994).

Loubet, J.-L., *Citroën, Peugeot, Renault et les autres* [Citroën, Peugeot, Renault, and the Others] (Paris, 1995).

Lung, Y., and Jetin, B., 'Un Ré-examen critique de la relation entre variété et modèles industriels à partir de l'industrie automobile' (forthcoming).

Midler, C., *L'Organisation du travail et ses déterminants. Enjeux économiques et organisationnels des réformes de restructuration des tâches dans le montage automobile* [The Organization of Work and its Causes: Economic and Organizational Challenges in the Reform and Restructuring of Tasks in Automobile Assembly], Thèse de 3ème cycle, University of Paris 1 (Paris, 1980).

Midler, C., *L'Auto qui n'existait pas* [The Car that did not Exist] (Paris, 1993).

Monnet, J.-C., 'La Politique d'innovation et de recherche', in M. Freyssenet and P. Fridenson (eds.), *Les Données économiques et sociales de Renault* [Economic and Social Data on Renault] (Paris, forthcoming).

Pointet, J.-M., 'Cohérence de la stratégie produit de Renault', in *Gérer et Comprendre*, 48 (1997).

Poitou, J.-C., *Le Cerveau de l'usine. Histoire des bureaux d'études Renault, de l'origine à 1980* [The Brain of the Factory: A History of the Renault Design Department, from the Early Days to 1980] (Université de Provence, Aix-en-Provence, 1988).

Renault, *Renault-Inter*, 162 (1973).

Womack, J. P., Jones, D. T., and Roos, D., *The Machine that Changed the World* (New York, 1990).

# 15

# *From British Leyland Motor Corporation to Rover Group: The Search for a Viable British Model*

ANDREW MAIR

The trajectory of Rover Group, formerly British Leyland Motor Corporation (BLMC), British Leyland (BL), and Austin-Rover, encountered and was strongly influenced by a series of different external industrial models between 1968, when BLMC was formed, and the late 1990s. These included the principles of Sloanism, Fordism, and Taylorism as observed at Ford of Britain and elsewhere, a Japanese model in the form of the company's erstwhile partner Honda, and from 1994 the particular German model of its new owner BMW. Over the course of these encounters, what can reasonably be considered the 'indigenous' craft-British and mass-British industrial models that had dominated the company (recognizing that these too incorporated elements borrowed from the wider global industrial environment) were moulded and shaped.[1]

From one perspective, it might be thought that the indigenous models were changed out of all recognition as a succession of new models was imported, each one replacing the previous with a more or less difficult transition. From the perspective adopted here, however, the lineaments of the indigenous models can be followed right through their encounters with the external models into the 1990s.

From this subtler and richer viewpoint, the trajectory from BLMC to Rover Group traces a series of adaptations, hybridizations, and partial borrowings in the search to create a viable British model capable of surviving in a competitive environment that was first Europeanizing and then globalizing. The mass-British model was represented by the descendants of Austin and Morris (combined from 1952 in British Motors Corporation (BMC)), while the craft-British model was exemplified by Triumph, MG, Rover, and Land Rover. From these two starting-points in the mid-1960s the trajectory into BLMC and beyond followed two strands, the first of which was continuous, the second of which had, except for a small pocket of stubborn resistance at Land Rover, virtually disappeared by 1980, only to be revived in new guises after 1986 and interwoven into the first.

Following the first strand, there was a series of attempts to restructure and revitalize the mass-British industrial model which had been successful from the 1920s to the early 1960s (Tolliday and Fujimoto 1998) but much of which had evolved incoherently by the 1960s so that its internal inconsistencies already posed severe threats to its competitiveness. BMC struggled to engage in mass car production on the foundations of a mass-British industrial model which simply could not manage annual output levels above a certain point and yet which paradoxically continued to pursue scale as a means to solve its problems. The story of this strand between 1968 and 1985 is one of largely failed attempts (measured in terms of commercial success) to adopt new practices and organizational forms which British managers based on their own understanding of Sloanism, Fordism, and Taylorism, resulting in the virtual collapse of the company.

Meanwhile, the second strand reveals the starvation or crushing of potentially internally consistent and competitively viable evolutionary developments of the craft-British model between 1968 and 1980, by the same processes of industrial restructuring which comprised the first strand (Whipp and Clark 1986, Whisler 1995). In 1968 companies like MG, Triumph, Rover (including Land Rover) were still trying to develop the craft-British model along its own, very specific trajectory manufacturing upmarket cars, sports cars, and four-wheel-drive vehicles in small, yet rising, quantities, some already experiencing considerable success, others with a strong base upon which to build. With the exception of Land Rover, by 1980 their efforts had disappeared, thanks to corporate BLMC and BL managers whose strategic and operational perspective was prisoner to the 'ideals' of the Fordism with which they were determined to replace the mass-British model.

Then, from 1986, Rover Group—essentially the remnants of Austin, Morris, and Land Rover—gradually began to create a new hybrid British model, on the basis of three elements: the major sites and legacies of the mass-British model; lessons from management's understanding of Japanese management practices; and now, reversing the dynamic, several traditions of the craft-British model, not the least of whose contributions was significant and vital financial support (cross-subsidization) from Land Rover, the growing competitive success of which seemed to reveal what might have been.

By the late 1990s Rover Group faced the new challenge of interaction with a thoroughly and particularly German industrial model in the shape of BMW Group, which purchased the company in 1994. How would a company based on a solid, successful, and confident industrial model approach its new subsidiary, still struggling to realize a new British model which promised much but which, after the trials of the previous three decades, remained weak and uncertain of itself?

## 15.1. TWO INDUSTRIAL MODELS AT BRITISH LEYLAND MOTOR CORPORATION, 1968

The year 1968 marked the culminating merger of British-owned motor vehicle companies into one large conglomerate, when British Motor Holdings (BMH: the outcome of BMC's 1966 purchase of Jaguar), owners of Morris, Austin, MG, and Jaguar, merged with Leyland Motor Corporation, a truck and bus manufacturer which also owned Rover and Triumph together with other smaller firms. The companies which made up the new conglomerate were organized on the basis of two closely related industrial models.

The craft-British model still formed the organizational foundation for a series of specialist producers such as Rover (including Land Rover), Triumph, and MG (Whipp and Clark 1986, Whisler 1995), as well as Jaguar and the Leyland truck, bus, and industrial vehicles companies. Work in the factories was organized by groups of workers represented by their own 'shop steward'. This trade union representative, who belonged to one of many unions depending on his particular craft, bargained with management over pay rates in a piecework system in which the workforce was paid for what it produced rather than the hours it worked. With factories often divided into segments separated by work-in-progress, a group could work at its own pace and go home once the daily quota was completed. With the shop stewards controlling the organization of work based on the practical skills of the workers, Taylorist procedures of measurement and knowledge formalization were entirely absent.

Production processes were largely labour intensive, and break-even points were low. Indeed the annual production of the craft-British companies often remained between 10,000 and 30,000 vehicles. In assembly, bodies on jigs were moved from one workstation to another by hand. Thus workers were directly disciplined neither by management nor by machine. The general absence of capital equipment meant that manufacturing was highly flexible in terms of product designs. MG, for instance, was able to introduce new products in only two weeks. Even production at the larger Triumph was only partly mechanized.

Products were designed by entrepreneurs and mechanics with no formal training, who had often risen from the factory floor or from the ranks of skilled workers where they had served long apprenticeships. Products were normally 'robust', that is, the basic design was retained for many years, with periodic updating. The craft-British companies focused on highly engineered or well-designed products for niche markets. Thus MG and Triumph vied for domination of British exports to the USA, where their sports cars were successful during the 1950s and 1960s. Rover developed a reputation for advanced engineering in the upper range of the British market, and the company deliberately kept a waiting list of customers as a measure of the appeal of the Rover car. Rover's Land Rover four-wheel-drive products concentrated on niche overseas markets (principally the former British Empire).

The craft-British producers typically had low levels of vertical integration. MG's engines came from the ordinary range of BMC. When Rover needed a large engine in the early 1960s it purchased rights from the American company Buick to build an aluminium V8 engine (still used by Land Rover in the late 1990s). Triumph realized economies of scale by creatively building a wide range of models on the basis of the same chassis, engines, and mechanical parts.

The craft-British model was internally consistent and coherent. It flourished in its niche markets during the 1950s and 1960s. Significantly, at Rover in particular, the model was also modernizing with considerable success, with well-engineered products and a reasonably well-managed and coherent, if stressful, process of evolution towards larger-scale production (Whipp and Clark 1986).

The mass-British model was the legacy of the British model which had proved so successful for Austin and Morris in the 1920s and 1930s in the face of Ford's failed attempts to impose Fordism in the United Kingdom, and which was still successful in the 1950s (Tolliday and Fujimoto 1998). BMC developed into a high volume company, with annual output rising from 300,000 in 1952 to 900,000 in 1965. Importantly, many of the features of the craft-British model were unchanged and initially, at least, seemed able to cope with higher volumes. Work was still organized by shop stewards who bargained with management over piecework rates. Companies were still run, and products designed, by entrepreneurs and mechanics, 'practical men'. On the other hand, there had been certain attempts to rationalize production at BMC, which justifies the characterization of a mass-British model. A basic range of three engines had been developed to power all the products, specialized transfer lines had been built to make them, and there was very significant 'badge engineering', in which the Austin and Morris range products were virtually indistinguishable, with old marques like Vanden Plas, Wolseley, and Riley retained for further badge-engineered variants of the same models (Williams *et al.* 1994). Some of the BMC products were successful in generating sales, including the innovative small mini with its front-wheel-drive transverse mounted engine (1959) and the 1100/1300 (1961–2). BMC remained the market leader, ahead of Ford of Britain, in the United Kingdom's pre-European Community tariff-protected market.

Despite modernization, the mass-British model was rife with internal contradictions, as a result of which the company was barely profitable and by the late 1960s was experiencing considerable problems funding new product development despite its high market share and rising sales (Wood 1988). BMC never rationalized the majority of its production facilities, and owned dozens of sites scattered across England. Production of a single car model could take place at a number of sites in succession, for stamping, welding, painting, and assembly, with bodies transported back and forth across England between stages. Worse, Austin and Morris each manufactured their own versions of the same car at their own factories, and their factories fed separate distribution networks which had not been rationalized either. Production was so poorly organized that there were frequent bottlenecks and problems, a continual threat to the piecework wages of workers. During sales slumps management provoked and prolonged strikes by workers as the only way to reduce output. Local disputes were common, yet there were no collective industrial relations mechanisms for resolving them, only a vast array of local agreements with many unions.

Indeed it was so arduous for management to reach agreement with the hundreds of shop stewards whose workers were involved in making a new model that even the most obvious product improvements were not made, and agreements became fixed for years at a time. Simultaneously, work groups were making their work more efficient but refusing to share the benefits with the company. Consequently, by the late 1960s BMC was seriously 'overmanned', with many employees having finished their work and going home after four or five hours of a nominal eight-hour day.

# The Search for a Viable British Model

BMC management had adopted none of the tools of 'modern' twentieth-century management, such as Taylorism, Fayolism, or bureaucratic structures, which elsewhere were thought necessary for mass production. Management was by persuasion, or coercion, or was simply absent. The main management theories were rejected by inward-looking company cultures, as were university engineers, who were viewed with suspicion in an anti-intellectual culture which emphasized experience over theory. Hence BMC had no programme to recruit university graduates to any level in the company. There was no corporate planning department. There was no internal accounting system to permit the financial performance of the car models to be calculated. BMC only discovered that money was lost on every Mini sold when Ford of Britain's engineers stripped down a car and worked out its cost structure. By the late 1960s, the mass-British model had become ossified and moribund, unable to cope with high volume production, unable to change itself, and only surviving for as long as it was largely sheltered from companies with superior industrial models.

## 15.2. CONGLOMERATION AND PURSUIT OF FORDIST IDEALS, 1968–1977

Although BMH was the larger company when it merged with Leyland, and while during earlier talks it had appeared that BMH would be in control of Leyland, BMH's deteriorating financial position meant that it was Leyland which gained effective control over the new British Leyland Motor Corporation (BLMC) in 1968. When the still powerful voice of Jaguar was included on the new board, it was clear that the senior managers of the craft-British producers were to control and manage the new conglomerate. Yet ten years later, not only had the inherent problems of the mass-British companies dragged BLMC further into crisis as they struggled to impose what they took to be the ideals of Fordism, but the craft-British model and its vehicle producers had virtually disappeared, having collapsed either through lack of support or through attempts to force the same Fordist ideals on them too.

The new BLMC management team openly admitted having no experience of running a large company. Yet they were soon pursuing ever higher volumes to retain profitability. At the same time they hoped to improve the productivity of the existing workforce (Church 1994, Williams *et al.* 1994, Wood 1988). BLMC had inherited a total of forty-eight factories, including BMH's twenty-three major plants. A number of senior and middle managers were hired from Ford of Britain and were expected to provide a logical way to organize the conglomerate. They soon argued that employment had to be cut immediately from 190,000 to 160,000. However, to the new BLMC leadership, rationalization, whether of factories or labour, was viewed as an impossible strategy owing to the power of the trade unions. Localized strikes over work organization or pay rates were by now a regular occurrence, disrupting production on a daily basis, and directly accounting for 2–3 per cent of total man-hours during the early 1970s (even before including their indirect effects). A strike-free period of two weeks in 1970 was seen as highly unusual. In this context, to even hint at site-use rationalization, even without any loss of employment, caused strikes. Hence the strategy of output growth seemed the only option. Indeed with the economic growth of the late 1960s and early 1970s production continued to rise, and output peaked at 916,000 cars in 1972. Profitability, while low (2–5 per cent), was maintained.

At the same time there were attempts to reform both the mass-British model and the craft-British model at the level of the production system. In 1970 BLMC appointed its first Director of Industrial Relations. From 1971 to 1974 a system of 'measured day work' (MDW) was progressively adopted to replace piecework at both craft-British and mass-British producers, as a means to increase managerial control over work. Yet MDW failed to deliver the expected increases in productivity, because it was introduced without the necessary complementary changes. Management had virtually no experience or competency in work study, and few managers or engineers understood the details of production. While the foremen now had a pivotal role in ensuring continued production, there were few experienced foremen able to command respect from shop stewards and work groups. Thus while management was theoretically in control of the production process, in practice traditional bargaining was retained, simply switched from pay-per-piece to pieces-per-day, or 'standard work'. Not only did productivity fail to rise but in some ways management was now in a worse position. As late as 1978, workers at some BLMC factories were still going home after four hours of the eight-hour day for which they were being paid, as they had traditionally done. Yet they had only completed 60 per cent of their MDW output (Wood 1988).

In the absence of productivity improvements, the BLMC strategy required successful new products. In the years prior to 1968, BMC/BMH had been unable to maintain an adequate rate of model replacement. Yet BLMC was in no stronger position financially. Every new product had to succeed in order to generate the resources for the next new product. In 1968 the BLMC car makers were formed into two divisions, one based on the former BMC, the other a specialist cars division composed of Jaguar, Rover, and Triumph. The main worry of BLMC management was the former BMC and for three years after 1968 the specialists were permitted to continue their existing and separate product programmes while BLMC focused on revitalizing the Austin and Morris ranges. Adopting some elements of Sloanist brand differentiation, it was decided to shift from badge engineering to separate product lines and images for Austin and Morris. Austin was to retain a conservative image, building the kind of front-wheel-drive cars which had become BMC's hallmark during the 1960s. Morris, by contrast, was now to build more adventurous and sporty rear-wheel-drive cars.

This was a very ambitious strategy, since the two marques would no longer share their product ranges and the rate of new product development would have to be doubled. Moreover, a key element of a coherent Sloanism, the sharing of mechanical components (see Chapter 2) was dispensed with. Then, the first two cars introduced under the new strategy were judged to be worse products, in design, handling, and product quality, than the 1961/2 Austin/Morris 1100/1300 models they were to replace. The rear-wheel-drive Morris Marina, introduced in 1971, incorporated mechanical components from the 1948 Morris Minor, and other parts from Triumph. Expected to compete in mass markets with Ford of Britain's Cortina, the Marina did not achieve expected sales, though it became BLMC's best-selling model in the absence of other successes. The second new model, the Austin Allegro, introduced in 1973 after a five-year development period during which senior management indulged in their own design whims (including a square steering wheel), was slower, less attractive aesthetically, and had more components which were liable to break down, than the 1100/1300. Its annual sales were only half those of its predecessor, and to maintain sales volumes BLMC kept the older models in production for a period.

By 1972 the dual Austin-Morris model strategy had already been abandoned. In a desperate effort to provide a new larger model for both Austin and Morris the body of the 1964 1800 model was given the newer engine of the final BMC product, the 1969 Maxi. A new large car was not ready until 1975 (the 18/22, later renamed Princess), but by then demand for large cars had collapsed owing to the twin impacts of oil crisis and ensuing recession.

Indeed after five years of slow reform sheltered by a period of economic growth, BLMC was ill-prepared for the mid-1970s recession. Market share in the United Kingdom began to decline from its 40 per cent level in 1968, at first slowly with rising BLMC sales in a growing market, and then rapidly, to 30 per cent of a much smaller market in 1975. With entry into the European Community in 1973 and falling tariff barriers, the share of imports in the United Kingdom market rose sharply, from 8 per cent in 1968 to 33 per cent in 1973, with Volkswagen and Renault performing well. The Japanese producer Nissan also made its first significant entry. By 1975 BLMC car production had collapsed from its 1972 peak of 916,000 to 605,000. While sales of the ageing Mini and 1100/1300 models rebounded as customers switched from the newer Marina to smaller cars, each Mini and 1100/1300 model sold was a drain on profits. The company lost money in 1975, but was already effectively bankrupt by late 1974 as its cash flow dried up.

The United Kingdom government under the Labour Party agreed to take control of BLMC in 1975, placing it under the aegis of its new National Enterprise Board (NEB). The government had commissioned the Ryder Report, named after chairman Sir Don Ryder, to set out a BLMC recovery plan. Ryder became the first chairman of the NEB, and therefore retained control over BLMC. The Ryder Report painted an optimistic picture, projecting that output would recover to 900,000 by 1980. The new strategy was to create growth through major investment in a single new model range for a company which now merged both Austin-Morris and Rover-Triumph-Jaguar into a single entity, BL Cars. The strategy attempted to impose a clearly structured organizational plan on the existing varied and diffuse operations, with factories now given numbers instead of names. Ryder sought to resolve industrial relations problems by purchasing reform. Coordinated discussions with trade unions were set in train to find a more rational system to replace the existing 246 separate bargaining units whose agreements were renegotiated in nine different months of the year. However, rationalization was thought unnecessary, and overmanning was not dealt with directly. Essentially, then, Ryder attempted to impose modernization, homogenization, investment, and growth, a Fordist ideal, onto the existing industrial models as the means to resolve their inherent problems. This was a further deepening of the reform trend under BMC/BMH before 1968, and under BLMC after 1968, a trend which was interpreted not as inappropriate but as not having been pursued far enough.

Progress under Ryder was slow. Profitability returned in 1976–7 as output recovered. But union negotiations to create a more rational and participative industrial relations system advanced very slowly, and there were disruptive strikes during 1976 at nine separate factories. A damaging one-month strike by toolmakers in 1977 severely curtailed production and was seen by the NEB board as a signal that Ryder was doomed to fail.

Other problems associated with the mass-British model had barely improved under BLMC and Ryder. Large inventories of components were held to guard against strikes at other—often BLMC—factories. Staff at purchasing departments had inherited weak policies and had to guess appropriate prices to pay suppliers. With no financial information, they frequently paid more than Ford of Britain for the same parts. These problems in purchasing reflected the continuing and widespread difficulties BLMC faced in general management. Leyland Motors managers, typical of craft-British producers, had virtually no general management skills to offer BMH and tried to manage the conglomerate using traditional guesswork methods. The basic antipathy towards management systems remained. Hence while corporate strategy was based on the ideals of Fordism at a general level, many BLMC managers were openly hostile to the practices needed to make the system function properly. To the incoming Ford managers trained in systematic management methods, the two cultures were utterly incompatible. The chairman from 1968 to 1974, Donald Stokes, was a salesman who announced that he did not believe in planning, measuring, or budgetary controls. Indeed no management or control systems were shared across BLMC, which retained a multiplicity of local, implicitly understood, and incompatible practices. The Marina model, for instance, had again been planned, developed, and introduced with no detailed information about its cost structures and no knowledge about whether or not it was in fact profitable.

Hence the mass-British industrial model remained largely intact, thwarting the Fordist-inspired strategy that the corporation's top management sought to impose. Some of the Ford-trained managers believed that BLMC was incapable of competing directly with Ford in mass market products, and should seek to create space for itself in higher niches more appropriate to a potentially coherent modified mass-British model. Yet BLMC management focused squarely on market share and sought products which could compete head on with Ford. Ryder's own authoritarian approach, which prevented the rest of the NEB and senior BLMC management from contesting his plan, only compounded the problems.

How did the craft-British model fare under BLMC? During the first three years under BLMC, each company continued to develop its existing new product projects, although nominally Triumph and Rover had joined Jaguar under the 'specialist cars' division. Triumph launched five new models between 1969 and 1972 and planned to introduce a new large car, while Rover launched the Range Rover upmarket four-wheel-drive vehicle in 1970, and continued to develop a planned series of three upper mid-range models financially supported by the success of the 1962 2000/3500 models.

Then in 1971 the BLMC board embarked on a new strategy for developing large cars and sports cars. The strategy was driven by a combination of Fordist ideals and internal corporate politics. The existing Rover and Triumph large car projects were stopped suddenly—with Rover's P8 project, which had been designed to compete directly with Mercedes, abandoned only six months prior to launch—and the two companies were merged. The new entity was to start from scratch to design two new high volume models, one for the large car segment, one for the sports car segment. Jaguar's place on the BLMC board played a crucial role, as did Triumph's long-standing ambitions to become a mass manufacturer of sports cars. Jaguar viewed the range of new Rover cars as significant competitors in its own market niche. With Rover diverted, Jaguar now had the domestic 'executive' car market niche to itself. Triumph won an internal competition against MG to build the 'corporate sports car', a decision that spelled the beginning of the end for a still successful—but about to be abandoned—MG.

Rover was widely viewed as the best managed and most successful of the craft-British companies, and was already modernizing its management with considerable success, including a cooperative and participative industrial relations regime as a framework for managing future restructuring (Whipp and Clark 1986). Until this point operationally and strategically independent, Rover was now forced to cooperate with close rival Triumph. Moreover, BLMC was trying to impose an industrial model quite foreign to Rover—and indeed to BLMC management itself. While Rover had already planned to increase its car production capacity to 75,000, BLMC ordered a 150,000 capacity assembly plant for the Rover site at Solihull in the West Midlands, in the belief that high sales of profitable large cars were necessary to support corporate growth. Unlike previous Rovers, the new product, the Rover 2300/2600/3500 series, launched after five years in 1976, had no technical originality. While aesthetically attractive and generating high initial interest, the new car ran into a series of major problems. The company was unable to meet early demand, particularly elsewhere in Europe, owing to a series of strikes at internal BLMC components suppliers, including a long dispute at a new Birmingham body painting factory built expressly for the project. The project had incorporated a new assembly line, a new paint factory, two new engines, and a new transmission simultaneously, and each caused problems. The smaller engines, supplied by Triumph, proved unreliable and frequently broke down. Only the larger Rover (Buick) engine was successful. Electrical parts failed frequently. The quality of the paintwork was poor and new car bodies soon rusted. The general level of trim was poor, in part because BLMC had insisted on the cheapest parts available, and in part because the cars proved difficult to assemble because manufacture had been ignored by the designers. BLMC production managers brought in from Austin and Morris had concluded that Rover engineers knew nothing of mass production and were advancing the project too slowly because they were too concerned over quality, and so they pushed the car into production and planned to ramp up output quickly. Finally, the product was poorly timed, introduced into an uncertain market. The result was that, despite production capacity of 150,000, the car only sold 19,000 in the British market in 1977, and the capacity was never to be more than one-third utilized. Rover's evolving craft-British model had simply been crushed by the imposition of an inappropriate industrial model which was itself poorly understood and implemented. Simultaneously, while Rover's 1970 Range Rover had been a small-scale but profitable success, Rover was denied the small investment needed to introduce a four-door version to promote continued sales growth.

Meanwhile, the craft-British sports car producers suffered similarly (Whisler 1995). MG was simply starved of all investment and obliged to continue a single line of cars, versions of the MGB. The model continued to sell well, with output of the 1962 car not peaking until 1976, but resources were not devoted to the technical modifications needed to meet increasingly stringent safety and pollution legislation in the USA. Triumph had been given responsibility for the new 'corporate sports car', the TR7, which was launched in 1975. However, the TR7 was underpowered, poorly built (a process involving three separate factories in two cities), and failed in the USA, with its dedicated factory operating at only 27 per cent of its capacity in 1976.

## 15.3. RATIONALIZATION AND RESTRUCTURING, STILL DRIVEN BY AN IMAGE OF FORDISM, 1978–1985

By 1977, as the 2300/2600/3500 and TR7 failed like the Marina and Allegro before them, output had scarcely recovered from the 1975 slump, employee numbers were rising to 195,000, and the small profits were a function of infusion of Ryder plan government funds. A second collapse of sales turnover and production began in 1978, exacerbated by the impact of the second oil crisis the next year. BLMC car output fell steeply from 651,000 in 1977 to 396,000 in 1980, less than half the 1972 peak. Sales collapsed at home and abroad. Sales outside the United Kingdom, which had accounted for 30–40 per cent of production (including CKD production abroad, in Belgium, Spain, and Italy) during the 1960s and 1970s, fell in tandem with home sales to one-third of their 1969 peak by 1981. The company began to make significant operating losses, 10 per cent of turnover in 1980 and 8.5 per cent in 1981.

## The Search for a Viable British Model

In 1977 Michael Edwardes, chairman of the manufacturer Chloride and member of the NEB board, agreed to take over BLMC as Chairman and Managing Director (Williams *et al.* 1987, Wood 1988). In stages between 1977 and 1980 he launched what seemed to be an entirely new strategy for BL Cars, as he renamed the company. At first Edwardes made incremental changes. He reversed the Ryder organizational centralization to recreate two car divisions for mass and specialist cars. He began to allow employment levels to decline through retirements and non-replacement of leavers. In 1978 came the first closure of a car factory in the United Kingdom for thirty years, as manufacture of the failing TR7 was transferred from Speke in Liverpool to Triumph's Coventry factory.

As the extent of the second crisis became clear, Edwardes's strategy became more radical. While he retained the basic Ryder plan to replace the model range and to invest in new capital-intensive production facilities, the new strategy involved large-scale rationalization and centralization of production facilities and the establishment of a new balance of power with the trade unions. These were significant departures from BLMC and Ryder. And yet in terms of the implicit industrial model, Edwardes's strategy was not so much a radical change as a renewed attempt to impose a version of Fordism.

The Triumph, MG, and Rover factories were all closed, including the new Rover factory at Solihull, closures which marked the formal end of the craft-British model. Of passenger car production under the craft-British system, only Land Rover's tiny output of Range Rover models remained (Edwardes had made the still profitable Land Rover operation an independent company). Production of all MG and Triumph models ceased and Rover 2300/2600/3500 production was transferred to the former Morris factory at Cowley, Oxford. Total employment at BL declined rapidly, from over 190,000 in 1978 to 96,000 in 1984. By the time Edwardes returned to Chloride after five years at BL in 1982 there were only two car production factories remaining, the former Austin factory at Longbridge, Birmingham and the Cowley plant.

Manufacture of MGs ceased despite continuing demand because BL could not support short-term financial losses in North America owing to the high exchange rate of sterling which had resulted from the policies of the Conservative government elected in 1979. BL sales in the USA collapsed from 75,000 units in 1976 to 6,000 in 1981 and 1 in 1984. The overseas ventures in Europe were also terminated as the high value of the pound sterling made export of CKD kits temporarily unprofitable. The BLMC distribution network in Europe, starved of models, collapsed. Sales in Belgium, for instance—a European stronghold which had been served by an assembly plant—declined from 84,000 in 1978 (14 per cent of total BL sales) to 5,000 in 1981 when the factory closed. Overall sales outside the United Kingdom fell below 100,000 per year in 1983–4.

The combination of a change of government at national level and the clear depth of the crisis facing BL created an atmosphere in which Edwardes and his senior management were able to alter the balance of power with the trade unions, particularly with the shop stewards. Edwardes had presented a first phase of his rationalization plans to the workforce in 1979, receiving 87 per cent support from the 80 per cent of the workforce which voted. On this basis, the most militant shop steward at Longbridge was symbolically fired for opposing the policy, and in 1980 management presented employees with an ultimatum: returning to work after a public holiday implied acceptance of a new industrial relations and work organization regime in which shop stewards no longer had the right to reject proposed changes to work organization. The era of 'mutuality' was over. Management argued that there was a role for 'responsible' unionism, defined as bargaining over pay and general conditions, not over working practices. New consultation procedures were established, and the facilities available to shop stewards were severely curtailed. Their number at Longbridge, for instance, fell from 800 in 1980 to 400 in 1982. Edwardes was seeking an industrial relations framework compatible with high volume production, to create a decisive break with a key trait of the mass-British model.

There were still strikes as management attempted to enforce new working practices. The sudden disappearance of shop stewards frequently made it more difficult for managers to obtain agreement from workers who now lacked representatives. However, the number of hours lost due to strikes (including forced stoppages in related factories) declined steeply, from 5.9 per cent in 1977 to 1.6 per cent in 1982.

Under Edwardes's guidance, BL Cars continued to focus on creating a range of basic small and medium mass-market cars with substantial government financial support. Total government aid under Ryder and Edwardes amounted to £2.1 bn., equivalent to about 70 per cent of annual turnover in the early 1980s. The most advanced project under the Ryder plan was a replacement for the Mini. This, however, was scrapped by Edwardes in 1978 after £300 m. had already been spent, when consumer tests showed that virtually no consumers preferred it to Ford's new small car, the Fiesta. The platform of the Mini replacement was then used to quickly develop a larger small car, the Metro, for launch in 1980, resulting in a compromised design, and still using small 1950s BMC engines.

Even with funds now available, BL Cars no longer possessed sufficient competent engineering resources for more than one new model project at a time, and so the five-door Maestro, and the derived four-door Montego, replacements for the Allegro and Marina, were not launched until 1983 and 1984. Following the same market-share-oriented mentality of BLMC, all three products were conceived as mass market cars to compete directly with Ford and Vauxhall (General Motors) products in order to secure the volumes believed necessary for profitability.

Major investments were made in new facilities, especially in body welding, to produce the Metro at Longbridge and the Maestro/Montego at Cowley. Automation was viewed as a means to increase direct labour productivity and product quality. With its government financial support, BL was quick to adopt the first wave of automated body welding lines being introduced in the world automotive industry. At Longbridge, a £200 m. new factory was built between 1977 and

1980 for the new welding lines. By 1982 BL management was claiming that the new Metro facilities were the most productive in Europe. Yet the system selected, with rigid 'multiwelders', was inflexible in terms of the products that could be manufactured. BL used its investment fund to build a plant with a capacity of over 300,000 per year. Like Ryder, Edwardes was optimistic. The new model range was supposed to permit BL to double unit sales, returning the company to a total output of 950,000 per year by the end of the 1980s.

Was the Edwardes strategy successful (Williams *et al.* 1987)? Undoubtedly Edwardes was tactically astute and succeeded in changing power relations within BL. At the same time he acted swiftly to close excess capacity. However, BL managers in general, including Edwardes, remained very much focused on events and processes within the company, particularly on 'reasserting the rights of managers to manage'. Yet to make a strategy based on a Fordist-inspired industrial model function coherently, sales had to rise to match capacity. The Metro lines now required 85 per cent capacity utilization to make them economical, and their inflexibility prevented them from being used to make other cars.

The Metro was relatively successful in the United Kingdom. However, market fragmentation following the entry of several small cars produced by European and Japanese manufacturers meant that domestic Metro sales fluctuated around only 110,000 in the early 1980s. The disappearance of most of the European sales network made exports difficult, and so Metro production fluctuated around 150–160,000. Capacity utilization rates were very low, the implications of which were only partly hidden by the fact that the capital investment was paid by government grants. The Maestro and Montego fared no better. While sales were respectable, the fragmentation of the domestic market since the mid-1970s meant that large volumes for single models were more difficult to attain. By 1985 combined Maestro and Montego domestic sales reached only 131,000.

BL had narrowed its basic product range to three key models just as the market had fragmented in a way that suggested companies needed a wide range of models to achieve high levels of domestic sales. And BL found it difficult to export its now adequate products. Failure to achieve projected sales meant that, in terms of operating profits, BL continued to lose money, albeit at a lower level, with the exception of a small profit in 1983.

Yet while BL focused most of its efforts on securing a core set of mass market BL models, at the margins of the company the seeds of a potential new strategy were being unwittingly sown (Mair 1994: ch. 5). With the Maestro and Montego not to be launched until 1983–4, in 1979 Edwardes signed an agreement with the Japanese manufacturer Honda to supply BL with CKD kits of its new Ballade model (a twin of the Honda Civic). The Ballade was to be built by Triumph and sold as the Triumph Acclaim from 1981. With the closure of Triumph, the Triumph Acclaim was built at the Cowley plant between 1981 and 1984, boosting sales prior to the arrival of the Maestro and Montego. The success of the venture led to an agreement to build the second version of the Ballade at the Longbridge factory from 1984 to 1989, with higher 'local content' including

BL engines (still 1950s BMC engines) in one version. Honda and BL also agreed to jointly design and build a new large car, the 1985 Honda Legend and 1986 Rover 800, both built at Cowley with the latter replacing the 1976 Rover 2300/2600/3500. To retain consistency in marketing, the second Ballade was named the Rover 200, the Morris name was dropped, and the Metro, Maestro, and Montego were all given the Austin badge. By the mid-1980s, BL had two car ranges: its own new products under the Austin name, and its Honda-based products under the Rover name. The 'volume cars' division of BL (Jaguar was still owned, but run as a separate business, until 1984) was renamed Austin-Rover.

Meanwhile, and separate from the Honda relationship in product development, BL began superficial experiments with new concepts acquired from the early stages of diffusion of 'Japanese' management ideas into the West. The early 1980s had already seen substantial changes to work organization for skilled maintenance employees, as complicated multiple-level grade systems with clear demarcations were collapsed into two basic grades when the new welding lines were introduced at Longbridge and Cowley, under pressure from the new industrial relations regime (Scarborough 1984, 1986). Now, there were attempts to define 'work teams', each controlled by a foreman, and each containing both production operators and skilled workers, who were to be relatively autonomous like the work groups of the old British model but this time clearly under managerial control and able to call on outside help (Willman 1987). A crude factory-level financial bonus scheme was operated during the early 1980s in an attempt to boost productivity.

## 15.4. ROVERIZATION: REDEFINING A BRITISH MODEL, 1986–1994

Between 1982, when Michael Edwardes left BL, and 1986, company strategy remained unchanged. Except for the decision to pursue the Honda relationship and a 1984 decision to develop a new small engine (with vital government financial support), few new initiatives were taken, as the company awaited the promised recovery. By 1985, however, it was apparent that sales of the Austin models were not reaching targets. On the basis of a recovery due to high sales in North America, Jaguar, which had always been run separately, was privatized in 1984. It was now planned to further break up BL for sale to other car makers, since this was viewed by some as the only way to retain any economic activity and employment at all in the shell of BLMC. Negotiations were started with General Motors to purchase Land Rover and the Leyland truck business, and with Ford to purchase Austin-Rover (the remnants of BMC).

However, domestic political opposition to the sale of these 'national champions' to foreign companies curtailed the negotiations. Eventually, with considerable government aid to wipe out debts and ensure healthy cash balances, both Austin-Rover and Land Rover were sold to the aerospace industry 'national champion' British Aerospace (BAe) in 1988 as a means to preserve their 'Britishness'

## The Search for a Viable British Model 409

while permitting BAe to increase its asset base by 50 per cent to support financing for its large overseas defence contracts. Despite early talk of industrial synergies, BAe effectively left the car maker to its own devices operationally and strategically while keeping tight control over—not permitting—any substantial new investments during its tenure as owner, which lasted until 1994.

Meanwhile, in 1986 the government had brought in Sir Graham Day, previously manager of a nationalized shipbuilding company, as Chairman and Chief Executive Officer. Day had immediately embarked on an entirely new strategy for Rover Group, as he renamed Austin-Rover/Land Rover. Day's strategy was based on a coherent and consistent industrial model. Significantly, the strategy drew strongly on several of the principles of the craft-British model. Land Rover, the only surviving example of the craft-British model, which was still effectively, if gradually, modernizing itself by centralizing its operations onto one site and updating its products and management techniques (neither Jaguar nor the bus and truck operations were in fact modernizing and both collapsed when they encountered recession), soon became a paradigm for the whole Rover Group.

First, in 1987 and 1988 Day sold all of the vertically integrated BL parts makers (exhausts, fuel tanks, radiators, metal stampings, and so on), mostly to their own managers, retaining only engine and gearbox manufacture and major body stamping operations. Secondly, the bus making operation was sold to its own management, and the truck and mid-sized van operation was sold to the Dutch truck maker Daf. By 1988 Day had sold eighteen of the twenty remaining businesses, retaining only Austin-Rover and Land Rover. The government had injected over £1 bn., or one-third of annual Rover Group turnover, into BL to make the sales (including to BAe) financially viable. In 1989 the Austin-Rover and Land Rover organizations were formally merged.

Thirdly, Day switched the product focus of Rover Group away from the products of the Edwardes era, the mass market Metro, Maestro, and Montego, and away from the obsession with market share and direct competition with Ford and General Motors. Maestro and Montego, only 3 and 2-years-old respectively in 1986, continued in production at the Cowley plant until 1994, but were effectively ignored in strategic terms: indeed they were now sold with no badge to indicate their manufacturer. Under Day's new 'Roverization' strategy, the company sought to develop a range of new Rover models of superior quality and design in the top niche of each size class. Day's stated goal was the creation of a British BMW. The aim was to create a profitable company which would remain small scale: 500,000 to 600,000 per year, for the medium-term future. Day was quite explicit that Rover Group would no longer be a 'volume producer'. Again Land Rover became emblematic.

Fourthly, to realize the strategy, Day pursued the link to Honda which had already delivered the Rover 200 (1984–9) and the Rover 800 (1986–1998), a link previously considered by most BL managers as marginal to their focus on mass market cars. Day sought to fill the gap between the Rover 200 and Rover 800 with new models also derived from Honda products. A new Rover 200 was launched in 1989 at Longbridge (twinned with a Honda Concerto, also manufactured by Rover for Honda). Rover produced the 200 in six different body styles, one of which was named Rover 400. In 1993 the Rover 600, twin of a Honda Accord model, was launched at Cowley. This strategy involved no major investments, and gave Rover a product range based on Honda designs and shared development costs.

However, Rover by no means became dependent on Honda technologically. A number of projects kept the capacity to design and develop cars intact. The only wholly new Rover Group product in which those capabilities appeared together between 1984 and 1996 was the Land Rover Discovery, launched in 1989, a four-wheel-drive recreational vehicle which was successful and highly profitable despite annual volumes under 50,000. Discovery legitimated Roverization. Simultaneously, there were smaller projects, outcomes of an enforced low-risk and low-investment policy used to update and refurbish existing models which adopted the practices of the craft-British model (regular update of basic 'robust' designs with commonization of as many parts as possible). These revealed that product development competencies remained intact and were being rebuilt: a revised Metro in 1990 with a new body shell; the six body variants of the Rover 200, introduced at regular intervals after 1989, the 1992 updating and partial new body for the Rover 800, and the body design to differentiate the Rover 600 from the Honda Accord. New engine variants were also introduced on a regular basis. The 1984 government decision to fund development of a new small engine to meet proposed European Community emission legislation now proved to have been crucial. The innovative design of the small aluminium K-series engine, launched in 1989, became a successful platform for independent product development and potential profitability.

Rover Group thus spread its model range widely, accepting small volumes for each model. The range incorporated the Mini, Metro/100 (two body styles), 200 (five body styles), 400, 600, 800 (three body styles), Land Rover Defender (two body styles), Discovery (two body styles), Range Rover, and from 1994 a revived MGB model, a total of eleven models built on ten platforms, with average volume per model and per platform of only 40,000–50,000. The company now possessed a clear, coherent, and potentially viable competitive strategy for small-scale profitability by the early 1990s. The promise of the strategy was confirmed by Discovery, the success of which supported the company financially, although the key test remained whether the mainstream Rover cars could be sufficiently 'Roverized' in design and customer appeal, drawing in part on the image of the old Rover company. This in turn rested on two factors: whether the basic Honda designs were too constricting, and whether Rover designers could define a modernized version of a craft-British 'Britishness' for the cars. The 1992 ingenious addition of a plastic false grille to the front of the cars to help create a 'modern but traditional' image suggested both the possibilities and the frailties.

To what extent did Rover learn a new industrial model for product development from Honda (Mair 1994)? In the early years of collaboration for the

Ballade-based Acclaim and first Rover 200 (1979–84), no attempt was made to learn. While BL production and logistics managers were surprised by the ease of assembly and precision of components supply arrangements Honda made for the Acclaim kits (exact numbers of perfect parts were sent rather than the boxfuls that were BL practice), it was not recognized that BL ought to learn from Honda. Indeed the two companies became equal partners in the XX project in which they jointly developed the Honda Legend/Rover 800. The project proved extremely difficult, as the industrial models of the two companies were almost entirely incompatible. For instance, BL engineers found it hard to understand the methodical precision of Honda engineers who kept changing (improving) targets during the design phase. Rover was rebuked for its poor seat design by Honda's seat maker. The ambitious plan for each company to manufacture the other's products proved a fiasco. So many differences of design were necessary to permit the project to proceed that the Rover 800, especially with the poor quality parts sent from the United Kingdom, disrupted production at Honda's Japanese Sayama plant. Honda, and Honda's European dealers, were so critical of the quality of the Honda Legends built at Cowley that Honda refused to purchase many. Shared production was abandoned by 1987/8.

The projects that followed (1989 Rover 200/400/Honda Concerto, 1992 Rover 600/Honda Accord (European version), 1995 Rover 400/Honda Civic (European version)) were led by Honda. Individual Rover engineers involved in these projects began to study Honda methods closely. A growing number of product development managers began to reason that the only way to cooperate with Honda was to adopt Honda processes. Yet, in part reflecting an ingrained suspicion of outside industrial models and new ways of thinking—legacies of the 'practical' British models—it was not until the early 1990s, after more than a decade of partnership with Honda, that new team-oriented product development and introduction processes were being adopted, practised, and refined for the new Rover products to be introduced in 1995 and 1996.

Rationalization of resources remained a major focus under Day and his successors, as they reduced the break-even point of the company, first below 450,000 and then towards 400,000. The Cowley plant was entirely restructured, shrinking in size to a fraction of its former area when a modern assembly hall was constructed in the early 1990s to build the 800 and 600 models. Rover Group employment continued to fall, from 42,000 in 1988 to 33,000 in 1992. Meanwhile, there were various attempts to introduce further Japanese-style employment practices during the late 1980s, some of them learned in a superficial way by senior Rover managers who had visited Honda factories, such as the 'zone briefings' and 'zone circles' introduced in 1986 and 1987 to provide employees with better information, top-down, about the company and to create opportunities for involvement in 'continuous improvement' (Smith 1988). The zone briefings soon faltered when management persistently ignored requests from supervisors for information to answer worker questions. The failure of the zone circles was in part a result of opposition from trade unions and shop stewards, now prepared to permit managers and engineers to design production processes, but not ready to permit workers to help them.

From 1987 to 1991, however, a systematic, planned, and internally cascaded Total Quality Improvement (TQI) programme was implemented in order to introduce a standard new set of business processes. Consultants played a large part in this programme, which was not related to Honda. Similarly, in 1990 a new department, Rover Learning Business, was set up to promote individual 'learning' opportunities for all employees with the goal of improving flexibility and participation in improvements. By 1991 there were the first serious attempts to understand teamwork at Honda and learn how Honda managed its workforce in North America. In 1992 a significant 'New Deal' was signed with the trade unions, now reduced to three principal unions through inter-union mergers. The New Deal set out a framework for developing new forms of labour relations and work organization, in an attempt to recreate 'greenfield' conditions on the old sites. The company was no longer permitted to make workers redundant without their (paid) consent; in principle this was an effective 'job for life'. Single status employment conditions were introduced: company uniform, shared restaurants, monthly pay by bank transfer. Trade unions were still recognized as the official representatives of employees, now in a single bargaining unit for salaries and conditions, with decentralized negotiations at factory level over work organization. Teamworking principles were to be accepted across the company, and were introduced step by step from 1992 to 1994 as working groups formed into teams which elected their own team leaders from volunteers approved by management. Employees were now, in principle, to be entirely flexible, willing to accept all task assignments given by management. A rational internal labour market was planned to reduce worker resistance to making suggestions which improved productivity; displaced workers were no longer assigned to a floating 'pool' of workers.

In practice, some of the new principles proved difficult to implement (Mair, forthcoming). While the formal adoption of teamworking proceeded smoothly, senior management was allowing local level experiments to find the best formula for Rover Group rather than imposing a fixed model top-down, and this was necessarily a slow process. Moreover, there was resistance from some production managers to the development of rational integrated learning processes that might permit workers to actually utilize newly acquired skills. Stopping the production lines to permit learning activities seemed to take second place to output on lines where demand for products was high; though it was easier at Cowley where the plant still operated below capacity.

Increasingly believing that it had now found a coherent industrial model which linked a wide range of low volume, upmarket products, with spin-off designs and regular updating of robust designs, to a flexible and knowledgeable workforce which was increasingly designing its own work processes, management persisted with efforts to diffuse and deepen the new model into the mid-1990s. Significantly, a number of the principles of Roverization—including low vertical integration, the spin-off/updating product development process, more autonomous work organization, the market niche/low volume strategy, and the attempt to define specifically 'British-style' cars—bore a close resemblance to the craft-British model, as if it had been updated by the addition of new social and technological management techniques.

To what extent can Roverization be judged a successful strategy? By the mid-1990s the results were still unclear. Positively, Rover Group appeared to have created a viable model for new product introduction, with new methods for low-cost product design and development. Not only had a wide range of products been introduced and regularly updated at low cost, but new models introduced in 1995 and 1996 confirmed that Rover Group had indeed retained and expanded the capability to design and develop competitive new products without Honda input. To the new Rover 400 (two body styles) derived from the Honda Domani/Civic were added three new models all developed with no Honda collaboration: a new Rover 200 (two body styles), a new Range Rover to replace the original 1970 model, and a new mid-engined MG model (MGF). Moreover, the K-series engine range, which lack of funding had at first limited to small engines, was now expanded to include a larger four-cylinder engine, a V6 large engine, and an advanced variable-valve timing version (working on different principles to Honda's variable-valve engine). An advanced direct-injection diesel engine was also introduced—and sold to Honda.

Market share in the United Kingdom continued to decline towards 10 per cent, but this was now argued to be a deliberate strategy to distinguish Rover from Ford and General Motors (Rover stopped pricing vehicles to attract large-scale company 'fleet' purchases and did not cut prices during the early 1990s United Kingdom recession). Exports to the USA of the Land Rover Discovery were now accelerating, and the company had developed an important market for over 5 per cent of its output in Japan, particularly for the Mini, the Discovery (now also sold to Honda), and soon the MGF.

Yet from an economic perspective Rover remained partly reliant on Honda, with which it shared product development costs and many components suppliers for the 400 and 600 cars. Moreover, while the new industrial model might be increasingly coherent and powerful internally, from the perspective of competitive strategy, a key question remained whether the niche pricing strategy, so necessary for profitable low volume production, could be maintained, particularly as other manufacturers (Fiat and Renault, for instance) were adopting similar strategies and offering similarly attractive products based on far higher production volumes. In effect, the domestic and European markets were fragmenting further with several manufacturers now vying for what used to be the 'quality' niches above the 'mass' market.

While product quality was now very high, gross productivity under Roverization only increased marginally. Between 1988, when the new configuration of the company had been settled by disposals of other businesses, and 1994, gross productivity rose only 13 per cent, or 2 per cent per year on average (comparing years with very similar overall output levels). While the Solihull Land Rover factory was operating at well over designed capacity, the extensively rebuilt Cowley plant was still operating at considerably less than its reduced full capacity. The fact that employment began to rise again to cope with increased production in 1994 suggested that productivity improvements were reaching a limit.

Moreover, while the details were now hidden in the BAe accounts, profits remained below 2 per cent. Operating profits during the recession in 1991 and 1992 were negative. Roverization may have been the only possible strategy, and may have been responsible for a remarkable ability to survive a further decade. Yet its long-term competitiveness relative to the fast-evolving strategies and models of other automobile producers remained uncertain.

## 15.5. BMW OWNERSHIP AND BEYOND: 1994 ONWARDS

BMW Group purchased Rover Group from BAe in 1994 for £800 m., as soon as BAe was legally permitted to sell Rover under the terms of its privatization arrangement. Honda had declined to take a majority share.

Rover, with its apparently internally consistent new industrial model developing, now had to deal with yet another outside model, that of BMW, with its quite different trajectory of highly successful modernization of a specific German model, one which, thirty years previously, must have shared much in common with the craft-British model of the old Rover company. Would BMW understand, and how would it interpret, Roverization? What strategies would BMW adopt regarding Roverization? Arguably, if Honda was a good partner for the 1980s and 1990s, BMW was potentially the ideal owner for the 2000s, yet much would depend on a further articulation between a British and an external industrial model.

By 1996, a number of clues were emerging. All ongoing vehicle projects had been continued except for early plans to replace the Rover 800 independently. Hence the new Range Rover was launched in 1994, the 200 and 400 replacements and MGF in 1995. From 1995 a major investment programme was made at the Land Rover factory, the first of significance at BL/Rover group since 1981, for production of a new niche small Land Rover from 1997. Construction of a planned new research and development centre was sanctioned. Following lengthy internal debates on what would be the first major strategic departure for Rover Group, a new Mini was announced for 2001, with its cost reduced by sourcing the engine in South America. Simultaneously, a new plant on a new site near Birmingham to build a small-medium engine (post K-series) for Rover and BMW was announced.

In terms of industrial models, intriguing issues were now being faced. From the economic standpoint, there was once more—as in the 1970s, as in the 1980s—clear pressure to attain higher volumes, with a target of 750,000 now openly declared. The question was whether the target for recovery in volumes would finally be reached. Also economically, beyond quick savings in global distribution logistics, a clear Sloanist strategy for new product development was emerging for the medium term, with future Rover and BMW vehicles to share engines, mechanical parts, and especially electronics. A key question here would be over the internal division of responsibilities for the core commonized components.

Yet perhaps most significantly of all, it was quickly clear that BMW managers and engineers found the internal functioning of Rover far more opaque than they had initially thought. Not only may the scale of investment in necessary new production facilities have been underestimated, but deep-seated differences in industrial models now needed to be grappled with, and it remained unclear to what extent these were understood by either BMW or Rover. Of particular relevance, BMW unwittingly discovered and brought to the surface the resilience of features of the old British industrial models which had endured for decades, still infused Rover Group, and were not necessarily inconsistent with Roverization. A 'can-do' engineering attitude based on practice rather than theory; few internal management controls; significant unease with measurement as a management tool; a vital role played by implicit knowledge; distrust of outside models coupled with great uncertainty how to react to BMW methods; arrogance and parochialism at the successful Land Rover; these were just some of the symptoms which led to great frustration for BMW managers and engineers who had to cooperate with, and manage, Rover, for these elements were certainly contradictory to BMW's model. The open question here was whether BMW would try to impose a 'rational' and 'systematic' management model on a 'wayward' Rover, and what impacts this might have, or whether the new owner would develop the highly sophisticated approach to adaptation and hybridization that the history of the Rover trajectory suggested might well be more successful.

# NOTE

1. This chapter focuses on the companies within the 1968 BLMC which in 1986 became Rover Group: the inheritors of vehicle makers Morris (including MG), Austin, Rover (including Land Rover), Triumph, and the Pressed Steel body company. Space precludes discussion of other companies within BLMC which were sold separately during the mid-1980s, such as Jaguar, Leyland Trucks, Leyland Bus, industrial and military vehicle makers, and parts makers. The discussion after 1986 is based on interviews with over 150 employees of Rover, Honda, and BMW, to whom I am particularly grateful. The interpretation of industrial models—particularly controversial in the British case—is my own. Steven Tolliday provided helpful comments on an earlier draft. David Brown of Birkbeck College collated the statistics for the appendix.

## STATISTICAL APPENDIX 15: Rover

| Year | Total car production (units) | Total sales outside UK (units) | Total sales (£m.) | Operating profit (£m.) | Total employees |
|---|---|---|---|---|---|
| 1968 | 818,289 | 397,501 | 907 | 45.5 | 188,247 |
| 1969 | 830,874 | 408,465 | 970 | 45.7 | 196,390 |
| 1970 | 788,737 | 368,380 | 1,021 | 13.4 | 199,524 |
| 1971 | 886,721 | 385,837 | 1,177 | 46.7 | 193,703 |
| 1972 | 916,218 | 347,325 | 1,281 | 41.0 | 190,841 |
| 1973 | 875,839 | 347,998 | 1,564 | 58.2 | 204,149 |
| 1974 | 738,503 | 322,523 | 1,595 | 19.3 | 207,770 |
| 1975 | 605,141 | 256,672 | 1,868 | −38.1 | 191,467 |
| 1976* | 687,875 | 320,836 | 2,892 | 116.1 | 183,384 |
| 1977 | 651,069 | 293,316 | 2,602 | 56.1 | 194,610 |
| 1978 | 611,625 | 247,936 | 3,073 | 71.3 | 191,853 |
| 1979 | 503,767 | 200,215 | 2,990 | −46.2 | 176,790 |
| 1980 | 395,820 | 157,829 | 2,877 | −293.9 | 157,460 |
| 1981 | 413,440 | 126,249 | 2,869 | −244.6 | 126,267 |
| 1982 | 405,116 | 133,862 | 3,072 | −125.8 | 107,763 |
| 1983 | 445,364 | 96,897 | 3,421 | 4.1 | 103,216 |
| 1984 | 383,324 | 78,664 | 3,402 | −11.7 | 96,001 |
| 1985 | 465,104 | 112,783 | 3,415 | −34.6 | 77,849 |
| 1986 | 404,454 | 123,558 | 3,412 | −246.4 | 73,396 |
| 1987 | 471,504 | 165,708 | 3,096 | 16.8 | 49,207 |
| 1988 | 474,687 | 137,048 | 3,224 | 71.0 | 42,300 |
| 1989 | 466,619 | 138,540 | 3,430 | 73.0 | 39,900 |
| 1990 | 464,612 | 168,928 | 3,785 | 65.0 | 41,400 |
| 1991 | 395,624 | 187,873 | 3,744 | −52.0 | 38,300 |
| 1992 | 378,797 | 159,107 | 3,684 | −49.0 | 34,600 |
| 1993 | 406,804 | 161,932 | 4,288 | 56.0 | 33,100 |
| 1994 | 460,000$^e$ | n/a | 4,900 | 82.0 | 36,238 |
| 1995 | 475,000$^e$ | 228,000$^e$ | 5,600 | 91.0 | 40,100 |

*Notes*:
* 1976 = fifteen-month accounting period.
$^e$ = author's estimate.
Total production is for passenger cars (includes Land Rover's Range Rover and Discovery but not Land Rover utility vehicles; includes 20,000–30,000 Jaguar cars per year until Jaguar was sold in August 1984). Total sales outside UK includes direct exports and CKD production abroad of cars (CKD car production took place until 1983). Operating profit is before interest, tax, extraordinary expenditures, and transfers from reserves (from 1988 to 1993 rounded figures for operating profit are given by owner British Aerospace). Total employees is average employees during the year (except 1994 and 1995 when year-end data are given): overseas proportion is 12–20 per cent during the 1970s; varies between 9–28 per cent during the 1980s; data not available after 1987 but limited to sales companies.

Total sales, profits, and employment refer to the whole BLMC/BL company including Jaguar and divisions making buses and trucks, components, and industrial vehicles from 1968 to the mid-1980s. Production and sales abroad refer to cars only. It is difficult to establish the proportion of turnover due to car sales; a good estimate is 65–75 per cent during the 1970s and early 1980s, rising to 90 per cent in 1987 with divestments, and 95 per cent in the early 1990s (the remainder being Land Rover utility vehicles).

*Sources*: Society of Motor Manufacturers and Traders; Annual Reports for BLMC, BL, British Aerospace, BMW.

# BIBLIOGRAPHY

Church, R., *The Rise and Decline of the British Motor Industry* (London, 1994).
Mair, A., *Honda's Global Local Corporation* (London, 1994).
—— 'The Introduction of Teamwork at Rover Group's Stamping Plant', in J. P. Durand, J. J. Castillo, and P. Stewart (eds.), *Teamwork in the Automotive Industry: Radical Change or Passing Fashion?* (Basingstoke, forthcoming).
Scarborough, H., 'Maintenance Workers and New Technology: The Case of Longbridge', *Industrial Relations Journal* 15/4 (1984), 9–16.
—— 'The Politics of Technological Change at British Leyland', in O. Jacobi, B. Jessop, H. Kastendiek and M. Regini (eds.), *Technological Change, Rationalisation and Industrial Relations* (London, 1986).
Smith, D., 'The Japanese Example in South West Birmingham', *Industrial Relations Journal*, 19 (1988), 41–50.
Tolliday, S., and Fujimoto, T., 'The Diffusion and Transformation of Fordism: Britain and Japan Compared', in R. Boyer, E. Charron, U. Jürgens, and S. Tolliday (eds.), *Between Imitation and Adaptation: The Transfer and Hybridization of Productive Models in the International Automobile Industry* (Oxford, 1998).
Whipp, R., and Clark, P., *Innovation and the Auto Industry: Product, Process and Work Organization*, Pinter Frances (London, 1986).
Whisler, T. R., 'Design, Manufacture, and Quality Control of Niche Products: The British and Japanese Experience', in H. Shiomi and K. Wada (eds.), *Fordism Transformed: The Development of Production Methods in the Automobile Industry* (Oxford, 1995).
Williams, K., Haslam, C., Williams, J., Johal, S., and Adcroft, A., *Cars: Analysis, History, Cases* (Providence, 1994).
—— Williams, J., and Haslam, C., *The Breakdown of Austin Rover* (Leamington Spa, 1987).
Willman, P., 'Labour-Relations Strategy at BL Cars', in Steven Tolliday and Jonathan Zeitlin, (eds.), *The Automobile Industry and its Workers* (Cambridge, 1987).
Wood, J., *Wheels of Misfortune: The Rise and Fall of the British Motor Industry* (London, 1988).

# 16

## A Second Comeback or a Final Farewell? The Volvo Trajectory, 1973–1994

CHRISTIAN BERGGREN

The year of the first oil crisis, 1973, marked the end of an era for the Western automobile producers, and for the Swedish car makers Volvo and Saab in particular. During the 1950s and 1960s, they had enjoyed rapid growth and prospered in spite of being small-scale producers together accounting for only 1 per cent of global car production. Operating in a small open economy, Volvo and Saab were severely affected by the sudden transition to a low growth market. In 1974, Volvo announced its first deal in a strategy for continued expansion: the acquisition of the DAF company in the Netherlands. The same year, Volvo also inaugurated two new plants in Sweden, Kalmar and Skövde, both designed according to socio-technical principles, by means of which Volvo hoped to solve its severe personnel problems and safeguard future worker recruitment. These two events, the acquisition in the Netherlands and the high-profile departure from a Fordist plant design in Sweden, highlight two themes which continued to shape Volvo's agenda into the 1980s and 1990s. The company's struggle to survive as an independent automobile maker in a small, open economy has shaped its strategic orientation. Its efforts to respond to Swedish labour market pressures formed the basis for a distinctive production model.

The first theme in the chapter is Volvo's efforts to cope with its basic competitive predicament: its position as a small-scale producer in a capital-intensive industry, heavily exposed to international competition. One aspect of this exposure was the high import penetration of the Swedish car market. Indeed Volvo never enjoyed a sheltered home market. Before World War II, imported cars accounted for 90 per cent of total sales. In the 1950s, when the Swedish makers adopted modern mass production methods, they gradually increased their share. In 1956, the Volvo PV444 became the best-selling model, a position Volvo models have retained ever since. The peak level of domestic market control was reached in 1970. When the Swedish makers started to upgrade their vehicles and focus on more exclusive segments, their domestic share eroded, and by 1990 was down to 30 per cent. By contrast, in France, domestic makers controlled 60 per cent of the market in 1990; in Germany the corresponding figure was even higher. Another aspect of international exposure was the high proportion of Swedish car production which was exported. The export orientation commenced in the late 1950s, when Volvo followed the example of Volkswagen, Renault, and Fiat, and started shipping cars to the USA. The early European exporters were defeated, but Volvo persevered and, in 1980, the USA became the company's single most important market. In 1970, 60 per cent of the Swedish car output was sold abroad; twenty years later this share had increased to 80 per cent.

From the early 1970s to the early 1990s the peculiarities of the Swedish labour market constituted a second and central factor in the development of the Volvo trajectory. For nearly all of this period unemployment in Sweden was below 3 per cent: in the second half of the 1980s, below 2 per cent. High female workforce participation was a second distinctive trait in Sweden; in the late 1980s, 80 per cent of all women of working age were employed outside the home. Thirdly, the Swedish labour market was distinguished by the high and even union density. In 1986, the unionization rate among blue-collar workers was 87 per cent and union density among white-collar workers was only a few percentage points lower. The confederation of blue-collar workers pursued a policy of 'solidaristic wages'; wages were to be fixed according to the characteristics of the work, irrespective of the profitability of individual firms. This egalitarian wage structure made it difficult for the automobile firms to recruit and retain workers during high growth periods. As a major manufacturing employer, Volvo strove to solve the labour problems through a series of innovative workplace designs. This endeavour made Volvo the icon of workplace reform in Europe until the crisis in 1992, when the company decided to shut its assembly plants in Kalmar and Uddevalla.

The focus of this chapter is on the passenger car sector. The development and survival of the Swedish automotive companies cannot be understood, however, without reference to their heavy trucks divisions, Volvo Trucks and Scania. The importance of trucks for the Volvo Group has been gradually increasing, from 22 per cent of total revenues in 1973 to 39 per cent in 1994. In contrast to their positions in automobile markets, the Swedish truck-making divisions are no minor global players. Of the world's heavy truck producers, Volvo is second only to Daimler-Benz. Together Volvo and Scania account for more than 20 per cent of global production outside the former Communist countries, and they also command eminent positions in the heavy bus market. Moreover, compared to cars, Swedish trucks are both more international and enjoy a stronger domestic position. In 1990, Volvo Trucks and Scania sold 95 per cent of their products outside Sweden. At the same time, they basically controlled the Swedish heavy truck market. Another important difference with the automobile sector was the virtual absence of any effective Japanese competition in the market for heavy commercial vehicles.

The following section of the chapter presents an overview of Volvo's production model and product strategies at the start of the period under scrutiny. The next two sections follow the evolution of its strategy and production model during the 1970s and 1980s, including references to the cross-flow of influences between the car and truck divisions, and ending with an analysis of the contradictions of 'Volvoism' in the late 1980s. The final section deals with the crisis of the early 1990s, when the Swedish labour market suddenly imploded, and Volvo, after an aborted plan to merge with the French automobile producer Renault, decided on a strategy of maintaining independence. Two final questions address the future: of Volvo as an independent car maker, and of its remarkable original production model.

## 16.1. STARTING-POINT: VOLVO'S SWEDISH FORDISM

In 1970, Volvo produced more than 200,000 cars for the first time. Three years later production surpassed 250,000. This implied a thorough effort to implement Fordist principles in Sweden: a concentration on only two car models, investments in dedicated machinery including 'Detroit automation' in the machining shops, and minutely divided work in the entire production system. Volvo was a pioneer in the adoption of American time-and-motion studies, such as MTM (Method-Time-Measurement) which had been invented during the war. When Volvo opened a major new assembly plant in Torslanda outside Gothenburg in the mid-1960s, the company boasted that by using MTM, assembly layout, work content, and the pace of production had been completely determined before the hiring of a single worker! In contrast to the American Fordist companies, however, Volvo did not pay its employees any premium for the repetitive work and high production intensity. As a result it became difficult to recruit Swedish workers during the 1960s, and the assembly lines were increasingly operated by immigrant workers from Southern Europe.

From an international perspective Volvo was a paradox: a small-scale producer with a very strong belief in the merit of scale. In order to realize economies of scale, the company adopted a policy of infrequent model changes. The rationale was spelled out in the *Annual Report* of 1971: 'We have concentrated our resources on a limited line of products. This is our way of reaching a satisfactory volume and thereby profiting from rationalization in spite of our internationally small size'. The strategy was successful, and for several decades Volvo was able to double the total volume of each successive model. The first post-war model, PV444/544, was produced in 400,000 units. Output of its successor, Volvo 120 ('Amazon') had reached 700,000 when it was discontinued in 1970. The next model, the 140/160 series, was only produced from 1966 to 1974, but reached an accumulated volume of 1.4 million.

A second way to compensate for small scale was to invest heavily in a distinctive brand image and to focus on niche markets. This was closely related to Volvo's early export orientation, since a niche producer confined to a small domestic market would never be able to enjoy economies of scale. The brand image was based on a few central product characteristics, such as quality, safety, and longevity. The development of a safety profile was stimulated both by early and advanced Swedish legislation and the focus on the demanding American market. Volvo consistently focused on stable market segments, such as well-to-do suburban families (only in Sweden do its products have a broader market appeal). These families' need for convenience and space made Volvo an early pioneer in comfortable station wagons. The same basic strategy was at work in the truck business, where Volvo early on chose to focus on the heavy segment (above 15 tons) and not compete in the mass market for light trucks. This strategy drove a rapid internationalization. By the mid-1970s foreign markets already accounted for 80 per cent of sales and 50 per cent of truck assembly.

A third means of finding economies of scale was Volvo's components sourcing strategy. From its very beginning, Volvo adopted a policy of low vertical integration. In the post-war period, this increasingly implied sourcing components from manufacturers outside Sweden. In the 1980s Volvo purchased 65 per cent of its components from non-Swedish firms, first and foremost German ones (Elsässer 1995: 87). By contrast, Mercedes-Benz sourced 90 per cent of its components from German suppliers, and Renault sourced 70 per cent from French suppliers. By purchasing from international high-volume component specialists such as Bosch (engine control and fuel injection systems) or Aisin-Warner (automatic transmissions), Volvo has been able to enjoy the same economies of scale as its foreign competitors. It must be pointed out, however, that Volvo did not try to develop any close partnerships with its components suppliers; relations were basically of the Fordist type, based on volume and price.

## 16.2. SOCIO-TECHNICAL BREAKTHROUGH AND THE EMERGENCE OF A NEW MODEL FOR VEHICLE ASSEMBLY, 1970–1981

In 1973, the last year of the post-war boom and the starting-point of this trajectory, Volvo Cars recorded record sales and profits. This was accompanied by severe labour problems, however: rampant turnover, absenteeism, recruitment difficulties, and occasional wildcat strikes. Appointed to the top office in 1971, the new Chief Executive Officer P. G. Gyllenhammar pledged to solve these problems by innovations in work organization and plant design:

> Due to the advanced economic and social structure of Swedish society, we have encountered earlier than most countries new problems in the organization of jobs and the working environment. Therefore, at Volvo we do not look upon these problems as a threat. Our familiarity with this type of question could well lead to an improvement in our competitive ability. (Volvo 1971)

In 1974 these efforts started bearing fruit. Two innovative plants were inaugurated, an assembly plant in Kalmar, and an engine factory in Skövde. Volvo remained true to its belief in the overall merits of scale, but with a significant modification. According to its new orientation, labour-intensive assembly could be performed at least as effectively in small-scale plants located outside the overheated labour markets of the big cities.

Thus Kalmar was planned for only 30,000 vehicles per year on a one-shift basis (a second shift was never introduced). The plant caught the attention of an international audience deeply interested in enhancing the quality of work life. The traditional conveyor belt was replaced by a flexible system, based on automatically guided vehicles (AGVs). This made it possible to divide the production flow into distinct segments with their own teams, where each team enjoyed a certain autonomy and workers could expand the content of work (from four to thirty minutes). The new form of organization was supported by an innovative building architecture, designed to give each team its own spatial identity. By contrast to the dominant mass production model, Kalmar emphasized the integrative advantage of small scale and the quality advantage of worker identification with the product. At the engine plant in Skövde, two hours from Gothenburg, a major investment in automatic machining of components was combined with a flexible assembly system, using a similar AGV technology. In engine production, however, the overall importance of scale prevailed, and the new plant was a large complex in the Volvo world, with a capacity of 300,000 units per year.

Volvo also planned to bring this new wave of work design to the New World. In 1974 construction of an assembly plant designed along Kalmar lines commenced in Virginia, in the USA. The timing of the new plants and plans was unfortunate, however. Kalmar and Skövde had barely begun production, when sales started to decline. In four years the car division lost a third of its volume and by 1977 the company was deep in the red. The plans for a US factory were shelved. In none of the years between 1974 and 1980 did the output of Volvo's core production units reach 1973 levels, despite new plant capacity. With few exceptions, the profit margin was close to or below zero during each of these years.

The oil crisis in 1973–4 and the subsequent years of stagnation convinced Gyllenhammar that the days of steady expansion in Sweden had come to an end. Alliances and acquisitions were deemed necessary to increase volumes and production scale in the future. In 1974, the failing Dutch company DAF, located in Born, was acquired. To maintain its exclusive dealerships, Volvo needed a broader product range, but did not want any large-scale investments in new models. The Dutch car added to Volvo sales volume, but unfortunately also to its losses. After a deal with the Dutch government, Volvo reduced its commitment to a minority stake. Gothenburg continued to be responsible for products and strategies, but the Dutch cars remained completely separated from the Swedish models, without any common components. In the crisis year of 1977, Volvo made a second attempt to expand its automobile business by an acquisition. Its recently introduced 240 series had not yet become successful, and prospects for the future looked grim. Saab, the junior Swedish car maker, with an annual volume of less than 100,000 units, was even deeper in trouble. A government report reflected the widespread opinion in Sweden that its small car makers would not be able to develop new models for the next decade. Supported by the government and the metalworkers' trade union, Gyllenhammar negotiated a merger deal with the family which owned Saab. The deal was fiercely opposed by engineers and managers at Saab, however, and they eventually persuaded the owner to withdraw.

To supplement the drive for increased scale in the automobile business, management at Volvo, like many other automobile executives in the 1970s, also sought to diversify and invest in sectors with a different cyclical business pattern. A start was made by acquiring businesses in the leisure industry. Volvo never became a successful owner of sportswear and small boat firms, however. At the end of the 1970s, the focus was shifted to the energy sector, in particular to oil exploration and trading. Volvo joined forces with a new financial company in Sweden and acquired a portfolio of diverse industry and trading interests. Whereas the automotive divisions had accounted for 80 per cent of Volvo's sales in 1973, this share had fallen to 40 per cent in 1982. At the height of the second oil crisis, when markets were more volatile than ever, Volvo's trading group was highly profitable. Later it became overextended, however, and a few years after the acquisition the oil trading company reported record losses, and was discreetly dissolved.

Volvo's car business was not saved by dividends from oil exploration or trading, however, but by the strong performance of the commercial vehicles division. Unaffected by the oil crises, the company steadily expanded production and sales of heavy trucks. In 1970, the total output was 16,000, the majority of which was in the heaviest segment: 16 tons and above. Ten years later, Volvo produced 28,000 trucks. The smaller bus operation grew even more rapidly, and tripled its output during the decade. Trucks and buses became the Volvo Group's premier generator of earnings, with a profit margin of 11–14 per cent.

Volvo's trajectory of workplace reforms had started in the automobile sector, where capacity bottlenecks and personnel problems were greatest during the early 1970s. When these problems abated, the commercial vehicles operations took the lead in a 'socio-technical relay race'. In 1976 Volvo Bus seized the opportunity to go further than Kalmar at its new plant in Borås. With product variety sharply expanding, the bus operation suffered from increasingly serious line-balancing problems at its traditional facilities. The Borås factory design was an attempt to find a system which could cope simultaneously with the flexibility requirements of the marketplace and the demands for more challenging and enriched work emanating from the labour market. Parallel assembly modules replaced the conventional conveyor line, making teams of ten to twelve workers responsible for building a complete chassis. Work cycles were extended to two to four hours. Materials handling was integrated with assembly. Instead of maintaining a fixed pace, teams distributed tasks and selected a pace of work according to how many workers were present and what kind of chassis they were building. At the outset, managers feared that the extended work cycles would make training demands extensive and make it difficult for the plant to cope with staff turnover. During the 1980s, however, the plant proved itself capable of an outstanding performance even in the midst of personnel instabilities brought about by sharp rises in volume.

Five years later, Volvo Trucks inaugurated a new assembly operation, the LB plant, north of Gothenburg. In contrast to Kalmar, which was by and large a management initiative, there was a significant union input at LB, both in the planning of the plant and in the actual managing of work organization. The production design, using buffered flows and operator-controlled AGVs, was a compromise between a traditional line system and the integrated assembly of the bus plant. Work organization at LB was more sophisticated however, with work groups having considerable decision-making power. They were supported by a new wage system designed to promote individual skills as well as group responsibilities. While the stimulus behind the Kalmar plant was primarily the labour market, from the very beginning the Borås and LB plants were conceived in terms of a dual need for reform: the difficulties of coping with increasingly differentiated customer demands; and the requirement to make assembly work much more attractive. The engineers involved in the plant projects were deeply convinced that in the future nobody would want to work on an assembly line. At the same time, the rapidly expanding programme of custom options made it imperative to continue developing more flexible structures.

## 16.3. REBOUND IN CARS, AND A RENEWED CRISIS OF INDUSTRIAL WORK, 1982–1989

In 1980 the stock market assessed the value of Volvo's car business as being close to zero! Two years later, the tide had turned and profits and production volume were surging. Three major factors contributed to Volvo's comeback. First, the crisis had triggered comprehensive rationalization activities. In a few years, overall productivity improved by 20 per cent, inventories were reduced, and the company's quality problems were resolutely confronted. A second factor in the turnaround was an overhaul of the company's product line-up. Introduced in the mid-1970s, the 240 model had suffered from a very slow sales start. Now it was upgraded and re-engineered to reduce production costs. Soon it became Volvo's foremost cash cow. A new model, Volvo 700, was introduced just as the business cycle started to recover. The third factor was that Volvo's recovery was helped by the dramatic shift of American–Swedish currency exchange rates. In 1981–2 the value of the Swedish krona was depreciated, and shortly afterwards the American dollar started to climb. By 1985, Volvo was exporting more than 100,000 cars per year to the USA. The profitability of the car division (measured as the return on employed capital) exceeded 40 per cent, making Volvo the most profitable automobile maker in the world. Total production increased from 225,000 in 1977 to 400,000 eight years later. The problem for Volvo was not to sell cars but to produce them. Sweden, too, was booming and unemployment had virtually disappeared. Public welfare provisions became even more generous, including 100 per cent compensation from the first day of sickness. More than ever before, Volvo was confronted with the problems of the Swedish labour market. Combined with negligible wage differentials between sectors (see Table 16.1), the labour shortage made it very difficult for automobile firms to retain workers in unattractive production jobs.

TABLE 16.1. *The wages of automobile workers in relation to the manufacturing average: a comparison of hourly wages in selected countries, 1970 and 1980*

| Countries | 1970 | 1980 |
|---|---|---|
| USA | 1.31 | 1.49 |
| Germany | 1.20 | 1.21 |
| France | 1.25 | 1.16 |
| Japan | 1.13 | 1.23 |
| Sweden | 1.19 | 1.05 |

*Source*: Altschuler *et al.* (1984).

Young people turned their backs on manufacturing jobs almost completely. Short effective working hours worsened the labour shortage. Workers in this industry in Sweden were absent twenty-nine days per year on account of illness; the figures for Germany and Britain were eighteen and eleven, respectively. The generous compensation system contributed to high absenteeism, as did extensive workforce participation by women, including child-rearing women. Women were often deployed in the most repetitive tasks, and tools and equipment were seldom adapted to their different physical abilities. The result was a rapidly rising level of repetitive strains and cumulative traumas among women. In the public debate the automobile industry was heavily exposed. The metalworkers union developed an ambitious programme for 'solidaristic work', which strongly emphasized the need for a general up-skilling and gradual elimination of repetitive and fragmented jobs. High expectations were placed on the automobile industry to provide the lead and restructure traditional mass production jobs. Assembly line jobs in particular were targeted.

All these factors created a crisis of industrial work in Sweden. Volvo Trucks responded by shifting assembly overseas. During the 1980s, output of trucks doubled from 30,000 to 60,000 per year, but all of the increase took place outside Sweden. In 1989, the Swedish plants only assembled a quarter of Volvo's truck production, and the Belgian plant had become the division's major assembly complex. At the same time, however, there was an important qualitative development in Gothenburg. The LB plant, the compromise of the early 1980s, was complemented by a new facility, which completely departed from the assembly-line model. In this so-called dock assembly system, teams of ten to twelve workers built complete and customized vehicles in work cycles of eight hours or more. The first dock was operational in 1988, the second in 1990, and expansion continued in successive steps. In this way, Volvo Trucks developed a plurality of assembly concepts within its European operations. In Belgium, with its abundant labour supply, the company continued to invest in assembly-line plants, where it could manufacture a simpler product mix cost effectively. In Sweden, Volvo invested in flexible and holistic forms of production for the assembly of complex and customized products. The inspiration for the Uddevalla car plant (see below) was obvious, but Volvo Trucks was able to link its production strategy closely to the requirements of specific market segments, which rendered it much more robust.

Volvo Cars, too, stepped up its international operations. In the early 1980s, the majority of Volvo cars were still assembled in Sweden. At the end of the decade, overseas plants produced 50 per cent more cars than the Swedish plants combined. The factory at Torslanda, Gothenburg, however, remained the main plant. And Sweden's personnel problems began to have severe repercussions here. After a brief spell of stability, staff turnover soared from 12 per cent per year in 1982 to 27 per cent seven years later. The new 700 series signified a move upmarket and more exacting demands on quality. The unstable workforce at Torslanda made it very difficult for the plant to achieve its quality goals without sacrificing productivity. To solve this renewed Swedish labour problem, Volvo returned to the socio-technical agenda of the early 1970s, which had now been enriched by the company's own experiences, above all in truck and bus production.

In 1985, Volvo Cars decided to expand assembly capacity by building a new plant in Uddevalla. The government offered lavish subsidies, since it hoped that Volvo would rescue the local labour market after the closure of a state-owned shipyard. At first, Kalmar was the model for the new Uddevalla operation. But the ideas of the project group were soon radicalized. After a visit to the new Nissan factory at Sunderland in the United Kingdom, the recipe of lean production was also rejected. Instead, the project team decided to build on the experience of the new production structures in bus and truck assembly. The Borås and LB plants had to a varying degree been compromises between the traditional principle of series flow (a 'line') and parallel and integrated assembly. Uddevalla chose to implement a true parallel production design, consisting of small shops with a total of forty teams working in parallel. Each of them was responsible for building complete cars. As at Kalmar, the novel production design was manifested in the architecture of the new building. Uddevalla was to compete head-on with mass-production plants. To make that possible, a dual technology strategy was devised: simple and low-cost technical aides for the parallel teams, combined with a centralized semi-automated system for picking and placing materials and subassemblies in individual kits. These were delivered from the materials shop on the request of the assembly teams. Of the workforce, 40 per cent were women, and special efforts were devoted to the ergonomic adaptation of the assembly process. Because of the long work cycles, new forms of training were vital. These 'holistic' learning principles stressed the need to understand the whole—the entire car or the complete function—and to combine mental maps with manual dexterity as a basis for generalized and robust assembly skills (Ellegård *et al.* 1991). Production layout, team organization, technical strategy, and principles of learning constituted the basis of a new industrial model, which held the promise of transforming repetitive mass production jobs into a new type of modern craftsmanship. The metalworkers' union was deeply involved, although its local organization in Gothenburg did not really believe in this promise. Starting operations in 1989, Uddevalla was planned to reach an annual volume of 40,000 cars in three years, a rather insignificant share of total production. Volvo none the less regarded the plant as a strategic response to Swedish labour market conditions.

In 1974, Kalmar and the new engine plant in Skövde had embarked on the uncharted socio-technical route. Fifteen years later, Uddevalla inspired the design of the next-generation gasoline engine plant at Skövde. This factory was to produce Volvo's new modular engine family, comprising four, five, as well as six-cylinder engines. Previously, the machining departments had been automated, whereas the assembly sections remained manual. The new plant took the process further by automating the basic internal assembly of the engines in a very flexible process (the most flexible automated engine plant in the world, according to its management). The remaining manual assembly was designed as twelve parallel modules. Half of them adopted dock assembly systems, where each operator could build complete engines. In the other half flexible flow-lines were utilized. A similar plant was constructed for the new generation of diesel engines introduced the following year.

Meanwhile, at the Torslanda assembly plant, a series of technological and organizational reforms succeeded each other. In the early 1980s, process engineers were greatly inspired by Volkswagen's foray into automated assembly and tried to launch an automation strategy at Gothenburg too. A long list of possible processes for automation was produced and several experiments carried out, but the equipment turned out to be excessively expensive in relation to minor savings in labour costs. The existing Volvo models were in no way designed for automatic assembly, and the production system lacked the precision—in components, tools, and machine settings—required for an automated process to be reliable. Further, Volvo did not enjoy the economies of scale of Europe's mass producers. The next attempt at reform at Torslanda was a project for stepwise socio-technical change. The plant was to be divided into a series of separate 'product shops', each shop being free to rearrange its production design to increase autonomy and flexibility. In 1987, this approach was tested in the department responsible for engine dressing. Again, the results were not convincing. The more complex flow system was difficult to monitor and control, and the approach involved significant costs, in extra buffers and equipment, for only modest increases in job content and worker autonomy.

As a consequence, the reform process at Torslanda was radicalized. A new plan proposed a complete conversion of half of the plant into small assembly shops, where teams would build complete cars as at Uddevalla. This approach too was tested. In 1990, after scarcely a year, the 'whole-car project' at Torslanda was discontinued, the justification being that assembly times were too high and materials handling too expensive. The local metalworkers' union supported the decision. Given Torslanda's Fordist traditions, the experiment had been far too radical, and it was expelled as an alien body. At Volvo Trucks a composite assembly philosophy was developed in response to the divergent demands of the marketplace. At Torslanda such opportunities were not explored. The outcome could have been different, if the experiment in whole-car assembly had started with special vehicles, which were a nuisance on the assembly line anyway. Nor did the project team receive any support from the manufacturing engineers responsible for the new 800 series. On the contrary, the latter favoured a highly automated process for the 'marriage' of the car body and engines/subassemblies. This was difficult to undertake within the context of parallel assembly. In the end, most of the ambitious reform attempts at Torslanda came to naught. Reorganization at Torslanda had reached an impasse.

At the same time, the breaking in of Uddevalla suffered from technical and organizational glitches. The outcome was an effective European relocation of Volvo's assembly operations. In 1989, after sales volume had already fallen significantly, the company decided to expand capacity at its Belgian plant at Gent. Soon afterwards the crucial decision was taken to introduce the 800 model at Gent, and not at Gothenburg where all previous model introductions had taken place.

When Volvo began production at Uddevalla, the new plant had to assemble a 10-year-old product. This paradox illustrates a striking contradiction in the trajectory of Volvo during the 1980s. At the factory level, the company redesigned its production system, adapted technology to human demands, explored novel methods of training and organizing, and articulated its plant design with inspirational architecture. Yet these efforts remained compartmentalized and isolated. The company's basic organizational structure remained firmly in a traditional Fordist mould, with a strict division of labour between powerful departmental hierarchies: product planning, marketing, styling, design and development, purchasing, engineering, manufacturing, distribution and sales. Why was this structure so persistent? One reason was Volvo's long tradition of compensating for low annual volumes by infrequent model changes. During the crisis in the late 1970s and early 1980s, Håkan Frisinger, the energetic managing director of Volvo Cars, tried to break the tradition and promised a significant product novelty every six months. Engineers were sent to Japan to learn from Honda's agile design and manufacturing process. These efforts failed, however. After the devaluation of the Swedish currency and the rise of the dollar, Volvo enjoyed record earnings without having to reform its engineering culture. At the same time the 240 model, which had been introduced in 1974, became a major hit; it seemed that Volvo could sell this old car for ever. When it was discontinued in 1993 (10 years later than originally scheduled), its accumulated output totalled 2.8 million cars. This performance reinforced the belief in infrequent model changes, and the 700 series was expected to repeat the feat. The powerful president of the company, Gyllenhammar, took a deep interest in reforms of manufacturing work, but lacked commitment to the car business as such. His strategic thinking was focused on grand deals, alliances, and diversifications. The most distinctive 'car guy' among senior management, Håkan Frisinger, was forced to leave the company because of tensions with Gyllenhammar. After that, there was no one at senior level pushing for reform of the product development and delivery process. There was no lack of resources; spending on design and development increased continuously during the 1980s. The problem was low productivity, and lack of effective management. Several projects were started but never completed. Only in 1991, ten years after the 700 model had been introduced, was Volvo in Sweden ready to launch a major new model.

The new plants for bus and truck assembly had been planned to cope with the dual pressure of more demanding workers coupled with more complex products: hence their emphasis on increased work content combined with production flexibility. Uddevalla, on the other hand, was conceived too much simply as an answer to labour market pressure. There was no articulated effort to make use of its flexible and adaptive potential to develop Volvo's product strategy. The planning of Uddevalla did not even include the transfer of Volvo's production of special vehicles (police cars, ambulances, and so on) from Torslanda to the new plant. In 1992, when it was too late, a pilot study demonstrated that these special vehicles could be built at Uddevalla in less than half the time spent at Gothenburg. After the closure of Uddevalla (see below), its methods were introduced at the special car assembly department in Gothenburg. The lack of support from an innovative product strategy made Volvo's production reforms highly vulnerable when the boom years suddenly came to a halt.

## 16.4. CRISIS AND NEW DIRECTIONS, 1990–1994

In the second half of the 1980s, the Swedish economy passed from export-driven growth and full employment to overheating and cost inflation, fuelled by increasingly speculative domestic demand and stagnating productivity growth. In 1985–8, the annual growth in labour productivity (value added per hour) in the manufacturing sector was only 2 per cent. In 1989–90 productivity and GDP growth declined even further and in 1990 there was a virtual standstill (Erixon 1990). In the midst of a looming crisis of costs, the Labour government announced the adoption of hard-currency policies, with the linking of the Swedish krona to the ecu. Full employment was abandoned as a central political goal. The first of a series of austerity programmes was implemented to cut public spending and prune welfare benefits. The problems were exacerbated by the rigid monetary policy of the Conservative coalition that won the election in 1991. Swedish GDP declined for three years in a row. Manufacturing output and employment fell by 20 per cent. Two hundred thousand out of a total of 1 million jobs were eliminated from the manufacturing sector. A further contraction was prevented only by an involuntary shift to soft-currency policies in 1992, when the krona was set free to float. From a historic low of 1.2 per cent in 1990, unemployment skyrocketed to more than 10 per cent in 1993. Suddenly, manufacturing companies no longer had recruitment problems. When alternative jobs were no longer available, and the average length of service increased because of lay-offs and reduced hiring, personnel turnover fell sharply. As a result of the deteriorating labour market and cuts in the compensation system, the problem of endemic absenteeism also disappeared. By the early 1990s, all of a sudden the labour-market-induced reform drivers were no longer very important. With few exceptions, the unions also ceased to act as effective agents of workplace change.

### 16.4.1. The Closure of the Branch Plants

The crisis in the Swedish economy greatly affected the automobile industry. During the boom period, domestic sales had been buoyant and in 1988, total sales peaked at 350,000 cars. This record level was followed by a steep decline and in 1993 only 125,000 new cars were registered in Sweden. Volvo's profitability had peaked as early as 1984, with a profit margin of 20 per cent. In the following years, profits declined with monotonous regularity, and in 1990 the car division was losing money again. The situation of the other Swedish automobile maker, Saab, was even more serious. The company had rejected a merger with Volvo during the crisis in the 1970s, and then prospered by upgrading its products. After a few years of healthy profits, earnings suddenly nose-dived. In 1989, Saab lost 2,100 m. kr. on sales of 15,300 m. kr. Management was paralysed, and the owners started looking for a strong industrial partner. Finally, an agreement was reached with General Motors (GM) to form a new fifty-fifty jointly owned company, Saab Automobile. Saab was integrated into GM's international purchasing system, and the Opel Vectra platform was used to reduce the cost and lead-time for the development of an urgently needed new Saab model. Saab's recently opened assembly plant in Malmö, which had been inspired by Uddevalla's efforts to create alternatives to the assembly line, was closed and all production concentrated in the main plant in Trollhättan. Massive lay-offs helped bring about a new 'lean' culture. Outsourcing and rationalization cut total employment at Saab from 15,800 to 6,500 people.

The fate of Saab seemed to reveal the destiny of Volvo Cars. Sales of large Volvo cars plunged from 280,000 in 1989 to 210,000 in 1991 and losses in the car division mounted. In previous crises, strong performance by the truck division had balanced the haemorrhaging in the car business, but now Volvo Trucks, too, reported losses. The Swedish car market had collapsed and an overvalued Swedish currency increased the difficulty of exporting. After the seven fat years, Volvo was in disorderly retreat. In 1992, the two small branch plants at Kalmar and Uddevalla were sacrificed. Volvo did not need their capacity any more. Nor did Volvo need to develop a strategic response to an advanced and demanding labour market. The relocation of assembly to Belgium, which had started in the 1980s, accelerated.

The crisis also triggered a revival at the old Gothenburg assembly plant, notorious for its personnel and productivity problems in previous decades. The deep recession brought a new stability to the plant and labour turnover fell from 27 per cent in 1989 to 6 per cent in 1992. This made it possible to invest in worker training, and a revised wage system stimulated a strong shopfloor commitment to quality. A new approach to production was introduced, called the KLE strategy (for quality, delivery precision, and economy, in that order). Its basic idea was to concentrate all efforts on achieving quality directly in the process. Thereby the need for inspection, rework, and extra handling would be eliminated and productivity would improve as a bonus for the efforts made to improve quality. In the first phase of the change programme, the role of foremen and supervisors was significantly strengthened, but in the second phase, the plant put much more emphasis on teams and blue-collar team leaders. Torslanda also tried to learn from Kalmar's effective methods to involve workers in systematic improvement. These methods included regular production breaks for team meetings, to discuss problems, and find corrective actions. The change programme at Torslanda worked surprisingly well, and the plant rapidly improved its productivity and quality.

Twenty years previously, the inauguration of Kalmar had signified a new manufacturing strategy with a belief in the advantages of small-scale assembly and production pluralism. The closure of Kalmar and Uddevalla spelt a return to a traditional belief in concentration and large-scale production. From a strictly economic point of view Kalmar and Uddevalla had always been marginal. As symbols of work reform they were uniquely powerful, however. The closure decision unleashed an intense debate, and top management declared that the decision did not imply any criticism of lagging performance. In fact, the two small plants had demonstrated an impressive improvement record over a short time period. After a painful breaking-in phase, Uddevalla first took off in 1990. This was followed by a new stalemate, when the plant struggled to develop an organizational form to match its assembly system. In mid-1992 there was a second revival under a new manager, who introduced a process-oriented organization with only two hierarchical levels, moved the management centre into the production area, and devised a radical programme to close the cultural gap between hourly and salaried workers. Kalmar had started with an innovative production system but a traditional Tayloristic rationalization approach. In the late 1980s, there was a revival. Teamwork and technology were developed in the socio-technical tradition, but linked to a new participative management culture and demanding business objectives. An elaborate process was put in place for groups to set their own targets, check and revise them, devise counter-measures, and set new goals.

In terms of productivity, Uddevalla and Torslanda competed neck and neck during 1991–2. Under its new management, however, Uddevalla displayed a steeper learning curve than Torslanda. Quality performance showed a similar pattern, but here Kalmar had the fastest rate of improvement. According to the J. D. Power standard American quality statistics for 1993, 940 model Kalmar cars had a perceived quality close to that of Toyota's Lexus. Apart from this quality achievement, the differences between the three Swedish assembly plants were not great, measured by traditional performance parameters. The distinctive advantages of Kalmar and Uddevalla were their flexibility, customer responsiveness, and competency for in-depth cooperation. According to Volvo's engineering department, which was responsible for preparing new models for manufacturing, Kalmar was outstanding in terms of the technical competence of its engineers and operators and their ability to cooperate with product engineers in a creative way. Uddevalla had superior flexibility and capacity for customizing vehicles. The annual model changes in 1990–2 were introduced with 25 per cent lower tooling costs and 60 per cent lower training costs per car compared with the Gothenburg plant: despite Uddevalla's long work cycles. An argument against Uddevalla's parallel design, however, was that it would be difficult and expensive to introduce the radically new 800 series, where the product design required automated equipment for critical operations. If Volvo's business and product strategy had been different, with an emphasis on rapid introduction and flexible assembly of a variety of niche vehicles and customized cars, the small plants would have been precious assets. In 1992, however, there was no interest in such a strategy. Once again the contradiction of 'Volvoism', between its innovative production policies and its rigid product planning, was highlighted.

### 16.4.2. Volvo and Renault: A Grand Merger Plan that Failed

The closure of the two small assembly plants in Kalmar and Uddevalla plants could also be seen as preparation for a much bigger deal, the planned merger between Volvo and Renault, which did not have much space to accommodate idiosyncratic Swedish production systems. In the difficult years of the late 1970s, Volvo management had attempted to solve the problems of the car business by merging with Saab. In the late 1980s alliances and mergers reappeared on the agenda. Previously, Volvo and Renault had started cooperating on limited projects. In 1990 a comprehensive alliance was disclosed, implying a successive integration of purchasing, quality functions, and product development. Three years later a full-blown merger proposition was presented at a press conference in Paris. The main rationale for the merger was to achieve economies of scale in components supply and product development. Volvo had recently finalized the expensive development of the 800 model; including investments in new plant and equipment, the bill totalled 15 bn. kr. The cost for designing and engineering a successor to the ageing 700/900 series was seen as prohibitive.

However, Volvo engineers had another view. They were proud of the innovations of the 800 model, its driving performance as well as safety features, and were confident that this platform could provide the basis for several future car models. When Volvo's plan to merge with Saab had been aborted, Saab engineers had played a key role. Now the Volvo engineers enacted a similar role. Many feared that a merger with Renault would result in a decline of independent research and development in Sweden. The engineers' union gained support from previous senior managers who had been alienated by Gyllenhammar's autocratic rule. The CEO of Scania, which had always grown organically, pointed out that it is extremely difficult to translate the total sales volume of a merged company into economies of scale in production, components, spare parts, and new products. In fact, Volvo had never been able to integrate its DAF business. Further, key investors were deeply sceptical about merger with a state-owned French company. Volvo's president had never concealed his contempt for the shareholder community, and now its leading players struck back with a vengeance. In the end, the merger plans were shelved. P. G. Gyllenhammar left and a remarkable era ended. Soon afterwards the new board announced a reversal of Volvo's business strategy, a new focus on the Group's core business, and a much stronger emphasis on shareholder interests.

### 16.4.3. The Turnaround

In 1992, Volvo had been at its nadir. Both the car and truck divisions were losing money. Two years later, the Group reported record earnings of 16 bn. kr. The driving forces of this dramatic turnaround were strikingly similar to the turnaround in the early 1980s: a favourable shift in exchange relations, new vehicle models, and hard-nosed rationalization. The first factor was external: the unintended depreciation of the Swedish currency by 25 per cent in 1992. The second factor was the belated take-off in sales of the 800 series in 1993. The same year Volvo Trucks introduced a completely redesigned model line-up which immediately boosted sales. The third turnaround factor was a thorough rationalization of production and distribution in order to increase productivity, improve quality, and reduce inventory. In 1993, the car division produced the same number of cars as in 1990, but the workforce had been reduced by 22 per cent. In total, the car and truck divisions now employed 47,000 people, compared with 58,000 in 1990. At the same time the emphasis on 'quality first', a crucial part of the aforementioned KLE strategy continued. In the J. D. Powers report on consumer satisfaction in the USA for the model year 1993, almost all automobile makers had improved, but progress at Volvo was more rapid than at other firms. The 940 model scored as the best European import. Another aspect of the streamlining programme was to reduce inventory in the distribution system. Traditionally, the overwhelming majority of Volvo cars had been built according to a central scheduling system, which was based on successively updated dealer forecasts. During the crisis, Volvo revamped this system with the explicit goal of assembling cars for Europe on the basis of individual customer orders only. In this change Uddevalla played a pioneering role. Custom-order assembly had the additional advantage of making it much easier to sell the cars at sticker price, thereby avoiding various dealer discounts. In 1994, between 80 and 90 per cent of all cars for Europe were built on custom orders.

### 16.4.4. Strategic Reorientation

A central theme of the new Volvo strategy was to concentrate on the core automotive business. The new leadership announced a radical divestment policy. As a result of sell-off, the financial position of the Group strongly improved. In 1994, trucks, buses, and cars accounted for 94 per cent of Volvo's total sales. 'Concentration on the core' had the advantage of creating a strong management focus, but risk exposure was heightened and increased volumes became a crucial issue. Accordingly, capacity expansion, in both the car and truck divisions, was an important part of the new strategy.

In the worst year of the recession, Volvo Trucks produced only 47,000 vehicles. In 1995, the volume had increased to 77,000 trucks. By the year 2000, the division plans to operate globally and sell at least 100,000 trucks per year. In the 1970s, Volvo Trucks had expanded in Europe. In the 1980s, it turned to North America, acquired the White corporation, and took over GM's heavy trucks business. In the 1990s, the focus was to be on Asia, and new plants were to be constructed in India and China. Volvo Cars, too, disclosed an expansion programme in 1995, which signalled an increase in its capacity for large cars from 300,000 to 450,000. After the divorce from Renault, Volvo did not intend to remain entirely independent, however. A further important part of the new strategy was cooperation with Mitsubishi Motor Corporation in the Netherlands. During most of the 1970s and 1980s, the Dutch operation had been a drag on Volvo's financial performance. One reason was the low level of production, which never exceeded 120,000 cars per year. Another was the complete lack of components commonization with Volvo cars designed in Sweden. In 1991, the Dutch operation was reorganized as a joint venture between Volvo and Mitsubishi, called Nedcar. The deal had two strands: the joint development of a mid-sized car on the basis of a Mitsubishi platform, and the complete renovation and expansion of the former Volvo plant in Born, employing Japanese technology and production methods. By 1995, plant capacity had been expanded to 200,000 vehicles and the twin models Mitsubishi Carisma and Volvo S40 were launched. The Carisma model sources its engines from Japan. Volvo had previously used Renault engines in its Dutch cars, but the new model is powered by a higher-performing engine built in Skövde to differentiate the Volvo product. Despite sharing responsibility with Mitsubishi, Volvo has a stronger influence in design and operations in Nedcar than it used to have in the previous structure, when the Dutch company was in reality an autonomous business. Further, Volvo has used the joint development effort to improve its product engineering capability and reduce lead-times in prototype development. When the estate version of the 850 model was prepared, estimated time for a complete prototype was one year. The project succeeded in reducing this time to thirty weeks; in the preparations for the new Dutch-built car, prototypes were produced in twenty weeks.

The Nedcar venture is significant in another way, too, in that it highlights the way that Volvo's production centre had gravitated to the Benelux region. Production of power-trains for cars and trucks remained concentrated in Sweden, and the same was true for truck cabs for Europe. Assembly of cars and trucks was increasingly concentrated in Belgium, however. In 1995, the Swedish plants produced only 14,000 trucks, compared to 21,000 in Belgium. Volvo is also expanding its Belgian car production. In 1990 Gent assembled 80,000 cars, compared to 160,000 in Sweden; in 1994 the output in Gent had increased to 150,000 cars, whereas production in Sweden had fallen below 100,000. This trend is continuing. Including the Nedcar venture, Volvo now produced more than two-thirds of its cars in the Lowlands.

In the years of the crisis of industrial work in the 1980s, it would have been an economic advantage for Volvo to maximize its production outside Sweden. Instead the company invested in Uddevalla. The situation in the mid-1990s was different. There were no longer recruitment problems in Sweden: high unemployment was a long-term prospect, welfare benefits had been curtailed, and industrial discipline restored. In Sweden, Volvo's market share was 25 per cent; a Volvo had been the best-selling model every year since 1956. In Belgium, Volvo is only one of five foreign automobile makers, so far the smallest. The market share is 2 per cent, below the level in Thailand. Paradoxically, Volvo is moving labour-intensive assembly out of Sweden, when from a comparative perspective, Swedish wages are lower than they have been for the entire post-war period. Volvo's strategic manufacturing decisions seem to suffer chronically from poor timing. In the 1970s, Kalmar was inaugurated when sales were heading steeply downwards; in the late 1980s, Uddevalla came on stream when the capacity was no longer needed; and in the 1990s, production is expanded in Belgium when it no longer has a cost advantage.

### 16.4.5. The Future: a Fragile Turnaround

Two broad questions will end this chapter. The first concerns the future of Volvo Cars, the second the future of Volvo's production model. In spite of the recovery between 1992 and 1994 the prospects for Volvo Cars remained uncertain. In 1996, after a year of flat sales and declining profit margins, the company announced a new round of restructuring, including the retrenchment of 2,000 employees, the majority of them in Sweden. As a part of the new strategic focus, the company had promised to launch a new model every year, but so far the visible results have been few. In 1996, the 800 series was 5 years old, and the 900 series was based on a 15-year old platform. The company had completely missed the new trend towards sport utilities and minivans, so popular in Volvo's main markets, the USA and Canada. In the early 1980s, Volvo had been an early European mover in product upgrading and internationalization. Complacency during the 1980s, followed by preoccupation with the Renault merger, resulted in a loss of precious time. By the mid-1990s, Volvo was once more in a position of trying to catch up with competitors. The German high-end specialists Mercedes, BMW, and Audi, were ahead of Volvo in model renewal, range of product offerings, as well as volume. In the 1970s BMW and Volvo had been equal in size. In 1995, BMW both enjoyed a broader product range and a higher volume per product, with 200,000 cars per model compared to 100,000 for Volvo. Further, the Germans had taken the lead in internationalization. Whereas Volvo continued to invest in the high-cost Benelux countries, BMW and Mercedes constructed plants in the USA, Mercedes invested in Mexico and Brazil, and in China both Audi and Mercedes were expanding rapidly. Compared to the global strategies of its competitors—and of its own truck division—the manufacturing structure of Volvo Cars remained narrowly focused on continental Europe. The division could draw on strong financial resources and engineering capabilities, as well as support from the powerful truck business, but whether that would be enough to safeguard Volvo as an independent car maker was far from certain. Possibly its strategy could include a broadening of cooperation with Mitsubishi, which could also involve joint production in the USA.

The second question for the future concerns Volvo's distinctive production model. The reopening of Uddevalla was the most spectacular item when Volvo announced its new investment programme. The plant was to be operated by a joint venture in which the British racing firm TWR Engineering controlled 51 per cent. It was now devoted to producing a small series of exclusive vehicles, based on the 800 platform. Did this decision also imply a revival of the Volvo production trajectory? The answer was no. The Swedish corporate and societal context of 1995 was radically different from 1985, when the original planning of Uddevalla started. Mass unemployment had succeeded the overheated labour market of the 1980s. The previous generous welfare benefits had been significantly reduced. The relative living standard of Sweden as measured by GDP per capita had regressed from the top group of OECD countries to below the mean. Uddevalla Mark I was perceived as a strategic response to an ever more demanding labour market. It challenged the parameters of mass production, be they American or Japanese. This meant that Uddevalla had to match the productivity of mass production, which created an enormous pressure at the plant to improve efficiency constantly. If it had been allowed to continue to develop, the potential consequences could have been far-reaching. However, to develop such an innovative production model independently implied a substantial cost premium, since Volvo could not rely on the standard equipment and standard solutions available for other makers. Moreover, a consistent innovation strategy would have required the rebuilding of all the major plants as well as re-engineering its cars to suit the new production model. In the affluent 1980s, that might have been possible; in the lean 1990s, it was not. In 1992, the window of opportunity for an alternative trajectory in high-volume production was shut. The context of Uddevalla Mark II is very different from Mark I. The renovation of the factory included a body and paint shop, but volume was to remain low: 20,000 cars per year at a maximum. Uddevalla II, as well as the similar assembly of low-volume special cars in Gothenburg, is very much a niche business, clearly separated from Volvo's high-volume units. Even if the new Uddevalla is a success, it will hardly influence Volvo's mass production system.

At its Torslanda operations, Volvo now subscribed to the lean production model, but modified by American as well as Swedish influences. The Japanese management system combines exacting performance demands with long-term employment security. In its new, shareholder-oriented management style, however, Volvo adopted an American approach to personnel policy, emphasizing immediate adjustment of the workforce when sales decline, even if corporate earnings were still strong. On the other hand there were also important Swedish influences at play at the plant level. Team leaders were elected by workers and rotated, instead of being appointed by management. Ergonomics remained a very important priority. Moreover there was a distinctive socio-technical heritage at Torslanda, for example in sections of the body shop and in the subassembly of doors and engines. Union input was still highly valued and the tradition of labour–management dialogue continued to play an important role. As noted above, however, the main volume of Volvo cars was now produced in the Benelux countries. This would have major implications for the future of the Volvo production model. Nedcar was run by Mitsubishi management. Management at Gent was always highly critical of Volvo's Swedish trajectory, and of Uddevalla in particular. The Gent plant was a pioneer within Volvo in the adoption of Japanese methods such as Total Productive Maintenance, and could be expected to be quick to learn from the Nedcar experience. Volvo's 'industrial model' in the 1990s, if there were to be one, might be better studied in Belgium and the Netherlands than anywhere else.

## STATISTICAL APPENDIX 16: Volvo

| | Sales | | | Cars | | | | | Trucks | | | | | Employment | |
|---|---|---|---|---|---|---|---|---|---|---|---|---|---|---|---|
| | | | | Production (000s) | | Sales | Profit | | Production (000s) | | Sales | Profit | | Employment (000s) | |
| Year | MSEK | Cars (per cent) | Trucks (per cent) | Sweden | International[a] | International (per cent) | Margin | | Sweden | International | International (per cent) | Margin | | Sweden | International |
| 1973 | 9,000 | 57 | 22 | 184 | 68 | 72 | 9 | | 13 | 3 | 76 | 12 | | 42 | 9 |
| 1974 | 10,500 | 53 | 22 | 165 | 70 | 69 | 3 | | 14 | 7 | 78 | 11 | | 46 | 11 |
| 1975 | 13,700 | 55 | 24 | 158 | 130 | 73 | 1 | | 15 | 10 | 79 | 7 | | 46 | 17 |
| 1976 | 15,700 | 54 | 26 | 161 | 136 | 69 | 1 | | 12 | 13 | 78 | 9 | | 45 | 17 |
| 1977 | 16,200 | 51 | 28 | 125 | 100 | 73 | −1 | | 13 | 12 | 78 | 11 | | 41 | 15 |
| 1978 | 19,100 | 54 | 28 | 146 | 115 | 76 | 2 | | 12 | 12 | 83 | 10 | | 42 | 15 |
| 1979 | 23,400 | 54 | 28 | 173 | 147 | 78 | 3 | | 14 | 14 | 83 | 12 | | 44 | 17 |
| 1980 | 23,800 | 50 | 28 | 139 | 130 | 77 | −2 | | 15 | 15 | 83 | 14 | | 44 | 17 |
| 1981 | 48,000 | 28 | 19 | 152 | 140 | 79 | 4 | | 13 | 15 | 89 | 8 | | 52 | 16 |
| 1982 | 75,600 | 24 | 16 | 167 | 152 | 79 | 10 | | 14 | 21 | 92 | 7 | | 53 | 18 |
| 1983 | 99,500 | 26 | 13 | 185 | 187 | 82 | 19 | | 12 | 22 | 91 | 1 | | 53 | 19 |
| 1984 | 87,100 | 35 | 19 | 190 | 190 | 87 | 20 | | 14 | 30 | 93 | 5 | | 49 | 16 |
| 1985 | 86,200 | 40 | 21 | 200 | 197 | 85 | 18 | | 11 | 30 | 93 | 5 | | 47 | 18 |
| 1986 | 84,100 | 44 | 22 | 200 | 215 | 86 | 15 | | 14 | 30 | 95 | 6 | | 52 | 19 |
| 1987 | 72,500 | 42 | 22 | 196 | 230 | 83 | 12 | | 14 | 33 | 93 | 9 | | 54 | 19 |
| 1988 | 96,600 | 41 | 27 | 180 | 221 | 81 | 9 | | 14 | 47 | 93 | 12 | | 54 | 23 |
| 1989 | 90,900 | 47 | 32 | 172 | 242 | 83 | 5 | | 15 | 45 | 93 | 9 | | 55 | 21 |
| 1990 | 83,200 | 47 | 34 | 156 | 220 | 85 | −2 | | 13 | 42 | 94 | 6 | | 49 | 23 |
| 1991 | 77,200 | 47 | 36 | 111 | 167 | 85 | −5 | | 14 | 39 | 96 | 3 | | 44 | 21 |
| 1992 | 83,000 | 54 | 36 | 121 | 183 | 87 | −4 | | 9 | 38 | 98 | −2 | | 39 | 21 |
| 1993 | 111,200 | 52 | 35 | 97 | 194 | 87 | 4 | | 9 | 42 | 99 | 8 | | 37 | 22 |
| 1994 | 134,000 | 55 | 39 | 99 | 253 | 88 | 21 | | 12 | 57 | 98 | 25 | | 36 | 21 |
| 1995 | 154,500 | 54 | 38 | tot. 370 | — | 87 | 8 | | 14 | 63 | 98 | 25 | | 42 | 24 |
| 1996 | 154,200 | 54 | 34 | tot. 376 | — | 88 | 12 | | 13 | 50 | 97 | 10 | | 44 | 26 |

*Notes:*
[a] Volvo's way of calculating profit margin (operating profit before taxes/sales) has been changed in the early 1990s.
Sales only comprise the automotive divisions, not Volvo's at times significant businesses in other industries.
Sales of trucks include buses.
Since 1995 Volvo has ceased to publicize production figures separately for Sweden.

*Sources:* Volvo, Annual Reports.

# BIBLIOGRAPHY

Altschuler, A., Anderson, M., Jones, D., Roos, D., and Womack, J. (eds.), *The Future of the Automobile* (MIT, Cambridge, Mass., 1984).

Berggren, C., *Alternatives to Lean Production: Work Organization in the Swedish Auto Industry* (New York, 1992).

—— *Mästarprestationer eller mardrömsfabriker? En utvärdering av Volvos småskaliga monteringsfabriker i Uddevalla och Kalmar* (Stockholm, 1993).

—— 'NUMMI vs. Uddevalla', *Sloan Management Review*, winter (1994).

Brulin, G., and Nilsson, T., 'The Swedish Model of Lean Production: The Volvo and Saab Cases', Proceedings of Third GERPISA International Conference (Paris, 1995).

Ellegård, K., Engström, T., and Nilsson, L., *Reforming Industrial Work—Principles and Realities in the Planning of Volvo's Car Assembly Plant in Uddevalla* (Stockholm, 1991).

Elsässer, B., *Svensk bilindustri—en framgångs historia*, (Stockholm, 1995).

Erixon, L., *Produktivitetsproblemet* (Stockholm, 1990).

Huys, R., and Van Hootegem, G., 'Volvo-Gent: a Japanese transplant in Belgium or beyond?' in Å. Sandberg (ed.), *Enriching Production* (Aldershot, 1995).

Kim, C., and Fujimoto, T., *Product Development Performance* (Boston, 1991).

Volvo, *Annual Reports*, 1970–95.

# 17

# Lada: Viability of Fordism?

JEAN-JACQUES CHANARON

This chapter examines three hypotheses about the trajectory of Avtovaz, the Russian manufacturer of automobiles and derived commercial vehicles (exported under the Lada brand), before arriving at a fourth more general hypothesis.

The first hypothesis is that Avtovaz has followed a trajectory characterized by two traits which make it a unique model. The company's evolution has been 'autocentric', in a quasi-closed system, as a result of its integration into the Soviet planned industrial system. Moreover, it has occupied a quasi-monopolistic position in a market in which demand has been far greater than supply. The second hypothesis is that the organization of production and inter-company relationships at Avtovaz have been based on an industrial model which reflects the original Fordism, as adapted to a unique national economic, social, and political context. The third hypothesis is that the developmental trajectory of Avtovaz has only experienced one real rupture in its history, that of transition from a Soviet planned economy to a market economy at the start of the 1990s. Until that point, there had been no real crisis intrinsic to Avtovaz's automotive production system itself, except for an accumulation of problems inherent in the Soviet economy as a whole which had to be overcome step by step during this period. The fourth and final hypothesis is therefore that the original Fordist model still appears to be viable under certain restricted conditions concerning the product, the mode of production, the organization of work, the market, and so on. In other words, the Fordist model may survive if focused on the cheapest and least sophisticated niches of the market.

The first section of the chapter describes the origins of Avtovaz. The second section describes the Fordist characteristics of its industrial model. The third section analyses the historical evolution of the Avtovaz trajectory during the 1970s, 1980s, and 1990s. The final section then discusses the conditions required for the continued viability of the Fordist model.

## 17.1. A MODEL OF 'AUTOCENTRIC' DEVELOPMENT

### 17.1.1. A Political Project

When the giant Togliatti complex, located on the Volga River, near Samara, was founded in 1966, the Soviet authorities explicitly planned a self-contained industrial development. The objective was to equip the Soviet Union not only with the capacity to produce private automobiles capable of effectively satisfying popular demand, but above all with a national automotive industry, by freeing itself as quickly as possible from the control of the Italian car maker (which was supplying product and process engineering) and the need to import most capital equipment and a substantial proportion of components. Avtovaz was to be capable of designing, producing, purchasing, and selling its own products throughout the world. A second objective was clearly to show Western countries, along with satellite Eastern European countries, that the Soviet Union was quite capable of competing on equal terms with them, both economically and technically, not only in the heavy industries, defence and space, but also in consumer goods and household durables.

The strategy adopted by the state authorities resulted directly from this political project. Fiat, the automobile producer selected as the partner in launching Avtovaz, came from a country which, if not a friendly nation, was at the very least believed to be more politically correct than other countries in ideological confrontation with the Eastern bloc. It was also a country in which the Communist Party was strong, indeed the strongest in Western Europe, even if it had not pledged total allegiance to the Communist Party of the Soviet Union (likewise its leftist trade unions). The chosen partner company was also viewed as more politically correct than others in its relationships with workers and their unions. It had also displayed a certain willingness to contribute to the development of the non-industrialized countries.

## 17.1.2. A Quasi-monopolistic Position

The particular Fiat model transferred was the Fiat 124, produced at Togliatti from 1970 as a popular, indeed populist, family car: a four-door sedan/saloon the design of which was truly 'classic', if not already outdated when transferred to Avtovaz. The model was still being manufactured in the mid-1990s.

Avtovaz was soon positioned as the Soviet Union's leading producer in what was essentially a captive domestic market (totally closed to imports), a market in which effective demand far exceeded overall supply. With a very low rate of consumer durable ownership (45 cars per 1,000 individuals in 1990) combined with adequate levels of savings to generate strong demand and waiting lists equivalent to several years' production, the company's market was guaranteed far into the future. Under such conditions, central planning was able to establish thoroughly reliable production schedules. It was the physical capacity of supply—which reached its current level in the late 1970s (Fig. 17.1)—which determined the size of the market. In fact production was divided *ex ante* between exports, which brought in hard currency to the national reserves, and the domestic market, which was considered more as a residual outlet.

Between 1980 and 1995, Avtovaz produced 62 per cent of all private vehicles made in the Soviet Union/Confederation of Independent States. All other producers

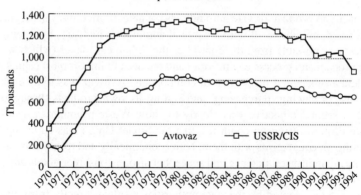

FIG. 17.1. *USSR/CIS and Avtovaz car production, 1970–1994*
*Note*: CIS stands for Community of Independent States.
*Sources*: Professional associations.

were smaller, either because their capacities were limited by a lack of investment (Moskvitch and Saporoshje ZAZ), or because they specialized in specific model ranges (Volga GAZ/UAZ).

### 17.1.3. *The Domestic/Export Dichotomy and the Economic and Financial Context*

Notwithstanding significant gaps and imperfections in the available data, a number of comments can be made regarding the economic environment of Avtovaz, at least in 1994. The Soviet system led to two particular features. The first was a significant gap between installed capacity, overly optimistic production plans, and actual output, which was always lower than forecast. The second was the clear dichotomy between the domestic market and export markets. The highest quality products, with components designed to meet specific regulations, were reserved for the latter. In 1994, 42.5 per cent of output was delivered abroad, with 60 per cent of this going to Western Europe.

As for the physical productivity of work (Table 17.1), always a questionable measure, the data are even more unreliable at Avtovaz, with its high level of vertical integration (67 per cent in 1994) and the difficulty of identifying its manufacturing employees (as opposed to those working in housing, shops, and services connected to the company).

Such a low productivity rate did not prevent the company from submitting accounts revealing profits and claiming profitability in press releases and in the rare documents circulated outside the factories. According to the official data, Avtovaz's gross profits were remarkable: 35 per cent of turnover in 1994 (Fig. 17.2).

## Lada: Viability of Fordism?

TABLE 17.1. *Labour productivity at Avtovaz in 1994*

|  | Number | Per capita Labour Productivity[a] |
|---|---|---|
| Employees and retirees | 220,000 | 3.0 |
| Employees | 175,000 | 3.8 |
| Togliatti plant | 126,000 | 5.2 |
| Production workers | 70,000 | 9.4 |

*Note*:
[a] Vehicles per head per year.

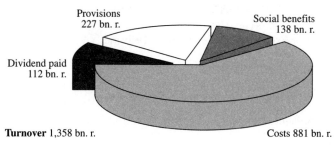

FIG. 17.2. *Avtovaz profitability in 1994*

### 17.2. AN EXAGGERATED FORDIST MODEL

#### 17.2.1. Fordist Technical Organization

Derived from the organization of work at Fiat's Mirafiori factory, designed in 1955 for the launch of the Fiat 500 model, the industrial model adopted by Avtovaz was based on the original Fordism. On the Togliatti site the company manufactured a very small range of models and versions on a moving assembly line, with the division of tasks pushed to the limit. The pace of the assembly line governed the pace of work and the effort of the workforce. The company's organization was highly centralized, hierarchical, and divided into departments according to classic functions and operations, with a quasi-omnipotent engineering department. It would seem that, contrary to Fiat's own advice (Fiat was prepared to provide improved technology), the engineers and political decision-makers involved in the strategic decisions over capital investment opted for a stable technology that the provider had fully mastered, so as to be able to blame any future malfunctions on Fiat.

TABLE 17.2. *Strategic versus operational variables*

| Domains | Strategic variables | Operational variables |
|---|---|---|
| Design | Range of models<br>Vehicles design<br>Process | Variants, options<br>Related technical choice<br>Implementation |
| Manufacturing | Overall organization<br>Annual volume | Definition of work posts<br>Daily volume<br>Quality control |
| Purchasing | Make or buy<br>Selection of suppliers<br><br>Modes of supply<br>Pricing | <br>Daily relationships (billing, complaints, and so on)<br>Logistics |
| Selling | Export/domestic division<br>Pricing | Deliveries |

## 17.2.2. The Specific Managerial Context

Avtovaz developed in a very specific context, namely that of a factory assigned to produce consumer durables, the organization and management of which were derived from the military-industrial complex model dear to the Soviet authorities. In this model, strategic management tasks, which were the responsibility of central ministries based in Moscow, were administratively separated from operational management functions, which were entrusted to functionaries. A formal plan for production objectives, linked to corresponding investment and workforce requirements, was imposed upon the latter, leaving them no room for manoeuvre except in the resolution of routine problems. Most of the so-called 'strategic' variables were the direct outcome of decisions taken in Moscow (Table 17.2).

## 17.2.3. The Stamp of the Political System

Management of the complex was clearly dependent on the prevailing ideology and social policy, as determined by the central authorities. Although the workers' union was entirely subject to the Communist Party and its party line, and therefore to its economic policy, the union had a broad range of powers, notably regarding workers' rights to welfare for which the company itself was responsible: social protection (health and pension benefits), supply of consumer goods, housing, holidays, cultural and sporting activities, and so on. Yet decisions on wage levels remained the responsibility of the political authorities.

# Lada: Viability of Fordism? 445

While the technical design of Togliatti was based on a classic Fordist model, the imprint of the political system remained the determining factor. The societal model in which the factory was placed had no mechanism to encourage productivity increases or divide them between capital and labour. If, in principle, there was coherence between the employment relationship of socialism and the organization of production imported from Fiat, since they were both based on simple planning, there was a major inconsistency as far as productivity gains were concerned, since Avtovaz had no mechanism to create them. This was clearly an 'imported' model, an attempt to graft a system designed for a capitalist market society onto a non-market context. Hence Avtovaz was a sort of 'transplant', but one created in a social system that was incapable of integrating it effectively, at least according to international criteria and standards.

## 17.3. THE TRAJECTORY OF AVTOVAZ

It is possible to identify three phases in the trajectory of Avtovaz. The phase from 1966 to 1975 was the start-up phase, spanning the construction of factories, the acquisition of basic know-how, and the increase in the rate of production towards installed capacity. The phase from 1976 to 1990 was a phase of stabilization, marked by the acquisition of knowledge and know-how related to the design of new models on the basis of existing platforms. The phase after 1990 was a period of uncertainty and search for a self-reliant path of development.

### 17.3.1. Launch, Increased Strength, and Learning, 1966–1975

The first four years were dominated by the construction of the new factory. On 22 April 1970 the first 2101 Avtovaz (Fiat 124) automobile, named the 'Jigouli', rolled off the assembly line. For the first three months the vehicles were based on completely knocked down (CKD) kits supplied by Fiat. From July 1970 the automobiles were 100 per cent Soviet. From 1970 to 1975 output increased progressively and rapidly (see Fig. 17.1). By 1973 cumulative production had already reached 500,000 vehicles.

This initial, euphoric phase focused on learning the techniques and organizational forms of mass production—a single model, a limited number of versions (four-door saloon/sedan, estate/station wagon), options, and colours—and of mass distribution, notably exportation to more than a hundred countries. In this period there were no major uncertainties; there was an abundant workforce for all types of tasks, attracted by a standard of living higher than the national average, with all requests for employment channelled by the omnipotent Communist Party. The financial resources to meet initial needs had been earmarked by the central authorities. The technologies (product and processes) were not difficult to master.

### 17.3.2. The Management of an Emblematic Industrial Complex, 1976–1990

Throughout this entire phase, which featured the disintegration of the Soviet Union and the abandonment of socialist dogma by the state, Avtovaz remained an industrial complex emblematic of the regime's capacity to manufacture products in line with the technological principles operating in Western countries but adapted to a socialist form of planned management.

Avtovaz was actually managed like all specialized complexes. Once the established capacity was attained (*circa* 800,000 cars per year), management operations were purely routine, responsible for achieving the plan's objectives while at the same time maintaining the plant, machines, and tooling in an acceptable condition. As Avtovaz was a rare example of a Soviet company capable of exporting to the West, it was closely scrutinized by the central authorities. They did not, however, invest the financial resources necessary for the regular renewal of the original production equipment. By 1990, the factory still largely operated with its original equipment, and was only just able to maintain production levels. Equipment had been in use for more than twenty years, and what equipment had been replaced was substituted by machinery similar in design to the original.

This phase also corresponded to the consolidation of knowledge and know-how, particularly in the learning of design-related skills. Despite limited resources, Avtovaz was able to design the Lada Niva, a small four-by-four vehicle (Lada 2121) launched in 1977. This was followed by the Samara/Spoutnik (the brand name in foreign and domestic markets respectively) (Lada 2109), launched in 1984, and then by the 'small' Oka (Lada 2105), on sale from 1989. Gradually, Avtovaz developed its capabilities to design, develop, test, and engineer products and processes, thus endowing the company with relative autonomy. However Avtovaz proved incapable of developing mechanical components (engines, gearboxes), and this considerably restricted its ability to adapt its products to meet the noise and pollution standards set by Western industrialized countries.

Moreover, a number of problems began to emerge, and specific solutions had to be adopted in a step-by-step manner. Thus the relative scarcity of unskilled workers was resolved by an influx of workers from neighbouring republics, attracted by wages higher than in their home regions. The rising level of absenteeism was resolved by chronic overmanning. The growing problem of suppliers unable to meet volumes and quality requirements was, in part at least, overcome by resorting to vertical integration. Yet the absence of financial resources necessary to replace and modernize production equipment and even increase capacity, or to diversify components supply (notably by buying diesel engines) was never resolved, with the central political authorities syphoning off the currency generated by exports without permitting the company to benefit.

In the domestic market, the company continued to benefit from a quasi-monopolistic position and a potential demand which greatly exceeded supply. Yet under these conditions, Avtovaz was unable to replace its model range and

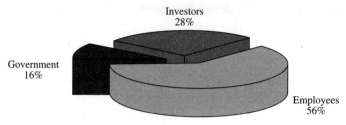

FIG. 17.3. *Shareholding of Avtovaz in 1994*

thereby stake its claim in the international marketplace. The company found a specific clientele drawn to the robustness and attractive prices of the export models (Jigouli and Samara). With a 'timeless' yet competitively priced product, Avtovaz occupied part of the niche market of working-class households and low-income employees who could thus obtain a small to medium-sized vehicle for the price of a small West European model.

### 17.3.3. *The Search for Appropriate Identity, Strategy, and Resources, Post-1990*

Perestroika brought a real break with the past, opening up an era of multiple changes and growing uncertainty for Avtovaz. As far as employment and the employee relationship was concerned, the chronic levels of low productivity could no longer be continued by the Moscow bureaucracy. Recognition of real prices and costs led to acknowledgement of massive overmanning, at least in relation to Western standards. A 1993 audit by Fiat estimated the labour force to be 40,000 higher than was needed if the company were to break even. Simultaneously, workers, white-collar staff, and managers resumed a combativeness long muzzled by the single union and the Communist Party, to demand the maintenance of their purchasing power, which was being whittled away by price rises.

In the managerial sphere, privatization, effective from January 1993, completely altered the conditions under which the company was managed; Avtovaz became a limited company with a chief executive and a board of directors appointed by shareholders (Fig. 17.3) instead of simply being nominated administratively. Henceforth, management would be answerable only to itself, the central authorities having abandoned their overriding control and admitted their inability to finance the necessary upgrading of production equipment and the product range.

Patent weaknesses in managerial skills became obvious, and the company was incapable of financing the appropriate training, which in any event did not exist in Russian schools and universities, nor the employment of foreign managers, nor the utilization of external consultants.

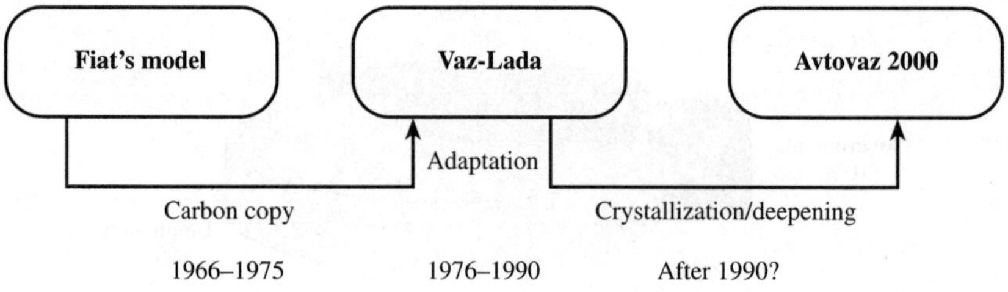

```
Fiat's model  →  Vaz-Lada  →  Avtovaz 2000
                  ↑Adaptation      ↑
   Carbon copy       Crystallization/deepening
   1966–1975    1976–1990    After 1990?
```

FIG. 17.4. *Avtovaz trajectory*

In the industrial and technological sphere, components shortages increased and components quality remained deplorable. Ageing production equipment was increasingly subject to breakdowns. Moreover, production equipment was technologically outdated, inflexible, and caused numerous quality problems. By early 1995, the company still possessed no CAD/CAM equipment, and only four computers in total, and no systematic quality control had been established. Lastly, the models produced were both technically and commercially outdated, and as a result did not conform to American and Japanese safety and pollution standards.

Financially, given the impossibility of generating funds internally, and with an inappropriate banking system, Avtovaz found it impossible to tackle its obvious deficiencies in terms of productivity, quality, and the technical and commercial 'modernity' of its products. Cash-flow shortages also made it impossible to purchase components and engineering services from foreign companies.

In the broader socio-political environment inflation was rampant. The rise of organized crime led to increased losses due to theft and the misappropriation of vehicles and components. There were mounting obstacles to the free movement of goods between Russia and its principal customers and suppliers since it now had to trade across the countries of the formerly Soviet Union. Transit duties, transport delays, and the risk purely and simply of the disappearance of goods all led to even higher transport costs. There was now a complete absence of industrial and technology policies, and the educational structures inherited from the Soviet system were inherently weak, notably in the management sciences.

In sum, by the mid-1990s the organizational form of production at Avtovaz remained typically Fordist. It was highly inconsistent with the employment relationship inherited from the socialist system, and which was far from having been effaced by liberalization. The company did not constitute a specific industrial model in its own right, unless we are prepared to accept that the adaptations linked to the 'Sovietization' of the original Fordist model are of a special and determinant nature which qualify it as a 'Soviet model' of mass production. The Avtovaz 'model' is in fact an importation of a particular form of work organization and technology which is now completely out of date.

## 17.4. WHAT FUTURE FOR AVTOVAZ AND ITS FORDIST MODEL?

With increasing uncertainty since the early 1990s, and notably financial difficulties, the company could not seriously envisage adopting a Toyotaist or Uddevallaist model so as to compete on an equal footing with its American, Korean, European, or Japanese rivals. The only available option was to revitalize the Fordist model, and this required a choice among three options.

The first option was for the company to ally itself with a foreign manufacturer in exchange for general aid. However, this would be at the cost of Avtovaz's autonomy, and might eventually lead to its demise as a Russian company. This is the 'voluntary' solution in which the company opens itself to international co-operation and the implementation of a real industrial partnership so as to simultaneously acquire new technologies, know-how, and resources. In the second option, the company would opt to develop its own way out of crisis, with a specific strategy based on further development and rationalization of the existing model. This is the 'traditionalist' solution based on an autonomous form of development, open only to step-by-step international cooperation through contracts and market relationships. Lastly, the company would seek a combination of the two solutions outlined above, mixing assembly work for another company with its own production. This was the solution which was negotiated in 1993 and 1994 for an agreement with General Motors to assemble the Opel Astra. This project was cancelled in 1995, with GM planning to switch to Yelaz, a new plant due to be set up at Yelabuga near the truck factory Kamez, by the end of the century, to which Avtovaz will only be a supplier of components.

In December 1995 Avtovaz changed its strategy with a subcontracting arrangement to Valmet to assemble 60,000 Lada Samara models in Finland over a three-year period. But the Russian company was still struggling to find a partner which would agree to license the assembly and distribution of a 'Western-style' model to Avtovaz.

Can the Fordist model remain viable in the context of the global automotive industry of the 1990s, marked by the widespread diffusion of organizational forms derived from Toyotaism? A revived Fordist model would be characterized by a search for all types of economies of scale. Its principal features are described in Table 17.3.

It is difficult to calculate the profitability thresholds for the mechanical components factories and assembly lines. The analyses available in the specialist literature are widely divergent, if not contradictory. With a production capacity of 800,000 units per year, it is possible to envisage a high level of profitability with a single platform, two basic models (as well as several derivative models, such as the estate/station wagon or small van), and two or three types of engine.

At issue is a possible long-term development perhaps leading to the emergence of a truly original model. This appears to be possible under the following conditions: maximum level of vertical integration, single product strategy, popular-car

TABLE 17.3. *Characteristics of a revived Fordist model*

| Area | Goals |
|---|---|
| Design | Minimize variants in mechanical components<br>Range reduced to a single platform: two models of vehicles, mid-sized, with a limited number of variants<br>Sharing of mechanical parts<br>Mechanized and automated processes |
| Manufacturing | Optimal size final assembly lines<br>Mechanical components plants of similar capacity<br>Maximum integration for economies of scale |
| Purchasing | All components for less important variants<br>Subcontracting for reasons of capacity and specialization |
| Selling | Fully owned integrated network<br>Cheapest prices in market |

market position, total reliability, simple technology, and minimum price. What if the original Fordist model, so disparaged as a system totally inappropriate to late-twentieth-century conditions of production (technology, organization) and demand (customization, technology, and so on), was viable under conditions of an extreme systematization and rationalization? This is the conclusion arrived at, though not conclusively proven, above. Yet several questions remain, and will not have conclusive answers for some time to come.

Does the conclusion reached by the analysis of Avtovaz contradict the central hypothesis of the IMVP studies (Womack *et al.* 1990) that the Toyota model is quasi-inevitable, or does it in fact simply occupy one possible niche, restricted to one place, and therefore in fact only temporary? Is there still a place for truly different, and not just hybridized, industrial models? In other words, can local specificities outweigh standardization? Is it possible to foresee such profound adaptations that they develop into real alternatives? If the above questions were answered in the affirmative, what would be the basic characteristics of the particular paths followed and solutions shaped? How would performance be measured?

No claims can be made to have responded to such fundamental issues, and yet the case of Avtovaz is instructive. There are multiple reasons for a solution to be found quickly, notably the fact that Eastern European countries and more particularly Russia given that it is by far the largest country in the region—cannot be isolated from the phenomenon of mass motorization. The solution must also recognize the weight of certain economic, financial, technical, social, political, and cultural irreversibilities. It seems out of the question, for political reasons, to make the Russian automotive industry a satellite to the large American, European, and Japanese multinationals.

It is likewise unthinkable to simply discard existing production equipment, given that it might still prove useful if the Fordist approach described above is fully applied with a deliberate choice to produce a cheap and simple car. In any case, it is possible to adopt aspects of the managerial methods which have evolved from recent developments, whether in Toyotaism or elsewhere—just-in-time, total quality, teamwork, quality circles, and so on—in so far as they are compatible with any forms of industrial organization, without requiring complete organizational changes but only marginal modifications.

## NOTE

Translated by Sybil Hyacinth Mair.

## BIBLIOGRAPHY

Womack, J. P., Jones, D. T., and Roos, D., *The Machine that Changed the World* (New York, 1990).

# 18

## *Conclusion: The Choices to be made in the Coming Decade*

MICHEL FREYSSENET, ANDREW MAIR, KOICHI SHIMIZU, AND GIUSEPPE VOLPATO

Many changes can be foreseen in the world automobile industry over the next few years: a geographical shift in the industry's quantitative centre of gravity, the liberalization and restructuring of international trade, the development of new electronic technologies in automotive products, growing attention to the ecological implications of vehicles, the emergence of new competitors, and the search for greater flexibility in the management of the workforce. In conjunction with these changes, automobile producers will have to modify the manner in which they internationalize their operations, adjust the structures of their supplier networks, change their product range strategies and planning of their products, and lastly, restructure their employment relationships.[1] Synthesizing and schematizing the future evolution of the automobile industry, it is possible to sketch out three scenarios.

### 18.1. FUTURE CHANGES IN THE INTERNATIONAL AUTOMOBILE INDUSTRY

*18.1.1. Two Types of Markets: Mature and Developing*

During the first century of the international automobile industry, there was a strong coincidence between the countries producing automobiles and the countries consuming them. Even as a new century begins symbolically, in practice change has already started. The new era will be characterized by a geographical shift in the pattern of demand for motorization, and as a consequence, a relocation of production towards countries which are crossing the threshold of wealth and social organization that permits and indeed requires motorization: countries such as Brazil, India, China, the countries of South-East Asia, and so on.

It is not easy to estimate the rate at which this change will occur, in as much as it depends upon rates of economic growth and on the qualitative characteristics of this growth in terms of consumption, investment, and import-export flows. However, a conservative hypothesis would suggest that annual global vehicle registrations by the year 2005 will exceed 65 million vehicles, 70 per cent of them passenger cars. Assuming that by 2005 North America, Western Europe, and Japan, the core regions of motorization, will absorb no more than 65 per cent of vehicles produced, we estimate that by the year 2005 the regions of recent take-off in motorization will be demanding close to 23 million vehicles, almost double the 13 million vehicles registered in 1995. If we also assume that the automobile companies will generally concentrate their core region factories on meeting core region demand, it will be necessary to build eighty additional factories, each producing 200,000 vehicles per year, in the emerging markets.

Qualitative changes are no less significant. In the past, the clear quantitative dominance of the three core regions led the automobile producers to view the rapidly emerging markets as marginal, residual markets that could be served by factories and car models which were obsolete by the standards of the core regions. With the marked shift in the industry's quantitative centre of gravity, it will no longer be possible to pursue this strategy.

### 18.1.2. The Liberalization and Restructuring of International Trade

The new factory location strategies of the automobile producers will be profoundly influenced by developments in frameworks for international trade. The producers will probably favour those countries which adopt legislation supporting the greatest flexibility in international trade: not only trade in cars and components, but more generally trade in terms of financial activity, technology diffusion, and the creation of joint ventures between local and foreign capital. However, liberalization may either be truly global, or it may be limited to multinational free trade regions such as EU Mercosur, APEC, and NAFTA.

The automobile producers and their suppliers will certainly have to discover new markets and labour relations[2] in the new growth countries. However, they will also have to face unstable market and labour relations in the developed countries where they already operate, given the problems of the established economic and social systems and the absence, as yet, of new ones.

### 18.1.3. Technological Change and Regulatory Norms

A third factor of change is the strong pressure of technological change and/or regulatory norms which increasingly concentrate on safety objectives: whether in terms of automobile use (active and passive protection systems) or environmental protection.

As far as the safe use of vehicles is concerned, the initiatives being undertaken by the automobile producers are manifested in a broad array of activities covering the entire spectrum of the technological frontier: the further diffusion of lightweight materials which deform in predictable ways (in accidents), the development of 'survival cells' within cars, the further diffusion and improvement of the air-bag, and so on. However, the most important tendency is the increased use of electronic components in the vehicle. New electronic components reduce damage in case of impact, but are also used to prevent accidents: such as the widespread diffusion of ABS (anti-lock braking systems) and ASC+T (automatic stability control and traction systems). Further developments can be expected with respect to air-conditioning, traffic information systems, navigation systems, head-up-display instruments, and so on.

These strategic technology requirements will play an important role in pushing the automobile makers to enter into collaborative agreements with specialist electronic components companies, to jointly develop the new equipment. Given the major R&D investments that this type of innovation entails, it is also likely that horizontal collaboration involving a number of automobile producers will be necessary in some cases.

At the same time, the most significant technological and organizational effort in product innovation is likely to concentrate on objectives related to ecological compatibility. While the automobile industry has already made significant advances over previous standards in the industry, ecological sensitivity is also growing apace among the population at large, as reflected in increased attention paid towards anti-pollution regulations. All this is translated into a strong thrust in industrial research towards the simultaneous reduction of engine-induced pollution and reduced fuel consumption (which in its turn further reduces pollution).

### 18.1.4. New Competitors

The fourth change will be the emergence of new competitors. The term 'new' has to be qualified, in the sense that in practice the financial, technological, and organizational investment required by automobile production leaves little room for latecomers. In reality, the new competitors are companies which are only partially 'new', like MCC Micro Compact Car AG, the fruit of an accord between the Société Suisse de Microélectronique et d'Horlogerie SA and Mercedes-Benz AG for the production of the 'Smart' car, or the South Korean automobile firms, which are already present in the market but which hope to develop further in ways that will bring them into closer competitive confrontation with the other international producers. Moreover, it is already clear that some new automobile companies will emerge through joint ventures between foreign and local capital in the emerging markets. In these cases, however, the local company will have objectives distinct from those of the participant automobile companies from the core regions. For instance, it is more than likely that the local company will expect the joint venture to export its products, not only to other developing countries, but also to the markets of the core regions in order to acquire financial resources in strong currencies.

### 18.1.5. The Transformation of Labour Markets and Labour Relations

A crucial dimension of future developments in the three core regions is the further raising of the average age of the automobile sector workforce. This ageing process appears to contradict the requirements for the production process to become more flexible, with the adoption of the JIT system, the creation of niche cars, and the customization of models. In the three industrialized poles the liberalization of international trade and changes in models of work organization have led to requirements for even greater flexibility. It remains unclear to what extent workforce flexibility must be increased further and whether this will be negotiated, or forced through unemployment, increased casual work, or the threat of factory relocation. The potential contradiction between flexibility and the ageing workforce is exacerbated if fewer new personnel are hired. In the past this type of conflict would have been overcome by increased automation. However, attempts to move in this direction have already revealed that the positive contribution of the workforce is absolutely fundamental if the levels of quality demanded by the clientele are to be attained.

## 18.2. THE CHOICES TO COME

### 18.2.1. Internationalization

The evolution of global automobile demand, augmented as it is by motorization in the emerging markets, appears to be promising for the industry as a whole. Individual companies may nevertheless face considerable risks, above all in terms of the specific countries in which they choose to invest and the timing of their internationalization. Indeed, a strategy of major investment in the emerging countries may entail considerable risks, as revealed by the difficulties that have confronted American and European producers in the past in Latin America. Schematically, there may be two main strategies. According to the first, it would be possible to continue to focus investment in the more developed countries, leaving to others the risks of opening up the emerging markets. In this scenario investment will focus on the development of new models and on internationalization within the countries where levels of motorization are already high. The second strategy, by contrast, focuses immediately on the emerging markets, which entails risks during the 'exploration' phase, but ensures a position of strength when demand growth stabilizes.

Of course, levels of internationalization already attained vary greatly. As evident from the analyses of the trajectories of the various automobile companies, there is no doubt that the major Japanese producers (Toyota, Honda, and Nissan) are the most advanced. As for the American companies, General Motors and Ford already possess an internationalized production base and have launched significant programmes as part of their strategies to increase integration and coordination, focusing on the countries of the NAFTA and European Union regions. The European automobile companies which specialize in the upper segments of the range like BMW and Mercedes are intensifying their internationalization within the countries of high motorization, with the construction of the factories at Spartanburg, South Carolina, by BMW for the production of the Z3 model, and that at Tuscaloosa, Alabama, by Mercedes, for the production of the All Activity Vehicle. As for the generalist European companies, they are relative laggards. While there are significant variations from company to company, the European generalists may be at a disadvantage in the race to internationalization, with the possible exception of South America. On the other hand, they have a number of advantages in the markets of Eastern Europe.

*18.2.2. Product Planning and Policy*

The automobile is a very complex product, and its producers have to take account of the innumerable specificities resulting from market variations. Above all, there are the cultural traditions of each country in terms of their form of motorization, and there is also a wide range of factors which are by their nature contested: various regulatory norms, the costs of managing a diverse range of automobiles, diverse traffic systems, and so on. Meanwhile, the advantages to be gained from standardization and economies of scale represent a significant factor in competition. Accordingly, a fundamental requirement over the next few years will be to create the best compromise between these two partly contradictory demands, resulting in specific products which still share components.

The numerous attempts, especially by American producers, to transfer automobiles from the USA to Europe and vice versa, have all proved unsatisfactory. The risk of trying to create an automobile which pleases everybody is that the product ends up pleasing no one. Some of the automobile producers have decided to offer two product ranges, one for the core regions, the other for emerging countries. Moreover, automobile demand in the high motorization countries is fragmented in ways that are increasingly pronounced, giving rise to a growing number of clusters of consumers with specific and distinct demands. While the market currently remains divided into mass demand and niche demand, it is possible that in the future it will be configured into a system composed solely of niches. This phenomenon is already evident in the USA and Japan, where the best-selling models account for under 2–3 per cent of total registrations, and it is beginning to appear in Europe, where the share of the best-selling model has fallen below 5 per cent. The flattening or reduction of the purchasing power of a significant part of the population, in particular the middle classes, has led some automobile companies to reduce the number of versions of each model and especially to reduce the number of platforms.

The issue is one of how to exploit the whole range of economies of uniformity: economies of scale and the shared production of component parts in ways that do not impede the diversity and personalization of the products; economies of scope in the planning of dual model ranges (for countries with high and low levels of motorization respectively) attained through the modularization of models; economies of replication in the creation of 'twinned' production sites in different countries, sometimes for assembly activities but especially for the production of component parts from a multiplicity of suppliers; and economies of learning when the more advanced factories transfer their personnel to the new emerging market locations. The combination of 'personalized' solutions created on the basis of nuclei of shared programmes, which permits various types of uniformity and therefore economies, is among the most promising areas for the application of industrial creativity. In this domain the positions of automobile producers in each of the three poles are quite different. The European companies have the advantage of greater experience with regard to stylistic aspects and therefore economies of scope, the Japanese firms have already demonstrated that they possess a notable maturity, particularly in the economies of replication and in the transfer of economies of learning, whereas the American firms can count on a marked advantage in the more traditional economies of scale.

As far as developments in the fundamentals of product development are concerned, it appears highly likely that the technological advances of the automobile producers will focus primarily on the improvement of conventional petrol and diesel engines. In fact, contrary to what might appear to be the case, the technology of internal combustion engines is still in its youth. Naturally, a century of technological innovation has resulted in considerable advances, but there remains significant room for improvement, above all if new standards of reference are employed, based not upon stable driving conditions but on the ecological efficiency of fuels under variable conditions of use, such as the successive accelerations and decelerations characteristic of driving in today's cities. As far as efficiency in conditions of operational flexibility is concerned, the most important technical innovations will focus on variability during use of three aspects of the engine: variability in the phasing of fuel intake valves, variability in the number of cylinders in use, and variability within the cylinders (expansion and contraction of the combustion chamber). Currently, only the first type of innovation has found its way to mass production cars, with the second type having reached the stage of working prototypes, while the third type is still at the research stage. In conclusion, it should be underlined that product innovation will be an essential element in the competitiveness of automobile firms for many years to come.

Government-financed research activities will play a vital role in the pre-competitive phase. An intelligent industrial policy to direct and channel the activities of automobile firms would appear to be a factor of great importance. Generally speaking, the Japanese and American industries are ahead in this regard, when compared with the European automobile companies, given the better focus and unambiguous goals of the policies adopted by the American and Japanese governments, in contrast to the tentative actions of the European Union. The American initiative to coalesce the nation's strengths around the 'New Generation Vehicle' project appears to be a significant example of this strategy in practice.

### 18.2.3. Partnership with Components Suppliers and the Sales Network

An important objective to bring to fruition is effective partnership between the automobile producers and their components suppliers. The internationalization process will require a major commitment to cooperation, whether to support investment by the major suppliers of sophisticated components in the emerging motorization markets, or to establish and develop new local suppliers of the more traditional components. The automobile producers need to develop real plans to guide the growth of local suppliers, with a continuous commitment of their own dedicated human resources inside supplier factories, and with the development of programmes for technological, organizational, and managerial training. Without a major organizational push towards integration between assembly factories and suppliers, an integration which is also essential for the organization of after-sales, repair, and spare parts services, the entire internationalization strategy runs the grave risk of failure.

A similar argument is valid for the networks of independent distributors not controlled by the producers. In this case the problem appears to be less pressing, in the sense that the growth in the 'parc' of vehicles will necessarily be gradual, whereas for suppliers their systems must function properly from the start. However, in the medium term the after-sales, repair, and spare parts services within the distribution networks will play a vital role in determining the success of different producers, and there are therefore issues to address given the particular types of relationships the producers maintain with independent distributors.

### 18.2.4. Flexibility and the Participation of the Workforce

As far as the management of employment relations is concerned, the strategies of the automobile companies will be oriented towards obtaining two results, which are partly contradictory. Above all, their objective will be to obtain the maximum commitment from the personnel. At the same time, they need to increase employment flexibility among their workers, given that demand will be increasingly marked by uncertainty and variability.

It is possible to identify a variety of strategies to promote flexibility. In some cases firms may adopt strong forms of flexibility in the utilization of the workforce, on the basis of the current phase of a weaker union presence in the factories. Other firms may prefer solutions of a more institutionalized nature, articulating new negotiated solutions. For instance, they may attempt to obtain a reduction in working hours combined with a reduction (if proportionately smaller) in salaries, with an acceptance that overtime will only increase work to a total of forty hours per week. Another strategy to increase the flexibility of work may be based on a form of duplication of production units between the developed and developing regions, with the personnel used in different ways in each region. The units in the industrialized countries could work systematically at full capacity, while oscillations in the market might be absorbed by the units located in developing countries.

However, the attempt to raise the employment flexibility of the workforce may raise questions about the attainment of higher levels of commitment, which will increasingly be the key factor in the achievement of higher quality standards. In this regard, the analysis of individual companies reveals a range of highly articulated positions, each with its advantages, although in this case favouring the Japanese firms, which have higher levels of participation and identification with company objectives.

None the less, the competitive challenge creates problems of adaptability for all the automobile producers. The European automobile companies clearly have the most difficult task ahead, given that they must achieve simultaneously a higher level of commitment on the part of the company personnel and greater flexibility in the use of their workforce. As far as commitment is concerned, in recent years the European producers have made significant progress, thanks to improvements in industrial relations with union organizations. Yet there remains much to be done in terms of experiments with organizational forms and new types of employment relations if the processes under way are to endure.

As regards the flexible utilization of the labour force, this is a requirement that will make itself increasingly felt in the future, and is an area in which new solutions need to be studied and experimented with. The issue in Europe over the next decade is the extent to which there will be a systematic and recurring short-term instability of demand, whether in individual markets or for individual models, even if the overall market context is characterized by gentle growth. In this case the automobile producers will require mechanisms to rapidly adjust potential supply to oscillations in demand. For example, it is in the interests of the European companies to measure working hours not on a weekly basis but on a monthly or annual basis, so as to be able to compensate over a longer period for potential fluctuations in requirements for labour. Hence the ability to increase or reduce the number of shifts would be a useful mechanism for adjusting productive capacity to demand for individual models. This type of complex problem may only be solved with significant creativity and with significant efforts to collaborate by company managers and union organizations. The issue is difficult, since the companies may simultaneously face situations of overmanning in the production of models whose sales are declining and situations of strong demand for overtime work on other models. In Japan this type of problem appears easier to resolve. The judgment of the Supreme Court of 28 November 1992 confirmed that Japanese workers were obliged to accept overtime work. But in Europe the traditional objectives of equity and fairness among workers, promoted by union organizations, may imply different approaches and major social issues.

Similar issues arise, although in different specific ways, for automobile companies operating in North America, where short-term fluctuations in demand have been a recurrent feature for fifty years. Company policies to make product supply more flexible are accepted to a greater extent, and the American legislative framework permits the temporary or systematic reduction of the workforce, or the utilization of overtime work, more expeditiously than in Europe. Yet the problem remains of how to achieve broader employee participation in company objectives. This is an important issue particularly with regard to factories located in regions with a long history of automobile production where levels of unionization are high. The experiment at the Saturn division appears very interesting, but the problems of General Motors in diffusing this model to other automobile divisions serve as a clear reminder of how difficult it is to generalize this type of culture into different contexts.

The Japanese producers may currently possess an enviable level of commitment among their personnel, a legacy of the history and culture of the country, yet the future of industrial relations at these companies is likely to include new problems to resolve, as the analysis of their respective trajectories has revealed. If much of the sentiment binding management and employee is based on the assumption of 'lifetime' dependence, the recent crisis following the bubble economy period has made this pact less certain. On the one hand, the evolution of relations with unions in Japan appears to be approaching the Western model, while on the other the relaxation of the 'lifetime' assumption may work against the traditional level of worker commitment. It seems likely that very innovative solutions will have to be introduced into the Japanese employment relationship, above all if the yen–dollar exchange rate remains high, reducing opportunities to export cars produced in Japan and increasing opportunities to import.

## 18.3. THE SCENARIOS

The survival and strategic position of companies will depend upon their strategies of internationalization, that is to say upon their capacity to face the multiplicity of uncertainties over markets and labour relations. Initially at least, the evolution of market and labour relations can be grouped into three main scenarios, as has been proposed for the Second International Programme of GERPISA, 'Between Globalization and Regionalization: What Future for the Automobile Industry?', with global homogenization at one extreme, regional diversification/ global 'commonization' at an intermediary level, and regional 'heterogenization' at the other extreme (Freyssenet and Lung 1996).

The global homogenization configuration corresponds to the case in which the liberalization of global trade prevails over the creation of more restricted economic spaces, with all automobile markets, or at least the most important, tending to homogenize, and with employment conditions converging under the impact of constraints which they all share. In this context the relevant strategic decisions would relate to the creation of a global product range, whether a traditional range or a niche range. This scenario assumes that the customs duties which have been reduced or eliminated are not in practice replaced by other restrictions on trade, notably a requirement that companies produce locally in order to obtain the right to sell. In this case, in the medium term at least, companies would have to invest where they wanted to sell, although even so the creation of a global product range is by no means impossible.

In this context, an internationalization strategy might consist of specializing each production operation and region in one segment of the range (either for components or for finished vehicles) in order to benefit from maximum economies of scale. If market variations for the same segment were to balance each other (the lower end of the mass market in the USA, the upper end of the mass market in Europe, for example), only a weak level of flexibility and polyvalency would be required in the factories. With each region specialized in one segment, each local subsidiary could have its own particular socio-productive system without this affecting the coherence of the whole, although if the differences were profound these might reflect significant variations in income distribution and therefore different automobile markets, thus representing a shift away from the hypothesis of homogeneous global demand.

The more moderate scenario of regional diversification/global 'commonization' assumes that, even in the absence of complete homogenization, both product markets and work processes would share certain fundamental traits across the different world regions, with differentiation only affecting secondary factors. In this framework a strategy of sharing major components and platforms (global commonization) and relatively similar policies towards the basic elements of employment relationships would allow companies to offer differentiated regional product ranges, and regionally specific employment relationships, which still met the same objectives in terms of flexibility and polyvalency. This scenario would lead to a centralization of platform design and the creation of an overall employment policy, and it would also lead to a globalization of components production and consequently of the supplier industry. At the same time, it would be accompanied by a regionalization of model design, with each model locally derived from global platforms, and of course a regionalization of their production. The determination of the concrete modalities of employment relationships would remain autonomous, as a function of the local context (rules on the determination of wages, for instance). Production could be transferred between regions if necessary, since the different models each factory would assemble would nevertheless share their platforms. Production programmes could therefore be planned at the global scale.

Finally, in the regional heterogeneity scenario, homogenization would clash more directly with the creation of relatively heterogeneous regional complexes. In this hypothesis the dynamic of regional integration prevails, each region including countries at different levels of development and with somewhat different competencies. This dynamic produces a specialization and concentration within the region which gives a boost to growth. However, regional integration may assume different forms, whether that of a simple free trade zone (in which economic and social power relationships are permitted to work themselves out) or true economic, political, and social integration (in particular, involving a controlled policy to increase the purchasing power of the entire population).

Markets and work would be so different, and permanently so, between world regions, that product ranges and employment policies would be specific to each region, with the exception of inter-regional trade in some niche markets. This scenario would tolerate the persistence of less internationalized companies, which would remain local players with an intricate knowledge of the regional market and great agility. However, given the handicaps this type of company would experience, the scenario seems to support a multi-regional configuration for the companies: regional subsidiaries would have great autonomy in terms of their product strategies, with the overall group exercising financial control and probably sharing knowledge and experience between the different regions.

# NOTES

Translated by Sybil Hyacinth Mair.
1. See Chapter 1, note 2.
2. See Chapter 1, note 1.

# BIBLIOGRAPHY

Freyssenet, M., and Lung, Y. 'Between Globalization and Regionalisation: What Future for the Automobile Industry?' *Actes du GERPISA*, 18 (1996).

# INDEX

Abbeglen, J. C. 112, 125
Abo, Tetsuo 149
absenteeism 15, 17, 20, 27, 96, 315, 374, 379, 421, 425
 *see also* work crisis
Adler, Paul 321
Agnelli, Gianni 313, 316
Agnelli, Umberto 316
Alfa Romeo 35, 100, 322, 325, 331
Allam, Delila 388
Altshuler, A. 32
American Motors Corporation (AMC) 36, 180
 purchase by Chrysler 194, 257, 258, 263
 PSA relationship 346
 Renault relationship 31, 187, 346, 381, 385
Andrea, D. J. 192, 207, 246
Asanuma, B. 68, 69
assembly line control (ALC) 76, 140, 158–9
Audi 26, 283–4, 285, 287, 302
Austin 17, 395–6, 398, 400–1, 408
 Nissan relationship 91, 92
Austin-Rover 395
 *see also* Rover Group
Auteri, E. 328
AutoAlliance 229–30
Autobianchi 15
AutoEuropa 230
Autolatina 230
automated guide vehicles (AGVs) 17, 39, 103, 126, 422, 424
automation 30
 Fiat 35–6, 317–18, 319–21, 335
 Ford 214, 221
 General Motors 188
 Honda 126
 Hyundai 5, 158–9, 160, 166–7, 170
 Mitsubishi 5, 145–6, 148, 150
 Nissan 96
 PSA 348, 350, 357
 Renault 380, 382, 383
 Toyota 64, 83
 Volkswagen 280, 291–3
 Volvo 420, 422, 424, 427
automobile public policy 13, 18, 21, 22, 23, 24, 30, 32, 34, 75, 85, 100, 111, 145, 181, 185, 190, 202, 249, 380, 410
 *see also* government and automobile firms relations
autonomization 64, 85
Auto Union 277, 278–80, 283–4

Avtovaz/Lada 6, 32, 440–51
 crisis 41, 447
 employment relations 446, 447, 448
 exports 442
 Fiat role 22, 441
 Fordist model 6, 440, 443–5, 449–50
 future outlook 449–51
 political system and 444–5
 production organization 443, 445, 448
 product policy 442, 447, 449
 supplier relations 446
 volume profit strategy 6, 32, 445, 449
 world car production 42–3

Babson, Steve 59, 238
Bélis-Bergouignan, Marie Claude 46, 74, 94, 95, 99, 116, 216, 234
Belo Horizonte factory, Brazil (Fiat) 332
Belzowski, Bruce 5, 36, 242
benchmarking 222, 303
Berggren, Christian 6, 17, 39, 418
Besse, Georges 352, 384, 385
Bieber, Owen 260
Big Three, *see* Chrysler Ford; General Motors
BMW 45, 194
 Rover purchase 6, 44, 45, 132, 396, 414–15
Bonazzi, G. 327, 331, 333
Borås factory, Sweden (Volvo) 423, 426
Bordenave, Gérard 5, 36, 211, 216, 233, 237
Bork, Mr 275, 284
Boyer, Robert 3, 4, 7, 9, 46, 91, 105, 150, 211, 215, 295, 366, 371
Bramalea factory, Canada (Chrysler) 258
Bresnahan, T. F. 55, 56
Briam, K.-H. 304
British Leyland Motor Corporation (BLMC), *see* Rover Group
Brumlop, E. 288
Buick (General Motors) 182, 183–4, 189, 195, 197

Cadillac (General Motors) 57, 183–4, 189
CAFE standards (Corporate Average Fuel Economy) 191, 250–1, 255, 256
Calvet, Jacques 352, 355, 361
Camuffo, Arnaldo 6, 36, 311, 318, 319, 326, 327
Cantarella, Paolo 322
Carrillo, J. 229
Castaing, François 261
Cerruti, G. 329, 331, 333
Chanaron, Jean-Jacques 6, 22, 43, 440
Charron, Elsie 373, 381

Chase, Herbert 59
Chevrolet (General Motors) 54, 182, 183–4, 189, 195
Chrysler 5, 6, 13–14, 54, 242–67
  American Motors Corporation purchase 194, 257, 258, 263
  competitiveness 191–2, 258–9
  crisis 28–9, 31, 36, 144, 244, 252–3, 263, 248–54
  distribution network 245
  diversification of business activities 257–8
  employment relations 200, 247–8, 251–2, 259–61
  finance 246, 248–9, 257–8, 262–3
  Fordist model 243–8
  future outlook 265–7
  globalization 245–6, 252, 267
  innovation and flexibility profit strategy 5, 14, 35, 36, 44, 254–6, 260
  minivans 187, 194, 256–7
  Mitsubishi relationship 140, 144–5, 147–8, 246, 249–50, 252
  national growth and redistribution modes and 28–9, 31, 32
  product development 261–2, 263–5
  production organization 247–8, 251–2, 259–61
  product policy 245, 249, 253
  recovery 254–62
  regionalization strategy 43–4
  statistics 268
  supplier relations 262, 264, 267
  union relations 31, 36, 189–90, 247–8, 259–61
Chrysler-Europe, purchase by PSA 343–4
  see also PSA (Peugeot Société Anonyme); Talbot
Chung, Ju-Young 155
Chung, Myeong-Kee 5, 23, 32, 154
Church, R. 399
Citroën 6, 30, 32, 37, 344
  crisis 28–9, 30, 340–1, 351
  distribution networks 359–61
  employment relations 342, 349, 352–3, 359
  exports 346
  manufacturing reform 356–9
  proposed relationship with Fiat 314
  purchase by Peugeot 338, 340–3, 379
  statistics 363
  training 359, 360
  wage system 349
  see also PSA (Peugeot Société Anonyme)
Clark, K. B. 322
Clark, P. 396, 397, 403
closing factories 5, 6, 14, 31, 103–4, 198–200, 254, 255, 405, 407, 431
Cole, R. E. 190, 207
Combe, Henri 351

commonization of platforms 13–14, 15, 17, 19, 26–7, 29, 35, 38, 76, 98, 103–4, 125–7, 196, 264, 282–3, 301–2, 347, 355–6, 368, 371, 378, 386, 456
company governance compromise:
  definition 3, 10, 46
  firms and:
    Avtovaz/Lada 444
    Chrysler 31, 253, 260
    Fiat 334
    Ford 31, 215, 218
    General Motors 12, 31, 189
    Honda 20, 25, 120–1
    Hyundai 172–3
    Mitsubishi 140
    Nissan 4, 92–3, 101, 106
    PSA 339
    Renault 37, 369–70, 373–6, 382, 388–9
    Rover 28
    Toyota 19, 25, 65, 80–1
    Volkswagen 16, 25, 26–7, 274–5, 282, 288–91, 298–9, 304–5
    Volvo 25, 421, 425–6
  see also company governance structure; employment relations; industrial models
company governance structure 6, 16, 91, 65, 112, 114, 118, 120, 139–40, 154, 163, 183, 217, 267, 283–4, 298–9, 300, 313, 339, 366, 369, 379, 384, 399, 414, 432–4, 441, 444, 447
completely knocked down (CKD) production, see knocked down (KD) production
Costa, F. G. 319
craft model:
  craft British model 395, 396, 397, 403–7, 409, 410, 414
  defined as:
    amalgam of several models 2
    old production process 51
  see also industrial models; piece work model; profit strategies (diversity and flexibility)
Cray, E. 182, 183
crisis of automobile firms 27
  Avtovaz/Lada 41, 447
  Chrysler 28–9, 31, 36, 144, 244, 252–3, 263, 248–54
  Citroën 28–9, 30, 340–1, 351
  Fiat 28–9, 30, 314–19
  Ford 28–9, 30, 219–20
  General Motors 28–9, 192–3, 200
  Honda 11, 38–9, 130, 133–5
  Hyundai 5, 32, 156, 163, 166
  Mitsubishi 23, 27, 144
  Nissan 23, 27–8, 103
  PSA 28–9, 350–1
  Renault 28–9, 383–4

Rover 28, 399, 401, 404–5
Toyo Kogyo/Mazda 23, 27, 30
Toyota 11, 38–9, 79–80, 85
Volkswagen 5, 11, 26, 38–9, 272, 284,
    297–8, 306
Volvo 11, 429–35
Cusumano, M. 49, 91, 92, 322
CVCC (Compound Vortex Controlled
    Combustion) engine (Honda) 115,
    118, 133, 135

Daewoo 32, 42, 157
Daf 26, 28, 409, 422, 433
Daimler-Benz 277–8, 283, 289, 389, 419
Dankbaar, Ben 148–9
Day, Sir Graham 409, 411
Dealessandri, T. 311
Decoster, Frédéric 388
Diamond Star Motors (DSM) 148–9, 258–9,
    261, 264
distribution networks 458
    Chrysler 245
    Fiat 325–6
    Honda 112–13, 118–19, 124–5
    Mitsubishi 141–2, 144
    Nissan 92, 96–7, 106
    PSA 359–61
    Toyota 63, 69–70, 75–6
    Volkswagen 302
diversification of business activities 30, 36,
    106, 192, 257–8, 294–5, 373, 376,
    378–9, 423, 430
DKW (Auto Union) 278
Dodge (Chrysler) 251, 255, 244–5, 258
Doleschal, R. 273
Dombois, Rainer 288, 293
Dreyfus, Pierre 367, 368, 370, 376, 379
Durand, Jean-Pierre 4, 211

Eaton, Robert 243, 263, 267
Edwardes, Michael 405–7, 408
Ellegård, Kajsa 17, 426
Elsässer, B. 421
employment relations:
    challenge in the future 454–5, 458–60
    definition 3, 10–11, 46
    firms and:
        Avtovaz/Lada 446, 447, 448
        Chrysler 200, 247–8, 251–2, 259–61
        Citroën 342, 349, 352–3, 359
        Fiat 35, 314–15, 319, 321, 327–30,
            333–4
        Ford 200, 214–15, 223–6
        General Motors 12–13, 184, 189–90,
            200–1
        Honda 39, 119, 121–4, 128
        Hyundai 162, 163–4, 165–70, 172–3
        Mitsubishi 140–1, 146

Nissan 20, 27–8, 35, 95–6, 101–2
PSA 37, 342, 348–50, 352–3, 358–9
Renault 37, 369–70, 373–6, 379, 381–2,
    387–9
Rover 37, 398–400, 406, 412
Toyota 19, 38–9, 64–8, 78, 80, 81–5
Volkswagen 27, 274–7, 281–2, 287–3,
    304–6
Volvo 27, 419–20, 424–5, 426
growth and redistribution modes of national
    income and 9
see also company governance compromise;
    industrial models; national labour
    relations; national labour compromise
Endres, Egon 284
Enrietti, Aldo 335
Erixon, L. 429
exports 23
    Avtovaz/Lada 442
    export competitiveness 9, 25–8
    Fiat 15, 43, 316, 332
    Honda 23, 25–6, 113–14, 118–19, 130
    Hyundai 161–3
    Japanese export strategy 23, 25–6
    Japanese exports to USA 77, 187, 191, 249,
        255
    limitation of Japanese imports 190, 202
    Mitsubishi 144, 149
    Nissan 27–8, 92, 94, 97, 98–9
    PSA 345–7
    Renault 15, 368–9
    Toyota 23, 25, 77
    Volkswagen 23, 26, 277
    Volvo 23, 26, 418–19, 424
    see also globalization; internationalization;
        regionalization

Fayolism 399
Ferrari 15, 134
Fiat 6, 15–16, 311–34, 350–1
    automation 35–6, 317–18, 319–21, 335
    crisis 28–9, 30, 314–19
    distribution network 325–6
    employment relations 35, 314–15, 319, 321,
        327–30, 333–4
    integrated factory concept 6, 326–30
    internationalization 19, 22, 31, 43, 316, 332
    lean production 330–2
    mass production 312–17
    national growth and redistribution mode 29,
        31
    organizational structure 313–14, 316–17,
        330, 334
    product development 322
    production organization 317–18, 320–1,
        323–4, 326–30, 331
    product policy 312, 332
    proposed relationship with Citroën 314

Fiat (cont.):
  recovery 318–21
  reorganization 316–17, 322
  role in Avtovaz factories, Soviet Union 22, 441
  SEAT relationship 314
  statistics 336
  supplier relations 323–5
  training 328
  union relations 313, 314–15, 319, 327, 329
  volume and diversity profit strategy 12, 15, 35, 44, 313
  wage system 315, 329
financing 21, 128, 156, 181, 249, 252, 253, 255, 263, 339, 341–2, 354, 377, 445, 448
flexifactory concept (Honda) 117, 122–3, 127, 133
Flynn, Michael 5, 15, 35, 179, 190, 192, 207, 246
Ford 5, 13–14, 52–4, 196–7, 211–36
  automation 214, 221
  competitiveness 191–2, 216, 222–3, 227–8
  cost reduction 220–2, 232
  crisis 28–9, 30, 219–20
  diversification of business activities 192
  employment relations 200, 214, 215, 223–6
  Fordist model 5, 36, 42, 213–19, 235–6
  globalization 5, 36, 231, 232–5, 236
  Hyundai relationship 155–6
  internationalization 215–16, 233
  Mazda relationship 223, 229–30, 233, 234
  national growth and redistribution modes 29, 30–1
  Nissan relationship 230
  production organization 213–14, 221, 223–4
  product policy 214, 217
  quality improvement 220–2
  recovery 226–8
  redefinition of boundaries 228–32
  revision of internal procedures 231–2
  statistics 239–40
  supplier relations 221–2, 230–1
  union relations 189–90, 214, 215
  Volkswagen relationship 230
  volume profit strategy 5, 35–6, 42, 214, 233
  world car concept 42, 232–4
Ford, Henry, I 213, 215, 256
Ford, Henry, II 215, 217
Fordism, defined as:
  type of macro-economic regulation 215
  model, see Fordist model
Fordist model:
  crisis of 159–64, 216–19
  defined as:
    centralized and unflexible production system 179–80, 183
    mass production 2, 164, 211, 213, 312–17, 338–9, 344

neo-Fordist model 235–6, 333
production paradigm 206
specific model 2, 12, 214, 365
Taylorist-Fordist model 2, 155, 164
difference from Sloanist model 16, 183
diffusion history 53–7
firms and:
  Avtovaz/Lada 6, 440, 441–3, 449–50
  Chrysler 243–8
  Fiat 312–17
  Ford 5, 36, 42, 213–19, 235–6
  General Motors 179–80, 192–3, 204, 206
  Hyundai 40, 42, 155, 159–64
  PSA 338–9, 344, 361
  Renault 366
  Rover 399–408
  Toyota 18
  Volkswagen 5, 15–16, 277–85
  Volvo 17, 420–1
transformations of 214–15, 333, 365
transplantation to other countries 2, 22–3, 40, 155
see also Ford Henry; industrial models; mass production; profit strategies (volume)
Frazier, Douglas 260
Freyssenet, Michel 3, 7, 9, 30, 40, 42, 46, 91, 105, 150, 211, 295, 365–7, 369, 372, 382–3, 388, 452, 460
Fridenson, Patrick 46, 366, 369, 379, 380, 382
Frisinger, Håkan 428
Fuji Heavy Industries 95, 106
Fujii, M. 96
Fujimoto, Takahiro 2, 9, 322, 395, 398
Fujisawa, Takeo 20, 110–16, 120
Fujisawaism 113, 115–17, 124, 134–5

Gardner, Glenn 264
Gautier, François 345
General Motors (GM) 5, 13–15, 179–206
  automation 188
  challenges 203–5
  competitiveness 182, 191–5, 204, 227–8
  crisis 28–9, 192–3, 200
  diversification of business activities 192
  early history 180–6
  employment relations 12–13, 184, 189–90, 200–1
  Fordist model 179–80, 192–3, 204, 206
  government relations 185, 202–3
  internationalization 43, 181, 188
  national growth and redistribution modes 29, 31
  negotiations with Hyundai 157–8
  oil crises and 186–91
  organization of 183–4, 195–9, 214–15
  production organization 196–7
  product policy 182, 187, 196

Sloanist model 5, 10, 12–15, 23, 179–80, 183, 198, 204
  statistics 208–9
  supplier relations 184–5, 190, 201–3
  union relations 184, 189–90, 201
  volume and diversity profit strategy 5, 10, 35, 43, 188
  wage structure 12, 184
  *see also* NUMMI
German model 414
  debate 295
  diversified quality production 295
  Volkswagen industrial model:
    definition and characteristics 273–7
    difficulties 277–85, 296–9
    renewing 285–95, 299–306
  *see also* industrial models; profit strategies (volume and diversity, quality and specialization)
Ghidella, Vittorio 320
globalization 5, 11, 36, 42–3, 120–30, 133, 147–9, 231, 232–5, 236, 245–6, 252, 267, 461
  *see also* internationalization; regionalization
GM, *see* General Motors
GMAD (General Motors Assembly Division) 183, 197, 204
Goeudevert, Daniel 298–9
Gorgeu, Armelle 383
government and automobile firms relations 31, 91, 157–8, 185, 198–7, 253–4, 274, 341, 353, 366, 381, 383–5, 401, 405, 409, 429, 440, 444, 447
  *see also* automobile public policy
Grieger, M. 273
growth and redistribution of national income 8–9, 10–11, 23–32, 41
  autocentred mode 9–10, 11, 28–32
  'developing' countries 11, 22, 32, 380
  export mode 9, 11, 23, 25–8, 418
  industrial models and 23–32
  invested-oriented modes 9
  modes confrontation 10–11, 23–32
  open country mode 9, 11
  speculative bubble 11, 32, 33–4, 35, 41, 77, 103, 134
  *see also* international context; profit strategies
Gyllenhammar, P. G. 421, 422, 428, 433

Halberstam, D. 251, 256
Hanada, Masanori 4, 20, 91, 93, 96, 102
Hanon, Bernard 376, 382, 383
Harbour, J. 193, 197
Hartz, P. 304–5
Hill, F. E. 211
Hillebrand, W. 302

Honda 4–5, 18, 20–1, 110–35
  in Asia 132
  automation 126
  crisis 11, 38–9, 130 133–5
  distribution networks 112–13, 118–19, 124–5
  employment relations 39, 119, 121–4, 128
  in Europe 127–9
  expert system 20, 111–12
  exports 23, 25–6, 113–14, 118–19, 130
  flexible mass production 110–14, 117
  globalization 120–30, 133, 231, 233
  innovation and flexibility profit strategy 4–5, 8–9, 18, 21, 25–6, 44, 114–15, 134
  in Japan 125–7, 130–1
  in North America 120–5, 131, 187
  peak production and profits 38
  production organization 112, 116–18, 119, 122–5
  product policy 114–15, 133–4
  product technologies 115
  regionalization strategy 43, 44
  research and development 20, 115–16
  Rover relationship 28, 44, 119, 127–9, 132, 134, 409–12
  statistics 136–7
  supplier relations 21, 118
  union relations 119, 122, 128
  wage system 20, 119, 122
  *see also* Hondaism
Honda of America Manufacturing (HAM) 38, 119, 120–1, 123–4, 129, 131
Honda Engineering 112, 117, 123
Honda, Soichiro 20, 110–11, 115, 120, 135
Honda of the UK Manufacturing (HUM) 128, 132
Hondaism, *see* Honda model
Honda model:
  amalgamated with Toyota model 11
  creation of 18, 20–1, 26
  definition as specific model 20–1, 110–13
  difference from Toyota model 2, 11, 117
  growth and redistribution modes of national income 2, 10, 25–6
  firms and:
    Chrysler 261–2, 265
    Honda 4–5, 26, 39, 43, 45, 110–13, 130–5
    Rover 410–12
  *see also* flexifactory concept; Fujisawaism; Fujisawa Takeo; Honda Soichiro; industrial models; mass production (flexible); profit strategies (innovation and flexibility)
Hounshell, D. A. 53

hybrid models:
    defined as:
        adaptation to local context 265, 333
        coexistence of models 396–9
        mix of models 172, 180, 243, 408–15
    firms and:
        Chrysler 243, 262–5
        Fiat 326–34
        General Motors 180
        Hyundai 165–8, 172–3
        Rover 396, 408–15
    other GERPISA book about 7
    *see also* industrial models
Hyundai 4, 5, 23, 154–73
    automation 5, 158–9, 160, 166–7, 170
    crisis 5, 32, 156, 163, 166
    employment relations 162, 163–4, 165–70, 172–3
    exports 161–3
    Ford relationship 155–6
    Fordist model 155, 159–64
    independent development 156–7
    international competitiveness 32, 165
    internationalization 32, 161–3, 165, 194
    Japanese management concepts 165–8
    knocked down production 155–6
    Mitsubishi relationship 149, 156, 158–9
    negotiations with General Motors 157–8
    operations in North America 161–2, 194
    production organization 156, 158–70, 165–8, 171–2
    product policy 170–1
    statistics 174
    supplier relations 171–2
    training 170
    union relations 163, 169–70
    volume profit strategy 40, 159–64
    wage system 163, 168–70, 172–3
    world car production 42–3

Iacocca, Lee 243–4, 252–4, 258–9, 262, 263
Imai, K. 100
industrial models:
    authors of *One Best Way?* and 6–7
    conditions for only one model 1
    definition 3, 10
    growth and redistribution modes of national income and 9, 10, 11, 23–32
    plurality 2, 11, 45, 211
    profit strategies and 10
    *see also* craft model; Fordist model; German model; Honda model; hybrid models; Japanese methods; lean production; mass production; modular production; performances; piece work model; profit strategies; reflexive production; Sloanist model; socio-technical model; Taylorist model; Toyota model

Ingrassia, P. 237
Innocenti 35
integrated factory concept, Fiat 6, 326–30
internationalization 29, 455–6, 461
    Chrysler 248, 252
    Citroën 22
    Fiat 19, 22, 31, 43, 316, 332
    Ford 215–16, 229–31, 233
    General Motors 43, 181, 188
    Honda 119–20, 127–30, 231, 233
    Hyundai 32, 161–3, 165, 194
    Mitsubishi 147–9
    Nissan 43, 92, 100–1
    PSA 29, 31, 194, 345–7
    Renault 22, 29, 31, 194, 369, 372, 378, 385
    Toyota 77–9, 85
    Volkswagen 43, 194, 277, 293–4, 297–304
    Volvo 421, 436
    *see also* exports; international context; globalization; regionalization
international context 23–5, 33–4, 41, 380, 453
    counter oil crisis 11, 32, 33–4
    custom duties 15, 18
    exchange rates 5, 10, 23, 24, 26, 29, 33, 75–6, 100, 194–5, 293, 365, 371, 377, 380, 424, 428–9, 433
    faltering of global growth 25–8
    international institutions 14, 15, 18, 29, 33, 41–2, 186, 216, 384, 401, 410, 453, 455
    oil crisis 10, 23, 24, 26, 27, 71, 179, 186, 202, 249, 284, 315, 340–2, 365, 382, 418
    trade conflicts and agreements 14, 24, 25, 75, 77, 85, 181, 190, 202, 216, 246, 255, 362, 453
    *see also* growth and redistribution of national income
Ishihara, Takashi 97
Isuzu 29, 134, 140, 181
Itami, T. 99

Jaguar 17, 396, 399, 400, 403, 408
    acquisition by Ford 228, 361
Japan 18
    company trajectories 4–5, 18–21
    competitiveness of companies 191, 194–5, 196
    export strategy 23, 25–6
    exports to USA 77, 187, 191, 249, 255
    growth and redistribution modes of national income 18
    limitation of US exports 190, 202
    transplants 34, 77–9, 100, 259
Japanese methods 1, 5, 18, 34–5, 186–7, 172, 186–7, 211, 243, 333, 351–61, 437
    *see also* lean production
Jeep 257, 258, 263
Jetin, Bruno 46, 114, 116, 237, 238, 368

## Index

Jordan, Maurice 339
Jürgens, Ulrich 6, 16, 273, 288, 292, 293, 304
just-in-time 19, 35, 63–4, 68–9, 83, 146, 171, 222, 324, 455

*kaizen* activities 64, 65–8, 70–1, 85, 86, 140, 146, 149, 150
Kamii, Y. 94
Kamiya, Shotaro 69
Karmann, Mr 278
Katz, Harry C. 224
Kawamata, President 92
Kawamoto, Nobuhiku 127, 133
Kawasaki factory, Japan (Mitsubishi) 140
Kawashima, Kiyoshi 120, 127
Keller, M. 203, 207
Kerkorian, Kirk 266–7
Kern, H. 292
Kia Motors 158, 160, 229
Klauke, P. 273
knocked down (KD) production 92, 119, 154, 155–6, 372, 407–8, 445
Knudsen, William 59
Koch, G. 275, 276
Kochan, T. A. 311
Kume, Tadashi 127

Labbé, Daniel 370, 374
labour shortage 15, 17, 18, 20, 21, 38, 80, 95, 96, 142, 365, 418, 419, 424, 439
  *see also* work crisis
labour turnover 15, 97, 80, 426, 430, 431
Lacey, R. 184
Lada, *see* Avtovaz/Lada
Lamborghini 258
Lancia 15, 322, 325, 331
Land Rover 28, 134, 395–6, 405, 408–9, 414
Lansard, Roger 345
Lapham, E. 266
lean production:
  apparent convergence to 2, 33, 34–7
  defined as:
    amalgam of Toyota and Honda models 2, 45
    Japanese methods 211
    optimal model for next century 1, 8, 45, 365
    Toyotaism 63, 173
  firms and:
    Fiat 330–2, 323–6
    Ford 212, 235
    Hyundai 154, 173
    Mitsubishi 149–50
    Renault 365
    Toyota 63, 87
    Volvo 426, 437
  *see also* industrial models; Japanese methods; Honda model; Toyota model

Lee, Y.-H. 170
Lefaucheux, Pierre 366
Levin, D. P. 254, 261, 266
Lévy, Raymond 384, 386, 387, 388, 389
Locke, R. M. 311, 313
Loderer 284
Lotz, Mr 283
Loubet, Jean-Louis 6, 16, 338, 368, 382
Lung, Yannick 4, 35, 42, 46, 74, 94, 95, 99, 116, 216, 368, 460
Lutz, Robert 261, 263, 264

Magnabosco, M. 311, 321
Mair, Andrew 2, 4, 17, 20, 110, 115, 121, 127, 128, 135, 410, 412, 452
market 9, 10, 12, 15, 24, 29, 452–3, 455
  Brazil 22, 32
  Canada 14
  emerging countries 22, 29, 32, 41, 365, 378, 452–3
  Europe 15, 24, 34, 296, 360, 377, 389
  France 350, 365, 371, 377, 383, 389–90
  Germany 273, 277, 282, 298
  Italy 315, 319, 321
  Japan 18, 24, 75, 95, 97
  Korea 23, 32, 34, 40, 155, 158–60
  Soviet Union 441
  Sweden 418, 422, 429, 436
  United Kingdom 397, 398, 401, 407
  USA 14–15, 24, 34, 156, 181–3, 186–8, 203, 244–5, 254–5
  *see also* growth and redistribution of national income
Maruyama, Y. 96
Maserati 258, 341
mass production:
  crisis of 314–17
  defined as:
    common to several models 2, 112
    Fordist model 51–3, 211, 338, 344
    profit strategy 365
    specific model 1
  diffusion history 53–7
  firms and:
    Chrysler 54
    Fiat 312–17
    Ford 213
    General Motors 54, 179
    Hyundai 157–9
    PSA 338–40
  kinds of:
    British 395–6, 398–9, 402–3
    diversified 63, 94–7, 365
    flexible 5, 110–4
  *see also* Fordist model; profit strategies (volume)
Mathieu, René 383
Matra 344, 355, 378, 383
Maxcy, G. 22

Mazda 27, 36, 181
  Ford relationship 223, 229–30, 233, 234
  *see also* Toyo Kogyo
McKinley, A. 225
Mercedes-Benz 45, 194, 303
  Mitsubishi relationship 149
Meyer, Alan 59
MG 17, 395–7, 403–5, 413
Midler, Christophe 376, 379, 380
Miller, Robert 263
minivans 34, 45, 131, 133, 134, 187, 193, 194, 256–7, 263, 264, 356, 378, 383, 385, 435
Mishina, Kazuhiro 44, 78
Mito, S. 116, 117, 135
Mitsubishi Heavy Industries (MHI) 21, 139–40
Mitsubishi Motors Company (MMC) 5, 21, 139–50
  automation 5, 145–6, 148, 150
  Chrysler relationship 140, 144–5, 147–8, 246, 249–50, 252
  crisis 23, 27, 144
  distribution network 141–2, 144
  early history 139–43
  employment relations 140–1, 146
  exports 144, 149
  globalization 147–9
  Hyundai relationship 149, 156, 158–9
  innovation and flexibility profit strategy 5, 21, 25–6, 27, 35, 36, 44, 114–15, 134
  Mercedes-Benz relationship 149
  Nedcar venture 148–9, 434–5
  passenger car sales growth 143–5
  production organization 142, 145–6, 147, 149–50
  regionalization strategy 43
  statistics 151–2
  supplier relations 141, 146
  training 141, 149
  transplants 148–9
  union relations 140, 149
  wage system 21, 141, 146–7
modular production 45, 456
Mommsen, H. 273
Monden, Y. 64, 70, 74, 98
Morris 17, 395, 396, 398, 400–1

Nabeoka, K. 322
Nader, Ralph 185, 283
national labour compromise:
  definition 46
  in different countries 9, 12, 15, 29, 32, 64, 215, 295, 454
  *see also* growth and redistribution of national income; national labour relations

national labour relations:
  definition 46
  in different countries 9, 12, 29, 454
Nedcar 148–9, 434–5, 437
Negrelli, S. 311
Nissan 4, 19–20, 91–107
  Austin relationship 91, 92
  automation 96
  crisis 23, 27–8, 103
  distribution network 92, 96–7, 106
  diversification of business activities 106
  employment relations 20, 27–8, 35, 95–6, 101–2
  exports 27–8, 92, 94, 97, 98–9
  Ford relationship 230
  internationalization 43, 92, 100–1
  mass production 94–7
  production organization 99, 104–6
  product policy 99
  statistics 108
  supplier relations 91–2, 98, 104
  union relations 92–3, 97–8, 104
  volume and diversity profit strategy 12, 18, 21, 27, 35, 36, 45, 139, 145, 150
  wage system 19–20, 93–4, 96, 101–2
Nomura, Masami 65, 67
Nonaka, I. 116, 333
Nordhoff, Mr 275, 280, 283
NSU 283
NUMMI (New United Motor Manufacturing Inc.) 5, 38, 43, 77–9, 180

Ohno, Taiichi 19, 49, 52, 64, 66, 82
Okamoto, H. 97, 99
Okuda, Hiroshi 85
Oldsmobile (General Motors) 183, 189, 197–8
Opel 5, 181, 193, 227, 303

Packard 56, 57, 180
Parayre, Jean-Paul 343–5, 347, 349, 350, 357
Parker, M. 78
Pascale R. T. 113, 135, 225
performances of automobile firms:
  break even point 5, 6, 8, 18, 21, 25, 28, 29, 31, 35–6, 38, 253, 296, 352, 371, 373, 382, 385, 397, 411
  capacity utilization 156, 157, 164, 314, 407
  market share 14, 28–9, 38, 72, 76–7, 97, 99, 103, 197, 226–7, 248, 255, 257, 314, 341, 350, 366, 371, 377, 380–1, 383, 388, 401, 413, 418, 435, 441
  productivity 15, 19, 66–8, 82–3, 86, 96, 97–8, 141, 164, 204, 218–19, 227, 334–5, 350–1, 353, 400, 413, 424, 431
  profit 14, 27, 29, 37, 38, 73–6, 85, 86, 88–9, 94, 100, 103, 108, 130, 136–7, 142–3, 151–2, 160, 174, 208–9, 227,

## Index

239–40, 248, 253, 256, 262, 268, 297–8, 308, 336, 338–9, 341, 351, 363, 366, 378, 384, 389, 391–2, 398–400, 402, 414, 416, 424, 433, 438, 442–3
  *see also* individual firms; industrial models
Peronnin, Jean 345
Perot, Ross 192
Pessa, P. 327
Peugeot, *see* PSA (Peugeot Société Anonyme)
Peugeot, Jean-Pierre 339
'piece work' model 17, 28
  *see also* profit strategies (diversity and flexibility)
Piëch 299
Pigozzi, Henri-Théodore 349
Pil, Frits 31
Pininfarina 322
Pointet, Jean-Marc 386
Poitou, J.-C. 376
pollution issues 13, 73, 74, 115, 118, 185, 202–3, 250, 454
Pontiac (General Motors) 54, 182, 183, 189
Porsche 273
Prince 91, 95, 106
product development 10, 16, 20, 26, 70, 115–16, 125–7, 144, 195–6, 204, 226, 231–2, 261–2, 263–5, 278–80, 283–4, 301–2, 322, 324, 345, 347–8, 355–6, 376, 388, 397, 410, 414, 445–6, 457
  *see also* production organization
production organization 3, 10
  Avtovaz/Lada 443, 445, 448
  Chrysler 247–8, 251–2, 259–61
  Citroën 356–9
  Fiat 317–18, 320–1, 323–4, 326–30, 331
  Ford 213–14, 221, 223–4
  General Motors 196–7
  Honda 112, 116–18, 119, 122–5
  Hyundai 156, 158–7, 165–8, 171
  Mitsubishi 142, 146
  Nissan 99, 104–6
  other GERPISA book about 7
  PSA 347, 356–9
  Renault 366, 369, 388
  Rover 397, 400, 405, 411–12
  Toyota 64, 69, 73–4, 76, 83–4
  Volkswagen 280–1, 291–3, 303–4
  Volvo 17, 420, 421–3, 426–7, 436–7
  *see also* automation; distribution; industrial models (definition); product development; supplier relations; team work
product policy 30, 456–7
  Avtovaz/Lada 442, 447, 449
  Chrysler 245, 249, 253
  Fiat 312, 332

Ford 214, 217
General Motors 182, 187, 196
Honda 114–15, 133–4
Hyundai 170–1
Nissan 99
PSA 347, 355–6
Renault 366, 369, 388
Rover 400, 401, 403, 405, 409
Toyota 75
Volkswagen 278, 282–3, 285–7, 298, 300–2
Volvo 428–9
  *see also* industrial models (definition); product development; profit strategies
product range:
  Avtovaz/Lada 443, 445–6, 449
  Chrysler 14, 244, 248–9, 254–7, 263–4
  Fiat 15, 35, 312, 332
  Ford 14, 36, 226–8
  General Motors 13–14, 35, 182, 187–9
  Honda 114, 125, 126, 127, 133–5
  Hyundai 156, 160
  Mitsubishi 141, 143, 145
  Nissan 19, 35, 91, 95, 99
  PSA 16, 338–9, 341, 355–6
  Renault 15, 367–8, 371, 378, 380, 384–5
  Rover 398, 401, 403, 406, 410
  Toyota 18, 63, 76
  Volkswagen 16
  Volvo 17
  *see also* commonization of platforms; product policy
profit strategies:
  conditions of viability 9–10, 12, 25
  definition 8, 35, 42
  types of:
    continuous reduction of costs at constant volume strategy and 8, 10
      PSA 6, 16, 35, 37, 339, 352–4
      Toyota 8–9, 18–19, 25, 26, 44, 67, 86
      *see also* Toyota model
    diversity and flexibility strategy and:
      Rover 17, 28, 397
      *see also* craft model; piece work model
    innovation and flexibility strategy and 8, 10, 20, 461
      Chrysler 5, 14, 35, 36, 44, 254–6, 260
      Citroën 340–1, 344
      Honda 4–5, 8–9, 18, 21, 25–6, 44, 114–15, 134
      Mitsubishi 5, 21, 27, 35, 36, 44, 45, 139, 145, 150
      Renault 44, 386–90
      Toyo Kogyo/Mazda 27
      *see also* Honda model
    quality and specialization strategy and 3, 8–9, 45, 461

profit strategies (*cont.*):
  Renault 35, 37, 386–90
  Rover 35, 37, 409–14
  Volvo 9, 26, 45, 420
  volume and diversity strategy and 2, 8–9, 12, 35, 461
    Chrysler 5, 12, 14, 245
    Fiat 12, 15, 35, 44, 313
    Ford 12, 14, 214
    General Motors 5, 10, 35, 43, 188
    Nissan 12, 18, 35, 43, 94–7
    PSA 12, 15, 44, 371–3, 377–80
    Renault 6, 12, 15, 44, 371–3, 377–80
    Volkswagen 8–9, 15, 26–7, 43, 285–7
    *see also* commonization of platforms; Sloanist model
  volume strategy and 460
    Ford 5, 35, 36, 42, 214, 233
    Hyundai 40, 159–64
    Lada 6, 32, 445, 449
    Volkswagen 16, 278
    *see also* Fordist model
  *see also* growth and redistribution of national income; industrial models
Proton 148, 149
PSA (Peugeot Société Anonyme) 6, 15, 16–17, 338–62
  automation 348, 350, 357
  continuous reduction of costs at constant volume profit strategy 6, 16, 35, 37, 339, 352–4
  crisis 28–9, 350–1
  distribution networks 359–61
  employment relations 37, 342, 348–50, 352–3, 358–9
  exports 345–7
  internationalization 29, 31, 194, 345–7
  Japanese model 351–61
  mass production 338–40
  product development 345, 347–8, 355–6
  production organization 347–8, 354–9
  product policy 347, 355–6
  purchase of Chrysler-Europe 343–4
  purchase of Citroën 338, 340–3, 379
  Renault relationship 339–40, 372, 379
  statistics 363
  structural reforms 345–7
  supplier relations 345, 354, 358
  training 359, 360
  union relations 349–50
  volume and diversity profit strategy 12, 15, 44, 371–3, 377–80
  wage system 349
  *see also* Citroën; Talbot

Raff, Daniel 2, 49, 54, 55, 56
recreational vehicles 27, 39, 44, 45, 84, 134, 145, 150, 410

reflexive production:
  defined as:
    specific and potential model 39–40
    neo-craft production 49
  Uddevalla factory, Sweden (Volvo) and 6, 27, 39–40, 419, 426–9, 430–2, 433, 435, 436
  Volvo and 9, 11, 27, 460–2
regionalization 11, 43–4, 460–2
Renault 6, 15–16, 365–90
  American Motors Corporation relationship 31, 187, 346, 381, 385
  automation 380, 382, 383
  competitiveness 371
  cost reduction 383
  crisis 28–9, 383–4
  diversification of business activities 371, 376, 378–9
  employment relations 37, 369–70, 373–6, 379, 381–2, 387–9
  exports 368–9
  innovation and flexibility profit strategy 44, 386–90
  internationalization 22, 29, 31, 194, 369, 372, 378, 385
  living car concept 383, 386
  growth and redistribution mode of national income 29, 31, 32
  operations in USA 381
  organizational structure 376–7
  Peugeot relationship 339–40, 372, 379
  production organization 366, 369, 388
  product policy 366, 369, 388
  proposed merger with Volvo 381, 384, 389–90, 432–3
  quality-based strategy 35, 37, 386–90
  recovery 383–9
  regionalization 43, 44
  Sloanist model 12, 15–16, 365, 366–71
  statistics 391–2
  supplier relations 383, 388
  training 374–5
  union relations 385
  volume and diversity profit strategy 6, 12, 15, 44, 371–3, 377–80
  wage system 366–7, 370, 374, 389
Renault, Louis 370
Riccardo, John 251, 252–3
Rieser, V. 327, 329, 331, 333
robots, *see* automation
Roos, D. 32
Rootes 246, 343
rotary engines 27, 283, 340–1
Rover Group 6, 17, 28, 395–9, 408–15
  British Aerospace relations 44, 128, 408–9
  craft-British model 397, 399–400, 403–5, 409–10, 414
  crisis 28, 399, 401, 404–5

employment relations 37, 398–400, 406, 412
Honda relationship 28, 44, 119, 127, 128–9, 132, 134, 407–8, 409–12
mass-British model 398–9, 402–3
production organization 397, 400, 405, 411–12
product policy 397, 400, 405, 411–12
purchase by BMW 6, 44, 45, 132, 396, 414–15
quality-based strategy 35, 37, 409–14
Roverization 408–14
statistics 416
union relations 397, 399, 412
Rubinstein, Saul 31

Saab 361, 418, 422, 430
Saehan Motors 157–8
*see also* Daewoo
Saga, Ichiro 101
Saillard, Y. 46
Sakiya, T. 112, 135
Salerno, Mario 230
sales distribution networks, *see* distribution networks
Samsung 173
Sartirano, L. 327
Saturn engine (Mitsubishi) 139, 143, 157
Saturn (General Motors) 5, 31, 43, 180, 189, 198, 201, 460
Schmücker, Mr 284
Schweitzer, Louis 389
SEAT 38, 296, 297, 302
  Fiat relationship 314
Schook, R. L. 122, 124, 225
Schumann, Michael 292
Shimada, H. 77, 78
Shimizu, Koichi 2, 4, 5, 19, 21, 38, 63, 65, 80, 81, 84, 93, 105, 139, 452
Shimokawa, Koichi 5, 21, 104, 106, 139
Shioji, Ichiro 93, 101
Simca 246, 343, 344, 347, 349
Skoda 38, 296, 302, 389
Slaughter, J. 78
Sloan, Alfred 12, 183, 214, 365
Sloanist model:
  conditions for viability 12–13
  convergence towards 12–23
  defined as:
    specific model 2, 10, 11, 12, 13, 365
    variant of Fordist model 179, 184, 188, 206
  differences from Fordist model 16, 183
  firms and:
    Fiat 12, 313
    Ford 12, 14, 214
    General Motors 5, 10–15, 23, 179–80, 183, 198, 204

Nissan 12, 18, 19–20, 35, 43, 94–7
PSA 12, 15, 16–17, 37, 43, 339, 345–50, 355–6, 361
Renault 12, 15–16, 365, 366–71
Rover 400–1
Volkswagen 8, 11, 15–16, 26, 38, 285–91, 302
growth and redistribution modes of national income and 10, 15–17, 26, 45
*see also* industrial models; mass production (diversified); profit strategies (volume and diversity); Sloan, Alfred
Smale, John G. 199
Smith, D. 411
Smith, John F., Jr 199, 203
Smith, Roger 188, 192, 198
socio-technical model 2–3, 83–4, 149, 375–6, 421–4, 427, 432, 437
  Kalmar factory (Volvo) 6, 27, 418–19, 421–2, 426–7, 430–2
  *see also* industrial models; reflexive production
Sorge, A. 295
Soviet Union 9, 22, 32, 448
Sperlich, Hal 244, 251, 252–3, 256, 261
sports-utility vehicles (SUVs) 193, 257, 263, 435
Ssangyong 173
Stalk, G., Jr 112, 125
Starkey, K. 225
Stempel, Robert 198–9
Stokes, Donald 402
Streeck, W. 295
strikes 12, 15, 16, 19, 20, 37, 67, 92, 163, 184, 216, 247, 282, 306, 319, 351, 353, 362, 374, 379, 381, 399, 400, 406, 421
  *see also* work crisis
Studebaker 57, 180
suggestions system 67, 165, 388
Sunbeam 343, 344
sunk investments 55–8
supplier relations 458
  Avtovaz/Lada 446
  Chrysler 262, 264, 267
  Fiat 323–5
  Ford 221–2, 230–1
  General Motors 184–5, 190, 201–3
  Honda 21, 118
  Hyundai 171–2
  Mitsubishi 141, 146
  Nissan 91–2, 98, 104
  PSA 345, 354, 358
  Renault 383, 388
  Toyota 26, 68–9, 79
  Volvo 421
Suzuki 29, 259
Suzuki, N. 78, 149

Tabata, T. 101
Takeuchi, Y. 105
Talbot 37, 344–6, 347, 351, 359–60
    see also PSA (Peugeot Société Anonyme)
Tanaka, T. 70
Taub, E. 225
Taylor, George 341
Taylorism, see Taylorist model
Taylorist model:
    defined as:
        specific model 2
        Taylorist-Fordist model 2, 155
    firm and:
        Hyundai 155, 164
        PSA 358
        Renault 366–7
        Rover 17, 397, 399, 400
        Volvo 431
    see also industrial models
team work 17, 19, 31, 34, 65, 84, 124, 146,
    149, 224–5, 296, 303–4, 326–7,
    330–1, 335, 350, 375–6, 379, 382–3,
    367–8, 398, 408, 412, 423, 425, 431,
    434, 437
    other GERPISA book about 7
Tidd, Joseph 2, 9
Toga, M. 252
Tolliday, Steven 2, 9, 215, 295, 398
total productive maintenance (TPM) 35, 76,
    101, 165, 172, 387, 437
total quality control 67, 92, 96, 121, 124, 140,
    159, 165, 221–2, 322, 383, 387
Townsend, Lynn 243, 244–6, 251
Toyo Kogyo 27, 30
    see also Mazda
Toyoda, Kiichiro 64
Toyoda, Sakichi 64
Toyoda, Tatsuro 85
Toyota 4, 18–19, 63–87
    automation 64, 83
    autonomization 64, 85
    continuous reduction of costs at constant
        volume profit strategy 8–9, 18–19,
        25–6, 44, 67, 86
    crisis 11, 38–9, 79–80, 85
    distribution network 69–70, 75–6
    employment relations 19, 38–9, 64–8, 78,
        80, 81–5
    exports 23, 25, 77
    humanization of work 83–5
    internationalization 77–9, 85
    peak production and profits 38
    production cost management 70–1, 75, 81–3
    production organization 4, 26, 63, 64, 69,
        73–4, 76, 83–4
    production planning 69–70, 73–5
    product policy 4, 26, 63, 64, 69, 73–4, 76,
        83–4

statistics 88–9
supplier relations 68–9, 79
training 83
union relations 65, 78
wage system 19, 38–9, 65–8, 78–9, 81–3
see also NUMMI; Toyota model
Toyotaism, see Toyota model
Toyota model:
    coherence and dynamism of model 26, 71
    creation of 18–19
    crisis and reorganization of 2, 45, 38, 80–6
    definition 18–19, 64
    difference from Honda model 2, 11, 117
    firms and 6, 14, 43, 63–90, 154, 351–61
    future uncertainties 85–6
    growth and redistribution mode of national
        income 10, 25
    hidden face of 4, 65–8
    history of 71–80
    Toyota Production System (TPS) 4, 26, 63, 64
    transplantation 77–9
    see also industrial models; Ohno, Taiichi;
        profit strategies (continuous costs
        reduction at constant volume);
        Toyota
Toyota Motor Kyushu (TMK) 82–3, 84
Toyota Motor Manufacturing Canada (TMC)
    77–8
Toyota Motor Manufacturing United Kingdom
    (TMUK) 77–9
Toyota Motor Manufacturing USA Inc. (TMM)
    77–9
Toyota Production System, see Toyota model
training 83, 141, 149, 170, 276, 328, 359–60,
    374–5
transplants 1, 22–3, 28, 34, 77–9, 100, 119,
    120–5, 127–9, 132–3, 147–9, 259
Triumph 395–7, 400–1, 403–5
TWR engineering 436

Udagawa, M. 92
Ueda, H. 68
union relations:
    Chrysler 31, 36, 189–90, 202, 247–8,
        259–61
    Fiat 31, 313, 314–15, 319, 327, 329
    Ford 189–90, 214, 215
    General Motors 12, 184, 201
    Honda 119, 122, 128
    Hyundai 163, 169–70
    Mitsubishi 140, 149
    Nissan 92–3, 97–8, 104
    PSA 349–50
    Renault 31, 385
    Rover 397, 399, 412
    Toyota 65, 78
    Volkswagen 275–6, 284
    Volvo 425, 426–7

# Index

United Kingdom:
  Honda operations 127–9
  Nissan factory, Sunderland 28, 35, 100
  Toyota Motor Manufacturing United
    Kingdom (TMUK) 77–9
USA:
  company trajectories 5
  imports 13, 24–5, 182, 249
  Japanese imports 77, 187, 191, 249, 255
  limitation of Japanese imports 190, 202
  monetary policies 23–5
  Renault operations 381
  Sloanist model, problems with 13–15
  Volkswagen operations 293–4

Valletta, Vittorio 313, 314, 316
Vernier-Palliez, Bernard 379, 380, 382
Volkswagen 5–6, 15, 16, 273–307
  automation 280, 291–3
  continuous improvement process 303–4
  crisis 5, 11, 26, 38–9, 272, 284, 297–8, 306
  distribution network 302
  diversified quality production 295
  diversification of business activities 294–5
  employment relations 27, 274–7, 281–2, 287–3, 304–6
  employment security 288–9, 304–6
  exports 23, 26, 277
  Ford relationship 230
  Fordist model 5, 15–16, 277–85
  German model 295
  governance structure 273–4, 283–4, 298–9, 300, 306
  internal competition 297–8
  internationalization 43, 194, 277, 293–4, 297–4
  peak production and profits 38
  production organization 280–1, 291–3, 303–4
  product policy 278–1, 284, 291–3, 298–9, 300–4, 306
  production in USA 293–4
  Sloanist model 8, 11, 15–16, 26, 38, 285–91, 302
  statistics 308
  team work 303
  training 276
  union relations 275–6, 284
  volume and diversity profit strategy 8–9, 15, 26–7, 43, 285–7
  volume profit strategy 16, 278
  wage system 282, 289–91
Volpato, Giuseppe 6, 36, 311, 316, 318, 326, 327, 331, 334, 452
Volvo 6, 17, 418–37
  automation 420, 422, 424, 427
  crisis 11, 429–35

diversification of business activities 423, 430
employment relations 27, 419–20, 424–5, 426
exports 23, 26, 418–19, 424
Fordist model 420–1
future outlook 435–7
innovation strategy 436
internationalization 421, 436
Nedcar venture 148, 434–5
production organization 17, 420, 421–4, 426–7, 436–7
product policy 17, 420, 421–4, 428–9, 436–7
proposed merger with Renault 381, 384, 389–90, 432–3
quality and specialization profit strategy 9, 26, 45, 420
recovery 433
reflexive production model 9, 11, 27, 39–40, 460–2
statistics 438
strategic reorientation 434–5
supplier relations 421
union relations 425, 426–7
wage system 419
VTEC (variable valve timing and lift technology) engine (Honda) 126, 129, 135

wage system 9
  Fiat 315, 329
  General Motors 12, 184
  Honda 20, 119, 122
  Hyundai 163, 168–70, 172–3
  Mitsubishi 21, 141, 146–7
  Nissan 19–20, 93–4, 96, 101–2
  PSA 349
  Renault 366–7, 370, 374, 389
  Toyota 19, 38–9, 65–8, 78–9, 81–3
  Volkswagen 282, 289–91
  Volvo 419
Warnecke, H. J. 304
Whipp, R. 396, 397, 403
Whisler, T. R. 396, 397, 404
White, J. B. 237
Wiersch, B. 273
Wilkins, M. 211
Williams, K. 398, 399, 405, 407
Womack, J. P. 1, 8, 33, 49, 63, 87, 211, 296, 450
Wood, J. 398, 399, 400, 405
work crisis 4, 10, 11, 15, 17, 38–9, 79–80, 95–6, 184, 247, 314–15, 365, 373–6, 420–1, 424–5, 435
  *see also* absenteeism; labour shortage; labour turnover; strike; unions
worker involvement 20, 64, 96, 124, 214, 224, 303–4, 366, 458–60

workers:
  classification 35, 65–6, 78, 96, 140–1, 146–7, 166–7, 329, 374–5, 379, 381
  immigrant 95, 281–2, 342, 349, 352, 369, 373, 281–2
  polyvalent 16, 165
  skilled 51–2, 55, 141–2, 170, 276, 281, 374–5, 379, 388, 397
  temporary 64, 72, 80, 95, 167, 388
  unskilled 3, 52–3, 281, 350, 369, 374–5, 380, 446

working time 5, 10, 21, 38–9, 78, 80–1, 83, 117, 140, 164, 285, 305–6, 315, 319, 349, 351, 366, 374–5, 388, 459
world cars 36, 42–3, 232–4, 332

Yakushiji, T. 207
Yamamoto, K. 93

Zavodi Crvena Zastava (ZCZ) 314
Zuboff, S. 333